OXYGEN TRANSPORT TO TISSUE XXIII

ADVANCES IN EXPERIMENTAL MEDICINE AND BIOLOGY

Recent Volumes in this Series

Volume 504
MYCOTOXINS AND FOOD SAFETY
Edited by Jonathan W. DeVries, Mary W. Trucksess, and Lauren S. Jackson

Volume 505
FLAVONOIDS IN CELL FUNCTION
Edited by Béla A. Buslig and John A. Manthey

Volume 506
LACRIMAL GLAND, TEAR FILM, AND DRY EYE SYNDROMES 3: Basic Science and Clinical Relevance
Edited by David A. Sullivan, Michael E. Stern, Kazuo Tsubota, Darlene A. Dartt, Rose M. Sullivan, and B. Britt Bromberg

Volume 507
EICOSANOIDS AND OTHER BIOACTIVE LIPIDS IN CANCER, INFLAMMATION, AND RADIATION INJURY, 5
Edited by Kenneth V. Honn, Lawrence J. Marnett, Santosh Nigam, Edward Dennis, and Charles Serhan

Volume 508
SENSORIMOTOR CONTROL OF MOVEMENT AND POSTURE
Edited by Simon C. Gandevia, Uwe Proske, and Douglas G. Stuart

Volume 509
IRON CHELATION THERAPY
Edited by Chaim Hershko

Volume 510
OXYGEN TRANSPORT TO TISSUE XXIII: Oxygen Measurements in the 21st Century: Basic Techniques and Clinical Relevance
Edited by David F. Wilson, Sydney M. Evans, John Biaglow, and Anna Pastuszko

Volume 511
PEDIATRIC GENDER ASSIGNMENT: A Critical Reappraisal
Edited by Stephen A. Zderic, Douglas A. Canning, Michael C. Carr, and Howard McC. Snyder, III.

Volume 512
LYMPHOCYTE ACTIVATION AND IMMUNE REGULATION IX: Homeostasis and Lymphocyte Traffic
Edited by Sudhir Gupta, Eugene Butcher, and William Paul

A Continuation Order Plan is available for this series. A continuation order will bring delivery of each new volume immediately upon publication. Volumes are billed only upon actual shipment. For further information please contact the publisher.

OXYGEN TRANSPORT TO TISSUE XXIII

Oxygen Measurements in the 21st Century: Basic Techniques and Clinical Relevance

Edited by

David F. Wilson
Sydney M. Evans
John Biaglow
Anna Pastuszko
University of Pennsylvania
Philadelphia, Pennsylvania

Springer Science+Business Media, LLC

Library of Congress Cataloging-in-Publication Data

Oxygen transport to tissue XXIII: oxygen measurements in the 21st century: basic techniques and clinical relevance/edited by David F. Wilson ... [et al.].
p. cm. — (Advances in experimental medicine and biology; v. 510)
Based on the 29th annual meeting of the International Society for Oxygen Transport to Tissue held in Philadelphia from August 11 to 15, 2001.
Includes bibliographical references and index.
ISBN 978-1-4613-4964-8 ISBN 978-1-4615-0205-0 (eBook)
DOI 10.1007/978-1-4615-0205-0
1. Oxygen—Physiological transport—Congresses. 2. Oxygen—Physiological transport—Research—Methodology—Congresses. 3. Tissue respiration—Congresses. I. Title: Oxygen measurements in the 21st century. II. Wilson, David F., 1938- III. International Society on Oxygen Transport to Tissue. Meeting (29th: 2001: Philadelphia, Pa.) IV. Series.

QP99.3.O9 O9399 2002
572'.47—dc21

2002034177

Proceedings of the 29th annual general meeting of the International Society for Oxygen Transport to Tissue held in historic Philadelphia, USA, on the campus of the University of Pennsylvania from August 11 to 15, 2001.

ISBN 978-1-4613-4964-8

Originally published by Kluwer Academic Publishers in 2003
Softcover reprint of the hardcover 1st edition 2003
http://www.wkap.nl/

10 9 8 7 6 5 4 3 2 1

A C.I.P. record for this book is available from the Library of Congress

INTERNATIONAL SOCIETY ON OXYGEN TRANSPORT TO TISSUE 2001

Officers:

President:	David F. Wilson, USA
President-Elect (2002):	Maureen Thorniley, UK
Secretary:	Oliver Thews, Germany
Treasurer:	Peter Keipert, USA
Chairman of the Knisely Award Committee:	Duane F. Bruley, USA

Executive Committee:

Tiziano Binzoni
Giuseppe Cicco
Jeff Dunn
Simon Faithfull
Joe LaManna
David Maguire
Avraham Mayevsky
Mamoru Tamura
Martin P. Wolf

Local Organizing Committee:

David F. Wilson
Sydney M. Evans
Cameron Koch
Don Buerk
Sergei Vinogradov
John Biaglow
Isolde Dunphy

SPONSORS

ISOTT is a scientific society and as such is most grateful for the generosity of sponsoring organizations. However, it does not endorse any of their products or services. This applies also to any material that may be distributed in mailings or at its meetings.

Sponsors:

We would like to extend our appreciation to the following organizations for their generous financial support of ISOTT 2001:

Alliance Pharmaceutical Corporation
AstraZeneca Pharmaceuticals, Inc.
Institute for Environmental Medicine, University of Pennsylvania

Thanks are also due to the University of Pennsylvania Department of Biochemistry and Biophysics for administrative assistance, and to the volunteers from the School of Medicine who so graciously contributed their time and effort to making the meeting run smoothly.

Exhibitors:

Harvard Apparatus, Inc.
Oxford Optronix, Ltd.
Spherix/Reming Bioinstruments

PREFACE

The ISOTT 2001 local organizing committee was pleased to welcome over 140 delegates from around the world to the 29th annual general meeting of the International Society for Oxygen Transport to Tissue. The meeting was held in historic Philadelphia, USA, on the campus of the University of Pennsylvania from August 11 to 15, 2001. In the tradition of ISOTT, the conference was a total immersion experience. Attendees were encouraged to eat together and spend their evenings relaxing together in a style that maximized exchange of ideas and interactions of younger scientists with their more senior colleagues. Delegates participated in a total of 122 presentations including poster displays, selected oral presentations, seminars by invited speakers and a round table discussion. In choosing invited speakers and oral presenters, special emphasis was placed on methods for oxygen measurement in living tissue and application of these technologies to understanding physiological and biochemical basis for pathology related to tissue oxygenation.

All of the manuscripts contained in this volume underwent both an editorial and scientific review, and only those meeting both criteria have been published. However, while all efforts have been made to eliminate editorial errors, some have undoubtedly been overlooked, for which the editors apologize.

Production of this volume would not have been possible without the skillful shepherding and hard work of Isolde Dunphy, who played a major role in organizing the meeting and took primary technical responsibility for production of the Proceedings, or the copy-editing of Laraine Visser. The editors would also like to thank the many colleagues who reviewed manuscripts for scientific content. Without them, these Proceedings would not have been possible.

On behalf of ISOTT, we congratulate Fahmeed Hyder and Theresa Busch respectively as the Melvin H. Knisely and Dietrich W. Lubbers award winners of 2001, selected for their outstanding achievements in research related to oxygen transport to tissue.

ISOTT 2001 Editorial Committee
David F. Wilson — Sydney M. Evans
Anna Pastuszko — John Biaglow

CONTENTS

TUMOR BIOLOGY AND PHYSIOLOGY

BLOOD PHYSIOLOGY AND BLOOD SUBSTITUTES

OXYGEN MEASUREMENT METHODS AND APPLICATIONS

TISSUE HYPOXIA/ISCHEMIA

THE ROLE OF TUMOR OXYGENATION IN VASCULAR AND CLINICAL RESPONSE TO ANGIOGENESIS INHIBITION

Michael S. Gee and William M.F. Lee*

1. INTRODUCTION

The concept of treating tumors with inhibitors of angiogenesis[1] is increasingly accepted but has yet to find practical clinical application. Beyond the discovery and clinical testing of potentially active agents, which is proceeding at a rapid pace,[2] much remains unknown about its therapeutic mechanisms, effective application and potential complications. Our work has focused on understanding how antiangiogenic agents produce their antitumor effect and exploring ways to monitor effectiveness of therapy and, in this study, we examine decreased tumor oxygenation as a potential consequence of angiogenesis inhibition.

2. METHODS

K1735 murine melanoma cells were maintained in DMEM supplemented with 10% fetal bovine serum and penicillin/streptomycin. Female C3H/HeN mice were purchased from Harlan Sprague Dawley (Indianapolis, IN). For tumor studies, 10^6 tumor cells were injected subcutaneously in the lower left flank. Treatment with recombinant murine interleukin-12 (rmIL-12; Genetics Institute, Andover, MA) or CC-9088 (Celgene Corporation, Warren, NJ) was initiated when tumors were 2-3 mm in diameter. For hypoxia studies, 0.25 mL of 10 mM EF5 [2-(2-nitro-1H-imidazol-1-yl)-*N*-(2,2,3,3,3-pentafluoropropyl) acetamide] in 0.9% saline was given intravenously 3 hours prior to tumor excision. Flow cytometric analysis[3] and confocal microscopy[4] were performed as described previously. Immunohistochemical staining for EF5 (Cy3-conjugated ELK3-51 Ab, gift from C. Koch), apoptosis (Apoptag kit, Intergen, Purchase, NY) and

* Department of Medicine and the Cancer Center, University of Pennsylvania Medical Center, Philadelphia, PA 19104.

Oxygen Transport to Tissue XXIII, edited by D. Wilson *et al.*
Kluwer Academic/Plenum Publishers, 2003

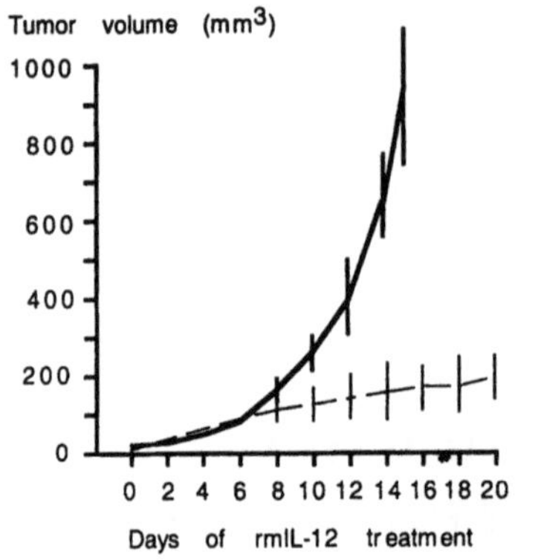

Figure 1. rmIL-12 suppresses growth of K1735 melanomas. Mice bearing established K1735 melanomas were treated with rmIL-12 on a 5 injections per week schedule for three weeks. Tumors were measured by calipers and size was estimated by the formula for volume of a spheroid [0.52 x $(\text{width})^2$ x length]. Average volumes for either vehicle (solid line) or rmIL-12 (dashed line) treated tumors are displayed. Error bars indicate standard deviation. N = 10 mice/group.

HIF-1α (H1α67 Ab, Novus Biologicals, Littleton, CO) was performed on frozen tissue sections (10 μm thick).

3. RESULTS

3.1. Angiogenesis inhibition produces tumor hypoxia associated with apoptosis

We began by studying the effects of the angiogenesis inhibitor rmIL-12 against K1735 murine melanoma tumors growing in syngeneic C3H/HeN mice. Our hypothesis was that rmIL-12 inhibition of vessel development would produce tumor cell ischemia. To detect ischemia, we measured tumor cell hypoxia by the binding of EF5, the hypoxia-activated pentafluorinated nitroimidazole.[5] An unusual and experimentally advantageous feature of K1735 tumors is that they are not normally hypoxic: Tumors between 3-12 mm diameter contained only 1%-2% severely hypoxic cells (< 1% O_2) by flow cytometric analysis of EF5 binding (calibrated by EF5 binding in K1735 cells grown *in vitro* at known O_2 levels). rmIL-12 treatment (0.125μg i.p./mouse/day, 5 days/week), led to significant tumor growth retardation (Figure 1) accompanied by the progressive development of tumor cell hypoxia during weeks 2 and 3 of therapy (Figure 2). These

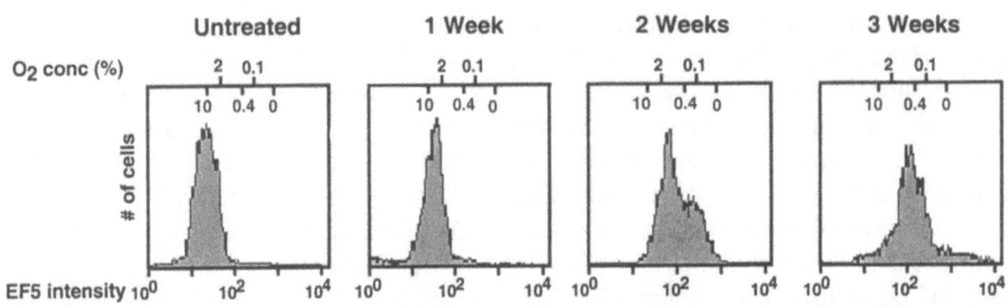

Figure 2. rmIL-12 treatment leads to progressive development of tumor cell hypoxia. The hypoxic marker EF5 was administered to mice bearing K1735 tumors treated for 0, 1, 2 or 3 weeks with rmIL-12. Cell suspensions prepared from excised tumors were stained for EF5. Flow cytometry histograms of representative tumors are shown. Lower horizontal axis indicates EF5 fluorescence intensity, while top horizontal axis indicates mean EF5 fluorescence intensity of K1735 cells under different O_2 levels *in vitro*. N = 5 mice/group.

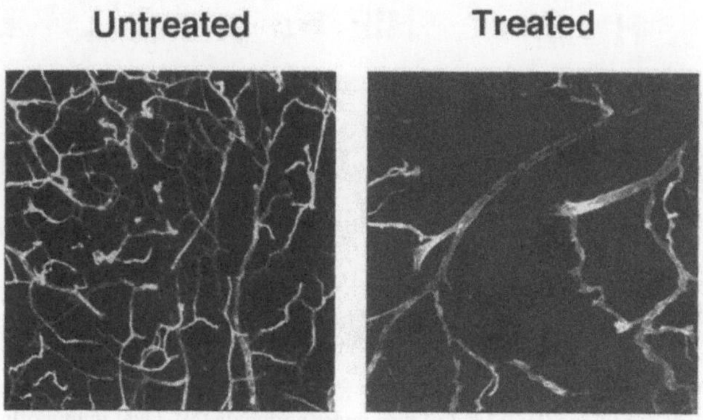

Figure 3. rmIL-12 decreases tumor vascularity. Mice bearing either untreated or rmIL-12 treated K1735 tumors were injected intravenously with a fluorescently-labeled lectin. Confocal microscopy (10x objective) was performed on 1 mm sections from excised tumors to visualize vasculature in three dimensions. Shown are representative projections (0.1 mm depth) from untreated (left) and 3 week rmIL-12 treated (right) tumors.

observations with EF5 were confirmed by polarographic oxygen electrode measurements (data not shown). Confocal microscopy revealed that rmIL-12 treatment led to pruning of tumor vessels and a decrease in vessel density (Figure 3). The development of hypoxia was accompanied by a significant increase in the percentage of K1735 tumor cells undergoing apoptosis. Apoptosis in the second and third weeks of therapy was due to severe tumor cell hypoxia because (a) apoptosis localized to regions of severe hypoxia (Figure 4) and (b) K1735 tumor cells grown under comparable hypoxia *in vitro* underwent significant apoptosis (data not shown).

The causal relationship between rmIL-12-induced tumor hypoxia, cell death and growth control is somewhat clouded by the complex nature of IL-12 as an antitumor agent. For example, rmIL-12 also induced significant tumor cell apoptosis after one week of therapy when severe hypoxia was absent. The high levels of IFN-γ detected in the sera of these mice after one week of rmIL-12 treatment were sufficient to induce

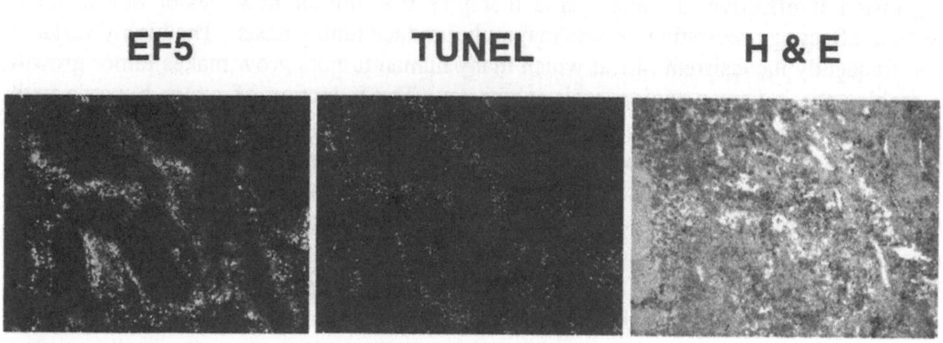

Figure 4. rmIL-12 therapy-induced hypoxia is associated with tumor cell apoptosis. Histologic sections from rmIL-12 treated tumors were simultaneously stained for hypoxia (EF5, left) and apoptosis (TUNEL, center). Shown are the staining patterns from the same representative high-powered field (100x magnification), with routine hematoxylin/eosin staining included (right) to show cell morphology. Note that the areas of apoptosis are confined within hypoxic tumor regions.

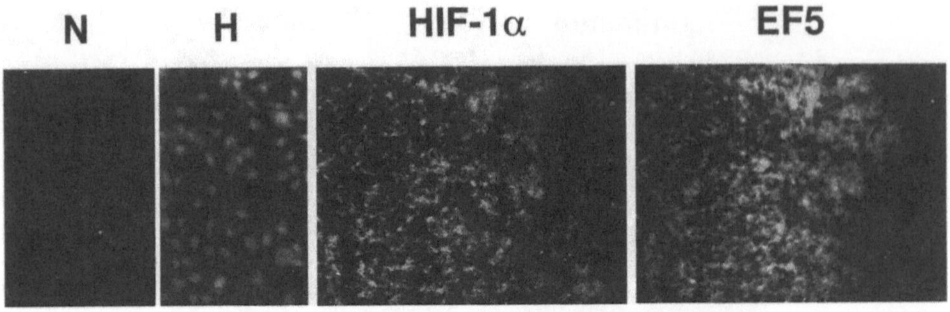

Figure 5. HIF-1α as a marker of tumor cell hypoxia. K1735 cells (left) exhibit more intense HIF-1α staining cultured under hypoxic conditions (H; 0.1% O_2) than under normoxia (N). Dual immunofluorescent staining of rmIL-12 treated K1735 tumors reveals areas of bright HIF-1α staining (middle) that co-localize with EF5 (right), indicating that HIF-1α can detect changes in tumor oxygenation with antiangiogenesis therapy *in vivo*.

K1735 apoptosis *in vitro*. This led us to conclude that this early phase of tumor cell death was unrelated to rmIL-12 antiangiogenic effects but, rather, was due to its induction of IFN-γ. The complexities of rmIL-12 as an antitumor therapeutic prompted us to examine the effects of a different antiangiogenic agent. We tested SelCID88, a structural thalidomide analog that was shown to inhibit angiogenesis in matrigel neovascularization assays. Treatment of K1735 tumors with this compound led to marked inhibition of tumor growth, development of regions of severe hypoxia and prevalent tumor cell apoptosis in the severely hypoxic regions (data not shown). Having confirmed these findings with another agent, we conclude that antiangiogenesis therapy produces tumor ischemia and that tumor control is likely the result of ischemic tumor cell death.

3.2 Assessment of tumor hypoxia by detection of HIF-1α protein

A problem confronting those testing antiangiogenic agents in clinical trials is determining treatment efficacy. Unlike cytotoxic chemotherapy which induces tumor regression if effective, antiangiogenic therapies that inhibit new vessel development without affecting preexisting vessels may only produce tumor stasis. The highly variable and frequently inconsistent rate at which many human tumors grow makes tumor growth a problematic indicator angiogenesis inhibition. The induction of tumor hypoxia with tumor angiogenesis inhibition led us to explore this as a way to monitor therapeutic efficacy. Agents such as nitroimidazoles that can identify hypoxic cells are being tested clinically[6] and may be useful for monitoring but, at present, their administration would complicate clinical trials of experimental therapeutic agents. We asked whether monitoring the reaction of tumor cells to hypoxia might be as effective for reporting angiogenesis inhibition. We examined HIF-1α as a potential monitoring tool, as it appears to be a principal mediator of hypoxic responses in a wide variety of cell types.[7] K1735 cells expressed low levels of HIF-1α protein under normoxic conditions that increased when the cells were incubated under hypoxic conditions (Figure 5). K1735 tumors, which do not exhibit detectable EF5 binding normally, also did not stain for HIF-1α.. rmIL-12 treated tumors, however, did exhibit areas of bright HIF-1α staining that co-localized with EF5 staining (Figure 5).

4. DISCUSSION

This study suggests histologic methods for visualizing the consequences of tumor angiogenesis inhibition. Cellular indicators of hypoxia such as HIF-1α offer the advantage of assessing tumor response to therapy. In this regard, HIF-1α may be an ideal marker because of its central role in mediating hypoxic responses in many cell types, although other oxygen-regulated proteins may prove similarly useful.[8] Bioreductively activated markers such as EF5 may be more useful in determining the degree of hypoxia achieved, as the hypoxic threshold for nitroimidazole binding may be less variable than that for HIF-1α induction among different cells. Currently, the most widely accepted indicator of successful angiogenesis inhibition is a decrease in tumor microvessel density. For this determination, tumor samples must be obtained before and after therapy. Thus, sample procurement may already be a part of clinical trials of angiogenesis inhibitors, setting the stage for validating this approach to monitoring antiangiogenic efficacy in human trials. Finally, the observation that antiangiogenesis therapy induces tumor cell hypoxia carries important physiologic implications regarding selection for cells resistant to other forms of therapy and potential therapeutic synergism between angiogenesis inhibitors and hypoxia-activated cytotoxic agents.

ACKNOWLEDGMENTS

We thank Genetics Institute for rmIL-12, Celgene Corporation for SelCID88 and Dr. Cameron Koch for EF5 reagents and expertise. Figures 1 and 2, published previously in *Cancer Research* (59: 4882-4889), were reproduced with permission from the American Association for Cancer Research. This work was supported by Medical Scientist Training Program (to M.S.G.) and NIH (R01 CA 77851 and CA 83042 to W.M.F.L.) awards.

REFERENCES

1. J. Folkman, Tumor angiogenesis: therapeutic implications, *N. Eng. J. Med.* **285**, 1182-1186, (1971).
2. N. J. Nelson, Inhibitors of angiogenesis enter phase III testing, *J. Natl. Cancer Inst.* **90**, 960-963 (1998).
3. M. S. Gee, et al., Hypoxia-mediated apoptosis from angiogenesis inhibition underlies tumor control by recombinant interleukin 12, *Cancer Res.* **59**, 4882-4889, (1999).
4. M. S. Gee, H. M. Saunders, J. C. Lee, J. F. Sanzo, W. T. Jenkins, S. M. Evans, G. Trinchieri, W. M. F. Lee, Doppler ultrasound imaging detects changes in tumor perfusion during antivascular therapy associated with vascular anatomic alterations, *Cancer Res.* **61**, 2974-2982, (2001).
5. C. J. Koch, S. M. Evans, and E. M. Lord, Oxygen dependence of cellular uptake of EF5: analysis of drug adducts by fluorescent antibodies vs bound radioactivity, *Br. J. Cancer* **72**, 869-874, (1995).
6. S. M. Evans, et al, Detection of hypoxia in human squamous cell carcinoma by EF5 binding, *Cancer Res.* **60**, 2018-2024, (2000).
7. G. L. Semenza, HIF-1: mediator of physiological and pathophysiological responses to hypoxia, *J. Appl. Physiol.* **88**, 1474-1480, (2000).
8. N. J. Beasley, C. C. Wykoff, P. H. Watson, R. Leek, H. Turley, K. Gatter, J. Pastorek, G. J. Cox, P. Ratcliffe, and A. L. Harris, Carbonic anhydrase IX, an endogenous hypoxia marker, expression in head and neck squamous cell carcinoma and its relationship to hypoxia, necrosis and microvessel density, *Cancer Res.* **61**, 5262-5267, (2001).

COMPARISON OF FLUCTUATIONS OF OXYGEN TENSION IN FSA, 9L, AND R3230AC TUMORS IN RATS

L. Isabel Cárdenas-Navia, Rod Braun, Kristina Lewis, Mark Dewhirst*

1 INTRODUCTION

Tumor hypoxia is known to reduce the sensitivity of tumors to radiation therapy.[1] Although hypoxia plays a major role in tumor response to therapy,[2,3] its exact causes are still unknown. In particular, the temporal characteristics of acute hypoxia have not been extensively explored. Previous publications have shown that PO_2 in tumors changes over time,[4,5] and that these changes may play an important and potentially radiobiologically significant role in tumor hypoxia. This study examines the kinetics of PO_2 fluctuations in FSA and 9L, and seeks to discover if there are differences in the power and frequency of these fluctuations between different tumor lines.

2 MATERIALS AND METHODS

2.1 Animals and Procedures

Twenty-three female Fischer 344 rats (Charles River Laboratories, Raleigh, NC) were used in this study. Eleven of the rats received subcutaneous implants of 1- to 2-mm^3 pieces of FSA rat fibrosarcoma tumors in the left hind limb; the other 12 received cell injections of 4-6 million 9L Glioma cells (K. Wheeler) subcutis on the left flank. After the tumors had reached 1-1.5 cm in diameter, rats were anesthetized with 50 mg/kg pentobarbital sodium, i.p. The femoral artery and vein were cannulated for later recording of BP and for venous access, respectively. A small portion (~4-10 mm^2) of the skin and tumor capsule was then removed to expose the surface of the tumor. This was

* Mark W. Dewhirst, Box 3455, Department of Radiation Oncology, Duke University Medical Center, Durham, NC, 27710, dewhirst@radonc.duke.edu

Oxygen Transport to Tissue XXIII, edited by D. Wilson *et al.*
Kluwer Academic/Plenum Publishers, 2003

kept moist by topical application of saline. A small incision was made in the left forelimb; a Ag/AgCl reference electrode was sutured into the subcutis. Rat body temperature was maintained by placing the rat on a temperature regulated water-heated blanket (K-module, Baxter Healthcare, Valencia, CA).

2.2 Oxygen Microelectrodes

Recessed-tip microelectrodes were produced using a previously published technique.[6] The resultant electrodes had tip diameters ~10 μm, with recess lengths on the order of 30 μm. The microelectrodes were pre- and post-calibrated and used as previously described.[7]

2.3 Intermittent Hypoxia Measurements

2.3.1 Experimental Setup

The rats were placed in a water-heated blanket and the left leg was stabilized on a rubber pedestal with tape. The arterial cannula was connected to a blood pressure transducer and amplifier (model 11-G4143-01; Gould Instruments, Valley View, OH) and the amplifier signal was digitized at 25 Hz and recorded using data acquisition software (AT-CODAS; Windaq, DATAQ Instruments, Akron, OH).

The electrode was inserted into the tumor and a non-zero PO_2 was found. The rats breathed air for 45 minutes, and were then switched to 100% O_2. After the experiment was completed, rats were sacrificed using euthasol, and in vivo zero-values for the microelectrodes were recorded.

2.4 Fourier Analysis

All Fourier analysis was performed using commercial software (CODAS, DATAQ Instruments, Akron, OH). The Fourier analysis was performed on records that contained 49,146 points per record (33 min length). The software averaged the records over every 5 or 6 points to obtain files that were 8,192 points in length. For the recordings during air-breathing, each record overlapped the previous record by 50%. For example, the first Fourier analysis of a PO_2 recording covered data from 0.0 to 32.9 min, the second from 12.0 to 44.9 min. For the O_2 breathing time a single analysis was measured from 57.0 to 89.9 min.

2.5 Analysis of Power Spectra

2.5.1 Power and Frequency of Fluctuations

The power of the Fourier power spectra (ie, the magnitude of the fluctuations) was analyzed in two different ways. The simplest method of comparing power between spectra is to look at the total power over a given frequency range.[8] Because most of the power in the present study was at very low frequencies, the range chosen for analysis was

0-2 cycles/min. The total power of a spectrum was obtained by a summation of all powers between 0 cycles/min and a given frequency. The total cumulative power was plotted as a function of frequency, and comparisons were made.

2.6 Statistics

All data were compared using non-parametric analysis. Differences among groups were tested using the Mann-Whitney U-test. Differences among paired data were tested using the Wilcoxon Signed-Rank test. Significance was achieved if $p<0.05$. All values presented in the text are means with standard deviations.

3 RESULTS

3.1 Air Breathing Results

3.1.1 Tumor PO_2 fluctuations

The duration of the PO_2 measurement period for FSA was 96.75 (2.55) min, and 91.61 (1.13) min (means) for 9L. Heart rate for all animals averaged 343 (10) bpm for the minute previous to beginning O_2 breathing, and 352 (4) bpm for the last 33 minutes of O_2 breathing. MABP averaged 113 (4) and 120 (5) mm Hg for air and O_2 breathing, respectively. Thus, systemic cardiovascular parameters were stable during these experiments.

An example of one FSA tumor recording is shown in Figure 1A. For this example, PO_2 slowly dropped from 8 mm Hg to almost zero during the air-breathing period. Shortly after the O_2 breathing was started, the PO_2 quickly increased to over 15 mm Hg. Further examination of the PO_2 trace reveals that the magnitude and frequency of fluctuations changed with the switch from air to O_2. During air breathing, the fluctuations varied less than ~1 mm Hg. After the rat began breathing O_2, the magnitude of the fluctuations increased to 4-5 mm Hg. For FSA overall, the average PO_2 during the last minute before O_2 breathing was 6 (2) mm Hg, and rose to 12 (3) mm Hg during the last 33 min of O_2 breathing. This increase in PO_2 was statistically significant.

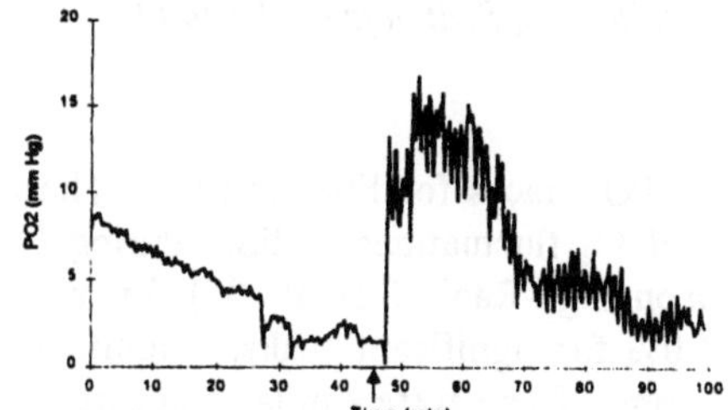

Figure 1A. Example PO_2 trace for FSA. Arrow indicates when O_2 breathing begins.

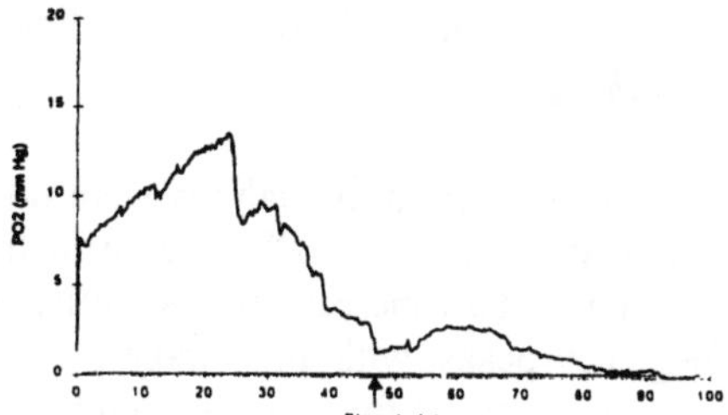

Figure 1B. Example PO_2 trace for 9L. Arrow indicates when O_2 breathing begins.

Figure 1B shows one trace of a 9L tumor recording. During the first 47 minutes of air breathing, PO_2 rose and dropped, varying from 3 to 14 mm Hg. Almost after the O_2 breathing began, PO_2 began to rise, and then gradually decreased to almost 0 mm Hg. The magnitude of fluctuations of PO_2 visibly changed when the rat was switched to O_2 breathing. During the air breathing, fluctuations vary 1-2 mm Hg; however, these fluctuations vary less than 1 mm Hg during the O_2 breathing. The average PO_2 during the last minute of air breathing was 9 (2) mm Hg, and was 10 (3) mm Hg for the last 33 minutes of O_2 breathing. This very slight increase in PO_2 was not found to be significant.

These FSA and 9L example traces showed characteristics of fluctuations of the distinct tumor types; however, there was no pattern for fluctuations for either tumor type.

Comparison of the baseline PO_2 for FSA and 9L showed there was no significant difference between the two.

3.2 Fourier Analysis of tumor PO_2 fluctuations

3.2.1 Air Breathing

Fourier analysis of the tumor PO_2 recordings showed that the fluctuations typically occurred at very low frequencies (<1 cpm) for both FSA and 9L. There were differences in the power spectra, however, indicating differences in the magnitude of the fluctuations. Figure 2 shows cumulative power vs. frequency of the power spectra for FSA and 9L over a frequency rage of 0-2 cpm. It is evident that the cumulative power of the 9L > FSA. The fluctuations in 9L PO_2 were largest of the tumors studied. FSA had a smaller magnitude of fluctuations, similar to that of R3230.[7]

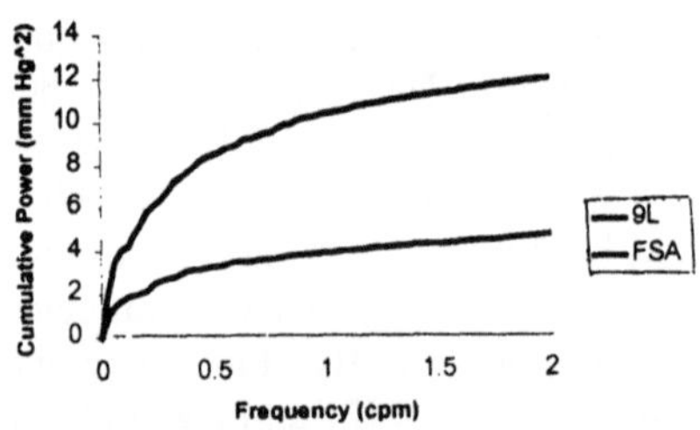

Figure 2. Comparison of PO_2 cumulative power for the same frequencies for FSA and 9L.

3.2.2 Changes in Frequency and Power in PO_2 following Switch from Air to O_2 Breathing

Fourier transform analyses were done for the PO_2 traces for FSA and 9L. These results showed an overall increase in frequency of O_2 fluctuations in FSA during O_2 breathing. This increase was significant (Wilcoxon Sign-Rank Test, $P<.05$) for many frequency peaks, with an average increase of 11.0% for significant peaks. Significant peaks occurred at frequencies from 0.23 to 1.42 cpm. There were similar comparisons for the changes in PO_2 power for FSA. A significant increase in magnitude of the fluctuations occurred over a large frequency range, 0.10-1.93 cpm, when O_2 breathing began. This change in magnitude increased by an average of 370.7% at the significant peaks. This is compatible with the visible increase in the magnitude of the fluctuations in figure 1A.

The relative frequency of fluctuations for 9L did not change when animals were switched from air to O_2. In contrast to FSA there was a significant decrease in the magnitude of the fluctuations for nearly all peaks in 9L, with an average 96.6% drop for the significant peaks, which corresponded to frequencies ranging from 0.06-2.35 cpm.

4 DISCUSSION

The dominant frequencies at which the fluctuations in both tumors occur are very low (<2 cpm); this range is too low to be explained by heart rate or influenced by microelectrodes, which have response rates of 40-100 ms.[9] Additionally, the fluctuations in PO_2 in FSA, 9L, and R3230[7] all occur at similar frequencies, with only the magnitudes varying, suggesting that these fluctuations are caused by a common mechanism in all three tumor lines.

Previous work has shown that red blood cell flux can cause measurable changes in PO_2, and can cycle in periods of 20-30 min[4]. Fluctuations seen in PO_2 could be caused by changes in red cell flux, which could be affected by variations in distribution at bifurcation points in tumor blood vessels or vasoactivity in these vessels.

The results of this study also show that power and frequency of PO_2 are impacted by hyper-oxic breathing, but the two tumors show different reactions to the O_2 breathing. After the O_2 breathing is turned on, the power of 9L fluctuations decreased significantly; whereas the power of FSA fluctuations increased significantly, as did the frequency, corresponding with an increase in PO_2. These effects suggest that the frequency and magnitude of fluctuations may play an important, although incompletely defined, role in changing tumor oxygenation. Future work in this laboratory will focus on further exploration of the importance of these low frequency fluctuations.

Previous studies of transient hypoxia have been conducted in a manner that could not assess the role of frequency and magnitude of PO_2 fluctuations in transient hypoxia. Studies of transient hypoxia using a dye mismatch method, for example, cannot measure these parameters.[10] Similar studies evaluating red cell flux using laser Doppler flowmetry probes do not directly measure PO_2.[11] Fourier transform analysis adds considerable information regarding the kinetics of intermittent hypoxia that may be helpful in optimizing O_2 transport in tumors.

Acknowledgements

This work was supported by a grant from the NIH/NCI CA40355.

REFERENCES

1. R. H. Thomlinson, and L.H. Gray, The histological structure of some human lung cancers and the possible implications for radiotherapy, *Br. J. Cancer* **9**, 539-549 (1955).
2. M. Hoeckel, C. Knoop, K. Schlenger, B. Vorndran, E. Bauβann, M. Mitze, G. Knapstein, and P. Vaupel, Intratumoral PO_2 predicts survival in advanced cancer of the uterine cervix, *Radiother. Onc.* **26**, 45-50 (1993).

3. D. M. Brizel, S.P. Scully, K.M. Harrison, L.J. Layfield, J.M. Bean, L.R. Prosnitz, and M.W. Dewhirst, Tumor oxygenation predicts for the likelihood of distant metastases in human soft tissue sarcoma, *Cancer Res.* **56**, 941-943 (1996).
4. H. Kimura, R.D. Braun, E.T. Ong, R. Hsu, T.W. Secomb, D. Papahadjopoulos, K. Hong, and M.W. Dewhirst, Fluctuations in red cell flux in tumor microvessels can lead to transient hypoxia and reoxygenation in tumor parenchyma, *Cancer Res.* **56**, 5522-5528 (1996).
5. D. J. Chaplin, P.L. Olive, and R.E. Durand, Intermittent blood flow in a murine tumor: radiobiological effects, *Cancer Res.* **130**, 171-182 (1992).
6. R. A. Linsenmeier, and C.M. Yancy, Improved fabrication of double-barreled recessed cathode oxygen microelectrodes, *J. Appl. Physiol.* **63**, 2554-2557 (1987).
7. R. D. Braun, J.L. Lanzen, and M.W. Dewhirst, Fourier analysis of fluctuations of oxygen tension and blood flow in R3230Ac tumors and muscle in rats, *Am. J. Physiol.* **277**, H551-H568 (1999).
8. R. D. Braun, R.A. Linsenmeier, and C.M. Yancey, Spontaneous fluctuations in oxygen tension in the cat retina, *Microvas. Res.* **44**, 73-84 (1992).
9. G. Schneiderman and T.K. Goldstick. Oxygen electrode design criteria and performance characteristics: recessed cathode. *J. Appl. Physiol.* **45**, 145-154, 1978.
10. M. W. Dewhirst, B. Klitzman, R.D. Braun, D.M. Brizel, Z.A. Haroon, T.W. Secomb, Review of methods used to study oxygen transport at the microcirculatory level, *Int. J. Cancer* **90**, 237-255 (2000).
11. D. J. Chaplin, and S.A. Hill, Temporal heterogeneity in microregional erythrocyte flux in experimental solid tumors, *Br. J. Cancer* **71**(6), 1210-1213 (1995).

EFFECT OF CARBOGEN-BREATHING ON REDOX STATUS OF THE RIF-1 TUMOR

Govindasamy Ilangovan, Haiquan Li, Jay L. Zweier, and Periannan Kuppusamy*

1. INTRODUCTION

Tissue redox status and pO_2 are two factors that are hypothesized to be different in normal and malignant tissues. Methods that can detect subtle differences in the above physiological parameters would greatly aid in devising appropriate treatment strategies[1,2]. We have previously used *in vivo* electron paramagnetic resonance (EPR) spectroscopy and imaging techniques and showed that tumor tissues are highly reducing and hypoxic compared to normal tissues[3]. The purpose of the present study was to obtain spatially resolved redox data from tumor tissue and to examine the effect of oxygenation on the tumor redox status.

2. MATERIALS AND METHODS

2.1. Chemicals

The EPR oximetry probe, LiPc, was prepared as reported[4]. The nitroxide probes 3-CP (3-carbamoyl-2,2,5,5-tetramethylpyrrolidine-N-oxy) and ^{15}N-TPL (Tempol, 4-hydroxy-2,2,6,6-tetramethyl piperidine-^{15}N-oxyl) were purchased from Aldrich (Milwaukee, WI) and C/D/N isotopes (Quebec, Canada), respectively. Solutions of the nitroxide were freshly prepared at a stock concentration of 150 mM in saline.

2.2. EPR Spectroscopy and Imaging Measurements

The EPR spectroscopic and imaging measurements were performed using L-band EPR imaging instrumentation[3]. The EPR spectral acquisitions and processing were performed using custom-developed data acquisition software. Typical calibration curves obtained for LiPc microcrystals at L-band EPR spectrometer using the topical resonator

* EPR Center, Division of Cardiology, Johns Hopkins University, School of Medicine, Baltimore, Maryland 21224, USA. Email: kuppu@welch.jhu.edu

Oxygen Transport to Tissue XXIII, edited by D. Wilson *et al.*
Kluwer Academic/Plenum Publishers, 2003

showed linear response from 0-100 mmHg of oxygen partial pressure with slope 8.9 mG/mmHg. Using these values the experimentally measured EPR spectral line widths are converted into pO_2 data.

2.3. Mice and Tumor Growth

Female C3H mice (weight: 25±5 g) were used. The mice were injected subcutaneously in their right hind leg with a single cell suspension of tumor cells (10^6 cells in 0.1 ml) in PBS. Tumors were allowed to grow to a size of about 8-10 mm in the greatest dimension.

2.4. Animal Preparation and Implantation of the LiPc Probe

The mice were anesthetized with isoflurane-air mixture and about 25 μg of the LiPc powder was implanted into tumor using a 21-gauge needle loaded with LiPc in the tip, and a wire stylus. All the *in vivo* EPR measurements of LiPc were made 48 hrs after the implantation. The tail vein was cannulated with heparin-filled 30-gauge catheter for infusion of nitroxide probes. The animal was placed on top of the resonator such that the tumor or the normal muscle was in direct contact with the active surface of the resonator[3]. Isoflurane (1.5% mixed with the breathing gas mixture, carbogen or room air) supply was maintained during EPR measurements using a Vasco Anesthesia veterinary anesthesia system (ProTech Medical, Hazel Crest, IL). A thermistor rectal probe was used to monitor body temperature. The body temperature was maintained at 37±1°C using an infrared lamp.

3. RESULTS AND DISCUSSION

3.1. Redox Pharmacokinetics of Nitroxides at Different Oxygenation Levels

In order to study the effect of oxygenation on the redox status of tumor tissue we performed simultaneous measurements of tumor pO_2 and redox status. Since the LiPc implanted in the tumor can report the oxygen level, the kinetics of the depletion of the exogenously infused nitroxide in the tumor tissue can be simultaneously followed and the effect of oxygenation on the redox reaction can be inferred from the difference in the rate constants[5]. Conventional nitroxide probes with ^{14}N nucleus will give a triplet EPR signal, in such a way that the middle peak of the triplet signal overlaps with the LiPc signal. Hence we used a ^{15}N nitroxide since such a nitroxide gives only a doublet, and the LiPc signal can be obtained in the middle of the doublet without any significant distortions, enabling one to simultaneously measure both the LiPc line width and the nitroxide signal intensity. However, the EPR experimental conditions, mainly the microwave power and modulation amplitude, used to obtain the undistorted LiPc signal are different from those used for the nitroxide signal. Thus, the LiPc and nitroxide signals were recorded using appropriate microwave power and modulation amplitude in separate acquisitions one after another continuously. In the present study we used ^{15}N-TPL as a redox probe. First the oxygenation (pO_2) of the tumor was measured from the EPR line width of the

implanted LiPc signal and found to be 2.3±0.4 mmHg (N=5). Then, ^{15}N-TPL (185 mg/kg body weight) nitroxide solution was infused through the tail vein and the EPR spectral acquisition was started simultaneously. The EPR spectrum obtained immediately after infusion showed only the LiPc signal at 446 G (the middle peak in Figure 1A). Subsequently a composite spectrum, containing both LiPc and nitroxide signals, was obtained (Figure 1A). It had a doublet peak at around the field value 432.5 (low-field peak) and 454.5 G (high-field peak) corresponding to the nitroxide and a strong signal in the middle from the LiPc in the tumor. The intensity (double integral of the low-field peak) of the ^{15}N-TPL signal was used to follow the pharmacokinetics and the LiPc signal (recorded at low power and modulation amplitude) was used to estimate the tumor pO_2. The decrease in the nitroxide signal intensity measured at the tumor site as a function of time is plotted in the Figure 1B (plot a). The LiPc spectrum did not show any change during the pharmacokinetic measurements indicating that the tissue pO_2 remained constant. In another set of experiments, tumor bearing animals were allowed to breathe carbogen for 20 minutes. After 20 minutes, first the LiPc spectrum was recorded and the line width was found to be higher and the corresponding pO_2 was measured to be 98±2 mmHg (N=5). Then, the same dosage of nitroxide was infused and the intensity of the nitroxide and the line width of the LiPc signals were followed.
The pharmacokinetic data obtained at this hyperoxygenation condition are presented in the Figure (plot b). The LiPc line width remained more or less constant during these

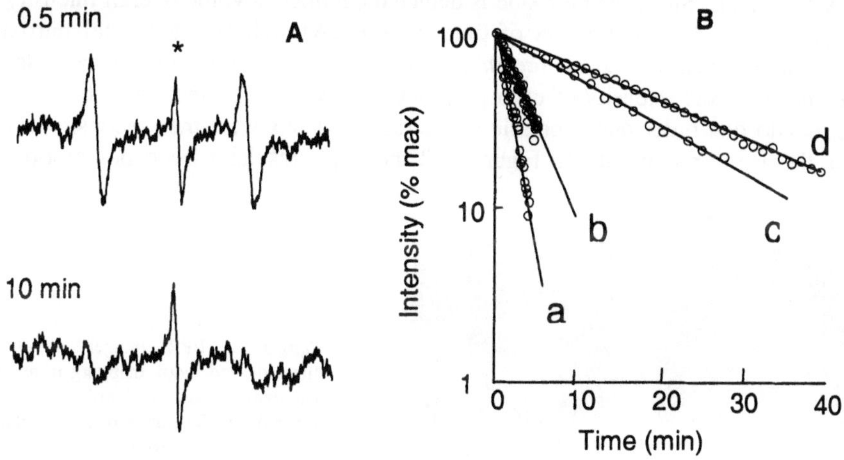

Figure 1. (A) Co-registration of the EPR spectra of implanted LiPc (oximetry probe; marked by *) and ^{15}N-TPL (redox probe) in the tumor at different times as indicated. The ^{15}N-TPL signal (doublet) decreased with time. (B) Pharmacokinetics of ^{15}NTPL (a & b) and 3-CP (c & d) measured from the tumor tissue of room air-breathing (a &c) and carbogen-breathing (b & d) mice.

pharmacokinetic measurements as well. As can be seen from Figure 1B, the relative intensity (the maximum was taken as 100%) of the ^{15}N-TPL changed with time at different magnitude when room air or carbogen was breathed. However, due to the very fast reduction rates of TPL, no significant differences between air and carbogen-breathing mice were observed (N=5). Hence the dependence of the tumor redox status on the oxygenation was studied by using another nitroxide spin label, namely 3-CP (with ^{14}N), which is reduced at relatively slower rate in normal and tumor tissues compared to TPL[3]. It was observed that the depletion rate was considerably reduced when the tumor was hyperoxygenated, as can be seen from Figure 1B (plots c & d). However, in this case we did not evaluate the pO_2 since the LiPc EPR signal merged with the middle peak of the triplet signal obtained for ^{14}N 3-CP. The nitroxide clearance rate constants (pseudo first order) were calculated to be 0.068±0.005 to 0.048±0.005 min^{-1} for normal and carbogen-breathing mice (N=5), respectively. It should also be noticed that compared to the TPL, 3-CP is cleared at slower rates, which is consistent with previous reports[3]. Such a slow clearance can be used advantageously for the EPR imaging applications, to obtain spatially resolved information of the redox status and its dependence on tumor pO_2.

3.2. EPR Imaging and Redox Mapping in RIF-1 Tumors

The pharmacokinetics data obtained using EPR spectroscopy denotes global evaluation of the redox activity in the whole tumor. In order to further delineate the spatially resolved differences in the tumor redox status, we used EPR imaging based pharmacokinetics and redox mapping in tumors. The spatial EPR images were collected at various timings after the infusion of the 3-CP probe through tail vein as previously reported[3]. The 2D images were collected at regular time intervals of 90 sec per image. The images were reconstructed from projections using filtered back projection methods as 128 x 128 pixels. Since the nitroxide is depleted continuously, the overall intensity of the image decreased as a function of time. Figure 2A1 indicates a typical nitroxide intensity map obtained from an air-breathing mouse. Each pixel in the image was indexed and the change in the intensity of each pixel was followed as a function of time, fitted using a pseudo first order rate expression. The rate constants were then reconstructed to obtain a 2D map of rate constants. Figure 2A2 shows the resultant frequency plot of the

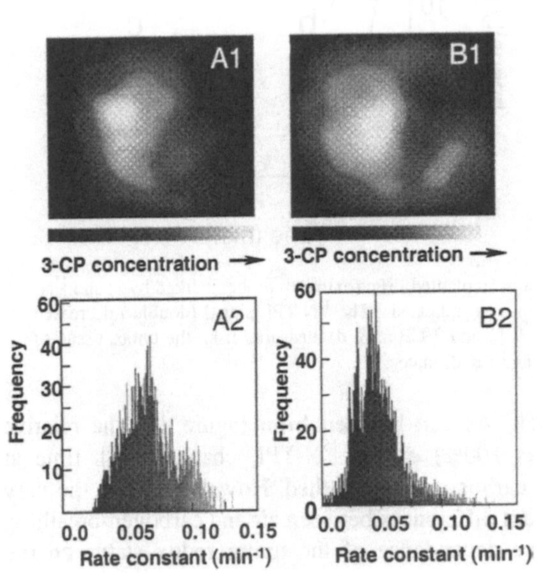

Figure 2. EPR images of 3-CP nitroxide and redox mapping in tumor. The images show the distribution of the nitroxide in the tumor of air-breathing (A1) and carbogen-breathing (B1) mouse, measured 5 min after infusion of the probe. Panels A2 and B2 show the frequency plots of nitroxide reduction rate constants obtained from pharmacokinetics data corresponding to A1 and B1 respectively. The distribution of rate constants, in the carbogen-breathing mouse (B2) is significantly narrowed and left-shifted compared to the air-breathing mouse (A2).

rate constants obtained from the image A1. It is clear from this frequency plot that the rate constants are distributed widely with a median value of 0.06 min^{-1}, which is close to the value obtained from the global EPR spectral pharmacokinetics data. Similar analyses of EPR imaging pharmacokinetics were made when the mice were breathing carbogen. The distribution of rate constants, in carbogen-breathing mice as seen in the Figure 2 B2, is significantly narrowed and less variable than that in air-breathing animal. Also the median value (0.04 min^{-1}) was significantly lower compared to the room air-breathing mice. The results indicate that the distribution of rate constants, and hence the tumor tissue redox state, is more heterogeneous in the case of air-breathing animal. On the other hand the distribution is significantly narrowed down in the case of carbogen-breathing suggesting that carbogen-breathing decreases the overall reducing capacity of the tumor tissue and it is less heterogeneous.

CONCLUSIONS

The present investigation demonstrates that tumor hyperoxygenation induced by carbogen-breathing significantly decreases the nitroxide reduction rate, which is a measure of tumor tissue redox state. The EPR imaging-based pharmacokinetics experiments indicate that the heterogeneity of tumor redox state over the entire tumor volume is considerably decreased by this hyperoxygenation approach.

ACKNOWLEDGMENT

This work was supported by NIH grant CA 78886.

REFERENCES

1. J. B. Mitchell, and A. Russo, The role of glutathione in radiation and drug induced cytotoxicity, *Br. J. Cancer* Suppl. **8**, 96-104 (1987).
2. I. J. Stratford, G. E. Adams, J. C. Bremner, S. Cole, H. S. Edwards, N. Robertson, and P. J. Wood. Manipulation and exploitation of the tumour environment for therapeutic benefit, *Int. J. Radiat. Biol.* **65**, 85-94 (1994).
3. P. Kuppusamy, M. Afeworki, R. A. Shankar, D. Coffin, S. M. Hahn, M. C. Krishna, J. B. Mithchell, and J. L. Zweier, *In vivo* electron paramagnetic resonance imaging of tumor heterogeneity and oxygenation in a murine model. *Cancer Res.*, **58**(7), 1562-1568 (1998).
4. G. Ilangovan, H. Li, J.L. Zweier, and P. Kuppusamy, Electrochemical preparation and EPR studies of lithium phthalocyanine: Evaluation of the nucleation and growth mechanism and evidence for potential dependent phase formation, *J. Phys. Chem.* B, **104**(17), 4047-4059 (2000).
5. Y. Miura, H. Utsumi, and A. Hamada, Effects of inspired oxygen concentration on *in vivo* redox reaction of nitroxide radicals in whole mice, *Biochem. Biophys. Res. Commun.*, **182**(3), 1108-1114 (1992).

EVALUATION OF MICROREGIONAL VARIATIONS IN TUMOR HYPOXIA FOLLOWING THE ADMINISTRATION OF ENDOSTATIN

Bruce M. Fenton, Scott F. Paoni, Brian K. Beauchamp, Baohuong Tran, Li Liang, Brian Grimwood, and Ivan Ding*

1. INTRODUCTION

Endostatin, a fragment of collagen XVIII, has been shown to potently inhibit both angiogenesis and the growth of experimental tumors, primarily through inhibition of endothelial cell migration and proliferation with minimal direct effects on tumor cells. Recent studies have also demonstrated that endostatin can enhance the antitumor effects of ionizing radiation, when administered before and during radiotherapy[1]. In order to optimize such combination therapies, an understanding of the accompanying changes in tumor pathophysiology, *i.e.*, oxygenation and blood flow, is clearly of prime importance. The primary aim of the current work was to implement an improved method for determination of the effects of endostatin on the microregional relationship between tumor perfusion and hypoxic marker uptake. By combining information from multiple images taken from the same frozen sections, additional pathophysiological information could be deduced that was unobtainable from the individual images.

2. METHODS

2.1. Animal Model

Cells from MCa-4 murine mammary carcinomas (3×10^6) were inoculated into the mammary fat pads of 6-8 week-old female C3H/HeJ mice (The Jackson Laboratory, Bar Harbor, ME). Tumor volumes were measured by calipers and the formula: volume = ½

*Bruce M. Fenton, Scott F. Paoni, Brian K. Beauchamp, Baohuong Tran, Li Liang, and Ivan Ding, Department of Radiation Oncology, University of Rochester School of Medicine, Rochester, NY 14642; Brian Grimwood, NY State Department of Health, Albany, NY 12201

Oxygen Transport to Tissue XXIII, edited by D. Wilson *et al.*
Kluwer Academic/Plenum Publishers, 2003

a^2b, where a and b are the minor and major tumor axes, respectively. Guidelines for the humane treatment of animals were followed as approved by the University Committee on Animal Resources.

2.2. Drug treatments

When tumors had grown to between 200-400 mm^3, recombinant murine endostatin (rmNYendo) was administered *i.p.* at 20 mg/kg for three consecutive days. Tumors were frozen at either day 4 or day 7, and 9.0 μm frozen sections were cut for later immunohistochemistry and image analysis.

2.3. $DiOC_7$ perfusion marker and EF5 hypoxic marker

To visualize blood vessels open to flow, an intravascular injected stain, $DiOC_7$, was used as described previously[2]. This agent has been shown to provide optimal visualization of tumor blood vessels by preferentially staining cells immediately adjacent to the vessels[3]. Localized areas of tumor hypoxia were assessed in frozen tissue sections by immunohistochemical identification of sites of 2-nitroimidazole metabolism as described previously[3]. A pentafluorinated derivative (EF5) of etanidazole was injected *i.v.* one hour before tumor freezing, at which time the EF5 has been shown to be well distributed throughout even poorly perfused regions of the tumor[4]. Regions of high EF5 metabolism were visualized immunohistochemically using a fluorochrome (Cy3, Amersham) conjugated to the ELK3-51 antibody. This antibody is extremely specific for the EF5 drug adducts that form when the drug is incorporated by hypoxic cells[5].

2.4. Immunohistochemistry and image analysis

Immediately following cryostat sectioning, tumor slices were imaged for $DiOC_7(3)$ perfusion staining using a Nikon microscope, digitized, background-corrected, and image-analyzed using Image-Pro software (Media Cybernetics, Silver Spring, MD) with a 800 MHz Pentium computer, as previously described in detail[3]. Color images from 16 adjacent microscope fields were automatically acquired and digitally combined. Each section was scanned under two different staining conditions. First, epi-illumination images of the fluorescent green $DiOC_7(3)$ staining were obtained immediately after the sections were sliced on the cryostat. Following the immunohistochemical staining procedures, the tumor section was returned to the same coordinates on the microscope stage, and fluorescent red-orange images were acquired showing the distribution of the EF5/Cy3 hypoxic marker staining. To account for variation in the intensity of the 100 W mercury lamp, calibrations images were obtained each day, using a hemocytometer filled with a reference concentration of Cy3 dye, as described in detail by Evans et al.[6]. Pixel intensities were then corrected for each image based on the ratio of the mean red intensities of the corresponding calibration images.

To obtain estimates of overall tumor hypoxia, fluorescent image montages of the EF5/Cy3 staining were quantified using the Image-Pro "histogram" tool to determine the mean intensity of the individual image pixels. To further quantitate microregional

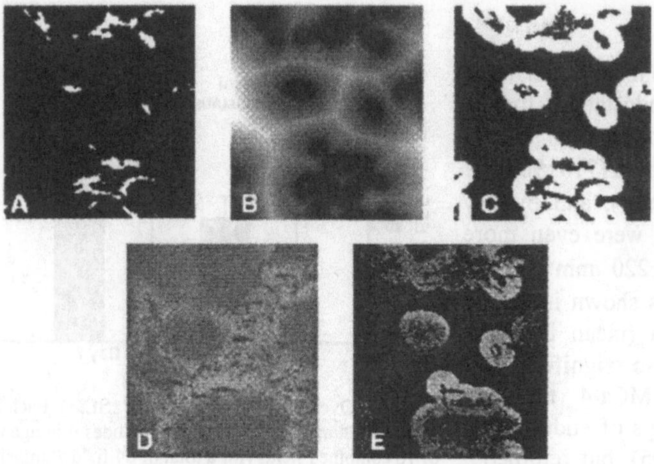

Figure 1. Image analysis procedures A) binary image of perfused vessels, B) distance filtered image, C) mask from thresholded distance map (includes gray levels from 1-20), D) EF5/Cy3 intensities (lighter intensities correspond to increased hypoxia), E) product of images C and D (selects only those EF5/Cy3 intensities within approximately 20 microns from the perfused vessels.

EF5/Cy3 intensities variations as a function of distance from perfused blood vessels, methods somewhat similar to those of Rijken et al.[7] were utilized. As summarized in Figure 1, $DiOC_7(3)$ images were first enhanced using the Image-Pro "color segmentation" tool to identify perfused blood vessels[3]. Specific colors were interactively selected and accumulated to obtain optimal discrimination between vessels and stroma, and a binary image of the selected colors was created (Figure 1A). This image was next inverted and a "distance filter" applied, which replaced the intensity of each pixel with an intensity proportional to the distance of that pixel from the nearest perfused vessel (Figure 1B). Thus, pixels immediately adjacent to the vessels were assigned intensity 1 in the new image, and the intensities of more distant pixels were increased by one gray level for each one pixel increase in distance from the vessel. This distance filtered image was then successively "thresholded" and binarized to select regions of the image within specific distances around perfused vessels. For example, thresholding between 1-20 gray levels selected a zone from 1-20 pixels away from a vessel, 21-40 selected a zone from 21-40 pixels away from a vessel, and so on. Figure 1C illustrates the binary image resulting from thresholding between gray levels of 1-20 (which selects regions within approximately 1-20 microns from the nearest vessel, since each pixel is roughly one micron across). This binary "mask" was then multiplied by the corresponding EF5/Cy3 image (Figure 1D) to obtain an image in which only those regions of the EF5/Cy3 image within the inner zone were included (Figure 1E). Finally, mean EF5/Cy3 intensities within this zone were determined using the Image-Pro "histogram" tool. EF5/Cy3 intensities of successive concentric zones at further distances from the perfused vessels were quantitated similarly and plotted as a function of distance from nearest perfused vessels.

3. RESULTS

At day four (one day after three daily doses of endostatin), treated tumors were significantly smaller than untreated controls (710±90 mm^3 versus 970±100 mm^3), and by day seven, differences were even more substantial (1670±220 mm^3 versus 1010±80 mm^3). As shown in Figure 2, overall hypoxia (mean EF5/Cy3 intensity) was also significantly reduced in the MCa-4 tumors following three days of endostatin treatment (p = 0.05), but returned to control levels by day 7.

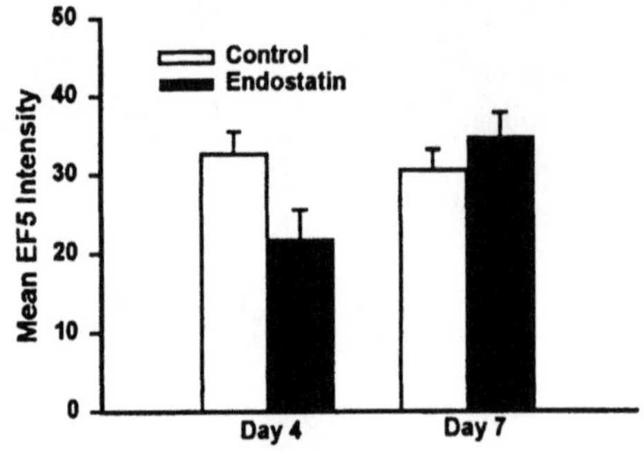

Figure 2 Overall EF5/Cy3 intensity (±SEM). Each bar is the mean of 6 tumors, each of which includes 4 image montages of 16 combined fields (for a total of 64 fields/tumor).

Figure 3 presents EF5/Cy3 intensities as a function of distance from the nearest perfused blood vessel, each point corresponding to the mean EF5/Cy3 intensity within a specific 20 micron wide zone (as detailed in Methods). EF5/Cy3 intensities within the first 20 micron zone are most closely related to adjacent intravascular oxygen levels, and the slopes of the curves are reflective of oxygen consumption rates in the surrounding tumor cells. Steeper slopes, which are indicative of a rapid increase in hypoxia with increasing distance from the perfused vessels, therefore correspond to increased consumption rates.

Despite the reduction in overall EF5 intensities at day 4 following endostatin (Figure 2), EF5 intensities within the zone closest to the nearest perfused vessel were not significantly reduced in relation to untreated tumors (Figure 3A), indicating that intravascular oxygen delivery was not substantially changed by treatment. In addition, the minimal difference in the slopes of these two curves suggests, at most, a slight decrease in oxygen consumption rate in the endostatin tumors. At day seven (Figure 3B), EF5 intensities of the zone closest to the blood vessels remain equal for endostatin and

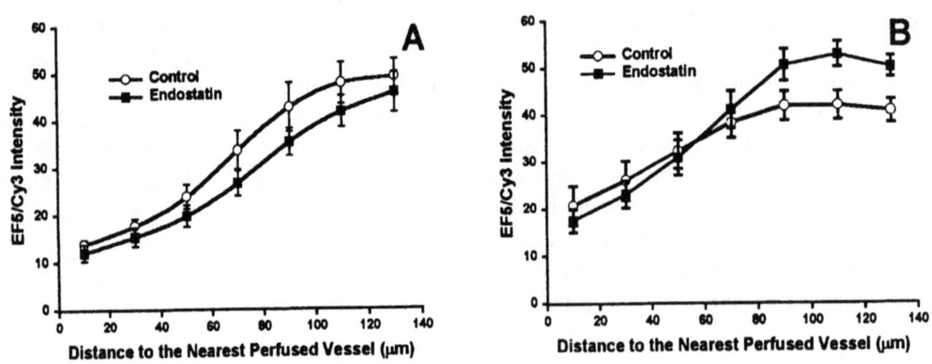

Figure 3 Mean EF5/Cy3 intensity within concentric zones surrounding perfused blood vessels. Each zone is approx. 20 microns in width. A) Day 4 (1 day following completion of treatment), B) Day 7 (4 days following treatment).

untreated tumors, suggesting little difference in intravascular oxygen delivery following treatment. At this time, however, the slope of the endostatin curve is increased in relation to controls, resulting in a significant increase in EF5 intensities for the zones furthest from the blood vessels. This is indicative of an increase in consumption in the endostatin tumors.

4. DISCUSSION

Although commonly used[8,9], changes in overall EF5/Cy3 intensities provide a somewhat oversimplified appraisal of the effects of endostatin on tumor hypoxia and can be misleading if large regions of necrosis are present (which do not metabolize EF5). The zonal analysis of EF5/Cy3 intensities as a function of distance to the nearest perfused vessels permits the acquisition of much more comprehensive information and can be limited to regions immediately surrounding perfused blood vessels, where necrosis would be less likely. From this microregional analysis, differences in tumor cell oxygen consumption rates and relative changes in intravascular oxygenation (which are related to vascular functionality) can both be estimated.

Since all images are based on two-dimensional slices through the tumors, vessels outside of the image plane are ignored. Thus EF5/Cy3 intensities could be locally reduced despite the absence of visible perfused vessels. Although this could potentially alter the relationship between EF5/Cy3 intensity and distance to the nearest vessel, such occurrences are expected to follow the distribution of visible vessels to some extent. Regions with high vascular densities will more likely have accompanying out of plane branches, which will be less likely for regions at large distances from any visible vessels. Although such out of plane branches would result in a decrease in absolute EF5/Cy3 intensities in these regions, the effects should be similar for different treatment groups and are not expected to mask relative differences among groups.

Previous reports of increased endothelial cell apoptosis[1] and decreased angiogenesis[10] following endostatin treatment suggest that this agent could result in a compromised vascular supply and a subsequent increase in tumor hypoxia. Somewhat surprisingly, overall tumor hypoxia was instead reduced following acute administration of endostatin. Since neither oxygen delivery nor consumption appear substantially changed, these results suggest that the relative proportion of functioning vessels may be increased in the endostatin treated tumors. This agrees with the recent proposal that antiangiogenic therapies could serve to improve vascular efficiency through the selective pruning of immature or less functional capillaries[11]. If confirmed in additional experiments, these results could provide an additional rationale for the reported improvement in response observed when combining endostatin with radiotherapy[1]. Further more comprehensive studies are needed to interpret these somewhat puzzling findings. Future experiments will investigate not only whether these increased oxygen levels are transient in nature, but also whether response to fractionated radiotherapy can be enhanced by extended endostatin administration.

ACKNOWLEDGMENTS

Financial support was provided by NIH Grant CA52586 and DOD Grant DAMD17-00-1-0420.

REFERENCES

1. N.N. Hanna, S. Seetharam, H.J. Mauceri, M.A. Beckett, N.T. Jaskowiak, R.M. Salloum, D. Hari, M. Dhanabal, R. Ramchandran, R. Kalluri, V.P. Sukhatme, D.W. Kufe, and R.R. Weichselbaum, Antitumor interaction of short-course endostatin and ionizing radiation, *Cancer Journal* **6**:287-293 (2000).
2. M.J. Trotter, D.J. Chaplin, and P.L. Olive, Use of a carbocyanine dye as a marker of functional vasculature in murine tumours, *Br.J.Cancer* **59**:706-709 (1989).
3. B.M. Fenton, S.F. Paoni, J. Lee, C.J. Koch, and E.M. Lord, Quantification of tumor vascular development and hypoxia by immunohistochemical staining and HbO_2 saturation measurements, *Br.J.Cancer* **79**:464-471 (1999).
4. B.M. Fenton, E.M. Lord, and S.F. Paoni, Effects of radiation on tumor intravascular oxygenation, vascular configuration, hypoxic development, and survival, *Radiat.Res.* **155**:360-368 (2001).
5. E.M. Lord, L. Harwell, and C.J. Koch, Detection of hypoxic cells by monoclonal antibody recognizing 2-nitroimidazole adducts, *Cancer Res.* **53**:5721-5726 (1993).
6. S.M. Evans, C.J. Koch, K.M. Laughlin, W.T. Jenkins, T. Vanwinkle, and D.F. Wilson, Tamoxifen induces hypoxia in MCF-7 xenografts, *Cancer Res.* **57**:5155-5161 (1997).
7. P.F.J.W. Rijken, H.J.J.A. Bernsen, J.P.W. Peters, R.J. Hodgkiss, J.A. Raleigh, and A.J. van der Kogel, Spatial relationship between hypoxia and the (perfused) vascular network in a human glioma xenograft: A quantitative multi-parameter analysis, *Int.J.Radiat.Oncol.Biol.Phys.* **48**:571-582 (2000).
8. B.M. Fenton, Influence of hydralazine administration on oxygenation in spontaneous and transplanted tumor models, *Int.J.Radiat.Oncol.Biol.Phys.* **49**:799-808 (2001).
9. S.M. Evans, S.M. Hahn, D.P. Magarelli, P.J. Zhang, W.T. Jenkins, D.L. Fraker, Hsi, RA, W.G. Mckenna, and C.J. Koch, Hypoxia in human intraperitoneal and extremity sarcomas, *Int.J.Radiat.Oncol.Biol.Phys.* **49**:587-596 (2001).
10. M.S. O'Reilly, T. Boehm, Y. Shing, N. Fukai, G. Vasios, W.S. Lane, E. Flynn, J.R. Birkhead, B.R. Olsen, and J. Folkman, Endostatin: An endogenous inhibitor of angiogenesis and tumor growth, *Cell* **88**:277-285 (1997).
11. R.K. Jain, Normalizing tumor vasculature with anti-angiogenic therapy: A new paradigm for combination therapy, *Nat Med* **7**:987-989 (2001).

EFFECT OF HYDRALAZINE ON BLOOD FLOW, OXYGENATION, AND INTERSTITIAL FLUID PRESSURE IN SUBCUTANEOUS TUMORS

Tomaž Jarm, Blaž Podobnik, Gregor Serša, and Damijan Miklavčič*

1. INTRODUCTION

Many experimental and clinical solid tumors are known to exhibit poor oxygenation[1-3] and high interstitial fluid pressure[4, 5] (IFP) in comparison to normal surrounding tissues. Oxygenation and IFP in addition to abnormal blood flow influence the response of tumors to various therapies and are also important for development and progression of malignant growth.[1-5] Hydralazine (HYZ) is an arteriolar vasodilating drug which is used for treatment of hypertension in humans.[6] HYZ can decrease both tumor blood flow[7-10] and tumor IFP.[9, 11] These two changes may have opposite effects on drug delivery and oxygenation in tumors. In our study we evaluated the effect of HYZ on blood flow, oxygenation, and IFP in a murine tumor model. Oxygenation was measured by a relatively new method which has recently become available, the OxyLite system.[12,13]

2. MATERIALS AND METHODS

2.1. Tumor model and anesthesia

The study was performed on solid subcutaneous SA-1 fibrosarcoma tumors (Jackson Lab., Bar Harbor, U.S.A.) growing dorsolaterally in the right flank of A/J mice (Rudjer Boskovic Inst., Zagreb, Croatia). Tumors were inoculated by injection of 5×10^5 viable SA-1 cells. Measurements were performed 8-10 days after inoculation when the tumors reached a volume of approximately 100 mm^3. Mice were anesthetized during experiments by isoflurane (concentration 1.7%) delivered via a miniature face mask in a mixture of O_2 and N_2O. Physiological temperature of mice was maintained by a regulated heating pad. At the end of experiments the mice were euthanized under anesthesia by cervical

* Tomaž Jarm, Blaž Podobnik, and Damijan Miklavčič, Faculty of Electrical Engineering, University of Ljubljana, Tržaška 25, Ljubljana, Slovenia SI-1000. Gregor Serša, Institute of Oncology, Zaloška 2, Ljubljana, Slovenia SI-1000.

Oxygen Transport to Tissue XXIII, edited by D. Wilson *et al.*
Kluwer Academic/Plenum Publishers, 2003

dislocation. Experimentation on mice was conducted in accordance with the pertaining legislation and was approved by the Veterinary Administration of Ministry of Agriculture, Forestry and Food of Slovenia (permit number 323-02-156/99).

2.2. Measurement methods

Relative blood flow in tumors was measured by laser Doppler flowmetry (LDF). A two-channel instrument OxyFlo (Oxford Optronix, U.K.) was used in our study with invasive bare optical fiber sensors (diameter 0.2 mm).

Partial pressure of oxygen (pO_2) in tumors was measured by a novel luminescence-based fiber-optic technique. This method employs a ruthenium luminophore, which is incorporated in a silicone cap fixed to the tip of a fiber-optic sensor.[12, 13] Pulses of blue light carried via the fiber excite pulsatile fluorescence of the luminophore. The life-time of this fluorescence is inversely proportional to local pO_2 in tissue in contact with the probe tip. The advantage of this method over the well established polarographic oximetry is that the fiber-optic pO_2 sensor does not consume oxygen and should therefore enable continuous monitoring of pO_2. A two-channel instrument OxyLite with combined pO_2 and temperature sensors (diameter 0.3 mm) were used in our study (Oxford Optronix, U.K.).

Interstitial fluid pressure (IFP) in tumors was measured by the so-called "wick-in-needle" technique.[11, 14] The system used in our study consisted of needle probes (diameter 0.5 mm), TSD104A pressure transducers, a DA100A amplifier, and an MP100 data acquisition unit (Biopac Systems, U.S.A.).

All probes were inserted into tissue approximately 5 minutes after the start of anesthesia in mice. The IFP probe was inserted into tumor from the caudal side. Both pO_2 and both LDF probes were inserted into tumor from the cranial side. All probes were slightly withdrawn after insertion to minimize tissue compression.

2.3. Hydralazine treatment

Hydralazine (Sigma Chemical, U.S.A.) was dissolved in sterile physiological saline and injected *i.v.* at a dose of 2.5 mg/kg of mouse weight. Control mice were injected with physiological saline only. Injections were made during measurements and only if all recorded signals had been stable for at least 10 minutes prior to injection.

3. RESULTS AND DISCUSSION

After injection of hydralazine (HYZ) blood flow started to decrease in all measurement locations. On average blood flow decreased by 50% within 5-10 minutes (Fig. 1a). No change was observed in control tumors. The results are in good agreement with the previously documented effect of HYZ on tumor blood flow.[7-10]

Two distinct types of pO_2 signals were observed after insertion of a probe into tumor (Fig. 2). First, pO_2 decreased within one minute to near zero level in all cases (the *decrease phase*). In about 1/3 of measurements this near-to-zero pO_2 (the *zero pO_2 phase*) was maintained during the whole period of observation (the *type II* signals). In the remaining 2/3 of measurements (the *type I* signals) pO_2 started to increase after 5-20 minutes (the *increase phase*) and finally stabilized at a new level (the *plateau phase*) 20-30 minutes after insertion of the probe. This plateau phase pO_2 varied among different

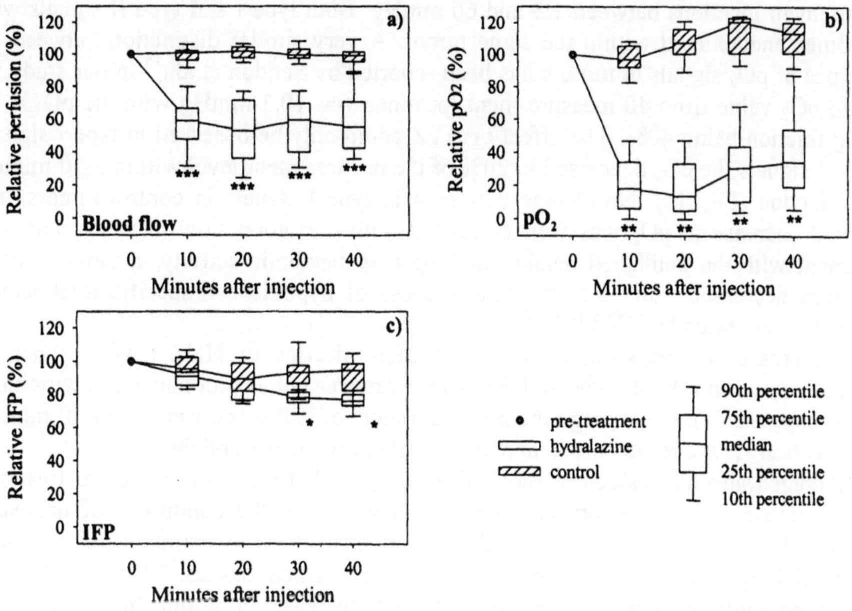

Figure 1. The effect of hydralazine on blood flow (a), oxygenation (b), and interstitial fluid pressure (c) in tumors. All individual values measured after injection were expressed relatively with respect to the pre-treatment value (100%) before averaging. Statistical significance of differences between HYZ-treated and control tumors was calculated by Mann-Whitney rank sum test (*p<0.05, **p<0.01, ***p<0.001).

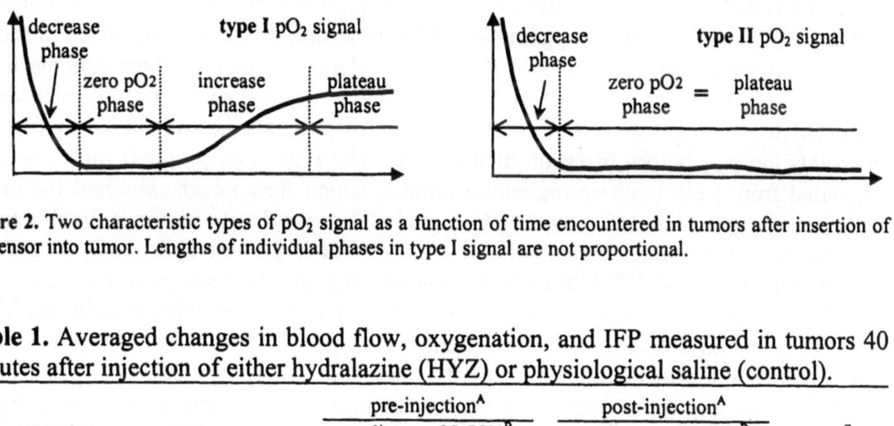

Figure 2. Two characteristic types of pO$_2$ signal as a function of time encountered in tumors after insertion of the sensor into tumor. Lengths of individual phases in type I signal are not proportional.

Table 1. Averaged changes in blood flow, oxygenation, and IFP measured in tumors 40 minutes after injection of either hydralazine (HYZ) or physiological saline (control).

			pre-injection[A]		post-injection[A]		
parameter	group	n	median	25-75%[B]	median	25-75%[B]	p[C]
blood flow (BPU)	HYZ	17	**245**	120-431	**138**	88-331	**<0.001**
	control	8	**540**	341-790	**545**	338-766	**0.938**
pO$_2$ (mmHg)	HYZ	13	**17.7**	2.9-36.6	**4.6**	0.4-11.0	**<0.001**
	control	6	**22.2**	13.6-32.7	**27.3**	14.9-36.8	**0.109**
IFP (mmHg)	HYZ	11	**10.5**	6.6-13.0	**8.2**	5.2-9.7	**<0.001**
	control	6	**10.1**	6.3-11.5	**8.8**	6.0-9.6	**0.219**

[A] Pre- and post-injection means just before injection and 40 minutes after injection respectively.
[B] The spread of data around the median value is given with the 25th and the 75th percentile limits.
[C] Statistical significance of change from the pre- to the post-treatment value (Wilcoxon signed rank test).

measurement locations between 1.5 and 60 mmHg. Both type I and type II signals were sometimes encountered within the same tumor. A very similar distinction between the two types of pO_2 signals in tumors has been reported by Seddon et al..[15] In our study the median pO_2 value from 40 measure-ment locations was 10.3 mmHg with the $pO_2 < 2.5$ mmHg fraction being 40%. The effect of HYZ could only be observed in type I signals. In type I signals the pO_2 decreased to 20% of the pre-treatment level within 5-10 minutes after injection (Fig. 1b). No change was seen in type I signals in control tumors. The observed decrease in pO_2 was well correlated with the blood flow decrease and is in agreement with the published results showing that metabolic activity in tumors due to HYZ was decreased[16] and that the effectiveness of hypoxic cell-specific treatment in tumors was increased by HYZ.[8, 10, 17]

A decrease was also observed in IFP after injection of HYZ (Fig. 1c) but the difference between HYZ-treated and control tumors was not as pronounced as with blood flow or oxygenation. On average the minimum level of IFP was observed 30-40 minutes after injection. These results are in agreement with previously published data.[9, 11]

Averaged absolute values of blood flow, pO_2, and IFP in tumors before treatment and 40 minutes after treatment are shown in Table 1. In the conditions of decreased systemic blood pressure caused by HYZ the organism tries to maintain normal blood flow in vital organs by reducing the blood flow to less critical tissues.[16] This "steal" effect is most probably responsible for the observed decrease in tumor blood flow and subsequently for the decrease in tumor pO_2.

The initial decrease of pO_2 immediately after the insertion of the probe is due mostly to the consumption of oxygen brought into tissue during the insertion. The multiphase behavior of type I pO_2 signals during the stabilization period after this initial decrease was most probably a result of tissue damage caused during the insertion of probes into tissue. Torn and occluded capillaries[18, 19] and vasoconstrictive reactions[15] may lead to decreased blood flow and deoxygenation of tissue in immediate vicinity of the insertion track and at the probe tip (the *zero pO_2 phase* in type I signals). Afterwards the tissue may be reoxygenated (the *increase pO_2 phase*) due to gradual restoration of microcirculation but it is not yet clear how well the level of pO_2, reached in the *plateau phase* represents the pO_2 before insertion of the probe. The type II pO_2 signals most probably originated from truly hypoxic regions in tumors. Blood flow never exhibited the multiphase behavior encountered in type I pO_2 signals. The most likely reason is that the tissue sampling volume of the LDF method is larger than that of the new optical pO_2 sensor.

The relatively high IFP in tumors in comparison to normal tissue was associated with poor oxygenation and resistance of tumors to radiotherapy, and with insufficient blood flow and impaired delivery of chemotherapeutic drugs to tumor cells.[4,5] Elevated IFP in tumors can contribute to development of hypoxia in tumors by decreasing the blood flow.[4, 5] Decreasing the elevated IFP may therefore improve tumor oxygenation. Application of HYZ in our study, however, decreased both the IFP and the pO_2 in tumors.

4. CONCLUSIONS

Blood flow, oxygenation, and IFP in SA-1 tumors in A/J mice were decreased by HYZ. The pronounced decrease in blood flow indicates that delivery of a chemotherapeutic drug and oxygen would be severely impaired by HYZ at the dosage used. The measured decrease in oxygenation proves that. Even though the new luminescence-based fiber-optic oximetry proved to be useful for detection of oxygenation changes, further investigation is needed to verify the absolute pO_2 values measured by this method.

ACKNOWLEDGEMENT

This study was supported by the Ministry of Education, Science and Sport of the Republic of Slovenia (grant J2-2222-1538) and by the European Commission and the 5th Framework Programme (grant QLK 3-99-00484, CLINIPORATOR project).

REFERENCES

1. M. R. Horsman, Measurement of tumor oxygenation, *Int. J. Radiat. Oncol. Biol. Phys.* **42**, 701-704 (1998).
2. M. Höckel, B. Vorndran, K. Schlenger, E. Baußmann, and P. G. Knapstein, Tumor oxygenation: a new predictive parameter in locally advanced cancer of the uterine cervix, *Gynecol. Oncol.* **51**, 141-149 (1993).
3. P. Vaupel, D.K. Kelleher, and O. Thews, Modulation of tumor oxygenation, *Int. J. Radiat. Oncol. Biol. Phys.* **42**, 843-848 (1998).
4. L. T. Baxter and R. K. Jain, Transport of fluid and macromolecules in tumors 1. Role of interstitial fluid pressure and convection, *Microvasc. Res.* **37**, 77-104 (1989).
5. M. F. Milosevic, A. W. Fyles, and R. P. Hill, The relationship between elevated interstitial fluid pressure and blood flow in tumors: a bioengineering analysis, *Int. J. Radiat. Oncol. Biol. Phys.* **43**, 1111-1123 (1999).
6. R. F. Albrecht, E.T. Toyooka, S. L. H. Polk, and B. Zahed, Hydralazine therapy for hypertension during the anesthetic and postanesthetic periods, *Int. Anesthesiol. Clin.* **16**, 299-312 (1978).
7. M. J. Trotter, B. D. Acker, and D. J. Chaplin, Histological evidence for nonperfused vasculature in a murine tumor following hydralazine administration, *Int. J. Radiat. Oncol. Biol. Phys.* **17**, 785-789 (1989).
8. J. Kalmus, P. Okunieff, and P. Vaupel, Dose-dependent effects of hydralazine on microcirculatory fuction and hyperthermic response of murine FSaII tumors, *Cancer Res.* **50**, 15-19 (1990).
9. R. A. Zlotecki, L. T. Baxter, Y. Boucher, and R. K. Jain, Pharmacologic modification of tumor blood flow and interstitial fluid pressure in a human tumor xenograft: network analysis and mechanistic interpretation, *Microvasc. Res.* **50**, 429-443 (1995).
10. M. R. Horsman, K. L. Christensen, and J. Overgaard, Relationship between the hydralazine-induced changes in murine tumor blood supply and mouse blood pressure, *Int. J. Radiat. Oncol. Biol. Phys.* **22**, 455-458 (1992).
11. B. Podobnik and D. Miklavčič, Influence of hydralazine on interstitial fluid pressure in experimental tumors – a preliminary study, *Radiol. Oncol.* **34**, 59-65 (2000).
12. D. R. Collingridge, W. K. Young, B. Vojnovic, P. Wardman, E. M. Lynch, S. A. Hill, and D. J. Chaplin, Measurement of tumor oxygenation: a comparison between polarographic needle electrodes and a time-resolved luminescence-based optical sensor, *Radiat. Res.* **147**, 329-334 (1997).
13. J. R. Griffiths and S. P. Robinson, The OxyLite: a fibre-optic oxygen sensor, *Br. J. Radiol.* **72**, 627-630 (1999).
14. H. O. Fadnes, R. K. Reed, and K. Aukland, Interstitial fluid pressure in rats measured with a modified wick technique, *Microvasc. Res.* **14**, 27-36 (1977).
15. B. M. Seddon, D. J. Honess, B. Vojnovic, G. M. Tozer, and P. Workman, Measurement of tumor oxygenation: in vivo comparison of a luminescence fiber-optic sensor and a polarographic sensor and a polarographic electrode in the P22 tumor, *Radiat. Res.* **155**, 837-846 (2001).
16. P. Okunieff, C. S. Walsh, P. Vaupel, F. Kallinowski, B. M. Hitzig, L. J. Neuringer, and H. D. Suit, Effects of hydralazine on in vivo tumor energy metabolism, hematopoietic radiation sensitivity, and cardiovascular parameters, *Int. J. Radiat. Oncol. Biol. Phys.* **16**, 1145-1148 (1989).
17. D. J. Chaplin and B. Acker, The effect of hydralazine on the tumor cytotoxicity of the hypoxic cell cytotoxin RSU-1069: evidence for therapeutic gain, *Int. J. Radiat. Oncol. Biol. Phys.* **13**, 579-585 (1987).
18. F. Steinberg, E. Hildenhagen-Brüggemann, and M. A. Konerding, Oxygen electrode injury in tumour tissue, in: *Tumor oxygenation*, edited by O. Thews, D. K. Kelleher, and P. W. Vaupel (Gustav Fischer Verlag, Stuttgart, 1995), pp. 186-193.
19. U. Schramm, W. Fleckenstein, and C. Weber, Morphological assessment of skeletal muscular injury caused by pO_2 measurements with hypodermic needle probes, in: *Clinical oxygen measurement II*, edited by A. M. Ehrly, W. Fleckenstein, J. Hauss, and R. Huch (Blackwell Ueberreuter Wissenschaft, Berlin, 1990), pp. 38-50.

OXYGENATION OF CERVIX CANCERS: IMPACT OF CLINICAL AND PATHOLOGICAL PARAMETERS

Michael Höckel* and Peter Vaupel**

1. INTRODUCTION

Current knowledge on the oxygenation status of carcinomas of the uterine cervix is mostly based on pretherapeutic data obtained in pre- and postmenopausal women. Mean and median O_2 tensions (pO_2) obtained from >20,000 pO_2-measurements in more than 400 squamous cell carcinomas of the uterine cervix observed by several research groups were, on average, distinctly lower than in normal tissues[1]. Oxygen tensions measured in the normal cervix of nullipara patients revealed a median pO_2 of 42 mmHg, whereas in conscious patients with locally advanced cancers of the cervix (stage FIGO Ib-IVa), the median pO_2 values were 10-20 mmHg[1]. In patients under general propofol-anesthesia the median pO_2 values were 3-8 mmHg[1]. Oxygenation status was found to be independent of various patient parameters (e.g., age, menopausal and parous state, use of oral contraceptives, smoking habits, Karnofsky index) or tumor parameters [e.g., clinical tumor size, FIGO stage, growth pattern (endophytic vs. exophytic)][2]. From the studies on the oxygenation status of locally advanced cervix cancers there is also clear evidence that the pO_2 distribution is extremely heterogeneous (both in space and in time) with no characteristic topological allocation of O_2 partial pressures within the tumor (e.g., tumor center vs. tumor periphery)[1]. Furthermore, there was no clear correlation found between the tumor pO_2 and vessel length density ("vascular density", mm/mm^3): Tumor regions with low vessel length density always showed low pO_2 values whereas tumor areas with higher vessel length densities can exhibit either a high or low median pO_2[3]. Tumor-to-tumor variability in oxygenation is found to be greater than intra-tumor variability. In cervix cancers of moderately/severely anemic patients hypoxic areas are more frequently detected than in non-anemic (cHb > 12 g/dL) patients. Approximately 60% of cervical cancers contain hypoxic tissue areas with pO_2 values $\leq$ 2.5 mmHg and local recurrences of cervix cancers have a higher hypoxic fraction than the primary tumors (38% vs. 25%) and lower median pO_2 values (4 mmHg vs. 10 mmHg)[4].

---

* Dept. Obstetrics & Gynecology, University of Leipzig, 04103 Leipzig, Germany
** Inst. Physiology & Pathophysiology, University of Mainz, 55099 Mainz, Germany

Oxygen Transport to Tissue XXIII, edited by D. Wilson *et al.*
Kluwer Academic/Plenum Publishers, 2003

Table 1. Oxygenation of the normal cervix and pretherapeutic oxygenation status of locally advanced cancers of the uterine cervix (measurements using the pO_2 histography system, 1990-2000, adapted from Höckel & Vaupel[12]).

	n	median pO_2 [mmHg]	mean pO_2 [mmHg]	f(0-2.5 mmHg) [%]	f(0-5 mmHg) [%]
normal cervix (nullipara)	7	42	40	8	13
primary SCC	95	11	16	25	38
FIGO IB	16	10	13	34	44
FIGO II	45	10	15	24	40
FIGO III	29	14	17	22	34
FIGO IVA	5	10	17	29	37
G1	8	15	18	14	23
G2	35	10	14	33	43
G3	19	11	14	19	39
primary AC	26	12	18	16	32
recurrent SCC	46	4	12	38	51
recurrent AC	14	6	17	31	47

n= number of patients (tumors) investigated, f = fraction; SCC = squamous cell carcinomas; AC = adenocarcinomas

Multivariate analyses have shown that hypoxia is a powerful prognostic factor in primary and recurrent carcinomas of the uterine cervix.[5-7] More recently, other authors have confirmed this finding, indicating that tumor hypoxia is an independent and adverse prognostic factor in locally advanced cancer of the uterine cervix with regard to overall and disease-free survival.[8-10] At the same time, tumor volume, FIGO stage (International Federation of Gynecology and Obstetrics stage), pathological stage, grade and histology are known to be clinically relevant prognostic parameters.

Based on these data, the main object of this updated investigation was to determine whether there is an association between tumor oxygenation (extent of hypoxia) and pathological tumor stage, grade and histology.

2. MATERIAL AND METHODS

Patient population, characterization of tumors, assessment of the pretreatment tumor oxygenation status and data evaluation have been described in detail earlier[2,4-6,11]. The study was approved by a Medical Ethics Committee, and informed consent was obtained from all patients.

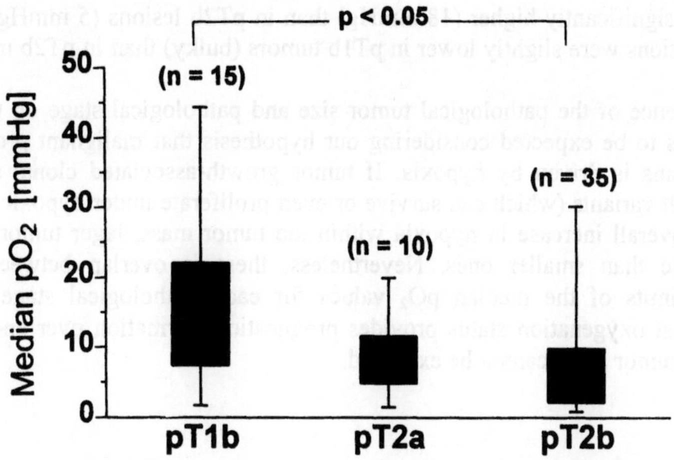

Fig. 1. Box and whisker plots of median pO_2 values of primary cervical carcinomas. Detailed pathological staging reveals stage-related differences in the oxygenation status of primary cancers of the uterine cervix.

3. RESULTS AND DISUSSION

Assessment of the pretreatment oxygenation status of locally advanced *primary cancers* of the uterine cervix in 121 patients revealed a median pO_2 value of 11 mmHg, a fraction of pO_2 values $\leq$ 2.5 mmHg of 24%, and a fraction of pO_2 values $\leq$ 5 mmHg of 37%. The percentage of patients with tumor pO_2 values $\leq$ 2.5 mmHg is 61%. All tumors of stage FIGO IVA are hypoxic.

The data characterizing the oxygenation status clearly indicate that there is no dependency on the maximum clinical tumor diameter[13,14] and on the clinical tumor stage according to the FIGO (see Table 1). In addition, oxygenation status and the extent of pretherapeutically measured hypoxic tissue areas are independent of the tumor histology (squamous cell carcinomas vs. adenocarcinomas), and histopathological grade (G1-G3; see Table 1). In contrast to these data, Lyng et al.[15] have reported a better oxygenation status of adenocarcinomas when compared with squamous cell carcinomas. In this latter study, the differences in pO_2 values between these two histologies were most pronounced at low pO_2. Our data also show a higher fraction of pO_2 values $\leq$ 2.5 mmHg in squamous cell carcinomas than in adenocarcinomas (25% vs. 16%, see Table 1). However, this difference was not statistically significant, and median and mean pO_2 values were similar in squamous cell carcinomas and in adenocarcinomas.

In a subgroup of 65 patients treated with primary surgery (with curative intent), pathological tumor staging (pT) was performed based on histopathological evaluation of the surgical specimens following radical hysterectomy or exenteration and lymph node dissection. This procedure identified a median maximum (histological) tumor diameter of 40 mm. In tumors with maximum extension < 40 mm (n = 37) the median pO_2 was 11 mmHg which was significantly higher than in tumors with a maximum diameter > 40 mm (n = 28, median pO_2 = 5 mmHg; $p < 0.05$ mmHg).[12] Median pO_2 values in stage pT1b tumors were significantly higher (18 mmHg) than in pT2b lesions (5 mmHg, $p < 0.05$). Hypoxic fractions were slightly lower in pT1b tumors (bulky) than in pT2b malignancies (Fig.1).

An influence of the pathological tumor size and pathological stage on the oxygenation status is to be expected considering our hypothesis that malignant progression of solid neoplasms is driven by hypoxia. If tumor growth-associated clonal selection of neoplastic cell variants (which can survive or even proliferate under hypoxic conditions) leads to an overall increase in hypoxia within the tumor mass, lager tumors should be more hypoxic than smaller ones. Nevertheless, there is overlap between the 95% confidence limits of the median pO_2 values for each pathological stage. Thus, the possibility that oxygenation status provides prognostic information over and above the pathological tumor stage cannot be excluded.

CONCLUSION

From these data it is concluded that (a) only a detailed histopathological tumor staging using surgical specimens enables detection of stage- and size-related differences in the oxygenation status of primary cancers of the uterine cervix. Clinical tumor dimensions and FIGO staging are probably not accurate enough for the demonstration of an influence on the tumor oxygenation status, (b) tumor oxygenation and the extent of tissue hypoxia are independent of the tumor grade (differentiation) and histological type (squamous cell cancers vs. adenocarcinomas), (c), tumor oxygenation is independent of various patient parameters, except hemoglobin levels, and (d) although our pO_2 data measured in the two histologies deviate to a certain extent from those published by Lyng et al.[15], the conclusions drawn are comparable.

REFERENCES

1. P. Vaupel and M. Höckel, Hypoxie beim Zervixkarzinom: Pathogenese, Charakterisierung und biologische/klinische Konsequenzen, *Zentralbl. Gynakol.* **123**, 192-197 (2001).
2. M. Höckel, K. Schlenger, C. Knoop, and P. Vaupel, Oxygenation of carcinomas of the uterine cervix: Evaluation by computerized O_2 tension measurements, *Cancer Res.* **51**, 6098-6102 (1991).
3. H. Lyng, K. Sundfor, C. Trope, and E.K. Rofstad, Oxygen tension and vascular density in human cervix carcinoma, *Brit. J. Cancer* **74**, 1559-1563 (1996).
4. M. Höckel, K. Schlenger, S. Höckel, B. Aral, U. Schäffer, and P. Vaupel, Tumor hypoxia in pelvic recurrences of cervical cancer, *Int. J. Cancer* **79**, 365-369 (1998).
5. M. Höckel, C. Knoop, K. Schlenger, B. Vorndran, E. Baussmann, M. Mitze, P.G. Knapstein, and P. Vaupel, Intratumoral pO_2 predicts survival in advanced cancer of the uterine cervix, *Radiother. Oncol.* **26**, 45-50 (1993).
6. M. Höckel, K. Schlenger, B. Aral, M. Mitze, U. Schäffer, and P. Vaupel, Association between tumor hypoxia and malignant progression in advanced cancer of the uterine cervix, *Cancer Res.* **56**, 4509-4515 (1996).
7. M. Höckel and P. Vaupel, Tumor Hypoxia: Definition and current clinical, biologic, and molecular aspects, *J. Natl. Cancer Inst.* **93**, 266-276 (2001).
8. A.W. Fyles, M. Milosevic, R. Wong, M.C. Kavanagh, M. Pintilie, A. Sun, W. Chapman, W. Levin, L. Manchul, T.J. Keane, and R.P. Hill, Oxygenation predicts radiation response and survival in patients with cervix cancer, *Radiother. Oncol.* **48**, 149-156 (1998).
9. T.H. Knocke, H.D. Weitmann, H.J. Feldmann, E. Selzer, and R. Potter, Intratumoral pO_2-measurements as predictive assay in the treatment of carcinoma of the uterine cervix, *Radiother. Oncol.* **53**, 99-104 (1999).

10. K. Sundfor, H. Lyng, C.G. Trope, and E.K. Rofstad, Treatment outcome in advanced squamous cell carcinoma of the uterine cervix, *Radiother. Oncol.* **54**, 101-7 (2000).
11. M. Höckel, K. Schlenger, S. Höckel, and P. Vaupel, Association between tumor hypoxia and malignant progression: The clinical evidence in cancer of the uterine cervix, in: *Tumor Hypoxia*, edited by P. Vaupel and D.K. Kelleher (Wissenschaftliche Verlagsgesellschaft, Stuttgart, 1999).
12. M Höckel and P. Vaupel, Prognostic significance of tissue hypoxia in cervical cancer, *CME J. Gynecol. Obstet.* **6**, 216-225 (2001).
13. P. Vaupel, D.K. Kelleher, and M. Höckel, Oxygenation status of malignant tumors: Pathogenesis of hypoxia and significance for tumor therapy, *Semin. Oncol.* **28**, 29-35 (2001).
14. P. Vaupel, O. Thews, and M. Höckel, Treatment resistance of solid tumors: Role of hypoxia and anemia, *Med. Oncol.* in press (2001).
15. H. Lyng, K. Sundfor, and K. Rofstad. Oxygen tension in human squamous cell carcinoma and adenocarcinoma of the uterine cervix, *Adv. Exp. Med. Biol.* **454**, 635-641 (1998).

QUANTITATIVE SPATIAL ANALYSIS OF HYPOXIA AND VASCULAR PERFUSION IN TUMOR SECTIONS*

Theresa M. Busch†, E. Paul Wileyto, Sydney M. Evans, and Cameron J. Koch‡

1. INTRODUCTION

The presence of tumor hypoxia is associated with poor patient outcome to cancer treatment, including radiation therapy[1] and surgery[2]. Hypoxia itself may indicate a more aggressive tumor phenotype[3], and tumor hypoxia has been associated with the probability of metastatic development[4]. Numerous methods are available to quantify hypoxia, including polarographic needle probes (*e.g.*, Eppendorf pO_2 histograph)[5], fluorescence-based needle probes (*e.g.*, OxyLite)[6], the comet assay[7], and nitroimidazole-based hypoxia markers (*e.g.*, pimonidazole, NITP, EF3, and EF5)[7-10]. A unique feature of hypoxia markers is that they permit the investigation of not only tumor-averaged hypoxia, but also the spatial distribution of oxygen within the biopsies sampled. Such data facilitate the investigation of the effects of intratumoral oxygen distribution on the microregional response to oxygen-dependent treatment modalities. For example, knowledge of the spatial distribution of oxygen is important in Photodynamic Therapy (PDT), which may target vascular-distant tumor cells, as well as the vascular endothelium itself[11]. The present manuscript describes a Matlab-based image-processing program that we have developed for the quantitative analysis of the spatial distribution of tumor hypoxia, as labeled by the hypoxia markers EF3 or EF5.

---

* This work was supported by NIH grants CA75285, CA85831, CA74071, and CA09677.

† To whom correspondence should be addressed at: B13 Anatomy Chemistry Building, 3620 Hamilton Walk, Philadelphia, PA, 19104-6072. (215) 573-3168, FAX: (215) 898-0090, buschtm@mail.med.upenn.edu.

‡ Theresa M. Busch, Sydney M. Evans, and Cameron J. Koch, Department of Radiation Oncology, and E. Paul Wileyto, Tobacco Use Research Center, School of Medicine, University of Pennsylvania, Philadelphia, PA 19104.

2. MATERIALS AND METHODS

2.1. Tumor model and hypoxia marker

Studies were performed using radiation-induced fibrosarcoma (RIF) tumors (5-7 mm diameter) grown intradermally on the shoulders of C3H mice[10]. EF3 (synthesized by Dr. Bill Dolbier, University of Florida, Gainsville) was administered *via* tail vein at 52 mg/kg for a 30 minute incubation. Mice were anesthetized (ketamine/xylazine at 175/10 mg/kg i.p.) during this incubation to allow the injection (*via* orbital plexus) of 30 mg/kg of Hoechst 33342 (Sigma, St. Louis, MO) at ~28.5 minutes and the rapid excision/freezing of the tumor at 30 minutes. To evaluate hypoxia and perfusion in tumors subjected to an oxygen consumptive condition, some tumors were treated with PDT. PDT was carried out as previously described[10], using the photosensitizer Photofrin (5 mg/kg *i.v.*) followed 24 h later by 630 nm illumination at 135 J/cm^2, 75 mW/cm^2.

2.2. Immunohistochemistry and fluorescence microscopy

Cryosectioning, immunohistochemistry and fluorescence microscopy for EF3 were performed as previously described[10]. To analyze the spatial relationship between hypoxia and blood vessel distribution, some sections were stained for both EF3 and CD31. CD31 staining involved a 1.5 h incubation in primary antibody (rat anti-mouse CD31; PharMingen, San Diego, CA) and a 45 min incubation in secondary antibody (Cy5-conjugated mouse anti-rat IgG; Jackson ImmunoResearch, West Grove, PA). Both antibodies were diluted 1:100 in antibody carrier[10]. CD31 staining was briefly fixed in 4% paraformaldehyde and then staining for EF3 was done as normal.

Fluorescence microscopy (Nikon LabPhot microscope with a 100-W high pressure mercury arc lamp and Photometrics Quantix CCD digital camera) was carried out to photograph Hoechst-perfused blood vessels, EF3 binding, and CD31 staining. Slides were then flooded with a solution of Hoechst (20 μM in PBS) to label cell nuclei, allowing a whole tissue image of Hoechst fluorescence. At the beginning and conclusion of each camera session, an image of hemacytometer-loaded calibration dye (Cy3 in 1% paraformaldehyde) was taken. The ratio of the exposure time to the mean fluorescence intensity of this image (x1000) was used to correct the fluorescence intensities of experimental images (based on their exposure time) to account for day-to-day variations in the lamp intensity. Background, determined from the lowest value on an image, was subtracted from all images.

2.3. Masking and image processing

Bitmap images (masks) were used to identify the location of tissue or features during processing and analysis. Two types of masks were made: a feature mask and an alignment mask. Feature masks highlighted the feature of interest in that particular image, while alignment masks highlighted a feature common to all the photos for a given section, *e.g.*, a hole in the tissue, to provide a reference point for alignment. Masks were created in Adobe PhotoShop (Adobe Systems Inc., San Jose, CA) by applying a series of functions and filters to the raw images. Alignment masks were made by applying the equalize function, followed by thresholding to highlight the hole or other feature common to all the images. A tissue feature mask was created from the image of Hoechst-flooded tissue; an equalize function was used to brighten the image, maximize and median filters were used to spread the nuclear-localized Hoechst fluorescence over the approximate diameter of a cell, and thresholding was performed. The edge feature mask traced the outline of contiguous tissue identified on the tissue feature mask. A feature mask of perfusion was made from the image of Hoechst-perfused blood vessels by applying the auto levels function to enhance contrast and thresholding. Similarly, a feature mask of CD31 was created using the auto levels and thresholding functions.

Images were processed and analyzed using a program written in the Matlab (The Mathworks, Inc., Natick, MA) programming language. In the first step of image processing, images were aligned. The tissue feature mask was used as a base and all other images were shifted in x and y coordinates until landmarks on the alignment masks were aligned. Images were cropped to maintain the largest area common to all.

Sampling was performed by digitally superimposing a grid over the image of interest, *e.g.*, the raw EF3 data. At the intersection of every 5th column and 5th row the pixel coordinates were checked against those pixels identified as containing tissue in the tissue feature mask. At tissue-containing points, the intensity value, *e.g.*, of EF3, was recorded as well as the distance traveled from that point to another designated feature of interest. In the final data set, the *shortest* distance traveled in any direction from a tissue-containing coordinate to the other feature of interest was saved. Typically 11,000 – 13,000 data points were calculated from each image.

All data was sorted and binned in Excel (Microsoft Corporation, Redmond, WA). Distances in pixels were converted to μm based on the photography-determine scale of 1.8 μm/pixel. Plots were created in SigmaPlot (SPSS Inc, Chicago IL).

3. RESULTS AND DISCUSSION

3.1. Sample feature masks

On each section, a digital image of EF3-stained tissue, Hoechst-labeled tissue, Hoechst-perfused blood vessels, and CD31-labeled blood vessels was acquired. Figure 1 shows the raw EF3 data (Figure 1A) and masks from a control tumor section. The masks include a tissue feature mask (Figure 1B), a perfusion feature mask (Figure 1C), and a CD31 feature mask (Figure 1D). Figure 2 depicts the raw EF3 data and masks of a section from a PDT-treated tumor. Utilizing these sections as examples, a method for the quantification of the distribution of hypoxia or other features of interest is described.

3.2. Hypoxia *versus* distance to the nearest perfused blood vessel

To describe the spatial distribution of hypoxia, EF3 binding was quantified as a function of distance to the nearest perfused blood vessel. These investigations necessitated carrying out multiple antibody-based staining protocols on the same section. To investigate the effects of the CD31 staining protocol on EF3 binding, the latter was compared between adjacent sections where one section was stained for only EF3 and the other was stained for both EF3 and CD31. Results indicated that the CD31 staining protocol may slightly reduce overall EF3 binding, however no evidence of a spatial

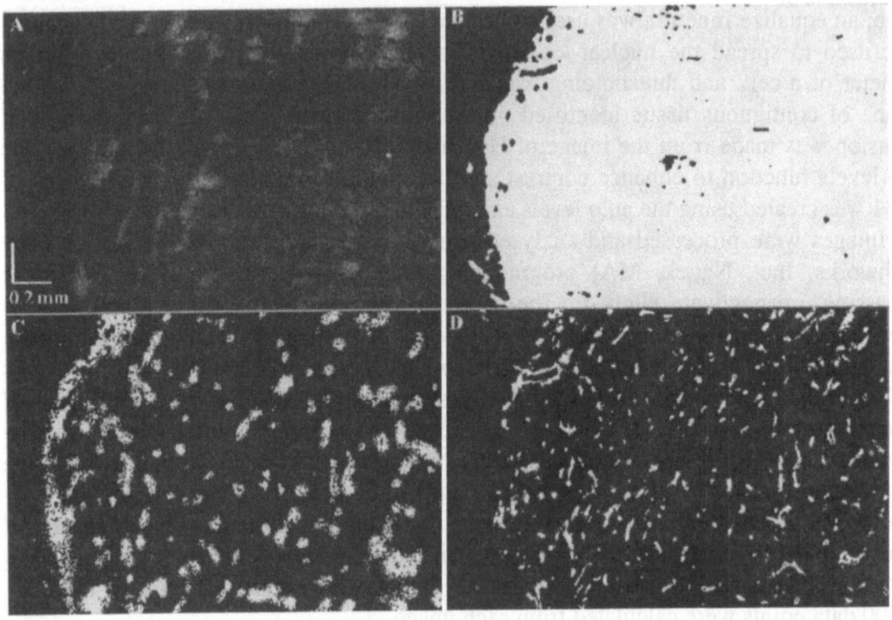

Figure 1. EF3 raw data (A), tissue feature mask (B), perfusion feature mask (C), and CD31 feature mask (D) of a section from a control RIF tumor.

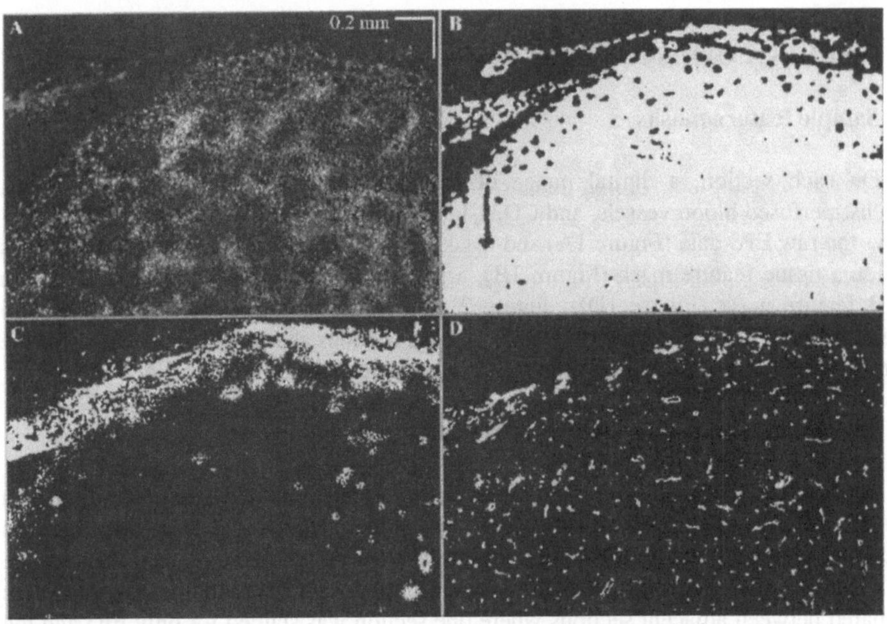

Figure 2. EF3 raw data (A), tissue feature mask (B), perfusion feature mask (C), and CD31 feature mask (D) of a section from a PDT-treated RIF tumor.

dependence to these effects was found. Thus, the relative distribution of EF3 was not affected.

To perform spatial analysis, first image alignment was carried out. The degree of alignment required is visible in Figure 2, where the perfusion mask needed to be moved 21 pixels down to align it with the other masks. The EF3 raw data was chosen as the image to be analyzed and spatial analysis was performed relative to perfused blood vessel structure. Perfused vessel structure was identified in Matlab by the creation of an overlay of the perfusion feature mask and the CD31 feature mask, as has been described by Fenton *et al.*[12]. Figure 3 plots averaged EF3 intensity values within 9 μm bins to the nearest perfused blood vessel. A small increase in EF3 binding at longer distances to the nearest perfused blood vessel was visible in the control tumor. A steeply increasing hypoxia gradient was found in the PDT-treated tumor, indicating that this method clearly detects spatial changes in EF3 binding. Furthermore, the overall higher levels of EF3 binding in the PDT-treated tumor indicate that more hypoxia was found in this tumor.

3.3. Perfusion *versus* distance to the tumor edge

Another type of spatial analysis that can be performed with the Matlab program is the quantification of the fraction of perfused tissue as a function of distance to the tumor edge. Figures 1C and 2C show that both of the sample tumors demonstrated significant perfusion along the peripheral edge. Yet, as shown in Figure 2C, PDT created a marked reduction in the perfusion of the tumor interior. For quantitative analysis, the perfusion feature mask was chosen as the image to be analyzed and spatial analysis was performed relative to the tumor edge identified on the edge feature mask. This mask is the outline of contiguous tissue on the tissue feature mask. Figure 4 plots the fraction of perfused tissue in 9 μm bins to the tumor edge. At distances within 50 μm of the tumor edge, more than half of the tissue demonstrated perfusion in both the control and the PDT-treated tumor. Both tumors showed a decrease in perfusion as distance to the tumor edge increased. The regular spatial distribution of interior blood vessels in the control tumor is indicated by the 10% perfused tissue fraction at distances >200 μm from the tumor edge. The sparseness of interior blood vessels in the PDT-treated tumor is evident as <2% of the tissue >200 μm from the tumor edge was perfused.

3.4. Conclusions

In summary, we have described the practical applications of a recently developed image analysis program that quantifies the distribution of hypoxia in a tumor section. The flexible design of this program facilitates performing spatial analysis on any one of a number of anatomical or physiological features. The present manuscript has described only several examples of the types of analysis that can be performed. Using the same techniques, for example, EF3 binding can be evaluated relative to blood vessel structure (regardless of perfusion) or relative to the tumor edge. Similarly, the spatial distribution of CD31-stained blood vessels can be quantified. The program is designed at present to be used for 2D analyses. A resulting limitation is that hypoxia or perfusion in the plane immediately above or below the section of interest is not considered. However, once a routine is added to facilitate image alignment in a third, z, direction, the same Matlab

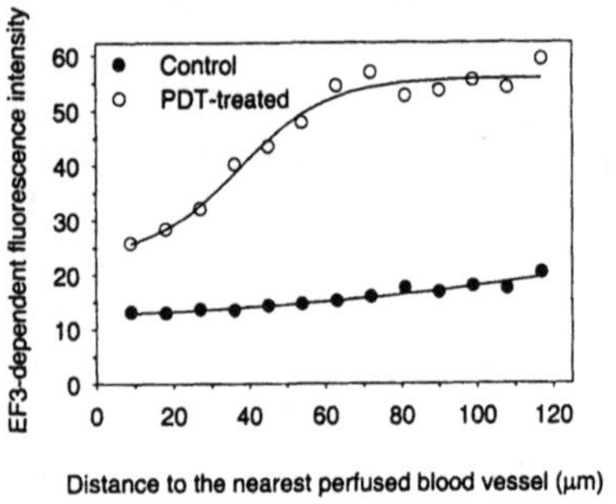

Figure 3. EF3 binding relative to the nearest perfused blood vessel. Data fit with a 4 parameter sigmoid curve.

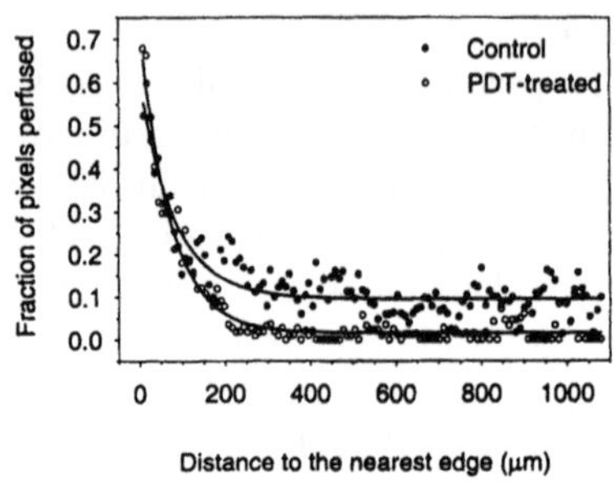

Figure 4. Perfused tissue fraction relative to the nearest tumor edge. Data fit with a negative exponential.

application can be used to calculate distance to the nearest feature of interest in 3 dimensions. In conclusion, the herein described program facilitates the routine analysis of the spatial distribution of hypoxia and ongoing studies utilize the resulting data to investigate the effects of tumor hypoxia on pre-clinical and clinical responses to cancer therapies.

REFERENCES

1. D.M. Brizel, R.K. Dodge, R.W. Clough, and M.W. Dewhirst, Oxygenation of head and neck cancer: changes during radiotherapy and impact on treatment outcome, *Radiotherapy & Oncology.* 53(2), 113-117 (1999).
2. M. Hockel, K. Schlenger, B. Aral, M. Mitze, U. Schaffer, and P. Vaupel, Association between tumor hypoxia and malignant progression in advanced cancer of the uterine cervix, *Cancer Res.* 56(19), 4509-4515 (1996).
3. A.C. Koong, N.C. Denko, K.M. Hudson, C. Schindler, L. Swiersz, C. Koch, S. Evans, H. Ibrahim, Q.T. Le, D.J. Terris, and A.J. Giaccia, Candidate genes for the hypoxic tumor phenotype, *Cancer Res.* 60(4), 883-887 (2000).
4. D.M. Brizel, S.P. Scully, J.M. Harrelson, L.J. Layfield, J.M. Bean, L.R. Prosnitz, and M.W. Dewhirst, Tumor oxygenation predicts for the likelihood of distant metastases in human soft tissue sarcoma, *Cancer Res.* 56(5), 941-943 (1996).
5. P. Vaupel, K. Schlenger, C. Knoop, and M. Hockel, Oxygenation of human tumors: evaluation of tissue oxygen distribution in breast cancers by computerized O2 tension measurements, *Cancer Res.* 51(12), 3316-3322 (1991).
6. J. Bussink, J.H. Kaanders, A.M. Strik, B. Vojnovic, and A.J. van Der Kogel, Optical sensor-based oxygen tension measurements correspond with hypoxia marker binding in three human tumor xenograft lines, *Radiat. Res.* 154(5), 547-555 (2000).
7. P.L. Olive, R.E. Durand, J.A. Raleigh, C. Luo, and C. Aquino-Parsons, Comparison between the comet assay and pimonidazole binding for measuring tumour hypoxia, *Br. J. Cancer.* 83(11), 1525-1531 (2000).
8. P.F. Rijken, H.J. Bernsen, J.P. Peters, R.J. Hodgkiss, J.A. Raleigh, and A.J. van der Kogel, Spatial relationship between hypoxia and the (perfused) vascular network in a human glioma xenograft: a quantitative multi-parameter analysis, *Int. J. Radiat. Oncol. Biol. Phys.* 48(2), 571-582 (2000).
9. S.M. Evans, S. Hahn, D.R. Pook, W.T. Jenkins, A.A. Chalian, P. Zhang, C. Stevens, R. Weber, G. Weinstein, I. Benjamin, N. Mirza, M. Morgan, S. Rubin, W.G. McKenna, E.M. Lord, and C.J. Koch,

Detection of hypoxia in human squamous cell carcinoma by EF5 binding, *Cancer Res.* 60(7), 2018-2024 (2000).
10. T.M. Busch, S.M. Hahn, S.M. Evans, and C.J. Koch, Depletion of tumor oxygenation during photodynamic therapy: detection by the hypoxia marker EF3 [2-(2-nitroimidazol-1[*H*]-yl)-*N*-(3,3,3,-trifluoropropyl)acetamide], *Cancer Res.* 60, 2636-2642 (2000).
11. T.J. Dougherty, C.J. Gomer, B.W. Henderson, G. Jori, D. Kessel, M. Korbelik, J. Moan, and Q. Peng, Photodynamic therapy, *J. Natl. Cancer Inst.* 90(12), 889-905 (1998).
12. B.M. Fenton, S.F. Paoni, J. Lee, C.J. Koch, and E.M. Lord, Quantification of tumour vasculature and hypoxia by immunohistochemical staining and HbO2 saturation measurements, *Br. J. Cancer.* 79(3-4), 464-471 (1999).

TUMOR OXYGENATION AND ITS RELEVANCE TO TUMOR PHYSIOLOGY AND TREATMENT

Peter Vaupel* and Michael Höckel**

1. INTRODUCTION

For many years, the identification of tumor hypoxia, its systematic characterization and the assessment of its clinical relevance were not possible due to lack of methods suitable for the routine measurement of intratumoral oxygen tensions in patients. In the late 1980s, a novel and clinically applicable standardized procedure was established enabling the determination of the tissue oxygenation status in primary tumors, local recurrences, and metastatic lesions in patients using a computerized polarographic needle electrode system.[1,2] Within a relatively short period of time, the significance of tumor oxygenation for therapy outcome became evident in numerous experimental and clinical studies.[3,4]

In this article, current knowledge on the oxygenation status of solid malignancies is compiled, the mechanisms causing tumor hypoxia are discussed, and emphasis is given to the significance of tumor hypoxia for tumor physiology, for tumor therapy and for long-term prognosis of patients.

2. PATHOGENESIS OF TUMOR HYPOXIA

Investigations have demonstrated that the presence of hypoxic tissue areas (i.e., areas with pO_2 values ≤ 2.5 mmHg) is a characteristic pathophysiological property of solid tumors and such areas have been found in a wide range of accessible human malignancies.[1-4] The hypoxic or anoxic areas arise as a result of an imbalance between the supply and consumption of oxygen. Whereas in normal tissues or organs the O_2 supply meets the metabolic requirements, in locally advanced solid tumors, the O_2 consumption rate of neoplastic as well as stromal cells may outweigh an insufficient oxygen supply and results in the development of tissue areas with very low O_2 levels.

---

* Inst. Physiology & Pathophysiology, University of Mainz, 55099 Mainz,. Germany.
** Dept. Obstetrics & Gynecology, University of Leipzig, 04103 Leipzig, Germany.

Oxygen Transport to Tissue XXIII, edited by D. Wilson *et al.*
Kluwer Academic/Plenum Publishers, 2003

Table 1. Causative factors of tumor hypoxia

A. Perfusion-limited hypoxia ("acute hypoxia")	
	abnormal structure of microvessels
	abnormal function of microvessels
	transient flow stasis (ischemia)
B. Diffusion-limited hypoxia ("chronic hypoxia")	
	enlarged diffusion distances
	adverse microvessel orientation
	adverse vascular/diffusion geometry
C. Anemic hypoxia	
	tumor-associated anemia
	therapy-induced anemia
D. Hypoxemic hypoxia	
	plasma flow only
	microvessels arising from venous side
E. Toxic hypoxia	
	HbCO formation (in heavy smokers)

Tumor hypoxia preferentially results from an *inadequate perfusion* due to severe structural and functional abnormalities of the tumor microcirculation.[5,6] Hypoxic (micro-)regions are heterogeneously distributed within the tumor mass and may be located adjacent to regions with O_2 tensions (pO_2 values) corresponding to those found in the normal tissue neighboring the neoplastic lesion. Perfusion-limited O_2 delivery leads to *ischemic hypoxia* which is often transient (up to 1 hr). For this reason, this type of hypoxia is also called "acute" hypoxia, a term which does not take into account the mechanisms underlying this condition (see Table 1).

Hypoxia in tumors can also be caused by an increase in diffusion distances, so that cells far away (> 70 μm) from the nutritive blood vessel receive less oxygen (and other nutrients) than needed. This condition is termed *diffusion-limited hypoxia*, also known as "chronic" hypoxia (see Table 1). In addition to enlarged diffusion distances, an adverse vascular/diffusion geometry and/or an adverse microvessel orientation can also cause hypoxia. Tumor-associated or therapy-induced anemia can lead to a reduced O_2 transport capacity of the blood, further contributing to the development of hypoxia (*anemic hypoxia*). Hypoxia is intensified especially in tumors or tumor areas that have low perfusion rates. A similar condition can be caused by carboxyhemoglobin (HbCO) formation in heavy smokers leading to a functional anemia, because hemoglobin blocked by carbon monoxide (CO) can no longer transport oxygen (*toxic hypoxia*).

Relatively often, tumor microvessels are perfused (at least transiently) by plasma only. In this situation, hypoxia around those vessels develops very rapidly because only a few tumor cells at the arterial end can be supplied adequately under this condition (*hypoxemic hypoxia*). There is abundant evidence for the existence of a substantial heterogeneity in the development and proportion (extent) of tumor hypoxia due to pronounced intra-tumor (and inter-tumor) variabilities in vascularity and perfusion rates. Whereas normal tissues can physiologically compensate for these deficiency states, locally advanced tumors (or at least larger tumor areas) cannot adequately counteract the development of hypoxia.

Table 2. Hypoxia modulated HIF-1 activity regulates gene products and protein functions in human cancer (selection)[9,10]

Adaptation process	Gene product	Protein function(s)
Metabolic adaptation	GLUT-1/3	glucose transport rate
	glycolytic enzymes	glucose turnover rate
	adenylate kinase-3	energy preservation
	carbonic anhydrase IX	H^+-elimination, pH-regulation
Adaptation of supply	erythropoietin	erythropoiesis, O_2-transport capacity
	VEGF; VEGF-R	angiogenesis
Growth adaptation	EGF, IGF-2	proliferation, cell survival
	Bcl-2 family	modulation of apoptosis

HIF-1 regulated gene products are likely to play key roles in tumor progression, aggressiveness, metabolic adaptation and may contribute to the increased resistance of hypoxic tumor cells to chemo- and radiotherapy. GLUT = glucose transporter, VEGF = vascular endothelial growth factor, VEGF-R = VEGF-receptor, EGF = epidermal growth factor, IGF-2 = insulin-like growth factor-2.

3. OXYGENATION STATUS OF HUMAN TUMORS

Squamous cell carcinomas of the uterine cervix and of the head and neck region have been the most extensively studied tumor types over the last 10 years with about 750 patients involved worldwide so far.[7,8] Consistently, all researchers have found that the oxygenation status is independent of clinical tumor size, stage, grade, histological type and of the location within the tumor, but is affected by the hemoglobin level and pathological tumor stage. Tumor oxygenation is compromised as compared to normal tissues and is heterogeneous both in space and in time. Tumor oxygenation and the extent of hypoxia are independent of various patient demographics and oncologic parameters, and exhibit substantial intra- and inter-tumor heterogeneity.[7,8] Local recurrences of squamous cell carcinomas have a higher hypoxic fraction than the primary tumors. 30-65% of locally advanced tumors contain hypoxic tissue areas, depending on the tumor type.

4. HYPOXIA AND ITS RELEVANCE TO TUMOR PHYSIOLOGY

Hypoxia can modulate gene expression, translation and protein degradation and thus can lead to substantial *proteome changes*. Overall, hypoxia can reduce the rate of protein synthesis, thus leading to a retardation or arrest of cellular growth resulting in cell quiescence. The ultimate cellular response to sustained, severe hypoxia is apoptosis or necrosis. In contrast, hypoxia may upregulate the synthesis of certain proteins (in the course of adaptation processes) in order to survive or escape the hostile metabolic microenvironment, such as glycolytic enzymes, glucose transporters, angiogenic molecules, growth and survival factors with a strong impact on tumor propagation. In addition, hypoxia-driven synthesis of resistance-related proteins may be responsible for the development of treatment resistance (see Table 2).

Tumor hypoxia (and other hostile microenviromental conditions) can additionally enhance malignant progression and increase aggressiveness through transient or persistent *genome changes and clonal selection* (often associated with subsequent clonal

Table 3. Tumor hypoxia and acquired treatment resistance to radio- and/or chemotherapy[a]

A. Direct effects
- through decreased generation of free radicals and thereby less DNA damage (bleomycin, anthracyclines)
- reduced "fixation" of DNA damage (X- and γ-rays)

B. Indirect effects via proteome changes[b]
- slowing of proliferation kinetics, increase in number of G_0 cells
- G_1/S-phase arrest (e.g., X- and γ-rays, vinca alkaloids, methotrexate)
- elevated levels of glutathione and related thiols (X- and γ-rays, alkylating agents, bleomycin, platinum compounds)
- elevated levels of resistance-related proteins and of DNA repair enzymes

C. Indirect effects via genome changes[c] and clonal selection
- loss of apoptosis and differentiation
- clonal heterogeneity
- "outgrowth" of resistant clonal variants
- increase in number of cells with aggressive phenotype

D. Secondary indirect effects via an intensified glycolysis with extracellular acidosis
- cell cycle effects
- activation of repair processes
- modification of cellular drug uptake and activation

E. Secondary indirect effects via a chaotic angiogenesis and microcirculation impacting intratumor pharmacokinetics
- impaired and uneven drug delivery
- arterio-venous shunt perfusion
- large diffusion distances

[a] selection of mechanisms; anemia acts as a factor worsening tumor hypoxia
[b] resulting from modulation of gene expression, post-transcriptional and/or post-translational effects
[c] through point mutations, gene amplifications, chromosomal rearrangements

dominance) due to a hypoxia-driven strong selection pressure. Hypoxia promotes genomic instability (through point mutations, gene amplification, and chromosomal rearrangements) thus increasing the number of genetic variants and thereby promoting clonal and intrinsic tumor cell heterogeneity. Outgrowth of resistant clonal variants in a "survival of the fittest" scenario and malignant progression are the final results.

Hypoxia-mediated clonal selection of tumor cells with persistent genomic changes can lead, *inter alia*, to a loss of differentiation and of apoptosis which can stabilize or further aggravate tumor hypoxia and which in turn again promotes malignant progression. Thus, hypoxia is involved in a *vicious circle* that is regarded as a fundamental biologic mechanism of malignant disease. Other consequences of hypoxia-induced malignant progression are an increased locoregional spread and enhanced distant metastasis.[4]

5. TUMOR HYPOXIA AND ACQUIRED TREATMENT RESISTANCE

Tumor hypoxia is a therapeutic problem, as it makes solid tumors resistant to sparsely ionizing radiation and some forms of chemotherapy. Hypoxia-associated resistance to radio-/chemotherapy is multifactorial with direct and indirect mechanisms involved (see Table 3).

6. TUMOR HYPOXIA AS AN INDEPENDENT PROGNOSTIC FACTOR

Multivariate analyses have shown that hypoxia is a powerful independent, adverse prognostic factor in locally advanced cancer of the uterine cervix[11], carcinomas of the head and neck[12] and in soft tissue sarcomas[13] with regard to disease-free survival and local control. Tumor hypoxia predicts for a worse outcome irrespective of the mode of treatment due to its impact on tumor propagation and malignant progression. Tumor hypoxia poorly correlates with clinical prognostic factors (e.g., stage, size, grade, nodal status). Whether an increased expression of angiogenic factors is an independent prognostic factor or just a surrogate parameter of tumor hypoxia is still an open question.

REFERENCES

1. P. Vaupel, K. Schlenger, C. Knoop, and M. Hoeckel, Oxygenation of human tumors: Evaluation of tissue oxygen distribution in breast cancers by computerized O_2 tension measurements, *Cancer Res.* **51**, 3316-3322 (1991).
2. M. Hoeckel, K. Schlenger, C. Knoop, and P. Vaupel, Oxygenation of carcinomas of the uterine cervix: Evaluation of computerized O_2 tension measurements, *Cancer Res.* **51**, 6098-6102 (1991).
3. P. Vaupel and D.K. Kelleher (eds), *Tumor Hypoxia.* (Wissenschaftliche Verlagsgesellschaft, Stuttgart, 1999).
4. M. Höckel and P. Vaupel, Tumor hypoxia: Definitions and current clinical, biological and molecular aspects, *J. Natl. Cancer Inst.* **93**, 266-276 (2001).
5. P. Vaupel, F. Kallinowski, and P. Okunieff, Blood flow, oxygen and nutrient supply, and metabolic microenvironment of human tumors: A review, *Cancer Res.* **49**, 6449-6465 (1989).
6. P. Vaupel, *Blood flow, oxygenation, tissue pH distribution and bioenergetic status of tumors.* (Ernst Schering Reasearch Foundation: Lecture 23, Berlin, 1994).
7. P. Vaupel, Durchblutung und Oxygenierungsstatus von Kopf-Hals-Tumoren. In: *Klinik des Rezidivtumors im Kopf-Hals-Bereich,* edited by H.D. Böttcher, T.G. Wendt, and M. Henke (Zuckschwerdt, München, 2001), pp. 7-23.
8. M. Höckel and P. Vaupel, Prognostic significance of tissue hypoxia in cervical cancer, *CME J. Gynecol. Obstet.* **6**, 216-225 (2001).
9. P. Vaupel, O. Thews, and M. Höckel, Treatment resistance of solid tumors: Role of hypoxia and anemia, *Med. Oncol.* in press (2001).
10. G.L. Semenza, Hypoxia, clonal selection, and the role of HIF-1 in tumor progression, *Crit. Rev. Biochem. Molec. Biol.* **35**, 71-103 (2000).
11. M. Höckel, C. Knoop, K. Schlenger, K.B. Vorndran, M. Baussmann, P. Knapstein, and P.Vaupel, Intratumoral pO_2 predicts survival in advanced cancer of the uterine cervix. *Radiother. Oncol.* **26**, 45-50 (1993).
12. M. Nordsmark and J. Overgaard, A confirmatory prognostic study on oxygenation status and loco-regional control in advanced head and neck squamous cell carcinoma treated by radiation therapy. *Radiother Oncol.* **57**, 39-43 (2000).
13. M. Nordsmark, J. Alsner, J. Keller, O.S. Nielsen, O.M. Jensen, M.R. Horsman, and J. Overgaard, Hypoxia in human soft tissue sarcomas: Adverse impact on survival and no association with p53 mutations. *Brit. J. Cancer* **84**, 1070-1075 (2001).

MECHANISMS UNDERLYING HYPOXIA DEVELOPMENT IN TUMORS

Mark W. Dewhirst*

1. INTRODUCTION

The purpose of this review is to critically evaluate physiologic mechanisms that lead to hypoxia in tumors. This review will challenge some of the current paradigms and present an alternative view. There are two types of hypoxia, as described radiobiologically. Long intercapillary distances between microvessels are believed to be causal for *diffusion limited or chronic hypoxia*. Intermittent vascular stasis is commonly believed to be responsible *for perfusion limited or acute hypoxia*. If one considers the multitude of factors that could influence oxygen transport, it becomes clear that these two classifications are gross oversimplifications of a more complex process. Factors that are known to affect delivery include microvessel density, vessel orientation, red cell flux, hemoglobin saturation and blood viscosity, to name a few. One must also consider the oxygen consumption rate, which can be influenced by cell proliferation rate and other factors, such as inflammation.[1,2] This review will present a comprehensive model including factors governing oxygen transport in tumors that provides a testable mechanistic underpinning for future research on this subject. The increased interest in tumor hypoxia because of its influences on gene expression and selection for more aggressive phenotype further emphasizes the need for more detailed study.[3]

2. FACTORS THAT INFLUENCE OXYGEN DELIVERY

It has commonly been reported that vascular density in tumors is lower than that in normal tissues.[4] Clearly, lowered vascular density can contribute to tumor hypoxia, particularly in situations where the intercapillary distance is larger than 300 μm. It was

---

* Mark W. Dewhirst, Box 3455 Duke University Medical Center, Durham, NC 27710, dewhirst@radonc.duke.edu

particularly in situations where the intercapillary distance is larger than 300 μm. It was first estimated that the maximum diffusion distance of oxygen was probably ≈150 μm based on work done on the 1950's by Thomlinson and Gray.[5] Seminal work using phosphorescence lifetime imaging in skinfold window chamber tumors validated this conjecture.[6] However, the orientation of vessels is also an extremely important factor as has been illustrated by Secomb, et al. by comparing predictions of oxygen tension distributions in tumor regions, based on the actual vascular geometries encountered vs. Krogh cylinder models with the same overall vascular length density.[7]

Another factor that can influence tissue oxygenation is vascular PO_2. Using microelectrodes in tumor bearing skin-fold window chamber tumors of rats we found that tumor-feeding arterioles had an average PO_2 of 30 mmHg, whereas downstream venules and capillaries averaged 10-12 mmHg. Approximately 25% of these vessels were perfused, but had PO_2 values near zero.[1] These tumors grow on one side of a fascial plane and the arteries feeding the tumor enter the tumor from that surface. Chamber tumors are 250 μm thick; thus, there is a relatively steep longitudinal O_2 gradient from the arteriolar side to the tumor surface. We have subsequently verified the presence of this gradient using phosphorescence lifetime imaging and hypoxia markers.[8,9] Metabolite and redox gradients also exist across the window tissue, consistent with a transition from aerobic to anaerobic metaolism.[9] These data suggest that there are not only longitudinal vascular PO_2 gradients within this chamber tissue, but there are also longitudinal tissue PO_2 gradients. Collectively, these data are consistent with the concept that hypoxic diffusion distances, zone sizes and severity of hypoxia will vary in tumors, based on intravascular PO_2 (Figure 1).

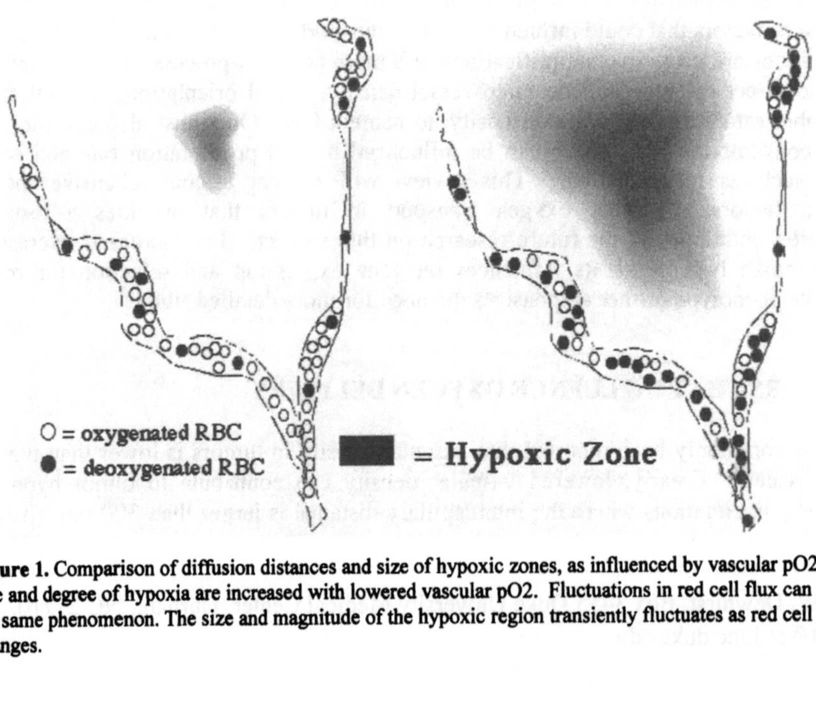

Figure 1. Comparison of diffusion distances and size of hypoxic zones, as influenced by vascular pO2. The size and degree of hypoxia are increased with lowered vascular pO2. Fluctuations in red cell flux can lead to the same phenomenon. The size and magnitude of the hypoxic region transiently fluctuates as red cell flux changes.

Another factor that becomes important in influencing oxygen transport is the influence of the microenvironment on blood viscosity. Both human and rat red cells undergo crenation when exposed to pH values of approximately 6.8 and $PO_2 < 2$ mmHg. Crenation increases red cell suspension viscosity by approximately 30 to 40%.[10] Drugs that restore red cells to their normal surface area/volume ratio restore viscosity to normal limits, increase red cell flux in hypoxic vessels and improve vascular PO_2[11] and radiation response.[12] Despite differences in viscosity, microvascular hematocrits are not significantly different in tumors vs. normal tissues.[13] However, the abnormal vascular geometry combined with altered reology contribute to the existence of plasma channels (vessels that carry no or few red cells). In the R3230AC mammary carcinoma model, we have observed that approximately 8 to 9% of microvessels are plasma channels.[14]

O_2 consumption rates of tumors are typically relatively low, compared to many normal tissues; nevertheless they are clearly out of balance with delivery suggesting that alteration in oxygen consumption rate might be a reasonable target for improving tumor PO_2.[15] Theoretical models support this hypothesis.[16] It has been demonstrated that the key factors that contribute to tumor hypoxia include limited arteriolar supply and irregular geometry; both of these factors contribute to steep longitudinal tissue O_2 gradients and hypoxia. The combination of irregular vasculature geometry and vascular hypoxia also leads to rheologic changes and plasma channels. All of these factors contribute to the overall lack of sufficient delivery.

The longitudinal oxygen gradients in tissues suggest that there may be tumor regions fed by relatively hypoxic vessels that are refractory to O_2 manipulation. We have coined the phrase **"Hypoxic Corner"** to describe these regions. These regions are physiologically similar to the "lethal corner" that was first described by Opitz in 1950.[17] We use the term hypoxic corner because this pathophysiologic scenario is by no means lethal to tumor cells. There are data to support this hypothesis. Recently, we combined hyperglycemia, to induce the Crabtree effect (to reduce oxygen consumption rate), with hyperoxic gas breathing. Overall, this led to a dramatic improvement in tumor PO2 (an average of 25 mmHg).[18] However, critical examination of these data suggests that those regions of the tumor that are most hypoxic are least effected by this manipulation. An important issue is whether these types of regions exist in three-dimensional tumors. One can argue that the skinfold window chamber tumor is a special case and that longitudinal gradients exist in only this type of tumor. Arguing against that are data with flank tumors, which we discussed above. However, critical examination of many Eppendorf studies illustrate this same phenomenon. In situations where measurements are repeated after some type of manipulation, there are typically large changes in median PO2 but less dramatic effects on the hypoxic fraction. This is true for preclinical studies as well as human trials, as we have reviewed recently.[18] It will be important to determine whether such hypoxic corners consist of viable cells or merely represent necrotic zones. More work needs to be done to clarify these issues clearly.

3. INTERMITTENT (ACUTE) HYPOXIA

The commonly accepted mechanisms for development of acute hypoxia assume that cessation of blood flow in vessels is prequisite for this to occur. It has further been speculated that this might occur as a result of vascular collapse, or plugging of vessels by tumor cells or leukocytes. We have compelling data to suggest that vascular stasis is not required for this to occur and that it is a much more common event than previously recognized. Using skin-fold window chamber preparations it has been found that red cell

required for this to occur and that it is a much more common event than previously recognized. Using skin-fold window chamber preparations it has been found that red cell flux in individual microvessels is nearly universally unstable and that changes in vascular pO2 are directly related to changes in red cell flux.[19] Complete vascular stasis, in contrast, occurred only 5% of the time and typically was of short duration (<1min).[14] We observed no evidence for vascular collapse and changes in red cell flux were not temporally related to vasomotion. Based on Greene's functions models we predicted that substantial proportions of tumor could be transiently hypoxic for as little as a two-fold change in red cell flux, which was commonly observed in these preparations.

Intermittent hypoxia has been further characterized by measuring PO_2 at single positions in tumors and normal muscle using 10 μm recessed tip polarographic electrodes.[20] PO_2 in these regions was also nearly universally unstable. Fourier transform analysis showed that fluctuations in both tumor and muscle PO_2 were relatively slow (<1-2cpm), ruling out vasomotion, respiration or heart rate as being causative. Similar frequencies of fluctuation have been seen in the 9L glioma and a rat fibrosarcoma, with variations in power between tumors. Differences between tumors are further illustrated by examining the number of times that PO_2 crosses a radiobiologically significant threshold of 10 mmHg over a period of 1 hour observation, for example. The fibrosarcoma line typically only has 1 or 2 events per hour, but these have a tendency to have a very long duration. Alternatively, the R3230AC mammary carcinoma has many more events per hour, but they are relatively short.[21] The 9L tumor tends to have the shortest durations of hypoxia averaging less than 1 minute per episode.

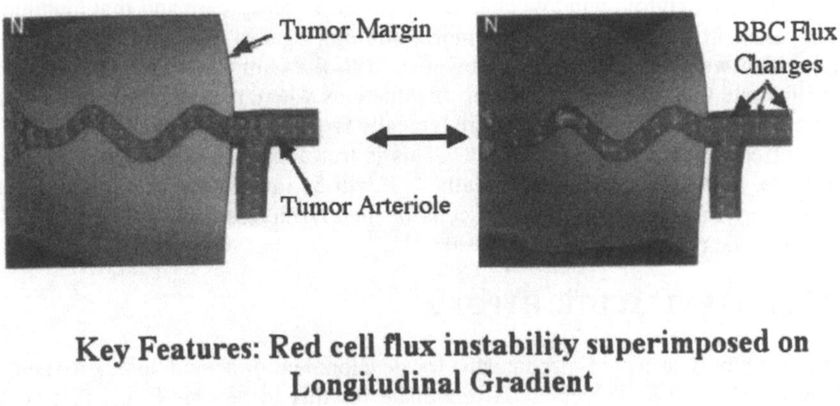

Figure 2. Main features of oxygen transport that govern development of hypoxia. Left panel – High red cell flux lengthens distance along the vessel where oxygen transport occurs efficiently. Regions distant from arteriolar source have shorter diffusion distance of oxygen. Lowered red cell flux shortens the length of the longitudinal gradient and the diffusion distance of oxygen.

of its design, there is no way to distinguish whether mismatch is occurring because of differences in frequency or magnitude of fluctuation. In addition, unless special care is taken in analysis, the results tend to emphasize differences in vascular stasis, as opposed to flow fluctuations.[24] What we have clearly shown is that complete vascular stasis is not a prerequisite for intermittent hypoxia and further that the phenomenon of intermittent hypoxia is an extremely common one. In three tumor lines, we have found it to be present in nearly 80% of the observations.

Thus, we propose a new model to explain the physiologic features that contribute to tumor hypoxia. It is based first on the premise that there are longitudinal oxygen gradients. As one moves away from the arteriolar source, microvascular PO_2 is reduced, leading to shorter diffusion distances and larger hypoxic zones between vessels. When red cell flux changes, there are temporary changes in the diffusion distance of oxygen. Near the arteriolar source, these fluctuations are of relatively little significance, but away from the source, they exert a greater influence on the PO_2 of tumor microregions (Figure 2).

3.1. Acknowledgements

This work was supported by a grant from the NIH/NCI CA40355. The author thanks Dr. Oliver Thews for his insights regarding the lethal corner concept.

REFERENCES

1. M. W. Dewhirst, Concepts of oxygen transport at the microcirculatory level, Semin Radiat Oncol. **8**, 143-50 (1998).
2. P. Vaupel, O. Thews, D. K. Kelleher, and M. Hoeckel, Current status of knowledge and critical issues in tumor oxygenation. Results from 25 years research in tumor pathophysiology, Adv Exp Med Biol. **454**, 591-602 (1998).
3. J. M. Brown, and A. J. Giaccia, The unique physiology of solid tumors: opportunities (and problems) for cancer therapy, Cancer Res. **58**, 1408-16 (1998).
4. M. Dewhirst, in Angiogenesis and blood flow in solid tumors/Drug Resistance in Oncology edited by B. Teicher (Marcel Dekker, Inc., New York, 1993), pp. 3-24.
5. R. Thomlinson, and L. Gray, The histological structure of some human lung cancers and the possible implications for radiotherapy, Br J Cancer **9**, 539-549 (1955).
6. G. Helmlinger, F. Yuan, M. Dellian, and R. Jain, Interstitial pH and pO2 gradients in solid tumors in vivo: high resolution measurements reveal a lack of correlation, Nature Medicine **3**, 177-182 (1997).
7. T. W. Secomb, R. Hsu, R. D. Braun, J. R. Ross, J. F. Gross, and M. W. Dewhirst, Theoretical simulation of oxygen transport to tumors by three-dimensional networks of microvessels, Adv Exp Med Biol. **454**, 629-34 (1998).
8. M. W. Dewhirst, E. T. Ong, R. D. Braun, B. Smith, B. Klitzman, S. M. Evans, and D. Wilson, Quantification of longitudinal tissue pO2 gradients in window chamber tumours: impact on tumour hypoxia, Br J Cancer **79**, 1717-22 (1999).
9. S. Walenta, S. Snyder, Z. Haroon, R. Braun, K. Amin, D. M. Brizel, W. Mueller-Klieser, B. Chance, and M. W. Dewhirst, Tissue gradients of energy metabolites mirror oxygen tension gradients in a rat mammary carcinoma model, Int J Radiat Oncol Biol Phys. *(accepted)* (2001).
10. B. Kavanagh, B. Coffey, D. Needham, R. Hochmuth, and M. Dewhirst, The effect of flunarizine on erythrocyte suspension viscosity under conditions of extreme hypoxia, low pH and lactate treatment, Br. J. Cancer **67**, 734-741 (1993).
11. M. W. Dewhirst, E. T. Ong, D. Madwed, B. Klitzman, T. Secomb, D. Brizel, J. Bonaventura, G. Rosner, B. Kavanagh, J. Edwards, and et al., Effects of the calcium channel blocker flunarizine on the hemodynamics and oxygenation of tumor microvasculature, Radiat Res. **132**, 61-8 (1992).
12. P. J. Wood and D. G. Hirst, Cinnarizine and flunarizine as radiation sensitisers in two murine tumours, Br J Cancer. **58**, 742-5 (1988).

13. D. Brizel, B. Klitzman, J. Cook, J. Edwards, G. Rosner, and M. Dewhirst, A comparison of tumor and normal tissue microvascular hematocrits and red cell fluxes in a rat window chamber model, Int. J. Radiat. Oncol. Biol. Phys. **25**, 269-276 (1993).
14. M. Dewhirst, H. Kimura, S. Rehmus, R. Braun, D. Papahadjopoulos, K. Hong, and T. Secomb, Microvascular studies on the origins of perfusion-limited hypoxia, Brit. J. Cancer **74**, S247-S251 (1996).
15. O. Thews, D. Kelleher, M. Hummel, and P. Vaupel, Can tumor oxygenation be improved by reducing cellular oxygen consumption?, Adv Exp Med Biol. **471**, 525-532 (1999).
16. T. W. Secomb, R. Hsu, E. T. Ong, J. F. Gross, and M. W. Dewhirst, Analysis of the effects of oxygen supply and demand on hypoxic fraction in tumors, Acta Oncol. **34**, 313-6 (1995).
17. E. Opitz, M. Schneider, Uber die sauerstoffversorgung des Gerhirns und den mechanismus von mangelwirkungen, Ergeb. Physiol. **48**, 126-260 (1950).
18. S. Snyder, J. Lanzen, R. Braun, G. Rosner, T. Secomb, J. Biaglow, D. Brizel, and M. Dewhirst, Simultaneous administration of glucose and hyperoxic gas achieves greater improvement in tumor oxygenation than hyperbaric gas alone, Int J Radiat Oncol Biol Phys (in press) (2001).
19. H. Kimura, R. D. Braun, E. T. Ong, R. Hsu, T. W. Secomb, D. Papahadjopoulos, K. Hong, and M. W. Dewhirst, Fluctuations in red cell flux in tumor microvessels can lead to transient hypoxia and reoxygenation in tumor parenchyma, Cancer Res. **56**, 5522-8 (1996).
20. R. Braun, J. Lanzen, and M. Dewhirst, Fourier analysis of fluctuations of oxygen tension and blood flow in R3230Ac tumors and muscle of rats, Am J Physiol. **277**, H551-H568 (1999).
21. M. W. Dewhirst, R. D. Braun, and J. L. Lanzen, Temporal changes in PO2 of R3230AC tumors in Fischer-344 rats, Int J Radiat Oncol Biol Phys. **42**, 723-6 (1998).
22. S. Hill, K. Pigott, M. Saunders, M. Powell, S. Arnold, A. Obeid, G. Ward, M. Leahy, P. Hoskin, and D. Chaplin, Microregional blood flow in murine and human tumours assessed using laser Doppler microprobes, Br. J. Cancer **74**, S260-S263 (1996).
23. D. Chaplin, R. Durand, and P. Olive, Acute hypoxia in tumors: implications for modifiers of radiation effects, Int. J. Radiat. Oncol. Biol. Phys. **12**, 1279-1282 (1986).
24. M. W. Dewhirst, B. Klitzman, R. D. Braun, D. M. Brizel, Z. A. Haroon, and T. W. Secomb, Review of methods used to study oxygen transport at the microcirculatory level [In Process Citation], Int J Cancer. **90**, 237-55 (2000).

LACK OF ASSOCIATION BETWEEN TUMOR HYPOXIA, GLUT-1 EXPRESSION AND GLUCOSE UPTAKE IN EXPERIMENTAL SARCOMAS

Oliver Thews, Debra K. Kelleher, Norbert Esser, Stephan Kraus, and Peter Vaupel*

1. INTRODUCTION

Hypoxia is a common phenomenon in experimental and human tumors which not only limits the efficacy of non-surgical treatment modalities such as irradiation and O_2-dependent chemotherapy but also affects the long-term prognosis of cancer patients. Numerous studies have clearly demonstrated that patients with more hypoxic tumors have a reduced probability for survival and disease-free survival[1]. For this reason it might be helpful to non-invasively identify patients with hypoxic malignancies.

In order to meet the energy demand, cells in hypoxic tumor regions increase their glycolytic activity. This leads to an enhanced formation of lactic acid and a pronounced extracellular acidosis. Since in this metabolic pathway the ATP yield is less (compared to oxidative glucose breakdown), these cells rely on an enhanced glucose uptake. For this reason it has been postulated that hypoxic tumor cells show an increased ability to take up glucose and therefore this physiological variable might be useful as a surrogate indicator for tumor hypoxia. In nuclear medicine, the cellular glucose uptake can be assessed using fluoro-deoxyglucose (FDG) which is taken up into the cells but not metabolized. In addition, it has been postulated that the expression of glucose transporter molecules (e.g., GLUT-1) also increases in tumor cells during severe hypoxia and is therefore correlated with the long-term prognosis of cancer patients[2].

The aim of the present study was to analyze whether tumor hypoxia has an impact on mechanisms of cellular glucose uptake *in vivo* and whether the uptake of FDG can be used as an indirect parameter for detection of hypoxia in solid tumors.

2. MATERIAL AND METHODS

2.1. Animals and tumors

Experimental DS-sarcomas were implanted onto the hind foot dorsum of male Sprague-Dawley rats by s.c. injection of DS-ascites cells (0.4 ml, 10^4 cells/μl). Tumors grew as flat, spherical segments and replaced the subcutis and corium completely. The

* Inst. of Physiology and Pathophysiology, University of Mainz, 55099 Mainz, Germany

Oxygen Transport to Tissue XXIII, edited by D. Wilson *et al.*
Kluwer Academic/Plenum Publishers, 2003

volume was determined by measuring the three orthogonal diameters of the tumor and using an ellipsoid approximation with the formula: $V = d_1 \times d_2 \times d_3 \times \pi/6$. Tumors reached the target volume of 0.4 - 3.0 ml approx. 6 to 14 days after inoculation. Animals were housed during the whole period of tumor growth either under normoxic ambient conditions breathing room air (21% O_2) or in a hypoxic atmosphere containing 8% O_2. All experimentation had previously been approved by the regional animal ethics committee and was conducted according to German federal law.

2.2. Measurement of the tumor oxygenation

Once tumors had reached the desired volume, oxygenation of the DS-sarcomas was assessed polarographically using O_2-sensitive needle electrodes (outer diameter: 300 µm) and pO_2 histography (Eppendorf, Hamburg, Germany). Animals were anesthetized with sodium pentobarbital (40 mg/kg, i.p.) and the Ag/AgCl reference electrode was placed in the lower abdomen between the skin and the underlying musculature. For tumor pO_2 measurement, the electrode was automatically moved through the tissue in pre-set steps of 0.7 mm. Approximately 80 to 100 pO_2 values were obtained from each tumor and the oxygenation status was described by the mean and median pO_2 and the fraction of pO_2 values ≤ 2.5 mmHg and ≤ 5 mmHg, respectively.

2.3. Measurement of GLUT-1 expression

For measurement of the expression of the glucose transporter 1 (GLUT-1) by DS-sarcoma cells *in vivo*, tumors were excised and disaggregated resulting in a single cell suspension. Expression of the GLUT-1 was determined by flow cytometry using polyclonal antibodies against the extracellular (type N-20, Santa Cruz Biotechnology, Santa Cruz, CA, USA) and the cytoplasmic domain (type C-20) of the GLUT-1 molecule. The measurement of both domains was necessary since GLUT-1 molecules are "stored" in the membrane of intracellular vesicles. On demand these vesicles fuse with the cell membrane resulting in the translocation of GLUT-1 transporters into the outer membrane. Quantifying the extracellular domain assesses the number of functional GLUT-1 molecules integrated into the cell membrane whereas the expression of the cytoplasmic domain correlates with the number of transporter molecules potentially available in the cell (including those "stored" in vesicles). In order to quantify the GLUT-1 expression, the measured fluorescence intensity was normalized to that of particles with a standardized fluorescence (Immuno-Brite, Beckman-Coulter, Krefeld, Germany).

2.4. Measurement of FDG-uptake

The fluoro-deoxyglucose (FDG) uptake of DS-sarcoma cells *in vivo* was assessed by injecting ^{14}C-labeled FDG i.v. (Amersham Pharmacia, Braunschweig, Germany) at a dose of 148000 Bq. After 80 min the tumors and a specimen of skeletal muscle was excised and the tissues were solubilized by incubating the samples with 1 ml BTS-450 (Beckman Coulter, Unterschleissheim, Germany) for 2 h at 40°C. Thereafter, 130 µl 30% H_2O_2 + 70 µl glacial acetic and 10 ml of a scintillation cocktail (Ready Organic, Beckman) was added. This solution was analyzed in a scintillation counter (LS 5000TD, Beckman). The variability of multiple measurements in tumors and various normal tissues (muscle, kidney, liver, spleen) was approx. 5% of the measured value. For quantification, the activity of the tumors was normalized to the reference tissue (skeletal muscle) resulting in a standardized uptake value (SUV) at $t = 80$ min.

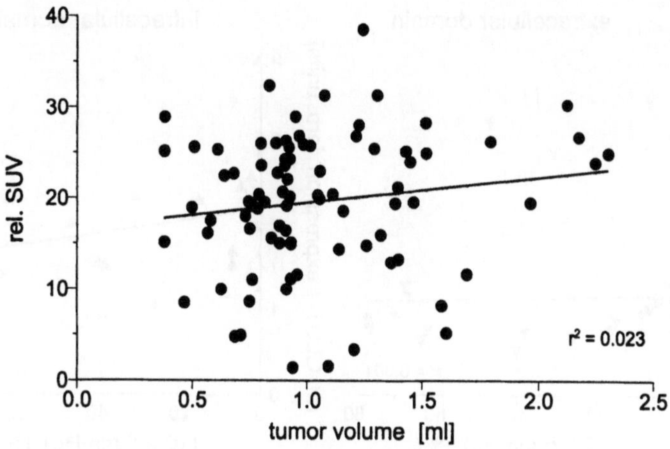

Figure 1. Correlation between tumor volume and ^{14}C-fluoro-deoxyglucose uptake (expressed by the standardized uptake value SUV)

3. RESULTS

The oxygenation of DS-sarcomas worsened profoundly with increasing tumor volume. The median pO_2 decreased significantly ($p = 0.011$) from 11 ± 4 mmHg in small tumors (volume < 1.0 ml, n=12) to 5 ± 2 mmHg in medium-sized tumors (1.0 ml -< 1.5 ml, n=11) and to 2 ± 1 mmHg in large tumors (≥ 1.5 mmHg, n=8), whereas the fraction of hypoxic pO_2-values ≤ 2.5 mmHg was 33 ± 6%, 44 ± 7% and 52 ± 3% in these three groups ($p = 0.020$), respectively. In parallel, the tissue glucose concentration decreased from 2.4 ± 0.2 to 1.9 ± 0.1 and 1.4 ± 0.1 mmol/l ($p = 0.007$) with increasing tumor volume indicating both a decreased availability and a higher glycolytic rate in larger tumors which is also reflected by a rising lactate concentration (6.1 ± 1.0, 10.2 ± 1.4 and 11.3 ± 0.8 mmol/l respectively, $p = 0.001$) in the tissue.

Although these biochemical parameters indicated a higher glycolytic rate in large tumors, the ^{14}C-FDG-uptake determined in the homogenized tumor tissue did not vary systematically with the tumor volume (Fig. 1). Since in DS-sarcomas the oxygenation status worsened with increasing tumor volume, these results indicate that the FDG-uptake seems to be independent of the extent of hypoxia within the tumor.

The expression of the GLUT-1 molecule also did not change with tumor hypoxia. In the DS-sarcomas investigated, neither the intracellular expression of GLUT-1 (which includes GLUT-1 transporters "stored" in the membrane of cytoplasmic vesicles) nor the expression of GLUT-1 on the extracellular surface of the cell membrane showed any systematic correlation with the extent of hypoxia (as expressed by the fraction of pO_2 ≤ 2.5 mmHg) (Fig. 2) or with tumor volume (data not shown). These findings were confirmed in tumors which grew under either normoxic (21% O_2) or hypoxic (8% O_2) ambient conditions. Reducing the oxygen content of the breathing gas during the whole period of tumor growth which leads to a pronounced worsening of oxygenation[3] had practically no impact on the GLUT-1 expression of DS-sarcoma cells *in vivo* (Fig. 3).

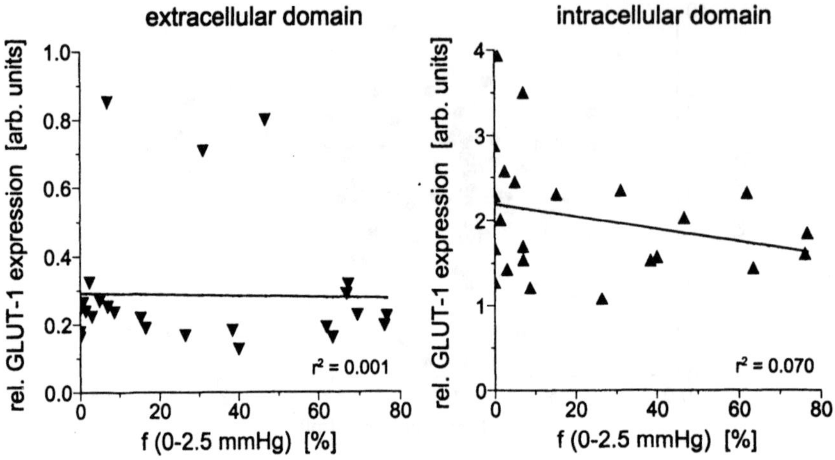

Figure 2. Correlation between tumor hypoxia (expressed by the fraction of hypoxic pO_2 values ≤ 2.5 mmHg) and the expression of the GLUT-1 molecule in DS-sarcoma cells *in vivo*

4. DISCUSSION

Fluoro-deoxyglucose (FDG) is accumulated in many tumor entities and can therefore be used to identify even very small primary lesions or distant metastases. The mechanism by which a specific accumulation in tumor cells occurs is still not fully understood. Since FDG is taken up into the cell via specific glucose transport molecules the cellular expression of these transporters (especially GLUT-1) might influence the FDG-uptake rate[4]. In heart muscle cells it has been demonstrated that FDG-uptake correlated well with the expression of GLUT-4 transporters[5]. Since hypoxia – which is a common

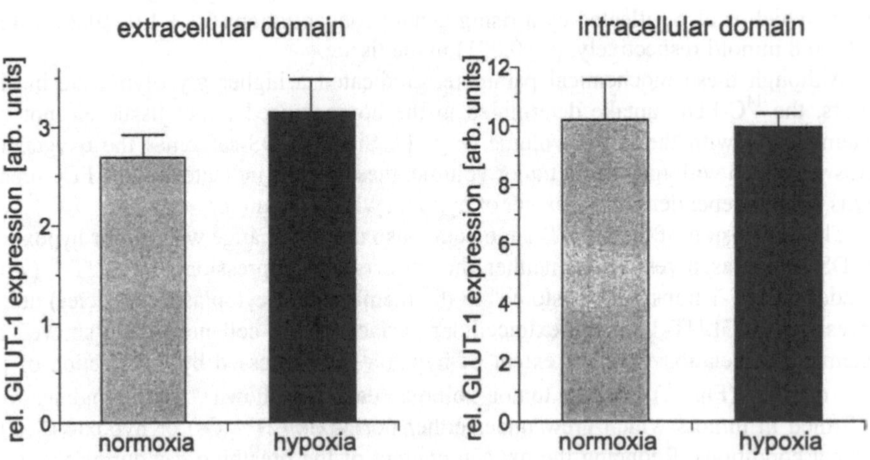

Figure 3. Impact of inspiratory normoxia (21% O_2; n=6) and hypoxia (8% O_2; n=10) during tumor growth on the expression of the GLUT-1 molecule

phenomenon in experimental and human tumors – augments the expression of GLUT transporters in tumor cells (either by improving the synthesis of the molecules or by translocating these molecules into the cell membrane) it was found that the FDG-uptake was increased by oxygen deficiency[6-8]. However, these results were only obtained under *in vitro* conditions. Since in some tumor entities it has been demonstrated that patients with tumors showing high FDG-uptake rates have poor long-term prognosis[9-11], it was postulated that hypoxia – which has been proven to be an independent prognostic indicator for human malignancies[1] – is responsible for the increased FDG-uptake in progressive tumors[4].

However, the results of the present study clearly show that in the experimental DS-sarcoma under *in vivo* conditions the GLUT-1 expression is not affected by the extent of hypoxia present in the tumor. No statistically significant variation in GLUT-1 expression could be found either in tumors of different volumes (in which the oxygenation status differs profoundly) or under inspiratory hypoxia which leads to a pronounced increase in tumor hypoxia[3]. These data are in good accordance with a clinical study analyzing the GLUT-1 expression semi-quantitatively in carcinomas of the cervix[2]. Since GLUT-1 expression is partly controlled by HIF-1 – a factor which is not only oxygen dependent – other microenvironmental parameters (e.g., pH) might play a role for induction. In the present study the FDG-uptake *in vivo* was also unaffected by the tumor size and thus by the oxygenation status. The weak correlation of these parameters seen in one study[10] might be the result of partial volume effects of the imaging technique due to its limited spatial resolution. Based on these results, it is unlikely that FDG-uptake (e.g., as determined by positron emission tomography) is a suitable indicator for indirect assessment of tumor hypoxia.

REFERENCES

1. M. Höckel and P. Vaupel, Tumor hypoxia: definitions and current clinical, biologic, and molecular aspects, *J.Natl.Cancer Inst.* **93**, 266-276 (2001).
2. R. Airley, J. Loncaster, S. Davidson, M. Bromley, S. Roberts, A. Patterson, R. Hunter, I. Stratford, and C. West, Glucose transporter Glut-1 expression correlates with tumor hypoxia and predicts metastasis-free survival in advanced carcinoma of the cervix, *Clin.Cancer Res.* **7**, 928-934 (2001).
3. D.K. Kelleher, O. Thews, and P. Vaupel, Hypoxyradiotherapy: lack of experimental evidence for a preferential radioprotective effect on normal versus tumor tissue as shown by direct oxygenation measurements in experimental sarcomas, *Radiother.Oncol.* **45**, 191-197 (1997).
4. E.K. Pauwels, M.J. Ribeiro, J.H. Stoot, V.R. McCready, M. Bourguignon, and B. Maziere, FDG accumulation and tumor biology, *Nucl.Med.Biol.* **25**, 317-322 (1998).
5. S. Egert, N. Nguyen, F.C. Brosius, III, and M. Schwaiger, Effects of wortmannin on insulin- and ischemia-induced stimulation of GLUT4 translocation and FDG uptake in perfused rat hearts, *Cardiovasc.Res.* **35**, 283-293 (1997).
6. P. Burgman, J.A. Odonoghue, J.L. Humm, and C.C. Ling, Hypoxia-induced increase in FDG uptake in MCF7 cells, *J.Nucl.Med.* **42**, 170-175 (2001).
7. A.C. Clavo, R.S. Brown, and R.L. Wahl, Fluorodeoxyglucose uptake in human cancer cell lines is increased by hypoxia, *J.Nucl.Med.* **36**, 1625-1632 (1995).
8. H. Minn, A.C. Clavo, and R.L. Wahl, Influence of hypoxia on tracer accumulation in squamous-cell carcinoma: *in vitro* evaluation for PET imaging, *Nucl.Med.Biol.* **23**, 941-946 (1996).
9. K. Dhital, C.A. Saunders, P.T. Seed, M.J. O'Doherty, and J. Dussek, [^{18}F]Fluorodeoxyglucose positron emission tomography and its prognostic value in lung cancer, *Eur.J.Cardiothorac.Surg.* **18**,425-428 (2000).
10. B. Nakata, Y.S. Chung, S. Nishimura, T. Nishihara, Y. Sakurai, T. Sawada, T. Okamura, J. Kawabe, H. Ochi, and M. Sowa, ^{18}F-Fluorodeoxyglucose positron emission tomography and the prognosis of patients with pancreatic adenocarcinoma, *Cancer* **79**, 695-699 (1997).
11. M. Oshida, K. Uno, M. Suzuki, T. Nagashima, H. Hashimoto, H. Yagata, T. Shishikura, K. Imazeki, and N. Nakajima, Predicting the prognoses of breast carcinoma patients with positron emission tomography using 2-deoxy-2-fluoro[^{18}F]-D-glucose, *Cancer* **82**, 2227-2234 (1998).

THE IMPACT OF TISSUE MICROENVIRONMENT ON TREATMENT SIMULATION

Alexandru Daşu and Juliana Denekamp*

1. INTRODUCTION

Theoretical modelling has been used for a long time for evaluating old treatments and for estimating the possible gain of new ones. Recently it has been considered for inclusion into treatment planning algorithms. For this purpose it is important to have both a theoretical model to describe accurately the radiation response and a realistic tissue model that takes into consideration the microenvironmental conditions that influence the radiation response.

It is well known that tumour vasculature is poor compared to that in normal tissues. This results in a unique environment in tumours characterised by an impaired supply of oxygen and glucose. The brief interruption of oxygen supply to the cells is known to result in increased radioresistance[1]. However, less known are the effects of nutrient deprivation. Poor supply of glucose to the hypoxic cells is accompanied on the long term by a decrease in the ability to repair DNA damage that renders the cells more radiosensitive[2-6]. Unfortunately little attention was paid to the clinical implications of such observations.

This study intends to investigate theoretically the influence of oxygen and nutrient deprivation on the tissue response to fractionated radiation treatments. The inclusion of the relationship between glucose deprivation and radiosensitivity represents a novel approach to theoretical simulation of radiation treatment.

2. CONDITIONS OF SIMULATION

The cellular response to radiation was calculated using two theoretical models. One of them is the linear quadratic (LQ) model that has been employed for many years in the theoretical evaluation of fractionated treatments and is the main candidate for inclusion into treatment planning algorithms. The other theoretical model is the linear quadratic model with inducible repair (LQ/IR). This model is a modified version of the LQ model that can explain the increased radiosensitivity of the cells to radiation doses below the

* Department of Radiation Sciences, Umeå University, 901 85 Umeå, Sweden

clinical dose range. The LQ/IR model assumes that unirradiated cells are capable of little or no repair, but that exposure to radiation results in a gradual induction of repair capacity with increasing radiation dose[7]. All this happens to doses below the clinical dose range, so that at doses above 1-1.5 Gy the repair capacity of the cells is fully activated. The fact that the two models are virtually identical in the clinical dose range led to the LQ/IR model being largely neglected in favour of the LQ model. However, some of the clinical implications of the differences between the two models are quite important as it has been shown elsewhere[8, 9].

Based on the experimental observations presented in the introduction, we postulate that the metabolic energy status of the cells influence their response to radiation. Thus, cells having low energy reserves are more or less unable to activate their DNA repair mechanisms and therefore are more sensitive than the cells with high energy reserves. The brief interruption in the oxygen supply characteristic to the short-term closure of some tumour blood vessels determines an increased radioresistance of the cells surrounding these vessels (acutely hypoxic cells). By contrast, cells situated far away from perfused tumour vessels (chronically hypoxic cells) have a poor supply of both oxygen and glucose, and this prolonged lack of nutrients results in a radiosensitisation. We postulate further that the magnitude of this sensitisation is given by the parameters of the LQ/IR model.

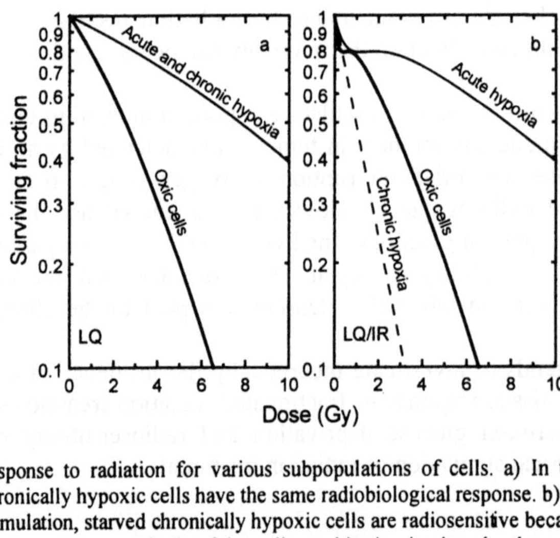

Figure 1. Predicted response to radiation for various subpopulations of cells. a) In the classical view on hypoxia, acutely and chronically hypoxic cells have the same radiobiological response. b) According to the new approach to treatment simulation, starved chronically hypoxic cells are radiosensitive because they are unable to induce the repair mechanisms. The magnitude of the radiosensitisation is given by the parameters of the linear quadratic model with inducible repair. Well-fed acutely hypoxic cells are radioresistant.

Figure 1 shows the shape of the survival curves for various subpopulations calculated using the classical view on hypoxia (panel a) and our new postulate on hypoxia and repair competence (panel b). According to the classical view, described by the LQ model that does not take into consideration the preparedness of the repair mechanisms and the possibility to induce them, there is no difference between the predicted response for acutely or chronically hypoxic cells. On the contrary, if the repair competence is taken into consideration, the response of the chronically and acutely hypoxic cells is quite well separated, the magnitude of the separation being given by the parameters of the LQ/IR model. Furthermore, if the extent of repair induction in the cells is greater than the radiochemical hypoxic protection (OER), as is the case shown in figure 1b, chronically hypoxic cells may even be more sensitive than the oxic cells rich in nutrients.

3. RESULTS AND DISCUSSION

Using the same cell parameters as in figure 1, we have calculated the response of the tumour (represented as a mixed population of oxic and hypoxic cells) predicted by the two theoretical models for fractionated radiation treatments. When calculating this response it was assumed that between each two consecutive radiation doses there is a process of redistribution between the various subpopulations that restores the initial fractions of cells. The results for a mixed tumour population were then compared to the response of a healthy normal tissue assumed to be a purely oxic population.

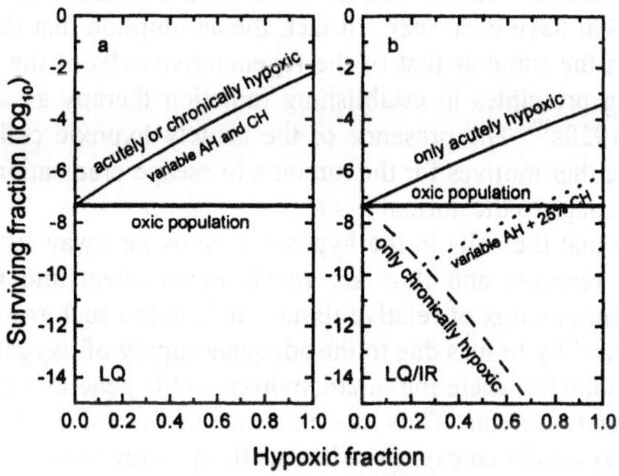

Figure 2. Predicted response to a standard treatment of 33 fractions of 2 Gy for tumours containing various subpopulations of cells. a) The classical view on hypoxia. b) The new approach on hypoxia and repair competence. The presence of acutely hypoxic cells worsens the response of the tumours. However, a small fraction of repair deficient chronically hypoxic cells counterbalances the effect of the acutely hypoxic cells rendering the tumour more sensitive than the normal tissue.

Figure 2 shows the calculated response to a standard radiation treatment of 33 fractions of 2 Gy as a function of the fraction of tissue hypoxia. A purely oxic population was considered to give the response of the normal tissue, while tumours were described as mixed populations with various degrees of hypoxia. For the ease of comparison, the response of the normal tissue is represented as a thick line across each panel.

If the response is predicted according to the classical view for which there is no difference between the radiobiological response of the two types of hypoxic cells (figure 2a), the presence of some hypoxia in the tumour worsens the response by increasing the overall radioresistance of the mixed population. Comparing the response of the tumour with the purely oxic population, it can be seen that the treatment results in less damage to the tumour compared to the normal tissue.

By contrast, if repair competence is taken into consideration (figure 2b), the response of the mixed population of cells is the result of a complex balance between the number of repair competent and repair incompetent cells. Thus, the presence of repair competent acutely hypoxic cells in the population worsens the response. However, the presence of the repair deficient chronically hypoxic cells with their increased radiosensitivity changes greatly the overall response of the mixed population. Thus, the presence of a small fraction of starved hypoxic cells makes the whole tumour more sensitive to radiation than the normal tissue even when some acutely hypoxic cells are present. This behaviour may provide an explanation for the clinical success of radiotherapy.

In curative radiotherapy, all the clonogenic cells in a tumour are destroyed, while the function of the adjacent healthy tissue is preserved. The differences between tumours and normal tissues with respect to tissue kinetics and architectural structure translate into the fact that normal tissues can tolerate much less cell kill than tumours. This means that a successful radiation treatment results in much larger tissue damage to tumours than to normal tissues. But the predictions of the theoretical modelling according to the classical view on hypoxia (figure 2a) are exactly the opposite. No consistent differences between the intrinsic radiosensitivities of tumours and normal tissues that could explain the difference in cell kill have been seen. In fact, the assumption that the radiosensitivity of the tumour cells is the same as that of the regenerative cells of the tissue of origin was one of the guiding principles in establishing radiation therapy as a successful form of treatment in the 1920s[10]. The presence of the acutely hypoxic cells and the increased proliferation are further motives for the tumours to escape eradication with the doses that give acceptable damage to the normal tissue.

The postulate that the cells in the hypoxic regions far away from the blood vessels have low energy reserves and thus are repair incompetent and radiosensitive could explain the apparent paradox of relative tissue cell kill. No such repair incompetent cells normally exist in healthy tissues due to the adequate supply of oxygen and nutrients. This is not the case in tumours where the microenvironment is generally characterised by poor levels of oxygen and nutrients. Thus, the existence of starved cells with their associated radiosensitivity may confer an overall radiosensitivity to the tumour population compared to the normal tissues. This translates into the fact that curative doses of radiation inflict more damage to the tumour, but not to normal tissues, thus allowing the preservation of their functions.

Another important implication of the postulated difference in the response to radiation of the two types of hypoxia is for predictive assays. Since various hypoxic cells have different responses, care should be taken when interpreting the results of various assays or when looking for correlations between them. For example, tumour oxygenation refers only to the oxygen availability and does not take into consideration the metabolic status of the cells. Radiobiological hypoxia on the other hand is a complex parameter given by the combined response of all the subpopulations in the tumour. Therefore no *a priori* correlation could be expected between these two parameters.

The modelling in this study was performed on mixtures of pure subpopulations. This combination is not encountered in reality. However, it allowed us to demonstrate the implications of the postulates we have proposed. If real distributions of oxygen and nutrients are taken into the modelling process, the conceptual results outlined in this article (or at least their direction) are not expected to change.

4. CONCLUSIONS

The modelling showed the clinical importance of distinguishing between the different types of tumour hypoxia based on their physiological and radiobiological characteristics. This distinction provided a better understanding of the tumour response to radiation treatments. Acutely hypoxic cells, having high energy reserves, are capable of inducing the DNA repair mechanisms and are thus radioresistant. By contrast, starved hypoxic cells have poor energy reserves and thus are more sensitive than the glucose-rich cells. The presence of a relatively small subpopulation of starving, chronically hypoxic cells in tumours could very much increase the cell kill. This does not happen in normal tissues, which do not have starving cells, and thus they can tolerate the radiation doses used clinically. It is therefore important to distinguish between the two types of hypoxia in predictive assays and other treatment simulations.

The modelling also showed that the LQ/IR model is more appropriate to describe tissue response to radiation treatment than the LQ model. Not only it describes better the response of cell to low doses of radiation, but the concept on which is based (induction of repair) allows an easy quantification of the radiobiological differences between acutely and chronically hypoxic cells.

ACKNOWLEDGEMENTS

This project was supported by grants from the Cancer Research Foundation in Northern Sweden. We would like to thank Prof. Bo Littbrand for help and encouragement and Prof. Jack F. Fowler for many simulative discussions.

REFERENCES

1. E. J. Hall, *Radiobiology for the Radiologist*, 5th edition (Lippincott Williams & Wilkins, Philadelphia, 2000).
2. E. J. Hall, J. S. Bedford, and R. Oliver, Extreme hypoxia; its effect on the survival of mammalian cells irradiated at high and low dose-rates, *Br. J. Radiol.* **39**, 302-307 (1966).
3. R. J. Berry, E. J. Hall, and J. Cavanagh, Radiosensitivity and the oxygen effect for mammalian cells cultured in vitro in stationary phase, *Br. J. Radiol.* **43**, 81-90 (1970).
4. W. A. Nagle, A. J. Moss Jr., H. G. Roberts Jr., and M. L. Baker, Effects of 5-thio-D-glucose on cellular adenosine triphosphate levels and deoxyribonucleic acid rejoining in hypoxic and aerobic Chinese hamster cells, *Radiology* **137**, 203-211 (1980).
5. I. J. Spiro, K. A. Kennedy, R. Stickler, and C. C. Ling, Cellular and molecular repair of X-ray-induced damage: dependence on oxygen tension and nutritional status, *Radiat. Res.* **101**, 144-155 (1985).
6. E. O. Pettersen and H. Wang, Radiation-modifying effect of oxygen in synchronized cells pre-treated with acute or prolonged hypoxia, *Int. J. Radiat. Biol.* **70**, 319-326 (1996).
7. M. C. Joiner and H. Johns, Renal damage in the mouse: the response to very small doses per fraction, *Radiat. Res.* **114**, 385-398 (1988).
8. A. Daşu and J. Denekamp, New insights into factors influencing the clinically relevant oxygen enhancement ratio, *Radiother. Oncol.* **46**, 269-277 (1998).
9. A. Daşu and J. Denekamp, Superfractionation as a potential hypoxic cell radiosensitizer: prediction of an optimum dose per fraction, *Int. J. Radiat. Oncol. Biol. Phys.* **43**, 1083-1094 (1999).
10. H. D. Thames and J. H. Hendry, *Fractionation in Radiotherapy* (Taylor & Francis, London, NewYork, Philadelphia, 1987).

EVIDENCE FOR AND AGAINST HYPOXIA AS THE PRIMARY CAUSE OF TUMOR AGGRESSIVENESS

Paul Okunieff, Ivan Ding, Peter Vaupel, and Michael Höckel*

In clinical trials, tumor hypoxia has consistently been associated with tumor aggressiveness. The evidence for an association between hypoxia and metastasis and more rapid tumor progression and death is seen in uterine cervical cancer,[1,2] and sarcoma of soft tissue.[3] Evidence is building in prostate,[4] vulva, head and neck,[5,6] and breast cancers.[7,8] A major question is whether hypoxia precedes tumor aggressiveness or whether aggressive tumors incidentally are also hypoxic.

INTRODUCTION

Cancer is frequently described as a genetic disease, the mechanism of which is the mutation of either anti-oncogenes or oncogenes to forms that promote immortal replication. Clear examples of the phenomenon are seen in certain pediatric cancers wherein single genetic constructs seem to be sufficient to cause malignant tumors. Unlike pediatric neoplasms, most adult tumors appear to require multiple mutations (perhaps hundreds) and cooperation between those mutations. In that case, oncogenic progression is a result of sequential changes that provide a selective advantage for tumor growth. The most famous example has been proposed for colon cancer by Vogelstein.[9] However, this model is now considered overly-simplistic.

Adaptation, which is required by all cells in the body, not just by malignancy, is physiologically achieved by up or down regulation of normal genes. For example, at the time of tissue injury, growth factors are produced in order to grow granulation tissue. The granulation response, which grows more briskly than virtually any tumor, turns off just as

*Paul Okunieff, MD, and Ivan Ding, MD. University of Rochester, Rochester, New York, 14642.
Peter Vaupel, MD. University of Mainz, Mainz, Germany 55099.
Michael Höckel, MD, PhD, University of Leipzig, Leipzig, Germany 04103.

briskly as it began. Hypoxia requires substantial adaptation, and includes many granulation-like signals.

Thus it is clear that tumors have a complex interaction with hypoxia. A better understanding of the involved mechanisms could lead to strategies for inhibition of tumor progression and oncogenesis. This article will explore the evidence that hypoxia dominates over genetics as the cause of malignant progression.

METHODS

In reviewing the literature, we will attempt to define the degree to which hypoxia causes oncogenic mutations or oncogenic progression. Alternatively, we will try and determine to what extent there is evidence that genetic changes that cause oncogenesis or progression incidentally cause hypoxia. If the data available fit most closely with Figure 1, then it will be argued that hypoxia dominates as the root factor causing tumor progression and/or oncogenesis. If the model depicted in Figure 2 predominates, then we will define genetic alterations as the pivotal factor. Likewise, if the process is interactive rather than being driven by a dominant factor (chicken and egg rather than cart and horse), as in Figure 3, then no definite conclusion will be drawn.

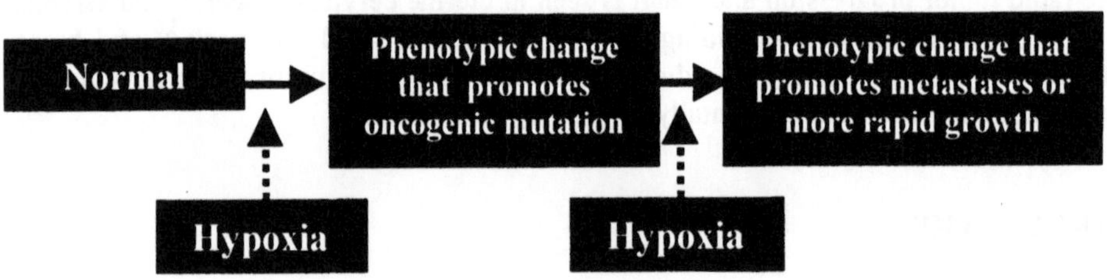

Figure 1. In this model, hypoxia is the dominant factor governing the oncogenesis and/or progression of the oncogenic phenotype. Here hypoxia causes microenvironmental changes that select for mutations and adaptations that initiate, promote, and advance oncogenesis.

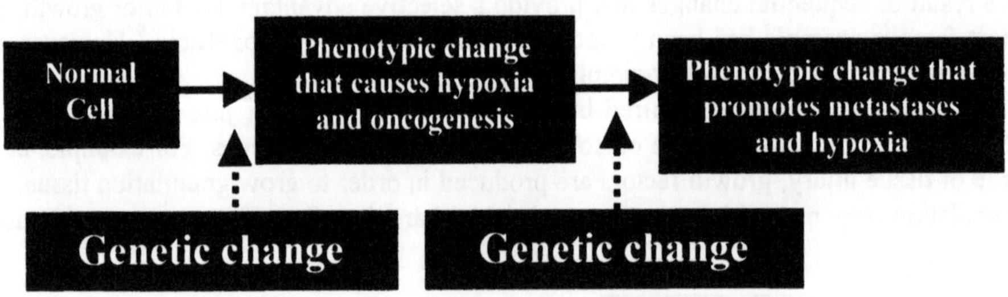

Figure 2. In this model, genetic mutation is the dominant factor governing the oncogenesis and/ or progression of the oncogenic phenotype. Here genetic mutations cause phenotypic changes in the tumor cells, which, in turn, cause hypoxia. Hypoxia is incidental and unimportant in the oncogenic progression.

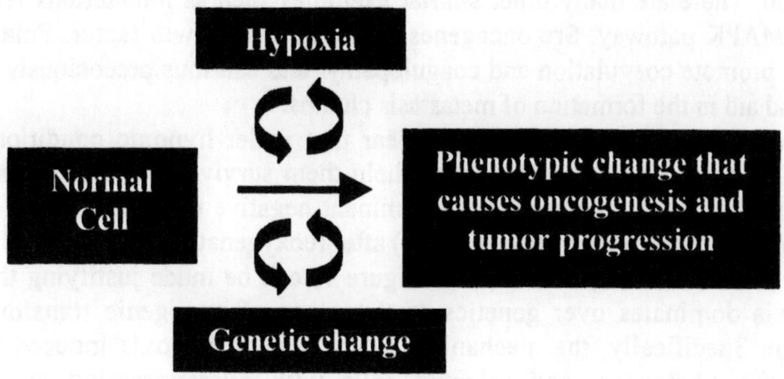

Figure 3. In this model, neither genetic mutations nor hypoxia dominate as the prime factor governing oncogenesis or oncogenic progression. This model suggests that the dominant factor cannot be determined, that hypoxia and genetic alterations are both responsible, and neither predominate.

RESULTS AND DISCUSSION

Evidence for hypoxia as a dominating factor for the selection of sublines important to tumor induction and progression

Evidence that hypoxia dominates as a mechanism of oncogenesis and tumor progression is substantial. Hypoxia can select, in a Darwinian sense, cells that are resistant to apoptosis. There are several examples of hypoxia selecting cells resistant to a hostile tissue microenvironment.[10-12] In order to survive adverse conditions, cells require the activation of their angiogenic process to compensate for an insufficient oxygen supply.[13-15] These factors include bFGF, VEGF, and EGF. These growth factors can themselves augment tumor aggressiveness, correlate with metastasis,[16-18] inhibit apoptosis,[19] and can be produced by the normal cells within the hypoxic tumor. Evidence of Darwinian selection was shown by Giaccia's group. They demonstrated this quite dramatically in an experiment in which a 1:1000 mix of P53/BCL2 mutant cells was created, cycled through 10 periods of hypoxia and reoxygenation, to yield a progeny of 1000:1 P53/BCL2 mutant cells. Since redox cycling causes DNA damage, apoptosis prevention allows mutations to stabilize and expand in a population of cells.[11,12,20] The end result being the potential for selection of a malignant clone. Thus, hypoxic stress selects antiapoptotic factor mutations in tumor cells. It may also select for mutations of many genes required for angiogenesis and growth, yielding malignant progression.

Hypoxia can also interact with oncogenes. For example, the angiogenic switch in Ras-transformed cells was promoted by the tumor microenvironment through VEGF being induced by hypoxia.[21,22] Ras transformed cells also produce high levels of VEGF, closing the positive feedback loop. Using dominant negative mutants of Ha-Ras and Raf-1, Koong et al. also showed that both H-Ras and Raf-1 inhibited hypoxia-induced NF-kappa B induction, and that low oxygen activates mitogen-activated protein kinases (ERK1 and ERK2).[23] Thus hypoxia can interact with oncogenes to suppress TNF induced apoptosis, while increasing proliferative stimuli. Seko et al.[24] reported that hypoxia can rapidly induce both Src and Raf-1 and ras, which led to activation of MAPK kinase signal transduction. There are many other similar examples such as interactions featuring Egr-1,[25,26] the MAPK pathway. Src oncogenes, and placenta growth factor. Finally, hypoxia might also promote coagulation and coagulopathy, and can thus precociously create more hypoxia and aid in the formation of metastasis clumps.[27]

Besides antiapoptotic signals, it is clear that under hypoxic conditions, ordinary adaptation by tumors activates genes that help them survive in a hypoxic environment. For example, Kim et al. reported that a dominant negative rac1 inhibits the intracellular production of reactive oxygen species (ROS) after reoxygenation in muscle cells.[28]

Thus, a full set of arguments (as in Figure 1) can be made justifying the statement that hypoxia dominates over genetics as the cause of oncogenic transformation and progression. Specifically, the mechanism exists through hypoxia induced increases in antiapoptotic substances, and selected cells with overexpression of antiapoptotic substances. Second, reoxygenation injury provides a powerful mechanism for creating new mutations in the cells that are apoptosis resistant. Finally, hypoxia at later stages appears to provide an environment for the phenotypic expression of oncogenes through antiapoptotic pathways, growth signals, angiogenic pathways, and coagulopathy.

Evidence for genetics as a dominating factor for the selection of sublines important for induction and progression of cancer

As previously discussed, cancer is a multistep process, usually consisting of an initial stage in which cells become immortal coincident with a step during which they are induced to proliferate. They must then induce angiogenesis and finally, must overcome several barriers to metastasize and grow in foreign tissue. While all these steps require signals, and many require mutations, none require hypoxia.

Common to all tumorigenesis is angiogenesis. Angiogenesis is a basic requirement for tumor growth, and features non-malignant cells, the endothelium.[29] Upregulation of oncogenes or reduced tumor suppressor genes are likely able to deregulate expression of both angiogenesis stimulators[30] and inhibitors in cancer cells.[31] Mutant ras expression, for example, is associated with increased production of vascular endothelial growth factor (VEGF) and MMP-related enzymes[32] and downregulation of thrombospondin-1 (TSP-1). More recently, Konishi et al.[33] reported that a mutant p53 induced the expression of VEGF mRNA, and that wild-type p53 down-regulated endogenous VEGF mRNA levels. In addition, VEGF status, K-ras status (and nodal status) were found to be significant indicators for prognosis in human lung cancer. Upregulation of VEGF and angiogenesis can then also be induced by constitutive activation of other oncogenic proteins (e.g., EGFR, Egr, Src, Raf, MEK, PI3K) acting at various levels on the Ras signaling pathway. In the early 1990's, our data showed that ras-transformed cells grown in nude mice had low PO_2 compared with the same histological type of spontaneous tumors. The ras transfected tumors grew faster and had a 10-fold higher rate of ATP consumption, and this caused a decreased PO_2.[34] Finally, experimental evidence strongly suggests that many of these oncogene-related properties can be abolished by effective application of the specific oncogene antagonists and signal transduction inhibitors.[30,35] In contrast, oxygen level has very little effect on cancer growth.

Thus data support the alternative hypothesis as well. Specifically, there is sufficient evidence to support the contention that oncogenic mutations precede hypoxia. Genetic mutations in oncogenes cause hypoxia through over-population and high metabolic rate. Since the mode of action of many carcinogens is unknown, the argument in favor of hypoxia as the primordial factor promoting oncogenesis and promoting tumor progression is stronger than the genetic alternative.

Evidence for other factors as key to tumor induction and progression

Overall, the microenvironment and genetics both play a critical role in tumor development and progression. The effects of both are commonly complementary. Consider: 1) Hypoxia has differential effects on angiogenic factor production.[36] For example, hypoxia strongly induces VEGF mRNA, but does not appear to increase VEGF-B or VEGF-C mRNA levels. Serum and growth factors, platelet-derived growth factor (PDGF), epidermal growth factor (EGF) and transforming growth factor-beta (TGF-beta) increased VEGF-C, but not VEGF-B mRNA expression. 2) Hypoxia and growth factors have distinct impacts on the microenvironment.[37] For example, transcriptional regulation of the VEGF promoter in glioblastoma by the growth factor EGFR appears to involve Ras/PI(3) kinase which is distinct from signals induced by hypoxia. 3) Hypoxia has dual effects on angiogenic and antiangiogenic factor expression.[38] For example, exposure of hypoxia to rodent cells with p53+/+ or p53-/- simultaneously suppressed TSP-1 and induced VEGF expression, regardless of the p53 genotype. Similarly, tumor tissue derived from the transformed p53+/+ and p53-/- cells overexpresses VEGF protein in regions of hypoxia, whereas TSP-1 protein was below the limits of detection in normoxic tumor tissue. 4) Combinations of hypoxia and genetic alteration regulates tumor-related growth factor expression.[39] For example, both ras and hypoxia up-regulated GLUT1 mRNA levels in Rat1-ras cells. High GLUT1 levels then allowed for anaerobic respiration even in normoxic conditions. More recently, Premkumar et al[40] reported that c-fos is essential for the activation of activator protein-1 transcription factor complex (AP-1) and subsequent stimulation of downstream genes. It is demonstrated that Scr and ras are affected by the hypoxia-induced c-fos gene expression.

Thus again, data exist that well support the third model in which oncogenic transformation and progression is a complementary interaction between the microenvironment and gene mutation without a dominant initiating event. However, the evidence for this pathway is less strong than that for the pathway depicted in Figure 1.

Clinical implications of hypoxia as the dominating selection factor of tumor induction or progression

We had expected, perhaps naively, to have arguments clearly distinguishing one pathway over the others. The implications, however, of our contention that the most common pathway is through redox stress are not small. For example, we have found that multiparous women have hypoxia in their normal cervix and an associated high risk of cervix cancer.[41] Others have shown that the prostate is hypoxic in older men with prostate cancer.[4] The latter appears to be reversed with Cox-2 inhibitors. Thus one could develop therapies that may prevent the oncogenic transformations.

Based on the above analysis it is clear that hypoxia is a very important factor, and is, for some tumors, the initiating event. Indeed if simple measures such as anti-inflammatory agents like Cox-2 inhibitors can reduce hypoxia, that might explain their success clinically in preventing gastrointestinal cancers, and for preventing metastogenesis in animal models. Regarding predictive assays, measurements of oxygen could provide a predictive assay both for identifying risk of developing malignancy and for predicting which tumors are likely to become aggressive.

REFERENCES

1. M. Höckel, C. Knoop, B. Vorndran, E. Baussmann, M. Mitze, P. G. Knapstein, P. Vaupel, Intratumoral pO2 predicts survival in advanced cancer of the uterine cervix, *Radiother. Oncol.* **26**, 45-50 (1993).
2. P. Vaupel, D. K. Kelleher, and M. Höckel, Oxygen status of malignant tumors: pathogenesis of hypoxia and significance for tumor therapy, *Sem. Oncol.* **28**(2S8), 29-35 (2001).
3. D. M. Brizel, S.P. Scully, J.M. Harrelson, L. J. Layfield, J. M. Bean, L. R. Prosnitz, M. W. Dewhirst, Tumor oxygenation predicts for the likelihood of distant metastases in human soft tissue sarcoma, *Cancer Res.* **56**, 941-943 (1996).
4. B. Movsas, J. D. Chapman, R. E. Greenberg, A. L. Hanlon, E. M. Horwitz, W. H. Pinover, C. Stobbe, and G. E. Hanks, Increasing levels of hypoxia in prostate carcinoma correlate significantly with increasing clinical stage and patient age: an Eppendorf pO(2), *Cancer*. **89**, 2018-2024 (2000).
5. D. M. Brizel, G. S. Sibley, L. R. Prosnitz, R. L. Scher, M. W. Dewhirst, Tumor hypoxia adversely affects the prognosis of carcinoma of the head and neck, *Int. J. Radiat. Oncol. Biol. Phys.* **38**, 285-289 (1997).
6. P. Okunieff, J. de Bie, E. P. Dunphy, D. J. Terris, M. Höckel, Oxygen distributions partly explain the radiation response of human squamous cell carcinomas, *Br. J. Cancer* **74**, S185-S190 (1996).
7. P. Okunieff, E. P. Dunphy, M. Höckel, D. J. Terris, P. Vaupel, The role of oxygen tension distribution on the radiation response of human breast carcinoma, *Adv. Exp. Med. Biol.* **345**, 485-492 (1994).
8. P. Okunieff, M. Höckel, E. P. Dunphy, K. Schlenger, C. Knoop, P. Vaupel, Oxygen tension distributors are sufficient to explain the local response of human breast tumors treated with radiation alone, *Int. J. Radiat. Oncol. Biol. Phys.* **26**, 631-636 (1993).
9. J. W. Arends, Molecular interactions in the Vogelstein model of colorectal carcinoma, *J. Pathol.* **190**, 412-416 (2000).
10. N. M. Mazure, E. Y. Chen, P. Yeh, Oncogenic transformation and hypoxia synergistically act to modulate vascular endothelial growth factor expression, *Cancer Res.* **56**, 3436-3440 (1996).
11. A. J. Giacca, Hypoxic stress proteins: survival of the fittest, *Semin. Radiat. Oncol.* **6**, 46-58 (1996).
12. T. G. Graeber, C. Osmanian, T. Jacks, D. E. Housman, C. J. Koch, S. W. Lowe, A. J. Giaccia, Hypoxia-mediated selection of cells with diminished apoptotic potential in solid tumours, *Nature* **379**, 88-91 (1996).
13. R.H. Ackerman, R. Subramanyam, J. A. Correia, N. M. Alpert, J. M. Taveras, Positron imaging of cerebral blood flow during continuous inhalation of $C^{15}O_2$, *Stroke* **11**, 45-49 (1980).
14. E. Berra, J. Milanini, and D. E. Richard, Signaling angiogenesis via p42/p44 MAP kinase and hypoxia, *Biochem. Pharmacol.* **60**, 1171-1178 (2000).
15. E. Berra, G. Pages, and J. Pouyssegur, MAP kinases and hypoxia in the control of VEGF expression, *Cancer Metastasis Rev.* **19**, 139-145 (2000).
16. F. G. Kern, S. W. McLeskey, L. Zhang, J. Kurebayashi, Y. Liu, I. Ding, S. Kharbanda, D. Chen, D. Miller, K. Cullen, S. Paik, R. B. Dickson, Transfected MCF-7 cells as a model for breast cancer progression, *Breast Cancer Res. Treat.* **31**, 153-165 (1994).
17. M. Nguyen, H. Watanabe, A. E. Budson, J. P. Richie, J. Folkman, Elevated levels of the angiogenic peptide basic fibroblast growth factor in urine of bladder cancer patients, *J. Natl. Cancer Inst.* **85**, 241-242 (1993).
18. J. Folkman, Angiogenesis and breast cancer, *J. Clin. Oncol.* **12**, 441-444 (1994).
19. P. Okunieff, M. Mester, J. Wang, T. Maddox, X. Gong, D. Tang, M. Coffee, I. Ding, In vivo radioprotective effects of angiogenic factors on the small bowel of C3H mice, *Radiat. Res.* **150**, 204-211 (1998).
20. C. Koumenis, R. Alarcon, E. Hammond, P. Sutphin, W. Hoffman, M. Murphy, J. Derr, Y. Tava, S. W. Lowe, M. Kastan, and A. Giacca, Regulation of p53 by hypoxia: Dissociation of transcriptional repression and apoptosis from p53-dependent transactivation, *Mol. Cell. Biol.* **21**, 1297-1310 (2001).
21. L. Hlatky, P. Hahnfeldt, C. Tsionou, and C. N. Coleman, Vascular endothelial growth factor: environmental controls and effects in angiogenesis, *Br. J. Cancer*. **27**, S151-156 (1996).
22. L. Hlatky, C. Tsionou, P. Hahnfeldt, and C. N. Coleman, Mammary fibroblasts may influence breast tumor angiogenesis via hypoxia-induced vascular endothelial growth factor up-regulation and protein expression, *Cancer Res.* **54**(23), 6083-6086 (1994).
23. A. C. Koong, E. Y. Chen, N. F. Mivechi, et al, Hypoxic activation of nuclear factor-kappa B is mediated by a Ras and Raf signaling pathway and does not involve MAP kinase (ERK1 or ERK2), *Cancer Res.* **54**, 5273-5279 (1994).

24. Y. Seko, K. Tobe, N. Takahashi, et al, Hypoxia and hypoxia/reoxygenation activate Src family tyrosine kinases and p21ras in cultured rat cardiac myocytes, *Biochem. Biophys. Res. Commun.* **226**, 530-535 (1996).
25. L. W. Lo, J. J. Cheng, and J. J. Chiu, Endothelial exposure to hypoxia induces Egr-1 expression involving PKCalpha-mediated Ras/Raf-1/ERK1/2 pathway, *J. Cell. Physiol.* **188**, 304-312 (2001).
26. C. J. Green, P. Lichtlen, and N. T. Huynh, Placenta growth factor gene expression is induced by hypoxia in fibroblasts: a central role for metal transcription factor-1, *Cancer Res.* **61**, 2696-2703 (2001).
27. N. C. Denko, and A. J. Giacca, Tumor hypoxia, the physiological link between Trousseau's syndrome (carcinoma-induced coagulopathy) and metastasis, *Cancer Res.* **61**, 795-798 (2001).
28. K. S. Kim, K. Takeda, R. Sethi, et al, Protection from reoxygenation injury by inhibition of rac1, *J. Clin. Invest.* **101**, 1821-1826 (98).
29. J. Rak, J. L. Yu, and G. Klement, Oncogenes and angiogenesis: signaling three-dimensional tumor growth, *J. Investig. Dermatol. Symp. Proc.* **5**, 24-33 (2000).
30. F. C. White, A. Benehacene, and J. S. Scheele, VEGF mRNA is stabilized by ras and tyrosine kinase oncogenes, as well as by UV radiation--evidence for divergent stabilization pathways, *Growth Factors* **14**, 199-212 (1997).
31. S. W. Lowe, Activation of p53 by oncogenes, *Endocr. Relat. Cancer* **6**, 45-48 (1999).
32. J. L. Arbiser, M. A. Moses, C. A. Fernandez, et al, Oncogenic H-ras stimulates tumor angiogenesis by two distinct pathways, *Proc. Natl. Acad. Sci. USA* **94**, 861-866 (1997).
33. T. Konishi, C. L. Huang, M. Adachi, et al, The K-ras gene regulates vascular endothelial growth factor gene expression in non-small cell lung cancers, *Int. J. Oncol.* **16**, 501-511 (2000).
34. F. Kallinowski, R. R. Friis, F. Van Roy, et al, Oxygenation of tumors derived from ras transformed cells, *Adv. Exp. Med. Biol.* **277**, 907-916 (1990).
35. E. Cohen-Jonathan, S. M. Evans, C. J. Koch, et al, The farnesyltransferase inhibitor L744,832 reduces hypoxia in tumors expressing activated H-ras, *Cancer Res.* **61**, 2289-2293 (2001).
36. B. Enholm, K. Paavonen, A. Ristimaki, et al, Comparison of VEGF, VEGF-B, VEGF-C and Ang-1 mRNA regulation by serum, growth factors, oncoproteins and hypoxia, *Oncogene* **14**, 2475-2483 (1997).
37. A. Maity, N. Pore, J. Lee, et al, Epidermal growth factor receptor transcriptionally up-regulates vascular endothelial growth factor expression in human glioblastoma cells via a pathway involving phosphatidylinositol 3'-kinase and distinct from that induced by hypoxia, *Cancer Res.* **60**, 5879-5886 (2000).
38. K. R. Laderoute, R. M. Alarcon, M. D. Brody, et al, Opposing effects of hypoxia on expression of the angiogenic inhibitor thrombospondin 1 and the angiogenic inducer vascular endothelial growth factor, *Clin. Cancer Res.* **6**, 2941-2950 (2000).
39. C. Chen, N. Pore, A. Behrooz, et al, Regulation of glut1 mRNA by hypoxia-inducible factor-1. Interaction between H-ras and hypoxia, *J. Biol. Chem.* **276**, 9519-9525 (2001).
40. D. R. Premkumar, G. Adhikary, J. L. Overholt, et al, Intracellular pathways linking hypoxia to activation of c-fos and AP-1, *Adv. Exp. Med. Biol.* **475**, 101-109 (2000).
41. M. Höckel, K. Schlenger, C. Knoop, P. Vaupel, Oxygenation of carcinomas of the uterine cervix: evaluation by computerized O2 tension measurements. Cancer Res. **51**:6098-102 (1991).

CHARACTERIZATION OF HUMAN Gc PROTEIN-DERIVED MACROPHAGE ACTIVATION FACTOR (GcMAF) AND ITS FUNCTIONAL ROLE IN MACROPHAGE TUMORICIDAL ACTIVITY

Saharuddin B. Mohamad, Hitoshi Hori, Hideko Nagasawa, Kenji Usui, and Yoshihiro Uto*

1. INTRODUCTION

Macrophages are essential for host defense and play an important role in orchestrating immune response of the host against threat signals. Macrophages are also known to have a critical role in antitumor immunity,[1] can infiltrate into tumor, and are found in most tumor sites.[2,3] Meanwhile, Gc protein (also known as vitamin D_3-binding protein) is a serum protein with multifunctional properties[4] and has been reported as a precursor for macrophage activation factor.[5] Gc protein can be converted by an inducible ß-galactosidase of B cells and neuraminidase of T cells to a potent macrophage activating factor (GcMAF), a protein with *N*-acetylgalactosamine (GalNAc) as the remaining sugar moiety.[6,7] Activated macrophages express tumoricidal activity by ingestion of tumor cells and release of reactive oxygen species (ROS) and reactive nitrogen species (RNS), or both.[5] We reported that *in situ* modification of Gc protein with ß-galactosidase and neuraminidase increased the release of superoxide in thioglycolate-elucidated mouse peritoneal macrophage.[8] Yamamoto et al. reported the possibility of using GcMAF as an immunomodulator for cancer treatment,[9] so it is important to provide an assay for GcMAF. Kanan et al. reported the quantitative analysis of GcMAF from human serum.[10] However, the sugar moiety of GcMAF has never been qualitatively studied. Here we present the study of qualitative analysis of GcMAF from purified human serum as well as its influence on the macrophage activity.

*Saharuddin B. Mohamad, Hitoshi Hori, Hideko Nagasawa, Kenji Usui, and Yoshihiro Uto
Department of Biological Science and Technology, Faculty of Engineering,
The University of Tokushima, Tokushima, 770-8506, Japan [E-mail: hori@bio.tokushima-u.ac.jp]

Oxygen Transport to Tissue XXIII, edited by D. Wilson *et al.*
Kluwer Academic/Plenum Publishers, 2003

2. MATERIALS AND METHODS

2.1. Materials

25-OH-Vitamin D_3 was a gift from Dr. Nobuto Yamamoto from Socrates Institute for Therapeutic Immunology and Albert Einstein Cancer Center, Philadelphia, USA. Other chemicals (biochemical grade) were purchased from Wako Pure Chemical Industries Co., Japan.

2.2. Purification of human serum Gc protein

The procedure was adapted from methods reported by Link et al.[11]

Human serum was diluted 1:1 with column buffer and applied to the 25-OH-D_3-Sepharose column at a rate of 0.4 mL/min. Absorbance at 280 nm was used to monitor protein elution from the column. After the sample was applied, the column was washed with 300 mL of column buffer at 2 mL/min. The protein remaining on the matrix was eluted with 6 M guanidine-HCl at 1 mL/min and 1 mL/fraction was collected. Fractions with protein peak of the guanidine eluted fraction were pooled and dialyzed with 10 mM sodium phosphate, pH 7.0 for the hydroxyappatite chromatography.

2.2.2. Hydroxyappatite chromatography

A 5-mL hydroxyappatite column (Econo-Pac HTP Cartridge 1. Bio-Rad) was equilibrated in 10 mM sodium phosphate, pH 7.0. The dialyzed sample from 25-OH-D_3-Sepharose chromatography was applied to the column (0.5 mL/min) and the column was washed with 50 mL 10 mM sodium phosphate, pH 7.0 (2 mL/min). A linear gradient from 10 mM sodium phosphate to 200 mM sodium phosphate, pH 7.0 was applied to the column. Fractions with protein peak were collected and the protein concentration was determined using the BCA method. Collected fractions were stored at -80 °C.

2.3. GcMAF preparation

Purified Gc protein (100 μg) was incubated with immobilized ß-galactosidase (1 Unit, in 0.5 mL 100 mM sodium phosphate buffer, pH 7.0) in a microcentrifuge tube at 37 °C by rotation movement for 1 hr. The immobilized enzyme was removed by centrifugation and pH of the supernatant was adjusted to pH 6.0 using 1M NaH_2PO_4. The supernatant (1.5~2.0 mL in 100 mM sodium phosphate buffer, pH 6.0) was incubated with immobilized neuraminidase (0.5 Unit) in a microcentrifuge tube at 37 °C by rotation movement for 1 hr. The immobilized enzyme was removed by centrifugation and the supernatant was made sterile by filtration and protein concentration was determined using the BCA method and then stored at -80 °C.

2.4. Isolation and culture of mouse peritoneal macrophages

Resident mouse peritoneal macrophages (female ICR mice, 7 weeks of age) were collected and after centrifugation at 1000 G at 4 °C for 10 min, the collected macrophages were cultured in 6-well plates with a concentration of 1~2 x 10^6 cells/ well in RPMI 1640 (serum free). Cultured cells were then washed 3 times with Hank's solution to separate adherent macrophages from non-adherent cells (T and B cells). The cultured macrophages were then treated as indicated for 3 hr, and superoxide generation assay was done as described below.

2.5. Superoxide generation assay

The method was modified from that reported by Johnston et al.[12] Briefly, after drug treatment for the times indicated, the plates were washed twice with PBS(-) and once with Krebs-Ringer phosphate buffer, and 1.5 mL of 50 μM cytochrome *c* in Krebs-Ringer phosphate buffer was added to the plates. Phorbol myristate acetate (PMA) was added to the final concentration 5 μg/mL in each well and cultured for 90 min at 37 °C/5% CO_2 in a humidified incubator. The reaction was stopped by means of an ice-bath, and the cultured medium was placed in an Eppendorf tube, and promptly cleared by centrifugation at 8000 G. The optical density of the supernatant was determined spectrometrically at 550 nm with reference at 540 nm (U-2000, HITACHI) using mixtures from plates without cells as blanks. The concentration of reduced cytochrome *c* was determined using the equation $\Delta E_{550\ nm} = 2.1 \times 10^4\ M^{-1}\ cm^{-1}$.

2.6. Phagocytosis assay

For phagocytosis assay mouse peritoneal cells were layered onto coverslips in a 24-well plate and incubated for 30 min to allow macrophage adherence. GcMAF was added to the culture to the final concentration of 10 pg/mL GcMAF and cultured for 3 hr and then assayed for phagocytic activity. Sheep red blood cells (SRBC) were opsonized by rabbit haemolytic serum (anti-sheep red blood C12HSB cells) purchased from Serotec Ltd. 0.5% opsonized-SRBC in RPMI 1640 (serum free) was overlaid on each macrophage-coated (monolayer) coverslip and cultured at 37 °C in a humidified 5% CO_2 incubator for 90 min. Noninternalized erythrocytes were lysed by immersing the coverslips in a hypotonic solution (1/5 PBS) for 5 to 10 sec. The macrophages were fixed with methanol, air dried and stained with Giemza stain. The morphological photograph of phagocytotic macrophage was taken using a phase contrast microscope (x 400).

2.7. Western blot and Lectin blot

0.1 and 0.2 μg protein for western blot and 1 μg protein for lectin blot were subjected to SDS-PAGE under reducing conditions followed by electroblotting onto a PVDF membrane. Nonspecific binding was blocked by incubation in Tris-buffer phosphate containing 0.1% Tween 20 and 3% BSA for overnight at 4 °C. For western blot, the blots were probed with anti-human Gc globulin (Code No. A 0021, DAKO) and after extensive washing, the blots were incubated with the secondary Ab (HRP-labeled anti-rabbit IgG). For lectin blot, the blots were incubated with HRP-labeled *Helix pomatia* lectin (Sigma-Aldrich Japan Co). The blots were developed using ECL Western blotting detection system (Amersham).

3. RESULTS AND DISCUSSION

Human serum was applied to 25-OH-D_3-Sepharose affinity and hydroxyappatite chromatography as described under Materials and Methods. Gc protein fraction was pooled and 0.1 and 0.2 μg of Gc protein was subjected to western blot. Fig. 1a shows the

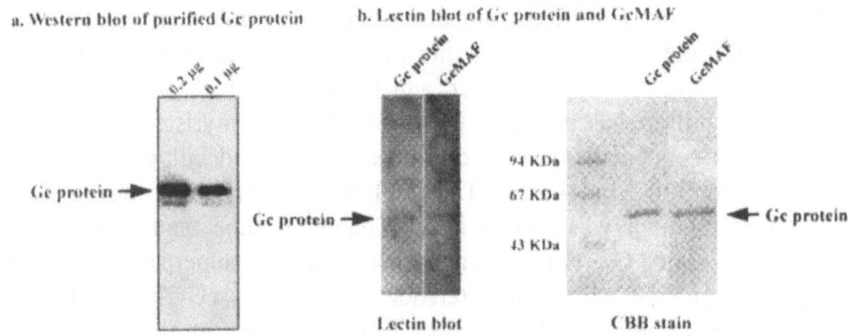

Fig. 1. Western blot and lectin blot analysis of Gc protein and Gc MAF. (1 μg protein was applied in each lane for lectin blot).

high degree of purity of Gc protein that was achieved after the two-step chromatography procedure. Lectins are good tools for exploring the interaction of carbohydrate because each lectin recognizes specific carbohydrates. *Helix pomatia* agglutinin, a lectin possesses particularly high specificity for terminal *N*-acetylgalactosaminyl residue.[13] 1 μg of purified Gc protein and prepared GcMAF was subjected to lectin blot (Fig. 1b). CBB stain of both Gc protein and GcMAF shows the detected band was about 52 KDa. Lectin blot analysis of both Gc protein and GcMAF gave a band at the same position as shown in CBB stain. As *Helix pomatia* lectin would be expected to specifically bind to terminal *N*-acetylgalactosaminyl residue, it would be possible that the GcMAF formation was not successful, or the band in Gc protein was an unspecific bound of lectin with the protein. We evaluated the influence of prepared GcMAF in macrophage activity using superoxide generation assay as described under Materials and Methods. Resident mouse peritoneal macrophages show higher superoxide generation activity at 10 ng/mL GcMAF treatment compared to the control (Fig.2). Increasing the GcMAF treatment was found to be inversely proportional to the macrophage activity. We also checked the phagocytosis activity of macrophages after 3 hr treatment of 10 pg/mL GcMAF (Fig.2). Fig. 2 shows the morphology of opsonized-SRBC phagocyted macrophages. 10 pg/mL of GcMAF enhanced the phagocytic activity of mouse peritoneal macrophage. Therefore, we consider that GcMAF was successfully prepared using the immobilized enzymes. The reason why an unexpected band was detected in Gc protein, is probably because of the

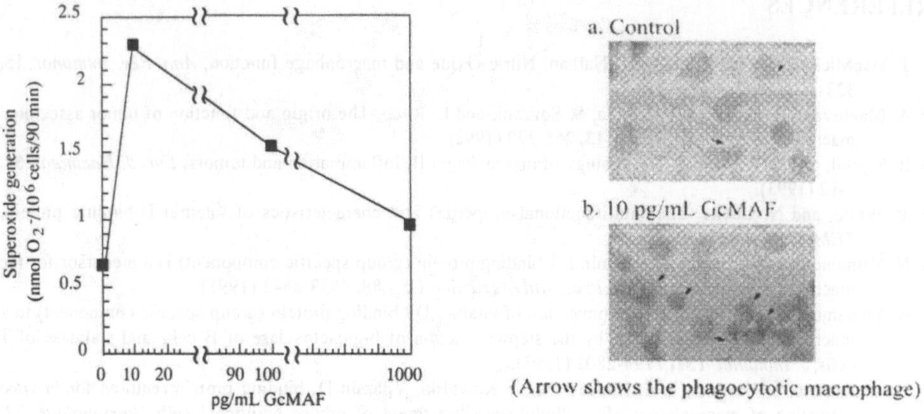

Fig.2. Superoxide generation assay of mouse peritoneal macrophage and morphology of opsonized-SRBC phagocytosed macrophage (Arrow shows phagocytotic macrophage).

unspecific bound of Gc protein to *Helix pomatia* lectin at the volume applied to the SDS-PAGE. We are revising the experimental methods of lectin blot to improve the quality of the results. GcMAF was reported to be involved in inflammation-primed macrophage activation, which finally leads to immune development. A lack of vitamin D_3-binding protein derived macrophage activation factor (DBP-MAF) was suggested to be involved in osteoperotic mutations because, DBP-MAF mediates bone resorption by activating osteoclast, which are responsible for bone resorption.[14] However, Odgren et al. reported that DBP-MAF production is not a universal characteristic of osteopetrotic mutations.[15] Swamy et al. reported data to support the essential role of GalNAc sugar moiety of GcMAF in Macrophage activation cascade using baculovirus expressed form of Gc protein, which is Glycosylated.[16] Furthermore, Swamy et al. also reported that the endogenous ligand of Gc protein (25-OH vitamin D_3) does not influence the activity of GcMAF to activate macrophage.[16] Because GcMAF has shown to be an excellent candidate immunomodulator for cancer treatment,[9] it is important to provide an assay system to characterize GcMAF formation. Here we provided important data and views to characterize GcMAF.

ACKNOWLEDGEMENTS

We thank Dr. Nobuto Yamamoto from Socrates Institute for Therapeutic Immunology and Albert Einstein Cancer Center, Philadelphia, USA for his technical advice and help in the study. This work was supported in part by a Grant-in-Aid for Scientific Research (c) No. 10672090 from The Ministry of Education, Science, Sport and Culture of Japan, and by the Sasakawa Scientific Research Grant from the Japan Science Society.

REFERENCES

1. J. MacMicking, Q. W. Xie, and C. Nathan, Nitric Oxide and macrophage function, *Ann. Rev. Immunol.* **15**, 323-350 (1997).
2. A. Mantovani, B. Bottazzi, F. Colotta, S. Sozzani, and L. Ruco, The origin and function of tumor associated macrophage, *Immunol. Today* **13**, 265-270 (1992).
3. R. Sejelid, and L. T. Busund, The biology of macrophage: II. Inflammation and tumors, *Eur. J. Haematol.* **52**, 1-12 (1993).
4. P. White, and N. Cooke, The multifunctional properties and characteristics of vitamin D-binding protein, *TEM.* **11**, 320-327 (2000).
5. N. Yamamoto, and S. Homma, Vitamin D3 binding protein (group specific component) is a precursor for the macrophage-activating signal, *Proc.Natl. Acad. Sci. USA* **88**, 8539-8543 (1991).
6. N. Yamamoto, and R. Kumashiro, Conversion of vitamin D3 binding protein (group specific component) to a macrophage activating factor by the stepwise action of ß-galactosidase of B cells and sialidase of T cells, *J. Immunol.* **151**, 2794-2802 (1993).
7. N. Yamamoto, S. Homma, J. G. Haddad, M. A. Kowalski, Vitamin D_3 binding protein required for *in vitro* activation of macrophages after alkylglycerol treatment of mouse peritoneal cells, *Immunology.* **74**, 420-424 (1991).
8. S. B. Mohamad, H. Nagasawa, Y. Uto, and & Hitoshi Hori, Tumor cell alpha-*N*-acetylgalactosaminidase activity and its involvement in GcMAF-related macrophage activation, *Comp. Biochem. Physiol. B, in press* (2001).
9. N. Yamamoto, Structural definition of a potent macrophage activating factor derived from vitamin D3-binding protein with adjuvant activity for antibody production, *Mol. Immunol.* **33**, 1157-1164 (1996).
10. R. M. Kanan, D. B. Cook, H. K. Datta,. Lectin immunoassay for macrophage-activating factor (Gc-MAF) produced by deglycosylation of Gc-globulin: Evidence for noninducible generation of Gc-MAF, *Clin. Chem.* **46**, 412-414 (2000).
11. R. P. Link, K. L. Perlman, E. A. Pierce, H. K. Schoes, and H. F. DeLuca, Purification of human serum vitamin D-binding protein by 25-hydroxyvitamin D_3-sepharose chromatography, *Anal. Biochem* **157**, 262-269 (1986).
12. R. B. Johnston, C. A. Godzik, and Z. A. Cohn, Increased superoxide anion production by immunologically activated and chemically elicited macrophages, *J. Exp. Med.* **148**, 115-127 (1978).
13. S. Hammarstrom, and E. A. Kabat, Studies on specificity and binding properties of the blood group A reactive hemagglutinin from *Helix pomatia*, *Biochemistry* **10**, 1684-1692 (1971).
14. N. Yamamoto, and V. R. Naraparaju, A defect in inducible ß-galactosidase of B lymphocytes in the osteopetrotic (*mi/mi*) mouse, *Immunology.* **88**, 604-610 (1996).
15. P. R. Odgren, S. N. Popoff, F. F. Safadi, C. A. MacKay, A. Mason-Savas, M. F. Seifert., and S. C. Marks Jr, The *toothless* osteopetrotic rat has a normal vitamin D-binding protein-macrophage activating factor (DBP-MAF) cascade and chondrodysplasia resistant to treatments with Colony Stimulating Factor-1 (CSF-1) and/or DBP-MAF, *Bone* **25**, 175-181 (1999).
16. N. Swamy, S. Ghosh, G. B. Schneider, and R Ray, Baculovirus-expressed vitamin D-binding protein-macrophage activating factor (DBP-MAF) activates osteoclasts and binding of 25-hydroxyvitamin D_3 does not influence this activity, *J. Cell. Biochem.* **81**, 535-546 (2001).

EFFECT OF HEMOGLOBIN SOLUTIONS AS HEMODILUENTS ON CAPILLARY OXYGEN TENSION

Roland N. Pittman, Kristin B. Miller, and Peter H. Tait*

1. INTRODUCTION

Isovolemic hemodilution is the replacement of red blood cells (RBCs) with an equal volume of hemodiluent (e.g., colloids, crystalloids or oxygen-carrying blood substitutes). Hemodilution reduces the systemic hematocrit, decreases the viscosity of blood, increases cardiac output, increases blood flow to organs and increases oxygen extraction by the tissues,[1–4] thereby providing beneficial consequences of hemodilution in clinical settings.

The limit for hemodilution is about 50% substitution of red blood cells (normal hematocrit ≅ 50%), because the transfusion trigger for packed red blood cells is 7 g Hb/dl blood, corresponding to a hematocrit of about 23%.[5] This hematocrit is the lower limit at which the tissue oxygen tension (PO_2), blood pressure, and functional capillary density (FCD) are not compromised.[5] The critical hematocrit, the value at which oxygen delivery to tissues is compromised, is usually agreed to be between 8 and 10%.[4, 6] The purpose of this study was to determine the effect of hemodilution on PO_2 values in the capillary network of skeletal muscle in the hamster, using three hemoglobin-based oxygen carriers and a non-oxygen carrying plasma expander as hemodiluents. Measurements of PO_2 were made *in vivo* at the arteriolar and venular ends of capillaries during stepwise isovolemic hemodilutions.

2. METHODS

2.1. Animal Preparation

Male Golden Syrian hamsters (Charles River) weighing between 80 and 120 grams (age = 42 ± 5 days, body weight = 98 ± 12 g, N = 47) were anesthetized with an initial injection of sodium pentobarbital (65 mg/kg BW) administered intraperitoneally. The anesthetized animal was placed supine on a heated Plexiglas platform, and the trachea

---

* Department of Physiology, Virginia Commonwealth University, Richmond, Virginia 23298

Oxygen Transport to Tissue XXIII, edited by D. Wilson *et al.*
Kluwer Academic/Plenum Publishers, 2003

was cannulated to ensure a patent airway. The right carotid artery and jugular vein were cannulated for the withdrawal and infusion of blood and hemodiluent, respectively. The right femoral vein was cannulated to administer supplemental anesthetic as needed.

The retractor muscle was prepared for intravital microscopic observations, as described by Sullivan and Pittman,[7] and placed dorsal side up on the heated platform. The muscle was covered with Saran film (Dow Corning) to minimize desiccation and gas exchange with the environment. The hamster's core and retractor muscle temperatures were maintained at 36 °C, using circulating warm water in the platform. Measurements were made using a Zeiss ACM microscope (Carl Zeiss, Hawthorne, NY) equipped with a Neofluar 40X/0.75 N.A. objective (Carl Zeiss) to observe individual capillaries.

2.2. PO_2 Measurements

The PO_2 of blood flowing through capillaries was measured using the phosphorescence quenching technique[8] implemented for intravital microscopy.[9] Prior to PO_2 measurements, a phosphor solution (concentration 10 mg/ml; dose 30 mg/kg body weight) was slowly injected into the jugular vein and the animal was allowed to stabilize for 15 minutes before PO_2 measurements were begun. The phosphor solution consisted of palladium meso-tetra (4-carboxyphenyl) porphine (MW=900 Da; Oxygen Enterprises, Philadelphia, PA) dissolved in distilled water and bound to albumin, with pH adjusted to 7.35-7.45.

After the phosphor was injected and the animal was allowed to stabilize, measurement sites from the arteriolar and venular ends of capillaries were identified. A square excitation region measuring 10 μm x 10 μm was placed over a selected capillary. A 10 Hz flash from a Xenon lamp was used to excite the phosphor. The phosphorescence signal was detected by a photomultiplier tube and its output was converted to a voltage before input to an analog to digital converter in a Gateway personal computer. Data acquisition and pre-processing were carried out with the software package LabVIEW 4.01 (National Instruments, Austin, TX). Twenty decay curves were averaged for each capillary site, and five arteriolar capillary and five venular capillary sites were measured at five different hematocrits in each animal. For each site and each hematocrit, the decay curves were averaged, normalized to the maximum phosphorescence signal, and fit using an equation for the decay time course of a Rectangular distribution of PO_2 to calculate the average PO_2 values:[9]

$$Y(t) = \exp\left[-(k_0 + k_q PO_2)t\right] \cdot (\sinh k_q\delta t)/k_q\delta t \qquad (1)$$

where $k_0 = 1/\tau_0$, τ_0 (= 546 μs) is the lifetime of the excited phosphor in the absence of oxygen, k_q (= $3.06 \times 10^{-4}\ \mu s^{-1}\ mmHg^{-1}$) is the quenching coefficient, δ is the half-width of the Rectangular distribution of PO_2, and sinh is the hyperbolic sine function. Equation (1) describes the time course of the normalized phosphorescence decay curve and the fitting of the data to this equation was programmed in Microcal Origin 6.0 (Microcal Software, Inc., Northampton, MA).

2.3. Hemodilution Solutions

In these experiments, four solutions were used as hemodiluents: Hespan (N=20; 6% hetastarch in saline; Dupont Pharmaceuticals, Wilmington, DE), DCLHb (N=9; diaspirin cross-linked hemoglobin), rHb1.1 (N=10) and 3261BR (N=8; all hemoglobin compounds supplied by Baxter Hemoglobin Therapeutics, Boulder, CO). Hespan is a non-oxygen-carrying plasma expander with an osmolarity of 310 mOsm/l (isosmotic with blood). DCLHb (MW=64 kDa; concentration 10.0 g/dl; P_{50}=32 mmHg; Hill coefficient = 2.4) is a hemoglobin-based oxygen carrier in which the two alpha subunits of DCLHb have been crosslinked with bis (3,5-dibromosalicyl) fumarate to prevent hemoglobin dissociation into dimers. rHb1.1 (MW=64 kDa; concentration 11.0 g/dl; P_{50}=32 mmHg; Hill coefficient = 2.2) is a recombinant human hemoglobin in which the two alpha subunits have been genetically crosslinked.[10] 3261BR (MW=64 kDa; concentration 9.1g/dl; P_{50}=14.6 mmHg; Hill coefficient = 2.15) is another genetically crosslinked human hemoglobin that was made by recombinant methods to have a lower P_{50} value. For comparison, hamster blood has a P_{50} of 29.3 mmHg and Hill coefficient of 2.2.

2.4. Hemodilution Procedure

Blood was withdrawn from the right carotid artery, while one of the hemodiluents was simultaneously infused into the right jugular vein. This isovolemic exchange was achieved using an infusion/withdrawal pump (Model 975, Harvard Apparatus Inc., Dover, MA) set at a rate of 0.39 ml/min. The relationship between systemic hematocrit and hemodilution time followed a monoexponential decay with a time constant of 9.1 min. A five-minute stabilization period followed each hemodilution step, and the hematocrit was determined by centrifugation of two 50 μl blood samples taken by a toe clip. For each group of animals, hematocrit was reduced from an average of 51% to 42% in Step 1, to 31% in Step 2, to 21% in Step 3, and to 12% in Step 4.

3. RESULTS

The relationship between systemic hematocrit and capillary PO_2 (arteriolar and venular ends; AC and VC, respectively) is presented in Figure 1 for hemodilution accomplished with the hemodiluents Hespan, DCLHb, rHb1.1 and 3261BR. This relationship for the four hemodiluents exhibited two different qualitative patterns. For Hespan and 3261BR, capillary PO_2 increased as the hematocrit was decreased from Control (Hct=51%) to Step 1 (42%), then there was a progressive decline in PO_2 as hematocrit fell to 12%, although the decline was much more pronounced for Hespan. For DCLHb and rHb1.1, capillary PO_2 decreased as the hematocrit was decreased from Control (51%) to Step 1 (42%), then rose between Step 1 and Step 2 (31%), before falling progressively to Step 4 (12%). Arteriolar capillary PO_2 for the three hemoglobin compounds remained above 30 mmHg through all four hemodilution steps, while PO_2 for Hespan fell precipitously below hemodilution Step 2.

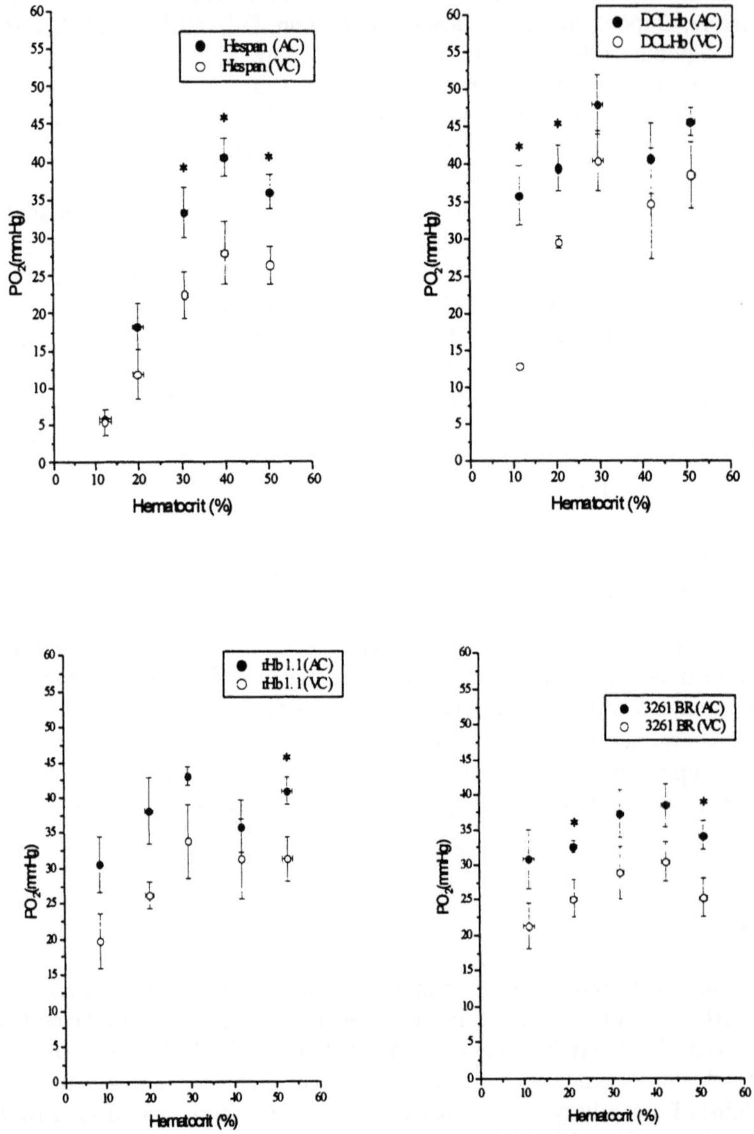

Figure 1. Plots of mean (± standard error) capillary PO_2 values versus hematocrit for hemodilution with Hespan, DCLHb, rHb1.1 and 3261BR. The filled circles represent data from the arteriolar ends of capillaries (AC) and the open circles represent data from the venular ends (VC). * indicates that there is a significant difference ($P<0.05$) between the arteriolar and venular capillary PO_2 at the given hematocrit.

4. DISCUSSION

4.1. Effects of Hemodilution on Capillary PO_2

The PO_2 at the arteriolar and venular ends of capillaries under control conditions are what would be expected in this preparation based on the results of Ellsworth et al.[11] for hemoglobin oxygen saturation (SO_2) of single RBCs: $S_{AC}O_2$=60% and $S_{VC}O_2$=40%, corresponding to $P_{AC}O_2$=35 mmHg and $P_{VC}O_2$=24 mmHg. These oxygenation values in capillaries are lower than standard "textbook" values due, in part, to less than optimal respiratory gas exchange usually observed in hamsters and to precapillary diffusive loss of oxygen in the microcirculation.[12]

For Hespan, arteriolar and venular capillary PO_2 stayed relatively constant down to a systemic hematocrit of about 30%, and when the hematocrit was reduced further to approximately 20%, there was a significant drop in PO_2 at both sites. When comparing arteriolar and venular capillary PO_2 at each hemodilution step, arteriolar capillary PO_2 was significantly higher than venular PO_2 except for hemodilution Steps 3 and 4. These results are consistent with the idea that oxygen diffuses from the blood in the capillaries to the tissues down to a systemic hematocrit of about 20%. At the lowest hematocrit (12%), there was no detectable difference in PO_2 at the two capillary sites, indicating that there was not a large enough PO_2 difference to drive oxygen from the blood to the tissue.

For each of the hemoglobin-based hemodiluents, there was no statistically significant difference among arteriolar capillary PO_2 values, indicating that these compounds were able to maintain PO_2 within a relatively narrow range (3 to 10 mmHg) over the entire range of hematocrits (i.e., 51 to 12%). As seen with Hespan, arteriolar capillary PO_2 was higher than the venular capillary PO_2. However, due to the presence of free hemoglobin in the plasma, the differences between PO_2 at the arteriolar and venular ends were not as large as for Hespan, and consequently did not achieve statistical significance except in a few cases where the variability in PO_2 was small. As a function of hematocrit, arteriolar and venular PO_2 tended to change in parallel.

4.2. Hemoglobin Concentration during Hemodilution

During hemodilution, the hemoglobin concentration of the blood can be calculated based on mass balances for hemoglobin contained within RBCs and that contained in the hemodiluent with the following result:

$$[Hb]_B = [Hb]_{RBC}\, H_f + [Hb]_{HD} \cdot (1 - (H_f / H_i)) \qquad (2)$$

where $[Hb]_B$ is the total hemoglobin concentration in the blood, $[Hb]_{RBC}$ is the hemoglobin concentration of a single RBC, $[Hb]_{HD}$ is the concentration of hemoglobin in the infused hemodiluent, H_i is the initial (i.e., control) hematocrit, and H_f is the final hematocrit. For Hespan, only the RBC contribution is pertinent, while for the hemoglobin-based solutions, both contributions need to be taken into account.[13] For the Hespan group, the transfusion trigger was reached between hemodilution Steps 2 and 3, thereby explaining the significant decrease in PO_2, which should lead to tissue hypoxia.

Note that for Hespan the difference between arteriolar and venular capillary PO_2 became smaller with decreasing hematocrit, and was undetectable at 12% hematocrit.

For DCLHb, rHb1.1 and 3261BR, the total hemoglobin concentration in the blood decreased as the hematocrit decreased, but at the lowest hematocrit of 12% the total hemoglobin concentration only reached 12 g/dl, well above the transfusion trigger of 7 g/dl. A total hemoglobin concentration of 12 g/dl is equivalent to a "RBC only" hematocrit of 35%. The curve in Fig. 1 for Hespan indicates that a hematocrit of 35% corresponds to a $P_{AC}O_2$ of about 35 mmHg, a value near those found with the three hemoglobin solutions at the lowest hematocrit. Thus, it appears that insofar as blood oxygenation in the capillaries is concerned, the total hemoglobin concentration is a major determinant of PO_2, whereas P_{50} is of minor importance, at least in the range of 15 to 32 mmHg.

ACKNOWLEDGEMENTS

The authors thank Dr. Aleksander S. Golub and Mr. Lei Zheng for their assistance with this study. This work was supported in part by Baxter Hemoglobin Therapeutics and by grant HL18292 from the National Heart, Lung and Blood Institute.

REFERENCES

1. L. Kuo, and R.N. Pittman, Effect of hemodilution on oxygen transport in arteriolar networks of hamster striated muscle, Am. J. Physiol. **254**, H331-H339 (1988).
2. D.R. Spahn, B.J. Leone, J.G. Reves, and T. Pach, Cardiovascular and coronary physiology of acute isovolemic hemodilution: a review of nonoxygen-carrying and oxygen-carrying solutions, Anesth. Analg. **78**, 1000-1021 (1994).
3. A.R. Pries, T.W. Secomb, M. Sperandio, and P. Gaehtgens, Blood flow resistance during hemodilution: effect of plasma composition, Cardiovasc. Res. **37**, 225-235 (1998).
4. P. Van der Linden, Anemic hypoxia. in: *Tissue Oxygenation in Acute Medicine*, edited by W.J. Sibbald, K. Messmer, and M.P. Fink (Springer-Verlag, Berlin, 1998) pp. 116-127.
5. M. Intaglietta, Microcirculatory basis for the design of artificial blood, Microcirculation **6**, 247-258 (1999).
6. D.R. Spahn, The optimal and critical hemoglobin in health and acute illness. in: *Tissue Oxygenation in Acute Medicine*, edited by W.J. Sibbald, K. Messmer, and M.P. Fink (Springer-Verlag, Berlin, 1998) pp. 263-275.
7. S.M. Sullivan, and R.N. Pittman, Hamster retractor muscle: a new preparation for intravital microscopy, Microvasc. Res. **23**, 329-335 (1982).
8. J.M. Vanderkooi, G. Maniara, T.J. Green, and D.F. Wilson, An optical method for measurement of dioxygen concentration based upon quenching of phosphorescence, J. Biol. Chem. **252**, 5476-5482 (1987).
9. L. Zheng, A.S. Golub, and R.N. Pittman, Determination of PO_2 and its heterogeneity in single capillaries,. Am. J. Physiol. **271**, H365-H372 (1996).
10. D. Looker, D. Abbott-Brown, P. Cozart, S. Durfee, S. Hoffman, A.J. Mathews, J. Miller-Roehrich, S. Shoemaker, S. Trimble, G. Fermi, N.H. Komiyama, K. Nagai, and G. Stetler, A human recombinant haemoglobin designed for use as a blood substitute, Nature **356**, 258-260 (1992).
11. M.L. Ellsworth, A.S. Popel, and R.N. Pittman, Assessment and impact of heterogeneities of convective oxygen transport parameters in capillaries of striated muscle: experimental and theoretical, Microvasc. Res. **35**, 341-362 (1988).
12. M.L. Ellsworth, C.G. Ellis, A.S. Popel, and R.N. Pittman, Role of microvessels in oxygen supply to tissue, News in Physiol. Sci. **9**, 119-123 (1994).
13. W.J. Sibbald, and A. Sielenkamper, Blood substitutes – Effects on the microcirculation. in: *Tissue Oxygenation in Acute Medicine*, edited by W.J. Sibbald, K. Messmer, and M.P. Fink (Springer-Verlag, Berlin, 1998) pp. 318-329.

DIFFERENT EFFICACY *IN VITRO* OF HEMOGLOBIN BASED OXYGEN CARRIERS AND RED CELLS

E. Bucci, T. L. Watts, H E. Kwansa, and A. Fasano[1]

A vast literature exists which either theoretically explores the mechanism of oxygen delivery from blood to the surrounding tissues [1-3], or simulates oxygen delivery and absorption by red cells suspension and cell free hemoglobin solutions using artificial capillary systems [4-6]. Very little is known about the physiologic efficiency of cell free hemoglobin solutions to sustain life of isolated organs and tissues. We thought that the oxygen dependent transport of metabolites across ileal mucosa, superfused in Ussing chambers, could be used to test the efficacy of hemoglobin based oxygen carriers [7].

1. METHODS AND RESULTS

The Ussing chambers system was modified by substituting the oxygenation chambers with membrane oxygenators. Rabbit intestinal tissues were mounted in Ussing cuvettes perfused with Ringer solutions containing either cell free sebacoyl crosslinked hemoglobin A(DECA) [8] or red cells. DECA is human hemoglobin intramolecularly crosslinked with a sebacyl residue between the two β 82 lysines of the partner β subunits across the β cleft. It does not dissociate into dimers and does not appear in the urine of infused animals. In 0.1M Tris-HCl buffer, at pH 7.4, and 37° C, its oxygen affinity has a P50 near 30 mmHg and an oxygen binding cooperativity with the Hill parameter n near 2.0.

In the Ussing chambers, transport across the ileal mucosa is produced by specific receptors, which take the metabolite inside the mucosal cells, from which they are released to the serosal side of the membrane together with Na+ ions. This generates a short-circuit

[1]Department of Biochemistry and Molecular Biology (EB,HEK), and Center for Vaccine Development (AF, TLW) University of Maryland, School of Medicine Baltimore, MD 21201, USA. Supported in part by NIH grants P01-48517 (EF) and R01-DK-48373 (AF).

Oxygen Transport to Tissue XXIII, edited by D. Wilson *et al.*
Kluwer Academic/Plenum Publishers, 2003

current (ISC) which is proportional to the amount of transported metabolite. Thus, the efficacy of transport is measured in μA/cm^2 of membrane surface [7].

Classically in order to maintain tissue metabolism, as monitored by metabolites transport, the perfusates are equilibrated with 95% oxygen, 5% carbon dioxide. A physiologically competent oxygen carrier should allow metabolite transport at much lower oxygen tensions.

We started experimentation with DECA and 30% oxygen. It should be stressed that already at 60% oxygen, Ringer alone does not allow transport. As shown in **Fig. 1**, DECA at 6.0 and 3.0 g/dL maintained identical transport, similar to that obtained with Ringer alone and 95% oxygen (not shown). DECA at 1.5 g/dL, was less efficient. Under 30% oxygen a suspension of freshly collected bovine red cells, diluted to a hemoglobin concentration of 3.0 g/dL, totally failed to maintain transport, **Fig. 2**.

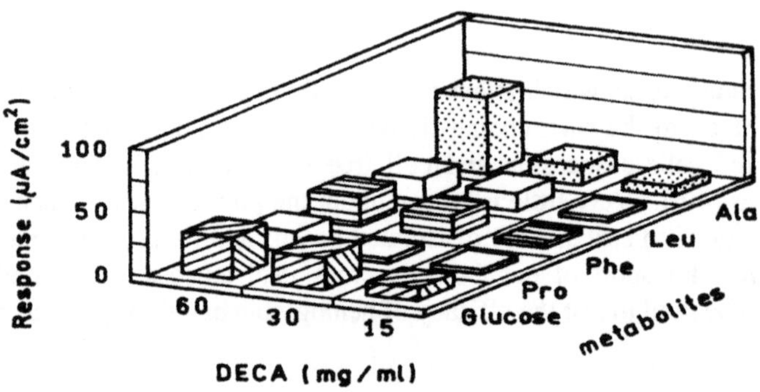

Fig.1. Effect of DECA concentration on metabolites transport after 4 h of perfusion at 37° C, flow rate 15 ml/min. The equilibrating gas contained 30% oxygen, 5% CO_2, balance nitrogen. Metabolite transport is monitored by the increase of the short-circuit current reported on the ordinates.

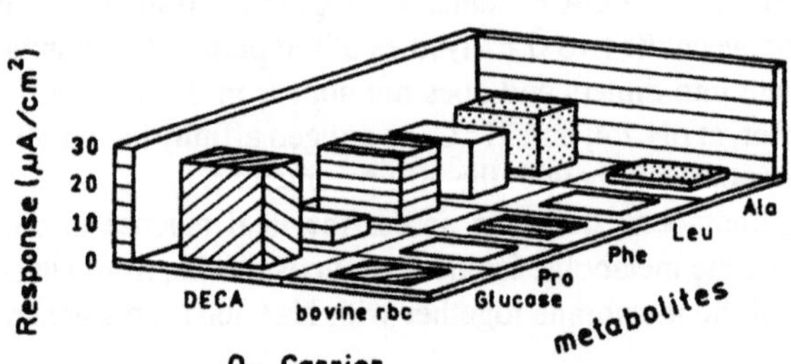

Fig.2. Comparison of metabolites transport (in μA/cm^2) after 4 h of perfusion by either DECA or bovine red cells. DECA is at 3 g/dl, bovine rbc provide a total Hb concentration of 3 g/dl. Perfusion at 37° C, flow rate 15 ml/min. The equilibrating gas contained 30% oxygen, 5% CO_2, balance nitrogen.

It should be stressed that the oxygen affinity of bovine red cells does not increase upon storage. In fact they do not contain 2-3-biphosphoglycerate and are regulated by chloride ions [9]. Thus DECA and bovine red cells had very similar oxygen affinity, with P50 near 30 mmHg at 37° C.

2. DISCUSSION

Useful to this discussion is **Table I**, which lists the relevant parameters of the used experimental conditions.

Table I

Comparison of oxygen content in Ringer solutions containing or not DECA. It should be noted that DECA has oxygen affinity with P50 = 30 mmHg at 37°C at pH 7.4 [8], similar to that of the red cells. Both DECA and the red cells are completely saturated with oxygen when exposed to a gas containing 30% oxygen (i.e. at the partial pressure of 228 mmHg). All data are normalized at 37°C.

Perfusate	Oxygen % in equilibrating Gas	Oxygen partial pressure (mmHg)	Oxygen in perfusate (ml/dL)	Total O_2 (ml/dL)	Transport
Ringer	95	722	2.3	2.3	Yes
Ringer	30	228	0.7	0.7	No
3g/dL DECA in Ringer	30	228	(DECA) 4.7 (Ringer) 0.7	5.4	Yes
3g/dL Hb inside red cells in Ringer	30	228	(Hb) 4.7 (Ringer) 0.7	5.4	No

The table shows that 2.3 ml/dL of free oxygen, in the perfusate equilibrated with 95% oxygen , are enough to sustain viability of the tissue. The total content of oxygen increases to 5.4 ml/dL in the DECA-Ringer perfusate, equilibrated with 30% oxygen. However, in this case, only 0.7 ml/dL of free oxygen are present in the perfusate. The rest (4.7 ml/dL) is bound to DECA. As 0.7 ml/dL of oxygen *per se* are not enough to sustain tissue viability, metabolites transport implies that also the oxygen bound to DECA was available to the tissue.

This situation prevails also for the red cells suspension. However in that case the oxygen of the red cells was not available to the superfused tissue.

A tentative explanation is that hemoglobin in solution wets the tissue creating a very small micro-environment between the hemoglobin molecules and the tissue wall, where the rate of oxygen consumption is matched by the rate of oxygen released from DECA. Instead, the red cells bounce from the tissue wall defining an average distance between them and the tissue. Thus the oxygen eventually released from the segregated hemoglobin must first transverse the cell membrane then diffuse through the intervening space before reaching the tissue. The outcome is that the rate of oxygen consumption exceeds the rate of oxygen release producing hypoxia. This hypothesis is consistent with the observation that in an artificial capillary oxygen is released much faster from cell free hemoglobin solutions than from red cells suspensions [6].

Also, at a partial pressure of oxygen of 228 mmHg, as shown in **Table I**, DECA is completely saturated with oxygen, which implies that it releases oxygen to the tissue against a partial pressure of oxygen which would keep it saturated at all time.

This suggests a different mechanism of oxygen transport between the red cells and DECA. Oxygen transport by red cells is regulated, as classically proposed, by oxygen partial pressure, oxygen affinity, rate of oxygen release from hemoglobin, rate of diffusion of molecular oxygen, and rate of oxygen consumption. Instead oxygen transport by cell free DECA is a phenomenon practically unregulated, subject only to the rate of metabolic consumption. This model probably applies to all of hemoglobin based oxygen carriers encouraging a revision of their design, so far based solely on the attempt of mimicking the oxygen affinity characteristics of blood.

REFERENCES

1. A. Dutta and A.. S. Popel, A theoretical analysis of intracelular oxygen diffusion. *J. Theor. Biol.*176,433-45 (1995).
2. A. Vadapalli, R. N. Pittman and A. S. Popel, Estimating oxygen transport resistance in the microvascular wall. *Am J. Physiol. Heart Circ. Physiol.* **279**,H657-71 (2000).
3. C. X. Eggleton, A. Vadapalli., T. K. Roy and A.S. Popel. Calculations of intracapillay oxygen tension distribution in muscle. *Math. Biosci.* **167**, 123-43 (2000)
4. D. D. Lemon, P.K. Nair, E. J. Boland, J. S. Olson and J. D. Hellum, Physiological factors affecting O2 transport by hemoglobin in an in vitro capillary system. *J. Appl.Physiol.* **62**,798-806 (1987).
5. T. C. Page, W.R. Light and J.D. Hellum, Prediction of microcirculatory oxygen transport by erythrocyte /hemoglobin solution mixtures. *Microvasc. Res.* **56**,113-26 (1998).
6. T. C. Page, W.R. Light and J.D. Hellum, Oxygen transport in 10 microns artuiicial capillaries. *Adv. Exp. Med. Biol.* **471**,715.21 (1999).
7. E. Bucci, A. Razynska, H.. E. Kwansa, B. Matheson-Urbaitis, M. O'Hearne, J. A. Ulatowski and R. C. Koehler, Production and characteristics of an infusible oxygen-carrying fluid based on hemoglobin intramolecularly cross-linked with sebacic acid. *J.Lab. Clin.. Med.* **128**,146-153 (1996).
8. Field M., Fromm D., McColl I. Ion transport in rabbit ileal mucosa. I. Na and Cl fluxes and short-circuit current. *Am. J. Physiol.* **220**:1388-1396 (1971).
9. C. Fronticelli,and E. Bucci, Solvent regulation of oxygen affinity in hemoglobin: sensitivity of bovine hemoglobin to chloride ions. *J. Biol. Chem.* **259**, 10841-10844 (1984).

A NOVEL BLOOD TRANSFUSANT CANDIDATE: INTACT HUMAN ERYTHROCYTES CONTAINING HEMOGLOBIN EXCLUSIVELY NITROSYLATED IN THE α-SUBUNITS

Antonio Tsuneshige and Takashi Yonetani*

1. INTRODUCTION

The transport of oxygen from lungs to peripheral tissues is carried out by hemoglobin (Hb), the main protein component inside red blood cells. However, the modulation of its physiological performance is regulated by the intracellular content of metabolites, mainly 2,3-diphosphoglycerate (DPG), which lower the oxygen affinity of Hb to optimal levels, so that an efficient oxygen delivery is accomplished within a narrow gradient in partial pressures of oxygen between lungs and tissues. However, intracellular levels of DPG in stored blood decay with time, which causes a gradual increase in oxygen affinity, thus imposing a limit to the lifetime of blood for practical use.

Our previous studies on the α-nitrosyl derivative of Hb — a tetrameric Hb in which only the heme groups of the two α-subunits were ligated with nitric oxide, or $\alpha(Fe\text{-}NO)_2\beta(Fe\text{-}O_2)_2$ — have shown that under acidic conditions and/or presence of allosteric effectors, such as DPG, this derivative shows extremely reduced affinity for oxygen in the β-subunits[1]. Promising results were also obtained from preparations of fresh erythrocytes containing α(NO)-Hb as previously reported[2].

In this study, we have prepared the α-nitrosyl Hb derivative, i.e., α(NO)-Hb, within expired intact human erythrocytes with an improved method over a previous protocol[2], using *S*-nitrosoglutathione (GSNO) as an NO donor. NO was incorporated accurately to a

* Department of Biochemistry and Biophysics, School of Medicine, University of Pennsylvania, Philadelphia, PA 19104-6059, U.S.A.

50% saturation of hemes and exclusively bound to the α-subunits, which were confirmed by EPR. By using GSNO, the formation of oxidized Hb, i.e., methemoglobin, was efficiently reduced without the addition of large amounts of reducing agents.

2. MATERIAL AND METHODS

2.1 Preparation of GSNO

Two hundred millimolar stock solutions of sodium nitrite and glutathione (GSH) containing 20 mM HCl were prepared in deionized water and placed in gas-tight containers sealed with rubber septa. Solutions were thoroughly deaerated by bubbling pure Ar. Aliquots of sodium nitrite and the acidified GSH solutions were mixed using gas-tight syringes in a sealed vial under Ar atmosphere, containing ammonium sulfamate to a final concentration of 10 mM. The reaction mixture was placed on ice for about 30 minutes. Once the reaction was completed, Ar was bubbled into the mixture to remove any excess of NO from the solution. The GSNO solution thus prepared was kept on ice and used as quickly as possible. Some portions were rapidly frozen and tested later for stability.

2.2 Titration of GSNO

A 250-fold dilution of freshly prepared GSNO solution in degased deionized water was prepared and kept in a sealed vial. Two milliliters of 10 μM Hb solution in 0.1 M potassium phosphate buffer pH 7.0 were placed in a cuvette with continuous stirring. A small amount of fresh sodium dithionite was added in slight excess into the solution to convert Hb to its deoxy form. Changes in absorbance at 432 nm were monitored after each addition of 10-μl aliquot of the diluted solution of the nitrosothiol. The nitrosothiol concentration was calculated from the inflection point, as this corresponded to the heme equivalents present in the Hb solution.

GSNO solutions stored frozen were tested, after thawing, for stability. Samples were always passed through a 0.22 μm filter and titrated before use. After titration, it was confirmed that GSNO was undistinguishable from the freshly prepared one for the purpose of present experiments.

2.3 Preparation of α(NO)-Hb-containing erythrocytes

Erythrocytes from expired blood obtained from the local branch of the American Red Cross were washed several times in isotonic saline solution (0.9% NaCl) and then resuspended in a 2-fold volume of 0.15 M sodium-potassium phosphate buffer, pH 5.5, containing 10 mM glucose. The phosphate buffer was prepared by mixing 0.15 M Na_2HPO_4 and 0.15 M KH_2PO_4 to adjust the pH to 5.5. The erythrocyte suspension was then transferred to a 300-ml Kjeldahl type flask sealed with a rubber stopper for deoxygenation with pure Ar at room temperature under continuous stirring. Once erythrocytes were confirmed deoxygenated judging by visual inspection, one equivalent on heme of sodium dithionite in deaerated deionized water was injected slowly into the flask with continuous stirring. After this, a volume of GSNO corresponding to a 50% saturation of hemes was added slowly in several small portions over a period of 1 h. Samples were taken at different time intervals to check distribution and concentration of nitrosylated Hb species by EPR as previously described[2]. A typical example of the determination of the concentration and quality of α(NO)-Hb within erythrocytes is shown in Figure 2.

2.4 Measurement of oxygen equilibrium curves

Oxygen binding isotherms of α(NO)-Hb-containing erythrocytes were determined using an automatic device previously described[2]. Briefly, the light beam originating from a modified Hitachi 557 dual wavelength spectrophotometer (On-Line Instruments, Bogart, Georgia) was guided into the oxygenation cell using an optical fiber. Scattered light was collected by a photomultiplier attached to the opposite side of the gas equilibration cell with variable lightpath length. Changes in absorbance were monitored at 560 nm and the reference light was fixed at 497 nm. The oxygen concentration in the sample was monitored using an oxygen electrode (O_2 Sensors, Malvern, Pennsylvania). Absorbance and oxygen concentration data were taken on-line and stored onto floppy disks for later processing. Measurements were carried out by suspending an equivalent of 60 μM heme of untreated erythrocytes or 20 μM heme of α(NO)-Hb-containing erythrocytes, in 0.15M sodium-potassium phosphate buffer at various pH values, 15°C.

3. RESULTS AND DISCUSSION

The major advantage of the present preparation protocol over the previous one relies on the use of GSNO as an NO donor that allows stoichiometrical addition of NO into the reacting sample and also the complete removal of unreacted nitrite. As seen in Figure 1A and B, a solution of GSNO can be easily titrated with deoxyHb in the presence of small amounts of sodium dithionite. The presence of sharp isosbestic points ensures the presence of two species. As GSNO is added, NO is promptly released and binds to deoxy hemes to form nitrosylHb. These two events can be monitored by a decrease at 432 nm and an increase at 414 nm, respectively.

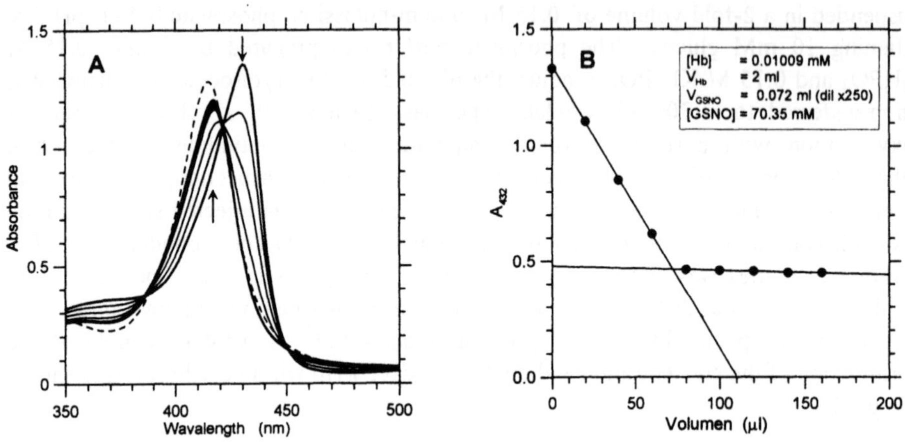

Figure 1. Titration of a solution of GSNO using deoxyHb as a trapping agent for NO. (A) Absorbance changes in the Soret region upon addition of aliquots of GSNO solution. *OxyHb (broken line)*, deoxyHb (*bold line*) and increasing amounts of nitrosylHb formed from NO released from GSNO binding to deoxyHb (indicated by the arrow at 417 nm). (B) Absorbance changes at 432 nm due to disappearance of deoxyHb and corresponding nitrosylHb formation. Conditions: 2 ml of approximately 10 μM Hb (on heme) in 0.1 M phosphate buffer, pH 7, at 15°C. DeoxyHb was obtained by adding a small amount of sodium dithionite into the solution. Aliquots of a dilute solution of GSNO were added rapidly into the cuvette with stirring, and absorbance in the Soret region upon each addition was recorded.

It is well known that nitrite is able to cross rapidly the erythrocyte membrane and enter into the intraerythrocytic environment oxidizing rapidly Hb to form metHb[3, 4], thus disabling it to perform gas transportation. The addition of ammonium sulfamate into the reaction solution of GSNO decomposes any unreacted nitrite. GSNO once formed is not able to enter erythrocytesm, but remains in the external medium. Under reducing conditions, such as in the presence of small amounts of sodium dithionite, GSNO decomposes releasing NO to form GSSG, according to: 2 GSNO → 2 NO + GSSG. The NO thus released from this reaction enters easily into erythrocytes and binds to unligated hemes. However, under acidic conditions and well below saturating conditions, the NO redistributes itself within the tetrameric Hb and ends up binding preferentially to the α-subunits. This process can be monitored by EPR as the intensity of the hyperfine structure around g = 2 increases with time (see Figure 2).

The formation of *S*-nitroso Hb, i.e., Hb in which the sulfhydryl group of the Cysβ93 residues are nitrosated, as previously reported[5] can be ruled out in our preparation procedure for several reasons: (1) GSNO does not penetrate the erythrocyte membrane

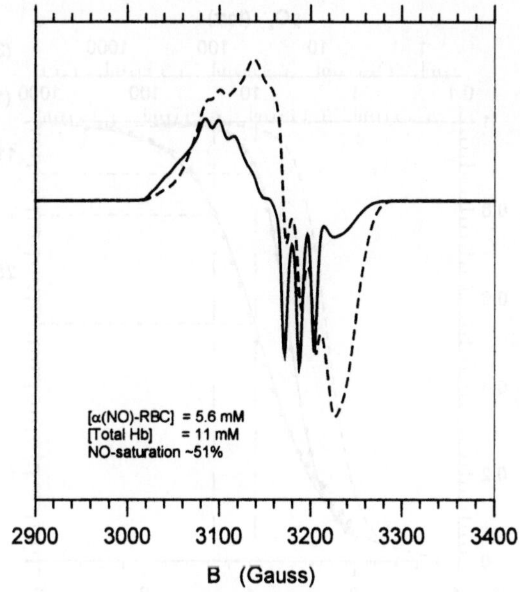

Figure 2. EPR measurement of α(NO)-Hb-containing eryhtrocytes (*solid line*) and fully NO-saturated erythrocytes (broken line). Aliquots of the erythrocyte suspension were taken directly from the reaction flask with a syringe, placed in quartz EPR tubes, and quickly frozen in liquid nitrogen. Total saturation of erythrocyte Hb with NO was achieved by adding a small amount of sodium dithionite and GSNO into the sample before freezing. Nitrosyl Hb concentration was calculated from the double integral of each spectrum and from nitrosylHb samples of known concentration. The intense triplet hyperfine structure indicates that NO is bound to the α-subunits.

and thence transnitrosation reaction with intracellular Hb is completely prevented; (2) the nitrosating agent that attacks sulfhydryl groups to form the corresponding *S*-nitrosothiol is N_2O_3, and not NO. N_2O_3 only forms in the presence of oxygen. In our procedure, oxygen is not available in the reacting mixture, and therefore N_2O_3 is not formed; (3) if *S*-nitroso Hb was indeed formed in our preparations, the oxygenation characteristics should have corresponded to those of a high affinity species, as reported by others[6, 7]; (4) finally, in the deoxy form the sulfhydryl residues from Cysβ93 are sterically hindered by a salt bridge between imidazole group of Hisβ146 and the carboxylic group of Aspβ94, and another salt bridge between the carboxylic group of Hisβ146 and the amino group of Lysα91 of the complementary dimer.

The oxygen delivery performance of erythrocytes containing α(NO)-Hb is shown in Figure 3. Previously, the $t_{1/2}$ for α(NO)-Hb in solution was determined as 7 h at 15°C. Since the determination of oxygen equilibrium curves takes approximately 1 h, isotherms were determined at this temperature to reduce the decomposition rate of this derivative.

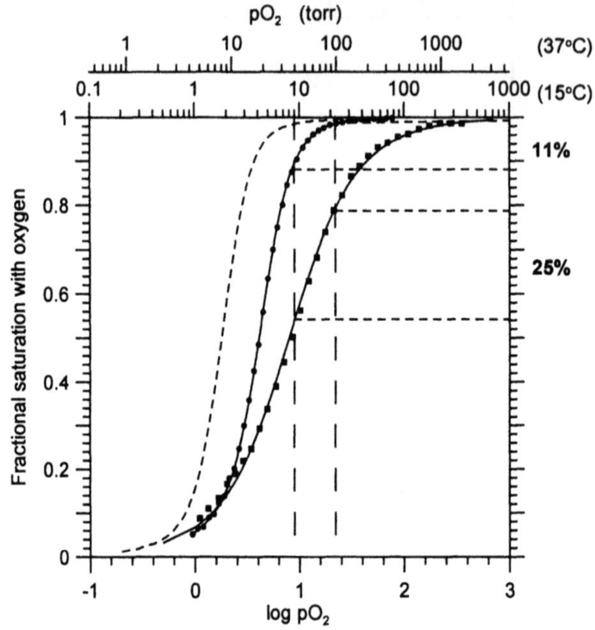

Figure 3. Oxygen delivery performance of intact α(NO)-Hb erythrocytes (!) and untreated erythrocytes (,), compared with that of Hb A in solution (*broken line*). Vertical lines correspond to partial pressures of oxygen at lung (right-hand) and tissue (left-hand), at 15°C. For comparison, a pressure scale corresponding to values at 37°C was added at the top. Measurements were done in isotonic phosphate buffer as indicated in the text.

This $t_{1/2}$ value could be even much larger in the present case because this derivative remains in the intracellular milieu. The P_{50} value extrapolated to 37°C was calculated to be about 37 torr. Despite the oxygen carrying capacity of α(NO)-Hb is reduced by 50% (as the α-subunits are ligated with NO, only the β-subunits are capable to transport oxygen), the oxygen delivery capacity exceeds twice the amount of oxygen delivered from untreated erythrocytes under the same conditions. Exchange of 25% of total blood volume with α(NO)-Hb-containing erythrocytes has been conducted in rats with no apparent adverse effects; microcirculation characteristics remained normal after observation for several hours (M. Suematsu, personal communication).

ACKNOWLEDGEMENTS

We thank Dr. Cameron J. Koch (Dept. of Radiation Oncology, University of Pennsylvania, School of Medicine) for donating the oxygen electrodes used in the present study and for his invaluable advice. This work was supported by grants HL 14508 from the National Institutes of Health.

REFERENCES

1. T. Yonetani, A. Tsuneshige, Y-x. Zhou, and X-s Chen, Electron paramagnetic resonance and oxygen binding studies of α-nitrosyl hemoglobin. *J. Biol. Chem.* **273**, 20323-20333 (1998).
2. A. Tsuneshige and T. Yonetani, T. Oxygenation properties of human erythrocytes containing exclusively α-nitrosyl hemoglobin: A promissing blood substitute candidate. *Artif. Cells, Blood Substit. Immobil. Bioltechnol.* **29**(5), 347-357 (2001).
3. E. Antonini and M. Brunori in: Hemoglobin and myoglobin in their interactions with ligands, Frontiers of Biology, Neuberger A. and Tatum, E. L., eds., North-Holland Publishing Company, Amsterdam, The Netherlands (1971).
4. E. E. Di Iorio, Preparation of derivatives of ferrous and ferric hemoglobin. *Methods Enzymol.* **76**, 57-72 (1981).
5. L. Jia, C. Bonavetura, J. Bonaventura, and J. S. Stamler, *S*-nitrosohaemoglobin: a dynamic activity of blood involved in vascular control. *Nature* **380**, 221-226 (1996).
6. R. P. Patel, N. Hogg, N. Y. Spencer, B. Kalyanamaran, S. Matalon, and V. M. Darley-Usmar, Biochemical characterization of human *S*-nitrosohemoglobin *J. Biol. Chem.* **274**, 15487-15492 (1999).
7. C. Bonaventura, G. Ferruzzi, S. Tesh, and R. D. Stevens, Effects of *S*-nitrosation of oxygen binding by normal and sickle cell hemoglobin, *J. Biol. Chem.* **274**, 24742-24748 (1999).

THE CONTRIBUTION OF HEMOGLOBIN TO OXYGEN TRANSPORT IN THE MICROCRUSTACEAN *DAPHNIA MAGNA*

A conceptual approach

Ralph Pirow*

1. INTRODUCTION

Problems of oxygen transport in microscopic organisms have been largely approached by theoretical considerations and diffusion models which have permitted the calculation of critical ambient oxygen concentrations, penetration depths for oxygen, and critical body sizes[1–3]. Such models provide valuable estimates when applied to unicellular organisms and early ontogenetic stages of metazoans such as eggs or embryos. However, fully developed small-sized animals of the advanced phyla are much more complex in anatomy, and their body forms often deviate from simple geometrical shapes. In addition, the presence of circulatory systems requires consideration of internal convective processes as well, which renders the theoretical approach much more difficult. To incorporate circulatory oxygen transport into a theoretical model, it is essential to have some knowledge of the circulatory architecture as well as information on key parameters such as perfusion rate, oxygen tension, and the concentration and oxygen-saturation of the blood oxygen carrier. In the past, small body size represented an almost insurmountable experimental obstacle which hampered the attempt to obtain adequate physiological information. Fortunately, with the advent of digital image processing and advanced optical techniques, expanding the methodical versatility of microscopy, it has become possible to gain this information from those animals which are distinguished by a high level of transparency[4–7].

The transparent microcrustacean *Daphnia magna* has proven to be a valuable model organism that has stimulated the development and adaptation of innovative optical techniques for studying oxygen transport in microscopic animals[6, 8–11]. Besides the interest in methodical advancement, it seemed also worthwhile to obtain a better understanding of the specific physiology of *Daphnia*, since this genus occupies an

* Ralph Pirow, Institute for Zoophysiology, University of Muenster, Hindenburgplatz 55, D-48143 Muenster, Germany, e-mail: pirow@uni-muenster.de, phone: +49-251-8323858, fax: +49-251-8323876

Oxygen Transport to Tissue XXIII, edited by D. Wilson *et al.*
Kluwer Academic/Plenum Publishers, 2003

important position in freshwater food webs and its representatives are frequently used as model organisms in ecological and toxicological studies. The application of these optical methods to *D. magna*, a species characterized by a striking variability in blood haemoglobin (Hb) concentration, has provided interesting information concerning the oxygen-transport capacity of the circulatory system[11, 12]. However, despite new insights into the circulatory physiology of this animal, the degree of utilization of Hb in circulatory oxygen transport still remains an open question that is difficult to tackle solely by experimental approaches. The purpose of the present paper is to introduce a conceptual approach to describe the diffusive-convective transport of oxygen in this species in a mathematically tractable way. For an animal with a given body size and physiological constitution, the proposed model makes it possible to predict the contribution of Hb to circulatory oxygen transport as well as the critical ambient oxygen tension (P_{O2amb}) at which oxygen supply to the tissue starts to become impeded.

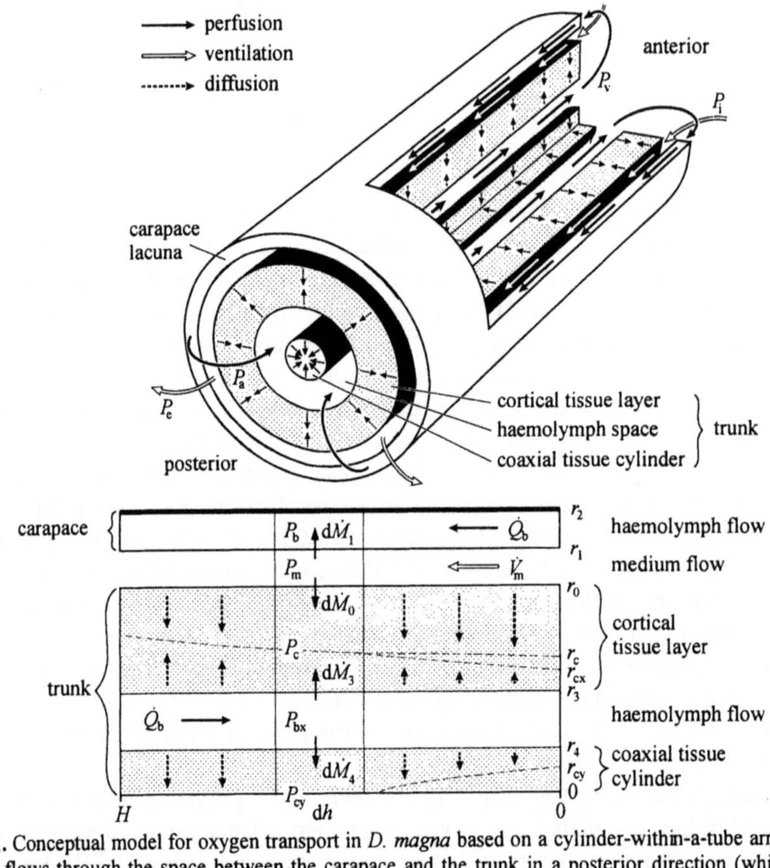

Figure 1. Conceptual model for oxygen transport in *D. magna* based on a cylinder-within-a-tube arrangement. Medium flows through the space between the carapace and the trunk in a posterior direction (white arrows) while oxygen is released both into the carapace lacuna and the cortex of the trunk. This cortical tissue layer is supplied with oxygen from the medium and from a truncal haemolymph space by diffusion (dashed arrows). Oxygenated haemolymph leaves the double-walled carapace and then enters the truncal haemolymph space (solid arrows). While flowing in an anterior direction, oxygen diffuses from this haemolymph space both into the coaxial tissue cylinder and the cortical tissue layer (dashed arrows). See text for details.

2. CONCEPTUAL MODEL FOR DIFFUSIVE-CONVECTIVE TRANSPORT

A mathematically feasible albeit hypothetical approach to model the diffusive-convective O_2 transport in a millimetre-sized animal with an open circulatory system is to assume a simple geometrical body such as a cylinder, which consists of a cortical tissue layer, a subcortical haemolymph space, and a coaxial tissue cylinder in the centre (Fig. 1). The cortex is supplied with O_2 from the ambient medium and the haemolymph space. From this haemolymph space, O_2 is also released into the central tissue cylinder. To take the effect of external convection and the role of the blood-perfused carapace as a site of O_2 uptake into account[10], the cylindrical trunk is further assumed to be enveloped by a double-walled hollow cylinder representing the carapace of the animal. Ambient medium flows through the space between the carapace and the trunk while O_2 is released into both the carapace lacuna and the cortex of the trunk. The differential amount of oxygen $d\dot{M}_0$ which is transported across the cuticle of the trunk is equal to the amount of O_2 respired by the cortical (hollow) cylinder element of height dh:

$$d\dot{M}_0 = a\pi\left(r_0^2 - r_c^2\right)\cdot dh \tag{1}$$

where a is the volume-specific O_2 consumption rate of the tissue, and r_0 and r_c represent the outer and the inner radius of the hollow cylinder element. The latter is obtained from the O_2 tension of the medium P_m by solving the following equation for r_c:

$$P_m = a/(2g_0r_0)\cdot\left(r_0^2 - r_c^2\right) + a/(4K)\cdot\left[r_0^2 - r_c^2 + 2r_c^2\ln\left(r_c/r_0\right)\right] + P_c \tag{2}$$

where P_c is the O_2 tension at r_c, K represents Krogh's diffusion constant for O_2, and g_0 is the diffusing capacity of the cuticle. Owing to the difference in O_2 tension between medium and haemolymph, P_m–P_b, the differential amount of oxygen $d\dot{M}_1$ diffuses into the carapace lacuna through the area element dA_1:

$$d\dot{M}_1 = g_1\left(P_m - P_b\right)\cdot dA_1 \quad \text{with} \quad dA_1 = 2\pi r_1 dh \tag{3, 4}$$

where g_1 and r_1 are the diffusing capacity and the radius of the inner carapace wall. As a consequence, the O_2 tension of the haemolymph increases by dP_b with:

$$dP_b = d\dot{M}_1/(\dot{Q}_b\cdot\beta_b) \quad \text{with} \quad \beta_b = \Delta[O_2]/\Delta P_b \tag{5, 6}$$

where $\dot{Q}_b$ is the perfusion rate, and β_b is the capacitance coefficient[13] for O_2 in the haemolymph which is defined as the increment in oxygen concentration $[O_2]$ per increment in P_b. For a given P_b, $[O_2]$ is calculated from:

$$[O_2] = \alpha_b P_b + C_{Haem}S \quad \text{with} \quad S = P_b^n/(P_{50}^n + P_b^n) \tag{7, 8}$$

where α_b is the physical solubility for O_2 in the haemolymph, C_{Haem} is the haem-based Hb concentration, and S represents the O_2 saturation of Hb. S is calculated from P_b, the half-saturation O_2 tension P_{50}, and the Hill coefficient n. The differential amount of oxygen $d\dot{M}_m$ that is released from the medium is given by Eq. (9). This release results in a reduction of the O_2 tension in the medium by dP_m with:

$$d\dot{M}_m = d\dot{M}_0 + d\dot{M}_1 , \qquad dP_m = -d\dot{M}_m / (\dot{V}_m \cdot \alpha_m) \qquad (9, 10)$$

where $\dot{V}_m$ represents the medium flow rate, and α_m is the physical solubility for O_2 in the medium. The differential amounts of oxygen $d\dot{M}_3$ and $d\dot{M}_4$, which are released from the truncal haemolymph space into the cortical tissue layer and the central tissue cylinder, respectively, are given by:

$$d\dot{M}_3 = a\pi\left(r_{cx}^2 - r_3^2\right)\cdot dh \quad \text{and} \quad d\dot{M}_4 = a\pi\left(r_4^2 - r_{cy}^2\right)\cdot dh \qquad (11, 12)$$

where r_{cx} and r_3 are the outer and inner radius of that tissue layer of the cortex which is supplied from the truncal haemolymph space, r_4 is the radius of the central tissue cylinder, and r_{cy} reflects the entrance depth for O_2 into the central tissue cylinder. The variable r_{cx} is obtained from the O_2 tension of the haemolymph P_{bx} in the truncal haemolymph space by solving the following equation for r_{cx}:

$$P_{bx} = a/(4K)\cdot\left[r_3^2 - r_{cx}^2 + 2r_{cx}^2 \ln\left(r_{cx}/r_3\right)\right] + P_c . \qquad (13)$$

Depending on whether or not the whole central tissue cylinder is sufficiently supplied with O_2, either Eq. (14) or Eq. (15) is used to determine r_{cy} for a given P_{bx}:

$$P_{bx} = a/(4K)\cdot r_4^2 + P_{cy} , \qquad r_{cy} = 0 \qquad (14)$$

$$P_{bx} = a/(4K)\cdot\left[r_4^2 - r_{cy}^2 + 2r_{cy}^2 \ln\left(r_{cy}/r_4\right)\right] \text{ with } 0 < r_{cy} < r_4 . \qquad (15)$$

P_{cy} represents the O_2 tension in the centre of this cylinder. As a consequence of the O_2 releases described in the Eqs. (11) and (12), P_{bx} changes by dP_{bx} with:

$$dP_{bx} = (d\dot{M}_3 + d\dot{M}_4)/(\dot{Q}_b \cdot \beta_{bx}) \qquad (16)$$

where the capacitance coefficient β_{bx} is obtained from Eq. (6) using P_{bx} instead of P_b. In Eq. (16), the right-hand term has no minus because flow is in the opposite direction compared to the flow orientation in the two other fluid compartments. These differential equations were solved separately for each cylinder element of height dh. The whole transfer of O_2 from the medium to the tissue was integrated numerically from the entering position of the cylinder at h=0, where P_m=P_i and P_b=P_{bx}=P_v, to the exit position at h=H, where P_m=P_e and P_b=P_a. P_i and P_e represent the inspiratory and expiratory O_2 tensions of the medium, whereas P_v and P_a refer to the O_2 tension of the haemolymph at the entrance and exit position of the carapace lacuna (Fig. 1). Given P_i, P_v was varied until the value for P_b at the position h=H was equal to that of P_{bx} at h=H.

Table 1. List of parameters used in the numerical evaluation. Data refer to 20 °C

Symbol	Value	Unit	Description	References & remarks
V	1.33	mm^3	body volume	
H	2.10	mm	length of the cylinder	
r_0	0.44	mm	radius of the trunk	
r_1	0.51	mm	radius of the inner carapace wall	
r_2	0.52	mm	radius of the outer carapace wall	
r_3	0.34	mm	inner radius of cortical tissue layer	
r_4	0.05	mm	radius of the coaxial tissue cylinder	
Φ	0.4		tissue fraction of body volume	Kobayashi (1983)[14]
a_0	21.1	$nmol\ h^{-1}\ mm^{-3}$	O_2 consumption rate (whole body)	Pirow et al. (2001)[11]
a	52.8	$nmol\ h^{-1}\ mm^{-3}$	O_2 consumption rate (pure tissue)	$a = a_0/\Phi$
Q_b	0.15	$mi\ h^{-1}$	perfusion rate	Pirow et al. (2001)[11]
V_m	4.00	$ml\ h^{-1}$	medium flow rate	Pirow et al. (1999)[9]
α_m	0.0137	$nmol\ mm^{-3}\ kPa^{-1}$	solubility coefficient for O_2 in water (0.2 ‰ salinity)	Gnaiger and Forstner (1983)[15]
α_b	0.0123	$nmol\ mm^{-3}\ kPa^{-1}$	solubility coefficient for O_2 in haemolymph (human plasma)	Christoforides et al. (1969)[16]
C_{Haem}	50, 340	$\mu mol\ l^{-1}$	haem-based Hb concentrations	Pirow et al. (2001)[11]
P_{50}	0.7	kPa	half-saturation oxygen tension	see Pirow et al. (2001)[11]
n	1.5		Hill coefficient	see Pirow et al. (2001)[11]
Δx_c	0.001	mm	thickness of the cuticle	see Pirow et al. (1999)[10]
Δx_e	0.005	mm	thickness of the epithelium	see Pirow et al. (1999)[10]
K_c	0.0034	$nmol\ h^{-1}\ mm^{-1}\ kPa^{-1}$	Krogh's constant of diffusion for O_2 in chitin	Krogh (1919)[17]
K	0.0581	$nmol\ h^{-1}\ mm^{-1}\ kPa^{-1}$	... in tissue	$2.2\times10^{-5}\ cm^2\ min^{-1}\ atm^{-1}$
g_0	3.44	$nmol\ h^{-1}\ mm^{-2}\ kPa^{-1}$	diffusing capacity of the trunk cuticle	$g_0 = K_c/\Delta x_c$
g_1	2.65	$nmol\ h^{-1}\ mm^{-2}\ kPa^{-1}$	... of the inner carapace wall	$g_1 = K_c \cdot K/(\Delta x_e \cdot K_c + \Delta x_c \cdot K)$

3. ASSUMPTIONS, PARAMETER VALUES AND PREDICTIONS

In the proposed model, the complex body shape of *Daphnia*[10] is reduced to a cylindrical trunk enveloped by a hollow cylinder representing the carapace. A plain circulation without any branching of haemolymph flow into subcircuits is assumed which is, of course, a simplification compared to the real situation in the animal[10]. Taking into account that not all the haemolymph ejected by the heart is pumped into the carapace lacuna, the perfusion rate used in the present model was assumed to be 50 % of the cardiac output of the animal. To further reduce mathematical complexity, a number of additional simplifying assumptions are made: (1) the tissue fraction of the trunk is regarded as a homogeneous O_2-consuming mass. (2) The volume-specific O_2 consumption rate of the tissue a is assumed to be constant throughout the tissue and independent of tissue O_2 tension. (3) Axially directed diffusion of O_2 is neglected. (4) Moving fluids are assumed to have uniform flow velocity profiles ("plug flow"). (5) Radial O_2 gradients in the haemolymph and the medium are neglected. (6) The outer wall of the carapace is assumed to be impermeable for O_2.

The values of physiological parameters used in the model were derived from Pirow et al.[11] and refer to a 2.65 mm long *D. magna* (Tab. 1). The length of the cylinder H was assumed to be 82 % of the body length of the animal, which represents the length of the carapace. The radii r_0 ... r_4 were calculated from the body volume V of the animal taking

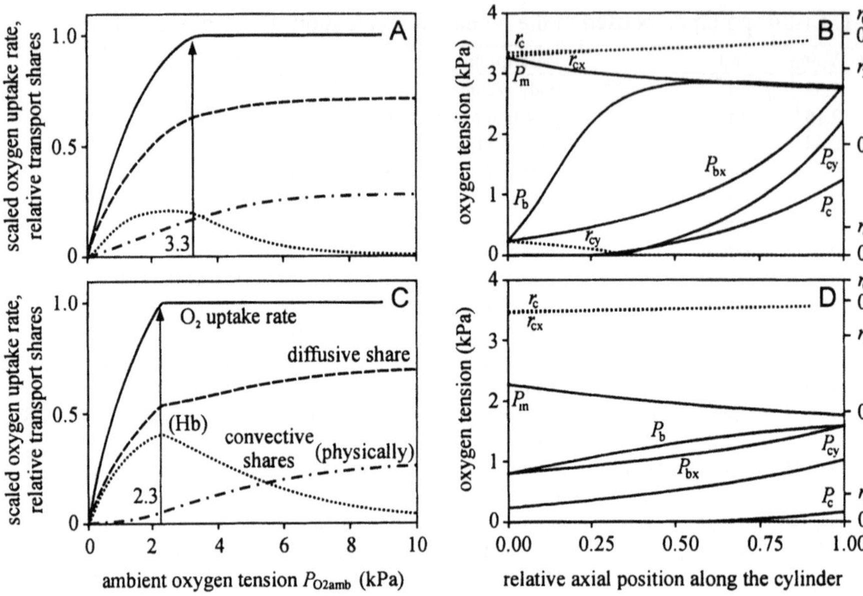

Figure 2. Results of the numerical evaluation of the model using a C_{Haem} of 50 µmol l^{-1} (Hb-poor state; A, B) and 340 µmol l^{-1} (Hb-rich state; C, D). Left: Dependency of O_2 uptake rate (solid lines) from P_{O2amb}. The vertical arrows and adjacent numbers indicate the critical P_{O2amb} at which the O_2 uptake rate decreased to 99 % of the maximum. The other lines show the relative diffusive and convective shares of tissue O_2 supply including those fractions which are transported by Hb and as physically dissolved portions. Right: Axial profiles of model parameters at the respective critical P_{O2amb} of 3.3 kPa (Hb-poor state; B) and 2.3 kPa (Hb-rich state; D).

the following conditions into account: (1) V comprises the tissual and haemolymph compartments, (2) the tissual compartment is 40 % of V (Kobayashi, 1983), (3) the medium flow velocity is 5 mm s^{-1} at $V_m = 4$ ml h^{-1}, (4) r_4 is 50 µm, and (5) the thickness of the carapace lacuna (r_2–r_1) is 10 µm. An estimate of V for a 2.65 mm animal was calculated by $V=(0.415\times\text{body length})^3$, where the shape coefficient of 0.415 was derived from Kobayashi[14] who reported a body volume of 1.12 mm^3 for a 2.5 mm *D. magna*. For Krogh's diffusion constant for O_2 in tissue K, a value similar to that reported for the frog satorius muscle[18] was chosen. The two C_{Haem} tested in the model were derived from Pirow et al.[11] who had measured 50 and 340 µmol l^{-1} in normoxia-acclimated Hb-poor and hypoxia-acclimated Hb-rich animals, respectively.

The numerical evaluation of the model for the Hb-poor and Hb-rich state yielded a critical P_{O2amb} of 3.3 and 2.3 kPa (Fig. 2A,C), respectively. These results indicate that the sevenfold increase in C_{Haem} from 50 to 340 µmol l^{-1} effected a 30 % reduction in the P_{O2amb} at which O_2 supply to the tissue started to become impeded. The reduction in critical P_{O2amb} was, however, not as large as expected from the experimental examination[11], which revealed a 73 % decrease from 4.8 kPa in Hb-poor to 1.3 kPa in Hb-rich animals. To bring model predictions into closer correspondence with experimental data, a critical reconsideration of the simplifying assumptions will be necessary when refining the model. For example, consideration of radial O_2 gradients in the medium and haemolymph would establish additional transport barriers for O_2, which shift the critical P_{O2amb} for the Hb-poor state towards higher values. In addition, a more elaborate concept of tissue perfusion combined with a shortening of diffusion distances in the tissual compartments would make the model more sensitive to elevated Hb concentrations and slight variations in Hb O_2 affinity, thus allowing the system to sustain an adequate tissue O_2 supply at a lower P_{O2amb}. Despite the slight differences in critical P_{O2amb}, the proposed conceptual approach represents a valuable complement to experimental approaches which aim to elucidate the role of Hb in *D. magna*. It permits to estimate that portion of the total O_2 required that is actually transported by Hb at a given P_{O2amb} (Fig. 2A,C). For cost-benefit analyses, which require information concerning the efficiency of Hb utilization, the model provides the percentage of total Hb that is involved in circulatory O_2 transport. For example, whereas 73 % of Hb was utilized in the Hb-poor state at the critical P_{O2amb} of 3.3 kPa, the degree of Hb utilization in the Hb-rich state at 2.3 kPa was 22 % (cf. P_{bx} profiles in Fig. 2B,D). This could suggest that the adaptation to hypoxia *via* increase in C_{Haem} is associated with a lower degree of Hb utilization. In conclusion, although this model may prove to be overly simplistic in some respects, nevertheless it does provide a suitable starting point for future research and could well invite further development. This might include a more elaborate concept of tissue O_2 supply, which better takes the specific circulatory patterns of the animal into account, as well as a consideration of axially directed diffusion and radial O_2 gradients.

REFERENCES

1. E. N. Harvey, The oxygen consumption of luminous bacteria, *J. Gen. Physiol.* **11**, 469–475 (1928).
2. R. W. Gerard, Oxygen diffusion into cells, *Biol. Bull.* **60**, 245–268 (1931).
3. A. Krogh, *The Comparative Physiology of Respiratory Mechanisms* (University of Pennsylvania Press, Philadelphia, 1941).
4. W. Burggren and R. Fritsche, Cardiovascular measurements in animals in the milligram range, *Brazil. J. Med. Biol. Res.* **28**, 1291–1305 (1995).
5. M. Colmorgen and R. J. Paul, Imaging of physiological functions in transparent animals (*Agonus cataphractus, Daphnia magna, Pholcus phalangioides*) by video microscopy and digital image processing, *Comp. Biochem. Physiol. A* **111**, 583–595 (1995).
6. R. J. Paul, M. Colmorgen, S. Hüller, F. Tyroller and D. Zinkler, Circulation and respiratory control in millimetre-sized animals (*Daphnia magna, Folsomia candida*) studied by optical methods, *J. Comp. Physiol. B* **167**, 399–408 (1997).
7. T. Schwerte and B. Pelster, Digital motion analysis as a tool for analysing the shape and performance of the circulatory system in transparent animals, *J. Exp. Biol.* **203**, 1659–1669 (2000).
8. R. Pirow, F. Wollinger, and R. J. Paul, Gas exchange and gas transport in a water breather of millimetre size, *Adv. Exp. Med. Biol.* **428**, 139–145 (1998).
9. R. Pirow, F. Wollinger and R. J. Paul, Importance of the feeding current for oxygen uptake in the water flea *Daphnia magna*, *J. Exp. Biol.* **202**, 553–562 (1999).
10. R. Pirow, F. Wollinger and R. J. Paul, The sites of respiratory gas exchange in the planktonic crustacean *Daphnia magna*: An *in-vivo* study employing blood hemoglobin as an internal oxygen probe, *J. Exp. Biol.* **202**, 3089–3099 (1999).
11. R. Pirow, C. Bäumer and R. J. Paul, Benefits of haemoglobin in the Cladoceran crustacean *Daphnia magna*, *J. Exp. Biol.* **204**, 3425–3441 (2001).
12. C. Bäumer, R. Pirow and R. J. Paul, Circulatory oxygen transport in the water flea *Daphnia magna*, *J. Comp. Physiol. B*, in press.
13. J. Piiper, P. Dejours, P. Haab and H. Rahn, Concepts and basic quantities in gas exchange physiology, *Resp. Physiol.* **13**, 292–304 (1971).
14. M. Kobayashi, Estimation of the haemolymph volume in *Daphnia magna* by haemoglobin determination, *Comp. Biochem. Physiol. A* **76**, 803–805 (1983).
15. E. Gnaiger and H. Forstner, *Polarographic Oxygen Sensors* (Springer Verlag, Berlin, 1983).
16. C. Christoforides, L. H. Laasberg and J. Hedley-Whyte, Effect of temperature on solubility of O_2 in human plasma, *J. Appl. Physiol.* **26**, 56–60 (1969).
17. A. Krogh, The rate of diffusion of gases through animal tissue, with some remarks on the coefficient of invasion, *J. Physiol.* **52**, 391–408 (1919).
18. M. Mahler, C. Louy, E. Homsher and A. Peskoff, Reappraisal of diffusion, solubility, and consumption of O_2 in frog skeletal muscle, with application to muscle energy balance, *J. Gen. Physiol.* **86**, 105–134 (1985).

MODIFICATIONS OF RED BLOOD CELL SHAPE AND GLYCOPROTEINS MEMBRANE CONTENT IN SEPTIC PATIENTS

Michael Piagnerelli [1], Karim Zouaoui Boudjeltia[2], Danny Brohee[2], Jean-Louis Vincent[1], Michel Vanhaeverbeek[2]

1. INTRODUCTION

The RBC must exhibit a remarkable ability to undergo cellular deformation, since its diameter (8 μm in humans) far exceeds that of the capillaries (2 to 3 μm) [1].

In sepsis, RBC deformability may be altered by extrinsic (interactions with leukocytes[2] and the release of oxygen free radicals[3]) and intrinsic factors (the increased production of nitric oxide [4], an increase in free intracellular calcium [5] and a decrease in adenosine triphosphate[6]). An increase in 2,3 diphosphoglycerate (2,3 DPG) concentration may also destabilize RBC membrane protein interactions [7].

RBC deformability is also influenced by the behavior of different proteins of the RBC membrane. An important transmembrane protein is glycophorin A which is highly glycosylated in the RBC. The two sialic acid residues (N-acetyl-neuraminic acid; SA) of each of these oligosaccharides, account for 60% of the negative charge of the RBC. Because of the negative charge on the RBC membrane surface due to SA, in physiologic conditions, RBCs normally repel each other and do not aggregate [1]. The importance of SA on RBC shape is demonstrated by the observation that neuraminidase-treated cells, which release their membrane SA content, undergo increased aggregation and display a reduced mean curvature [8]. Decreases in RBC membrane SA content have been demonstrated in RBCs of patients with diabetes mellitus [9] as well as in senescent RBCs [6]. The decrease in SA may explain the increase of RBC aggregability observed in diabetes mellitus [9] but effects of SA on alterations of RBC shape remain controversial.

In sepsis, several studies have demonstrated that RBCs have an increased aggregability and a decreased deformability [10,11]. To explain these alterations, we hypothesized that a decrease in the SA content of RBC membranes may occur in septic patients. We studied the RBC membrane SA content and its relationship with RBC shape in critically ill patients with or without sepsis compared to healthy volunteers. For estimation of RBC

1. Department of Intensive Care Medicine. Erasme University Hospital 1070 Brussels.
2. Experimental Medicine Laboratory. A. Vesale Hospital. 6110 Montigny-le-Tilleul. Free University of Brussels (ULB) Belgium

Oxygen Transport to Tissue XXIII, edited by D. Wilson *et al.*
Kluwer Academic/Plenum Publishers, 2003

shape, we adapted the technique described by Aminoff et al. using the Becton-Dickinson FACScan flow cytometer [12]. We expanded this method with the measure of the spherical index for RBC shape estimation.

2. PATIENTS AND METHODS

2.1. Patients

This study included 20 healthy volunteers and 39 adult critically ill patients with or without sepsis. Criteria for sepsis included clinical and bacteriological evidence of infection plus signs of a systemic response[13]. Exclusion criteria were RBC transfusion in the last 72 hours, acute bleeding, hematologic disorders, recent chemotherapy, history of diabetes mellitus, burns, cardiogenic shock, chronic renal failure, cirrhosis and pregnancy. The clinical characteristics of the subjects are summarized in Table 1.

2.2. Blood Sampling

In patients, blood samples were drawn once during the first 24 hours of sepsis (in the septic patients) or during the 24 hours following ICU admission (in the non-septic patients). Laboratory studies are summarized in the Table 1.

Table 1. Characteristics of the subjects studied

	Septic (n=19)	Non-septic (n=20)	Volunteers (n=20)
Age (years)	69±13[a]	66±16[a]	31±10
Hct	29.1±4.9[a b]	33.9±4.3[a]	41.2±3.1
Hemoglobin	9.7±1.6[a b]	11.4±1.4[a]	13.9±1.0
MCV	88.0±4.0	91.1±5.7	87.9±7.0
WBC	11.09±4.0[a]	10.2±2.8[a]	6.4±1.3
CRP	11.4±8.0[a]	7.2±7.8[a]	<0.5
Glucose	8.58±3.63	8.03±2.20	NR
Lactate	1.55±1.12	1.77±.82	NR

Hct: Hematocrit (%); hemoglobin (g/dL); MCV (fl.^{3}): mean corpuscular volume; WBC (/mm^3): white blood cell; CRP (mg/dL): C-reactive protein; Glucose (mmol/L); lactate (mmol/L); NR: not relevant
[a] p value < .05 vs control, [b] p value < .05 vs non-septic

2.3. Sialic Acid Content

Following total blood centrifugation at 4000 g for 10 minutes at 20 °C, the supernatant was removed and the RBC pellets were re-suspended and washed three times with 10 ml of phosphate-buffered saline (PBS; 10mM phosphate buffer, 145 mM NaCl, pH 7.4; 10^6 cells/ml) in isotonic solution. The erythrocyte ghosts were prepared by lysis with 50 ml of distilled water and washed three times with PBS with centrifugation at 5000 g for 60 min. at 20 °C. The ghost pellets were re-suspended with 300 µl of PBS and 10µl of Triton X-100 were added. The samples were agitated for 10 min. at 37 °C and then centrifuged at 100,000 g for 70 min. at 4 °C. The membrane proteins were collected in the supernatant. The protein concentrations were measured by the method of Lowry et al. using bovine serum albumin as standard[14]. The SA content of RBC membrane proteins was assessed by high-performance liquid chromatography (HPLC) at 25 °C. SA was specifically labeled with ortho-phenylenediamine (OPD) yielding stable fluorescent quinoxaline derivative. SA was released from glycoproteins in a $NaHSO_4$ solution (0.5 M; 80°C; 20 min.) and labeled in the same solution with the OPD (20 mg/ml; 80 °C; 40 min). The fluorescence for the OPD-N-acetylneuraminidic acid was measured at 420 nm after 236 nm excitation[15] Intra-assay and inter-assay coefficients of variation were 5% and 13%, respectively.

2.4. Flow Cytometry Studies

2.4.1. RBC shape analysis

Data were collected on a Becton-Dickinson FAC SCalibur® cytofluorimeter. Both the general principles of FACScan flow cytometer and the details used herein have been described elsewhere [12,16].The forward li ght- FAC SCalibur® scatter (FSC) channels were set on linear gain. Cell size is the principal component of the FSC signal. RBC FSC distribution parameters were defined on healthy volunteers, in isotonic (286 mOsm) PBS and at 25°C. In this group, as described previously by Rolfes-Curl et al[12], the FACScan clearly viewed the flow of ellipsoid, biconcave RBCs as essentially two populations of cells under isotonic conditions. The FSC histograms showed a typically bimodal distribution of RBC in isotonic buffer, and a monomodal distribution in hypotonic solution [12]

We determined two gates of interest in each FSC histogram: R1 and R2. The limit between R1 and R2 was the mean through value in the valley between the two modes. We computed the M2/M1 ratio, named the spherical index, between the FSC median values in R2 and R1 regions , and the moment (asymmetry of the histogram; moment = 3 x (mean-median)/ σ). A negative moment indicates a left asymmetry. A decrease in M2/M1 is correlated with the sphericity of the RBC. The cytometer settings were kept constant and the stability of the FSC channel was checked with calibrated size beads (BDIS CaliBrite Beads®) in each experiment (interassay variability: 3%).

2.4.2. Statistical analysis

Non parametric statistical tests were used, including the Kruskal-Wallis test, and the Dunnett test for pair-wise comparisons. The RBC modification shapes in isoosmolar and hypoosmolar solutions were evaluated by the Wilcoxon signed rank test. The correlations were evaluated by the Spearman test. Data are presented as mean ± SD. A p value < .05 was considered as statistically significant.

3. RESULTS

3.1.1. SA content

SA content was significantly lower in the septic patients than in both the non septic patients (1.98 ± .79, 2.20 ± .39 μg/100μg membrane protein, $p = .01$) and the healthy volunteers (2.71 ±1.00 μg/100μg membrane protein, $p < .001$).

3.1.2. RBC shape

The M2/M1 ratio was significantly reduced in all ICU patients (septic and non-septic), corresponding to a right shift of the events in R1 gate (right shift of the median channel in the FSC histogram). Also the moment decreased, reflecting the RBC spherocytation and FSC symmetry. The spherical index (M2/M1) was strongly correlated with the moment (volunteers: $r^2 = -.86$, $p < .001$; non-septic patients: $r^2 = -.74$, $p < .001$; and septic patients: $r^2 = -.74$, $p < .001$, respectively).

Correlation between flow cytometry study and SA content

There was a correlation between the M2/M1 ratio, the moment and the RBC SA membrane content for the ICU patients (septic and non-septic) (M2/M1 and SA, $r^2 = .17$, $p = .01$; moment and SA, $r^2 = .15$, $p = .015$, Spearman test).

4. DISCUSSION

Microcirculatory alterations in sepsis are complex. Disturbances of RBC and WBC rheologies may contribute to the impaired cellular oxygen supply [17]. Increased RBC aggregability and decreased RBC deformability have been found to be common in sepsis [2,3, 10]

Alterations in RBC deformability have been studied by different techniques including micropipette aspiration[1],fil tration procedure[10] or ektacytometry[1]. Of the available methods, micropipette approach provides the most comprehensive information concerning various membrane properties[1]. The filtration technique of a cell suspension through micro-pores has been used most frequently. However, the lack of selectivity (RBC versus WBC), sensitivity and/or reproducibility (homogeneity of filters) counterbalance the obvious advantages of this relatively simple and cheap technique. Ektacytometry is particularly useful for documenting decreased membrane integrity of the membrane due to protein defects[1]. The present study confirms that flow cytometry can also be used to evaluate RBC shape[12,16]. The FACScan system gave a bimodal distribution in contrast to a gaussian distribution obtained with the other flow cytometrie system (Coulter Epics 541). This difference is attributed to the difference in tube geometry and forward scatter detectors[12]. In our study we did not add fluorescently labeled agglutinins which may alter the RBC shape. Moreover, we applied very low shear stresses (RBC flow rate : 12 μL/min.) to allow the RBC to rotate in the flow without deformation.

Our results revealed that the RBCs of septic patients have a more spherical shape evaluated either by the spherical index or by the moment.

In physiologic conditions, senescent RBC are less deformable, which is apparently related to alterations in the cytoskeletal configuration, and which, in turn, may play a role in the distribution of integral glycoproteins, leading to their recognition and removal from the circulation[6]. Marikovsky et al[6] demonstrated on ATP-depleted RBCs, as well as on old RBCs, a relationship between the change in RBC shape and alterations in surface charge topography, suggesting that shape-change might influence the surface distribution of negative charges due to SA.

We demonstrated that the RBC SA membrane content is decreased in septic conditions and this is associated with modifications of RBC shape. Non-septic ICU patients had a somewhat lower RBC SA membrane content decrease than volunteers. In fact, this group contains patients with surgery and cerebral hemorrhage. These patients may present a similar inflammatory response even in the absence of infectious complications[18]. Non-septic ICU patients were therefore considered as an intermediate population with alterations in RBC shape and RBC membrane.

It is possible that in sepsis, like in the senescence process, a decrease in RBC membrane SA content may increase the RBC clearance by the reticulo-endothelial system. Simchon et al.[19] demonstrated in rats an increased entrapment, in selective regions (especially in the liver and the spleen) of RBCs treated by neuraminidase to decrease their SA membrane content. Further studies including measurement of sialidase activity are needed to understand the process of SA decrease in septic conditions.

A better understanding of the mechanisms of RBC rheologic alterations in sepsis and its effect on oxygen transport may be important in the development of new therapeutic strategies to improve cell oxygen availability and thereby reduce organ failure in severe sepsis and septic shock.

REFERENCES

1. Mohandas N., Chasis JA. 1993. Red blood cell deformability, membrane material properties and shape: regulation by transmembrane, skeletal and cytosolic proteins and lipids. Semin. Hematol. 30: 171-192.
2. Todd JC., Poulos ND., Davidson LW., Mollitt DL.1993. Role of the leukocyte in endotoxin-induced alterations of the red cell membrane. Am. Surg. 59 :9-12.
3. Goode HF., Webster NR. 1993. Free radicals and antioxydants in sepsis. Crit. Care Med. 21 : 1770-1776.
4. Graf J., Eichelbrönner O., Sibbald WJ. 1999 The red blood cell and nitric oxide. In : Vincent JL (ed). Yearbook of Intensive Care and Emergency Medicine. Springler-Verlag, Berlin Heidelberg New York Tokyo. 465-475.
5. Todd JC., Mollitt DL. 1995. Effect of sepsis on erythrocyte intracellular calcium homeostasis. Crit. Care Med. 23 : 459-465.
6. Marikovsky Y. 1996. The cytoskeleton in ATP-depleted erythrocytes : the effect of shape transformation. Mech. Ageing Dev. 86 : 137-144.
7. Waugh RE. Effects of 2,3 diphosphoglycerate on the mechanical properties of erythrocyte membrane. Blood 1986; 68 : 231-238.
8. Grebe R., Wolff H., Schmid-Schonbein H. 1988. Influence of red cell surface charge on redcell membrane curvature. Pflugers Arch. 413 : 77-82.
9. Rogers ME., Williams DT., Niththyananthan R., Rampling MW., Heslop KE., Johnston DG. 1992. Decrease in erythrocyte glycophorin sialic acid content is associated with increased erythrocyte aggregation in human diabetes. Clin. Sci. 82 : 309-13.
10. Baskurt OK., Gelmont D., Meiselman HJ. 1998. Red blood cell deformability in sepsis. Am. J. Respir. Crit. Care Med. 157: 421-427.
11. Astiz ME., DeGent GE., Lin RY., Rackow EC. 1995. Microvascular function and rheologic changes in hyperdynamic sepsis. Crit. Care Med. 23: 265-271.
12. Rolfes-Curl A., Ogden LL, Omann GM, Aminoff D. 1991. Flow cytometric analysis of human erythrocytes: Possible identification of senescent RBC with fluorescently labeled wheat germ agglutinin. Exp. Gerontol. 26: 327-345.
13. ACCP/SCCM Consensus Conference. 1992. Definitions for sepsis and organ failure guidelines for the use of vasoactive therapies in sepsis. Crit. Care Med. 20 : 864-874.
14. Lowry OH., Rosebrough JJ, Farr AL et al. 1951. Protein measurement with the folin phenol reagent. J. Biol. Chem. 193: 265-275.

15. Anumula KR. 1995. Rapid quantitative determination of sialic acids in glycoproteins by high-performance liquid chromatography with a sensitive fluorescence detection. Anal Biochem. 230: 24-30.
16. Gutowski KA., Hudson JL., Aminoff D. 1991. Flow cytometric analysis of human erythrocytes: I. Probed with lectins and immunoglobulins. Exp. Gerontol. 26: 315-326.
17. Hinshaw LB. 1996. Sepsis/septic shock : participation of the microcirculation : an abbreviated review. Crit. Care Med. 24 : 1072-1078.
18. Gabay C., Kushner I. 1999. Acute-phase proteins and other systemic responses to inflammmation. N.Engl.J. Med. 340: 448-454.
19. Simchon S., Jan KM., Chien S. 1988. Studies on sequestration of neuraminidase-treated red blood cells.Am. J. Physiol. 254: H 1167- H 1171

FIBER OPTIC IMMUNOSENSORS FOR CARDIO-VASCULAR DISEASE DIAGNOSIS: Quantification of Protein C, Factor V Leiden, and cardiac Troponin T in plasma

Hyun J. Kwon, Stephen C. Peiper, and Kyung A. Kang*

1. INTRODUCTION

Cardio-vascular diseases are principal causes of the mortality in the United States. These are closely associated with thrombophilia, and the most prevalent cause of the hereditary thrombophilia is the deficiency or dysfunction of hemostasis related plasma factors. Currently, there is an on-going effort to develop near real-time quantifying methods for the three important molecules related to cardio-vascular diseases in the Department of Chemical Engineering at the University of Louisville. They are: Protein C (PC), Factor V Leiden (FVL), and cardiac Troponin T (cTnT).

PC is a key anticoagulant and anti-thrombotic in plasma at the normal concentration of 4 μg/ml. Activated Protein C (APC) achieves its anticoagulant role by inactivating Factor V (FV) and Factor VIII. The heterozygote (less than 60% of the normal level) has an 8-10 fold higher risk of the thrombotic complications by the age of 40 years.[1] Currently, PC deficiency is diagnosed by enzyme linked immunosorbent assay (ELISA), which takes at least 6 hours to complete.

Individuals with the normal PC level may also develop thrombosis if they have FVL (or APC resistance). FV is present in the blood plasma at the concentration of 4-14 μg/ml. Activated FV serves as a procoagulant in the blood coagulation pathway but gets inactivated by APC. A single point mutation (G_{1691}➌A) in the FV gene leads to an amino acid substitution (Arg_{506}➌Glu) at the APC cleavage site, forming Factor V Leiden (FVL). FV and FVL are equally well activated by thrombin but *FVL does not get inactivated by APC*, while FV does.[2] FVL is the most prevalent cause of hereditary thrombophilia, accounting for 21-40% of individuals with thrombosis.[3] Current

* Hyun J. Kwon and Kyung A. Kang, Department of Chemical Engineering, University of Louisville, KY 40292. Stephen C. Peiper, James Graham Brown Cancer Center, University of Louisville, Louisville, KY 40202.

diagnostic methods are limited to DNA analysis (long, complicated, and expensive) or clotting time assay (not specific to FVL).

cTnT is found in cardiac muscles and is released into blood when cardiac muscle cells are damaged, frequently due to the insufficient oxygen supply. Therefore, cTnT is a highly sensitive marker in plasma following cardiac muscle damage or acute myocardial infarction. cTnT level more than 1.5 ng/ml usually shows 3-4 times higher mortality rate with acute myocardial infarction.[4] cTnT amount has been measured by ELISA and the commercially available biosensors, Elecsys (Roche Diagnostics, Inc.)[5] and TropT® (Roche Diagnostics, Inc.).[6] However, ELISA requires several hours to perform; Elecsys is rather expensive ($ 50,000-100,000); and TropT® is not quantitative.

The biosensor under development performs a sandwich type immunoassay on an optical fiber with high sensitivity and selectivity.[7] One type of monoclonal antibody against the analyte (1° mAb) is immobilized on the fiber sensor surface. For an assay, a sample is applied to the sensor chamber allowing the analyte to bind to 1° mAb. Next, another type of monoclonal antibody (2° mAb) linked with the fluorophore (Cy5) is applied to react with the bound analyte. Laser is applied to excite the fluorophore and the emitted fluorescence is measured to correlate with the analyte concentration. Advantages of this biosensor include the short assay time (~15 minutes), high portability, easier operation than ELISA or DNA analysis, and inexpensive cost per assay.

In this paper, the current developmental status of the three biosensors are described: (1) the sensing performance of the PC biosensor in the viscous plasma was studied; (2) monoclonal antibodies are being developed as the reagents for a FVL biosensor; and (3) the feasibility study for a cTnT sensor was performed.

2. MATERIALS AND METHODS

2.1. Sensor Preparation for PC and cTnT Detection

Optical fibers were purchased from Research International (Woodinville, WA). Either 6 cm (for PC sensing) or 12.5 cm (for cTnT sensing) of fiber was decladed and used for antibody immobilization. For the PC sensor, pure PC and two different types of monoclonal antibodies were donated by the American Red Cross (Rockville, MD). The 1° mAb (MAb 7D7) was biotinylated and immobilized on the fiber as described by Spiker and Kang.[7] The 2° mAb (MAb 8861) was linked with Cy5 (Amersham-Pharmacia). For the cTnT sensor, cTnT and two monoclonal antibodies were donated by Roche Diagnostics, Inc. (Switzerland). Anti-cTnT M7-IgG (1° mAb) was biotinylated and immobilized on the fiber and Anti-cTnT M11.7-IgG (2° mAb) was linked with Cy5. The cTnT sensor was constructed, following the same method as the PC sensor.

2.2. Biosensor Assay Procedures

For the static incubation, the measurement was performed as described by Spiker and Kang[7] with the incubation times of 5/3 minutes (PC/2° mAb) for PC sensing and 10/10 (cTnT/2° mAb) minutes for cTnT sensing. For the convective flow study, the sample was applied through the sensing chamber by a peristaltic pump (Ismatec) at a predetermined flow rate. The incubation times were 3/3 minutes for PC sensing and 10/10 minutes for cTnT sensing. All PC and cTnT samples were prepared in human serum albumin (HSA) solution (103 mg/ml) to emulate plasma viscosity. Our previous studies show that this solution behaves very similar to the plasma but is less expensive.

3. RESULTS AND DISCUSSION

3.1. PC Biosensor

In the previous PC sensor study, we have proven that viscous plasma samples produced less signal intensity due to the decrease in PC molecule diffusion to the fiber surface.[8] One way of enhancing mass transport rate of molecules in a viscous medium to the fiber surface was thought to be adding the convective transport mechanism. Therefore, sensing performance of the PC sensor was examined at various flow rates. The PC and 2° mAb were introduced to the chamber using a peristaltic pump at the flow rate of 0.1 and 0.3 ml/min. The sample incubation time was reduced from 5 (static) to 3 minutes (convective) to minimize the sample volume. Figure 1 shows that, compared to the stagnant sensing (⊖), at the flow rate of 0.1 ml/min (⦶) the signal intensity increased by 50-100%, and at the flow rate of 0.3 ml/min (**O**) by 130-240%. This result confirms that convection plays the main role in increasing the signal intensity. Now, this PC sensor is capable of quantifying the PC amount at the deficiency level (0.5-2.5 μg/ml) in viscous plasma within 10 minutes with high signal to noise ratio.

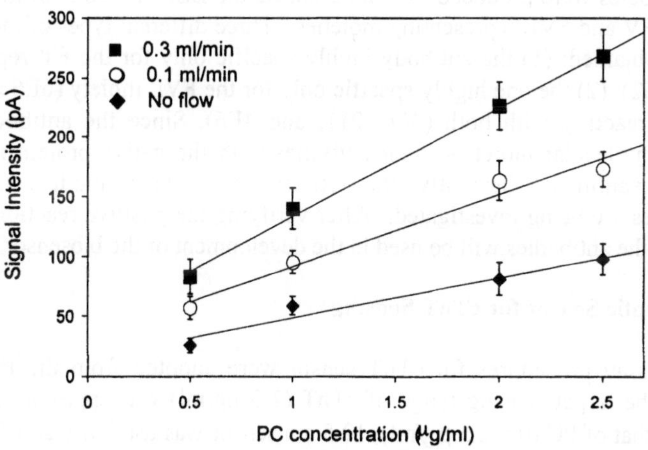

Figure 1. Signal intensity of the PC biosensor with static and convective sample application. For convective flow study, the PC samples and 2° mAb were applied at the flow rate of 0.3 and 0.1 ml/min. [Experimental conditions: 6 cm fiber; 5/3 min incubation for static conditions; 3/3 min incubation for convective flow study].

1.2. Antibody Development for FVL and FV Sensing

In order to develop a biosensor for the FVL diagnosis, it is critical to obtain the antibodies that differentiate FVL from FV because their amino acid sequence differs only by one. Therefore, we attempted to generate unique antibodies that recognize only FVL and also only FV. Two different molecular moieties (proprietary information) that

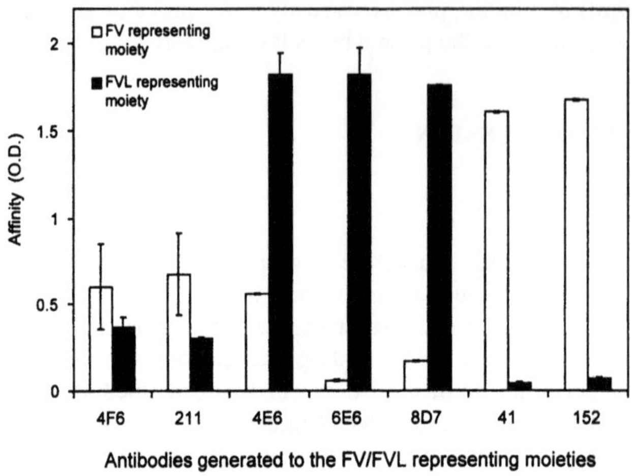

Figure 2. The activity of antibodies to FV and FVL representing moieties.

contain the mutation site of FVL and the corresponding region of FV were developed and injected to mice. Then hybridoma cell lines were generated from the mice, and monoclonal antibodies were produced. Figure 2 shows the ELISA results of monoclonal antibodies to the FV and FVL representing moieties. Three different types of monoclonal antibodies were obtained: (1) the antibody highly specific only for the FV representing moiety (41 and 152); (2) the one highly specific only for the FVL moiety (6E6 and 8D7); and (3) the one reacting with both (4F6, 211, and 4E6). Since the antibodies were generated to the molecular moieties, their activities with the native protein molecules should be also examined. Currently, the activities between antibodies and native FV/FVL molecules are being investigated. After verifying the positive reaction with the native molecules, the antibodies will be used in the development of the biosensor.

3.3. The Fiber Optic Sensor for cTnT Sensing

The initial assay procedures for cTnT sensor were adopted from the PC sensor. However, since the target sensing range of cTnT (1-3 ng/ml) was approximately 1000 times lower than that of PC (0.5-2.5 μg/ml), 12.5 cm sensor was used instead of 6 cm and the incubation times were extended from 5/3 to 10/10 minutes. The sample and 2° mAb were applied either with static incubation or convective flow (Figure 3). With the static condition, the signal intensity was approximately 40 pA at 3 ng/ml of cTnT, and the signal for the sample with 1 ng/ml was extremely low (Figure 3). As in the study with the PC sensor, to enhance the sensing efficiency, during cTnT and 2° mAb incubations, convective flow was applied to the sensing chambers at the flow rate of 0.3 ml/min. With the convective flow, the signal intensity was doubled. This result shows that the fiber optic sensor is capable of quantifying cTnT level in the range of 1-3 ng/ml within 25 minutes, sufficient to detect the threshold concentration (1.5 ng/ml).

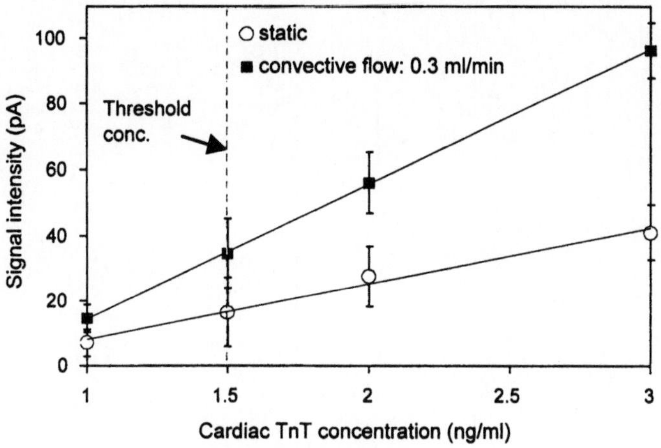

Figure 3. The performance of the fiber optic cTnT sensor under the static and convective flow incubation. [Experimental conditions: 12.5 cm fiber sensors; 10/10 min incubation for cTnT/2° mAb]

CONCLUSIONS

Fiber optic sensors are being developed for PC and cTnT detection. The PC sensor is capable of detecting PC in the range of the PC deficiency (0.5-2.5 μg/ml). To develop a biosensor for FVL diagnosis, monoclonal antibodies are being developed to quantify the amount of FV and FVL. In the feasibility study, the cTnT sensor is capable of quantifying the cTnT level (1-3 ng/ml) in plasma, discriminating the abnormal from the normal. The fiber optic sensors under development may provide a means of significantly faster and user-friendly sensing for three different molecules that are extremely important for proper oxygen transport in human body. The authors acknowledge the National Science Foundation CAREER Award (BES-9733207).

REFERENCES

1. Markis, M., Rosendaal, F. R., and Preston, F. E., Familial thrombophiloa:genetic risk factors and management, *J. Internal Medicine* **242** (Suppl. 740), 9-15 (1997).
2. Kalafatis, M., Bertina, R. M., Rand, M. D., and Mann, K. G., Characterization of the molecular defect in factor VR506Q, *J. Biol. Chem.* **270** (8), 4053-4057 (1995).
3. Appleby, R. D. and Olds, R. J., The inherited basis of venous thrombosis, *Pathology* **29**, 341-347 (1997).
4. Collinson, P., Stubbs, P., Moseley, D., Oakes, M., Greenwood, T., and Noble, M., Measurement of cardiac troponin T for risk stratification in unstable angina, *Clin. Chem.* **41**, 165 (1995).
5. Forest, J. C., Masse, J., and Lane, A., Evaluation of the analytical performance of the Boehringer Mannheim Elecsys 2010 immunoanalyzer, *Clin. Biochem.* **38**, 386-393 (1998).
6. Muller-Bardorff, M., Sylven, C., Rasmanis, G., Jorgensen, B., Collinson, P. O., Waldenhofer, U., Hirschl, M. M., Laggner, A. N., Gerhardt, W., Hafner, G., Labaere, I., Lenberger, R., Zerback, R., and Katus H. A., Evaluation of a point-of-care system for quantitative determination of troponin T and myoglobin, *Clin Chem. Lab. Med.* **38** (6), 567-574 (2000).
7. Spiker, J. O. and Kang, K. A., Preliminary study of real-time fiber optic based protein C biosensor, *Biotech. Bioeng.* **66** (3), 158-163 (1999).
8. Kwon, H. J., Balcer, H. I., and Kang, K. A., Protein C biosensor sensitivity for biological samples and sensor reusability, *Comparative Biochemistry and Physiology Part A*, In press.

SOFT GEL CHROMATOGRAPHY COLUMN ANALYSIS AND DESIGN FOR THE PRODUCTION OF HIGH MOLECULAR WEIGHT BLOOD FACTORS: MODEL MOLECULE PROTEIN C

Weiwei He and Duane F. Bruley*

1. INTRODUCTION

The purpose of this investigation is to examine how the mechanical characteristics of chromatographic resins influence the biomechanical design of chromatography columns. It is crucial for preparative chromatography that column scale-up takes into consideration the collapse potential of the gels because of the cost and inconvenience of such an episode. It is desirable to minimize process cycle-time considering resin mechanical properties to achieve economical preparative scale production of a desired protein. It is also true that a resin can have desirable mechanical characteristics but the surface chemistry for adsorption of the desired product might be inappropriate for the separation and purification of a specific product. Therefore, it is important to consider both the biomechanical and the biochemical characteristics of the resin systems to attain optimal process design and performance for reduced production cost.

The primary gel tested in this investigation is Sepharose 2B, which consists of Agarose, and dextrans. The beaded agarose matrix shows a low non-specific binding and cross-linked forms show excellent chemical and physical stability. Increasing agarose concentration decreases the matrix porosity, thereby altering the void fraction range while increasing rigidity.

The mechanics of flow through porous media is important in many scientific and engineering applications. When fluid flows through a porous, deformable medium there is a coupling between the flow and the deformation of the medium, which may lead to various interesting phenomena. For example, the strain distribution in a porous, deformable matrix under steady mechanical compression is uniform, but if the strain is

* University of Maryland, Baltimore County College of Engineering, ECS 314, 1000 Hill Top Circle, 21250 Baltimore, MD, USA

produced by fluid flow, the strain distribution can be highly non-uniform [1]. In general, the nature of the flow and resultant strain distribution depend upon the properties of the porous matrix of the fluid and their interactions as well as the boundary conditions, geometry, and applied pressure [2]. The study of flow through the gel particles has general application in the area of chromatography column design [3, 4].

This work is to establish theoretical and experimental relationships for the understanding of column behavior and the scale-up of Immobilized Metal Affinity Chromatography (IMAC) and Immuno Affinity Chromatography for the low cost production of large quantities of Protein C (model molecule). Gel matrices for Protein C [5] separation are being analyzed to determine the relationship between the mechanical characteristics of the gel, the flow conditions and the column collapse potential.

2. MATERIALS AND METHODS

Due to availability, the primary gel tested was Sepharose 2B laced with Protein C (PC) Monoclonal Antibody (Mab) donated by the American Red Cross (ARC) (This gel had previously been used for Protein C separation).

Necessary experimental studies were performed to determine flow rate and pressure drop correlations for a chromatographic column (MT 20 column), packed with Sepharose 2B. The fluid (water) flow rate and pressure drop (ΔP) was examined to determine how they influence column stability. Gel packing analysis includes two considerations; one is bulk compression, and the second is gel particle deformation.

Column packing

Specific MT 20 columns, 15mm diameter and 130mm length, were purchased from the Bio-Rad company. The gel slurry was equilibrated with water until all ethanol was removed. The gel was then placed in a flask under vacuum for thirty minutes. Then, the degassed gel was flowed in small quantities into the column, using a pipette, to a height of about 80mm. The column was allowed to settle until there was no further reduction in gel height. This step was done overnight, before doing a flow rate vs. pressure drop experiment, to allow the gel matrix to achieve most of the expected bulk compression.

Experimental measurements

A packed column was connected to a triaxial pressure panel and cell (manufactured by the Karo-Warner company). First, measurements were made at pressure increments starting from a minimum gauge pressure drop of zero psi. Using 5 psi increments, the gauge pressure drop reached a maximum (100psi) after which the pressure drop increments were reversed and data were recorded at 5 to 10 psi increments back to zero. The pressure drop across the column was read from a digital pressure gauge on the control panel. After each pressure increment the column was allowed to achieve steady state and then the time to flow 20ml (sometimes 10 ml) of water through the column matrix was recorded. The small range of pressure from 1psi to 10psi was also measured at 1 psi interval to get the permeability for the steady solution. The water flowed directly through a tube at the bottom of the chromatography column and into a burette at atmospheric pressure.

3. THEORETICAL MODEL

Biot's Consolidation Theory [6] (-a mathematical description of three-dimensional consolidation in a porous, elastic media-) was reduced to a single dimension and time to develop a set of systems equations for the microscopic theoretical analysis of the column. These equations describe the pore pressure distribution and the solid skeleton displacement in the porous media (gel).

Neglecting the initial effective stress and body forces and assuming a one-dimensional, time-dependent process along the axial z direction, the governing equations for the solid and fluid respectively are:

$$(\lambda + 2G)\frac{\partial^2 u}{\partial z^2} - \frac{\partial p}{\partial z} = 0 \tag{1}$$

$$\frac{\partial}{\partial z}\left[\frac{k}{\mu}\left(\frac{\partial p}{\partial z}\right)\right] = (1-\phi)C_s\frac{\partial p}{\partial t} + \frac{\partial^2 u}{\partial t \partial z} \tag{2}$$

A central finite difference for the second order variable and forward finite difference for the first order variable was utilized to obtain the equation solutions for the gel deformation (mm) and fluid pressure drop (psi). Solutions have been obtained for both the process steady state and the dynamic state. The solid displacement equation (1) and the fluid equation (2) are simplified to allow steady-state solutions.

$$(\lambda + 2G)\frac{d^2 u}{dz^2} = \frac{dp}{dz} \tag{3}$$

$$\frac{d}{dz}\left[\frac{k}{\mu}\left(\frac{dp}{dz}\right)\right] = 0 \tag{4}$$

Boundary conditions:

$$p(L,j)=\Delta p \ @\ z=L; \tag{5}$$

$$u(0,j)= 0 \ @\ z=0; \tag{6}$$

$$\frac{\partial u}{\partial z}(L,j) = 0 \ @\ z=L; \tag{7}$$

$$p(0,t) = 0 \ @\ z=0 \tag{8}$$

Initial conditions:

$$p(l,0) = 0 \ @ \ t=0 \qquad (9)$$

$$u(l,0) = 0 \ @ \ t=0 \qquad (10)$$

Experimental results

The average values for five tests of the target resin, Sepharose 2B in a MT 20 column were calculated. A plot shows that flow rate has a maximum value when starting from a minimum pressure drop and continuing to a maximum pressure drop. The results show that there is a hysteresis phenomenon resulting from the gel characteristics. The flow rate is consistently lower when starting from the maximum pressure drop. This hysteresis loop indicates that the gel behavior changes from elastic to a pseudo-plastic at high-pressure drops. This indicates that because of water and reabsorption over time the resin matrix has a viscous-elastic behavior.

After the flow rate was measured, Darcy's law was used to calculate the permeability.

Computer simulations

Steady-state solutions

The steady-state simulation was carried out for Sepharose 2B. It was found that the pressure drop decreases more near the chromatography column end. The single measurement for displacement at the top of the gel matrix compared well with the model solution.

Dynamic solutions

The first layer of the gel matrix was chosen to test the dynamic solution. The other layers showed a similar trend to the first layer. The steady-state time was chosen to simulate the pressure and displacement along the gel matrix column. It was shown that the pressure drop reaches a steady-state value of about 4 psi after approximately thirteen seconds for the first layer of the gel matrix. It was determined that displacement reaches 0.002 m after approximately twenty seconds which fits with the experimental data measured in steady-state.

4. CONCLUSIONS

The main purpose of this investigation was to study the optimal design of chromatography columns from a *biomechanical* perspective.

This investigation employs both experimental and theoretical analysis to determine chromatographic column design to avoid " column collapse".

After considerable research Biot's Consolidation Theory was selected for the theoretical modeling and simulation studies. Even though Biot's approach was developed primarily for soil analysis it was considered suitable for application to the flow and solid mechanics involved in the flow through the resin bed in a chromatographic column.

Darcy's law was selected for the column macro analysis since the Reynold's number was calculated to be 2.46*10-6.

The fact that the maximal flow rate of Sepharose 2B yielded a non-zero value (approximately 10psi) prompted us to establish a concept of "Pseudo Collapse" point; an unexpected and important result of the experimental studies.

Both steady and dynamic model have been used to simulate events in the macro and micro level of the gel matrix. The analytical solutions of the steady model are in agreement with the experimental results. The dynamic model has effectively simulated the micro behavior of the gel matrix inside the column. The steady-state solution of the dynamic model is almost the same as the steady solutions. Therefore, Biot's Consolidation Theory can be used as a good mathematical model for chromatography column analysis and design.

An important conclusion of this work is that Darcy's Law can be applied to soft-gel chromatography column design since the constitutive relationship is an integral component of Biot's Consolidation Theory. The application of Darcy's Law to column design could help with the low cost production of Protein C, the pivotal anti-coagulant in the human coagulation cascade.

ACKNOWLEDGEMENTS

We want to thank NSF Grant Number CTS-0090749, NSF Grant Number CTS-9904465, the Hillis-Carnes Corporation, the American Red Cross, and the Whitaker Foundation for financial and material support on this research.

REFERENCES

1. C. G. Caro, M. J. Lever, G. E. Morgan, K. H. Parker and C. P. Winlove. Studies of Transport in Deformable Porous Material With Relevance to Connective Tissue properties. J. Physical. 358:3-7 (1984).
2. K. H. Parkar, R. V. Mehta and C. G. Caro. Steady Flow in Porous, Elastically Deformable Materials. Transaction ASME. 54:794-796 (1987).
3. M. Rashidi, L. Peurrung, A. F. B. Tompson & T. J. Kulp. Advance in Water Resources, 19:163-180 (1996).
4. S. Rodriguez, C. Romero, M. L. Sargenti, A. J. Muller and A. E. Saez. J. Non-Newtonian Fluid Mechanics, 49:63-85 (1993).
5. D. F. Bruley, and W. N. Drohan. Protein C and related anticoagulants. Gulf Publishing Company (1990).
6. M. A. Biot. Theory of Elasticity and Consolidation for a Porous Anisotropic Solid. J. Appl. Phys., 26:182-185 (1955).

DEVELOPMENT OF AN ECONOMIC MINIANTIBODY PRODUCTION PROCESS FOR THE PURIFICATION OF PROTEIN C (ANTI-COAGULANT/ANTI-THROMBOTIC)

Lino K. Korah, Doh G. Ahn, and Kyung A. Kang*

1. INTRODUCTION

Protein C (PC) is a vital anticoagulant and antithrombotic in the blood coagulation cascade. Because PC is activated only when it is needed, it does not cause internal bleeding complications as does heparin or coumadin. PC deficiency can lead to blood clot formation in blood vessels, which can eventually lead to heart attack, stroke, skin necrosis, or even death. About 1,400 people worldwide currently die each day of septic shock and recently, PC has also been proven highly effective for septic shock treatment.[1] Therefore, an inexpensive supply of PC is important for PC deficient patients, sepsis patients, and those undergoing major surgeries.

Blood plasma contains several proteins homologous to PC and many of these are coagulants. Hence, traditional purification methods other than immuno-affinity chromatography cannot be employed for PC purification. Since current methods of PC purification usually use monoclonal antibodies (MAbs) produced by animal cells, it is very expensive. A new approach is to use single chain variable region of antibody molecule (mini-antibodies), which is the smallest fragment (MW=30kD) with a complete antigen binding function.[2] Mini-antibody (mini-Ab) consists of a heavy chain (V_H) and a light chain (V_L) of the variable region of the antibody and they are connected by a linker. Mini-Ab can be produced by recombinant *E. coli* and is secreted into the liquid culture medium.[2] Advantages of producing mini-Abs using *E. coli* over regular MAbs include less contamination problems during the production, less shear sensitivity of the host microorganism, less expensive, and easier scale-up.[3]

In this paper, an economic analysis for the production of the PC mini-Abs using *E. coli* is presented. Also, results of some of the optimization efforts in culture conditions are analyzed. The effect of the mini-Ab concentration on its dimerization is demonstrated.

---

*Lino K. Korah, Doh G. Ahn and Kyung A Kang, University of Louisville, Louisville, KY 40292

2. MATERIALS AND METHODS

Thirteen recombinant *E. coli* colonies producing mini-Abs against PC molecule (PC mini-Ab) by phage display technique were obtained from Dr. Michael Sierks at the Arizona State University.[2] One colony that showed the highest production, E4FX, was selected for further studies. The supernatant of E4FX culture broth contained approximately 200 μg/ml of the mini-Ab. The *E. coli* was grown in 2XTY media (*cell growth media*) with 1% glucose at 37°C in a shaking incubator (New Brunswick Scientific; Edison, NJ) at 300 rpm. When the optical density (O.D.) of the media reaches to 0.9, cells were separated using a centrifuge (Fischer Scientific; Pittsburgh, PA). They were then added to a new 2XTY media (*induction media*) containing 0.1% glucose, and the inducer 1 mM isopropyl β-D-thiogalactopyranoside (IPTG; ICN Biomedical, Inc.; Irvine, CA). They were cultured at 30°C in the incubator at 300 rpm for 24 hrs. During this period PC mini-Ab is produced. The cells were then separated using a centrifuge and the supernatant was concentrated to a desired concentration by an ultrafiltration (molecular cut-off of 10,000; Fischer Scientific; Pittsburgh, PA) using a centrifuge at 4000 rpm for 20 minutes. The concentration of the mini-Ab was measured by ELISA.

3. RESULTS AND DISCUSSION

3.1. Production Cost Analysis

Currently, as previously stated, purification of PC from blood plasma is performed using MAbs produced by animal cell culture. Therefore, a preliminary mini-Ab production cost is estimated compared to MAbs for the feasibility of its usage for the PC purification. The cost for screening the cell lines and for the antibody purification are assumed to be approximately the same for the MAb and the mini-Ab. In this estimation, the cost difference in bioreactors, the labor, and utility are not included. For the mini-Ab production, it takes two days to complete one batch culture. However, for MAb production it takes at least 4-6 days.[3] In our experiments, an average of 200 mg/l of the

Table 1. The cost comparison for the mini-Ab from *E .coli* and the MAb from animal cell culture

E. coli		Animal Cell	
Media	Cost / liter ($)	Media	Cost / liter ($)
2XTY	3.45	DMEM[a]	27.50
IPTG	1.82	Calf (bovine) serum	30.00
Ampicillin	0.59	Antibiotic	2.68
Glucose	0.23		
PMSF	1.58		
Total cost of medium/L	7.66	Total cost of medium/L	60.18
Mini-Abs produced/L	200 mg	MAbs produced/L	100 mg
Production rate (mole/day·dollar)	**4.35×10^{-7}**	**Production rate (mole/day·dollar)**	**2.8×10^{-9}**

[a] Dulbecco's modification of Eagle's Media

mini-Ab was obtained. For MAbs, the average production rate of 100 mg/l was reported.[4] The molecular weight of mini-Ab (MW=30kD) is five times less than that of MAb (MW=150kD). A rough estimation of the production rate of mini-Ab is 4.35×10^{-7} mole/$.day. For MAb, it is 2.8×10^{-9} mole/$.day, which is approximately 150 times higher. For this estimation, the affinities of both antibodies are assumed to be the same. Also, the number of available binding sites for MAb is assumed to be only one since the other site may not be available due to the steric hinderence once the one site is occupied.

3.2. Glucose Effect on Mini-Ab Production

In the production of the PC mini-Ab, two different media are used sequentially: the *cell growth media* (for the *E. coli* growth) and *induction media* (for the mini-Ab production). *E. coli*'s glucose metabolism produces acetate, and the presence of acetate can affect the mini-Ab production.[5] After the cell growth stage, the mini-Ab was produced in the *induction media* containing three different glucose concentrations, and also without changing media from the *cell growth media* (Figure 1). When 0.1% glucose concentration was used, the PC mini-Ab production level was the highest. When the glucose was not contained in the induction media, the mini-Ab concentration was decreased by 17%. Without glucose, IPTG is used as the substrate for cell growth and maintenance[6] and, therefore, the decrease in the mini-Ab production. At the concentration of 1% and when the media was not changed from the *cell growth media* to the *induction media*, the production levels were almost zero, possibly due to the presence of acetate in the media. From this observation, the glucose concentration 0.1% was chosen as the optimum concentration.

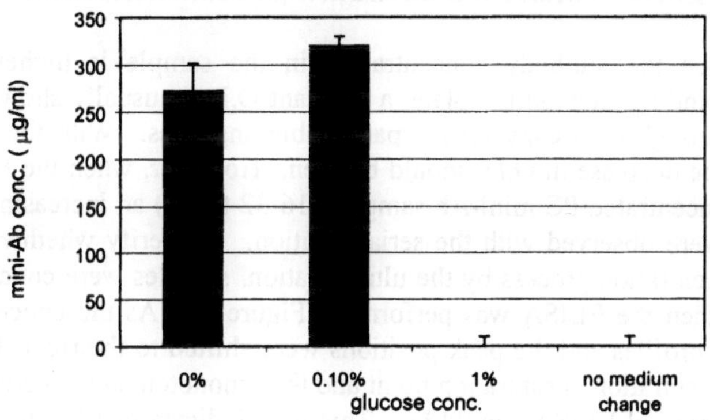

Figure 1. Effect of glucose in the *induction media* on the mini-Ab production.

3.3. IPTG Effect on Mini-Ab Production

The most expensive component in the mini-Ab media is IPTG. Hence, the IPTG concentration in the *induction media* was varied from 0.05 mM to 2 mM (Figure 2). At 0.1 and 1 mM of IPTG, the production of the mini-Ab was almost the same. However, at 0.05 mM and 2 mM of IPTG concentrations, there were approximately 44% of reductions in the mini-Ab production. Since IPTG is needed as an inducer for the PC mini-Ab production, it is natural that lower production was observed at lower

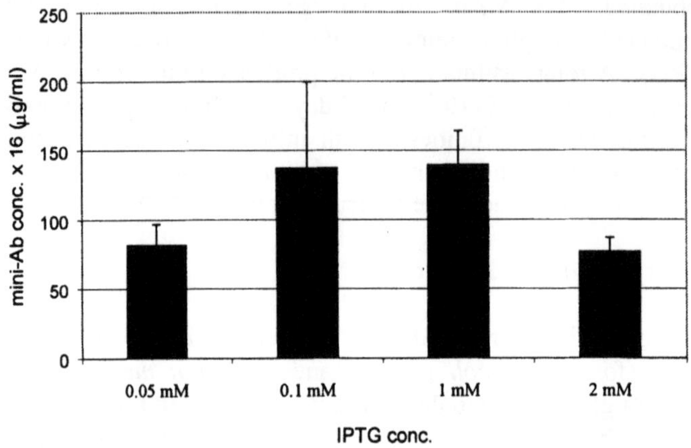

Figure 2. Effect of IPTG (inducer) on the mini-Ab production.

concentration. The reason for the lower production at higher IPTG concentration is currently being investigated.

3.4. Concentration Effect on Mini-Ab Dimerization (or Multimerization)

In ELISA, when the antibody concentration in the sample is higher than the maximum antigen binding sites on the plate, a constant O.D. is usually shown until the sample is diluted enough to occupy only a part of binding sites. With the increasing dilution, a monotonic decrease in O.D. should be seen. However, when the ELISA was performed using concentrated PC mini-Ab samples (16-32 times) an increasing and then decreasing O.D.'s were observed with the serial dilution. To verify whether this trend was due to the concentration process by the ultrafiltration, samples were concentrated at various ratios and then the ELISA was performed (Figure 3). As the concentration is increased, the O.D. profiles and the peak positions were shifted to the right. For a fresh sample without concentration, a saturation point and then monotonically decreasing O.D. is obtained, as the normal ELISA should be. This may indicate that the high mini-Ab concentration possibly causes the monomer to change to the dimer or multimer.[7] Usually the dimeric mini-Ab has stronger affinity for antigen than monomers.[7] However, for ELISA, the primary antibody (9E10) binds to the c-Myc part of mini-Abs, which is very close to the region of dimerization. Since the dimers usually have higher affinity for PC, it competes with monomer to bind to PC on the ELISA plate. If the dimer or multimer is bound to PC on the ELISA plate, the c-Myc part may not be accessible for 9E10. Hence, there may be a decrease in ELISA signal intensity at higher concentration.

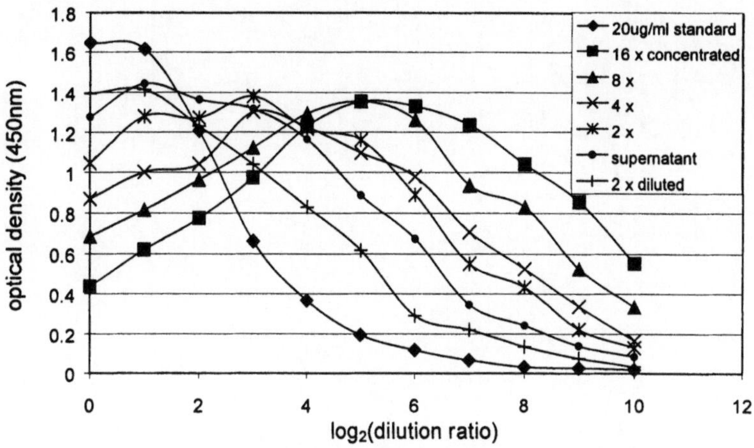

Figure 3. Concentration effect on dimerization of the mini-Ab.

CONCLUSIONS

Purified PC has very important applications in treating various thrombotic complications and also sepsis. A significant part of the PC purification cost is from the expensive regular MAb produced by animal cells. The PC purification cost may be reduced by using an alternative form of cheaper antibody, the mini-Ab. The production cost for the mini-Ab was estimated to be approximately 150 times less than that of the regular MAb. Production of the PC mini-Ab was increased in the lab scale using a lower glucose concentration (0.1%) in the *induction media*. The cost of production was reduced by using 0.1 mM of IPTG instead of 1 mM, which was previously used. It was also found that the mini-Ab appears to dimerize or multimerize at higher concentration. From our preliminary study results, the mini-Ab production appears to be an economically feasible process. The authors would like to thank The American Red Cross for the material support and Dr. M. Sierks at the Arizona State University for the supply of 13 *E. coli* colonies producing the PC mini-Abs.

REFERENCES

1. Eli Lilly, 2001, Indiana, (August, 2001); www.lilly.com.
2. Wu, H., et al., Artificial antibodies for affinity chromatography of homologous proteins: application to blood proteins, *Biotechol.* Progr. **14**: pp. 496-499 (1998).
3. Freshney, I. R., *Culture of Animal Cells. A manual of basic Technique*, 4th Edition (Wiley-Liss, New York, 2000), pp. 387-390.
4. Karu, A. E., Bell, C.W. and Chin, T. E., Recombinant Antibody Technology, *ILAR J* **37**: pp. 127-132 (1995).
5. Akesson et al., On Line detection of acetate formation in *Escherichia coli* cultures using dissolved oxygen responses to feed Transients, *Biotech Bioeng.* **64** (5): pp. 590-598 (1999).
6. Cohen, G.N. and Monod, J., Bacterial permeases, *Bacteriol. Rev.* **2**: pp. 169-194 (1957).
7. Arndt, K. M., et al., Factors influencing the dimer to monomer transition of an antibody single-Chain Fv fragment, *Biochemistry* **37**(37), 12918-12926 (1998).

NITRIC OXIDE SYNTHESIS IN BRAIN IS STIMULATED BY OXYGEN

Stephen R. Thom and Donald G. Buerk*

1. INTRODUCTION

The purpose of this investigation was to determine the impact of elevated partial pressures of O_2 on the steady state level of nitric oxide ($^{\bullet}NO$) in the central nervous system. Previous investigations have inferred that elevations in brain O_2 tension alter concentrations of $^{\bullet}NO$. However, conclusions were based on measurements of the products produced as a consequence of $^{\bullet}NO$ production, rather than direct measurements, and the results have been contradictory (1-4).

Interest in this area has been fueled by studies suggesting that elevated $^{\bullet}NO$ concentrations play a role in O_2 toxicity. Rats exposed to extreme pressures of oxygen (5 atmospheres absolute, ATA) were found to have increased intrasynaptosomal free Ca^{++} concentrations which secondarily increased NOS activity (2). NOS activity was assessed indirectly as cyclic GMP level. The occurrence of seizures, a manifestation of O_2 toxicity, was inhibited by a NOS inhibitor and also by a Ca^{++} channel blocker. More recently, Elayen *et al.* (1) reported four- to six-fold elevations in nitrite and nitrate in microdialysate perfused through the brains of rats exposed to 3 ATA O_2. They measured the K_m for O_2 of neuronal NOS (nNOS), found it to be 260 mm Hg, and implied that nNOS activation in brain may be a direct biochemical consequence of elevated O_2 tension. In contrast to these results, Demchenko *et al.* (3,4), reported a ~ 36 % decrease in nitrite and nitrate concentrations in brain microdialysate when rats were exposed to 5 ATA O_2. They proposed that $^{\bullet}NO$ was depleted in brain due to reactions with superoxide radicals ($O_2^{\bullet}$).

2. METHODS AND MATERIALS

Wistar male rats were anesthetized with ketamine (83 mg/kg ip) and xylazine (11 mg/kg ip), the scalp was removed, a 4 mm hole was drilled through the skull, and in an adjacent area the skull bone was thinned but not completely removed. A second dose of

* Departments of Emergency Medicine and Physiology, and the Institute for Environmental Medicine, University of Pennsylvania, Philadelphia, PA 19104-6068.

Oxygen Transport to Tissue XXIII, edited by D. Wilson *et al.*
Kluwer Academic/Plenum Publishers, 2003

ketamine/xylazine amounting to 3/4 of the initial dose was then given, and the animal placed in a hyperbaric chamber. Electrodes were placed on micromanipulator arms above the head and slowly advanced through the dura and ~1 millimeter into the cerebral cortex. In the adjacent area where skull bone had been thinned, the laser-Doppler flowmeter (a 3 mm probe attached to Model BLF 21, Transonic Systems, Ithaca, NY) was fixed using another manipulator arm. The hyperbaric chamber used in this study was rated for a maximum pressure of 3.0 ATA and has been described in a prior publication (5). Once in the closed chamber, animals were monitored for approximately 30 minutes until electrode and cerebral blood flow recordings became stable. During this time pure, compressed air was flowed through the chamber to remove exhaled gases, but no additional pressure applied. Where specified, rats were injected with 7-nitroindazole (12 mg/kg ip), n-acetylcysteine (40 mg/kg ip), or geldanamycin (0.3 mg/kg ip) 30 minutes prior to pressurization. Others received bovine erythrocyte (copper-zinc) superoxide dismutase [SOD] (25,000 U/kg) intravenously immediately prior to pressurization.

3. RESULTS

Elevation in brain •NO and O_2 concentration with hyperbaric oxygen: The dose-response relationship between O_2 partial pressure and brain •NO concentration is shown in a representative experiment (Figure 1). Anesthetized rats with •NO- and O_2- selective microelectrodes in their brains were placed in a hyperbaric chamber and monitored while breathing air for approximately 30 minutes to establish baseline values. The chamber gas flow was changed from air to 100% O_2 and monitored at 1 ATA O_2, followed by pressurization to 2 ATA, and then 2.8 ATA with pure O_2. There were significant elevations in •NO concentration in response to hyperoxia. The effect of pressure was assessed by changing the gas flow at 2.8 ATA from pure O_2 to a mixture of nitrogen and just 7.46 % O_2. As shown in Figure 1, this caused both the •NO level and cerebral blood flow to decrease, demonstrating that pressure *per se* was not responsible for the changes observed. The elevations in brain •NO associated with changes in O_2 tension could be fit to Michaelis-Menten kinetics (Figure 2). The apparent K_m for O_2 was 550 μM.

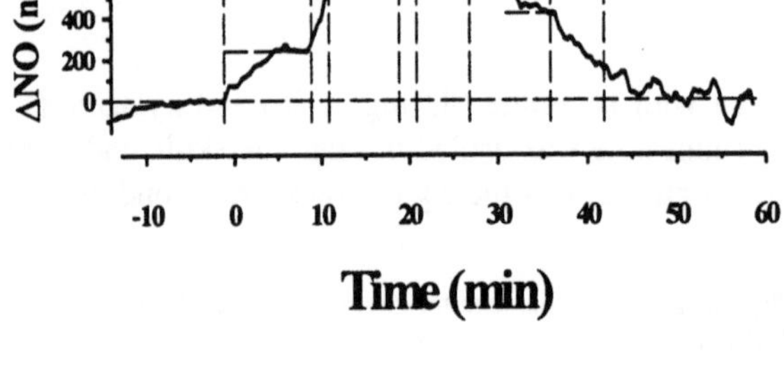

Figure 1. Representative experiment showing results of a rat with implanted O_2 and •NO-specific electrodes. Elevations in brain O_2 tension and •NO concentration occurred in response to changing the breathing gas from air to 100% O_2 at ambient pressure, to 100% O_2 at 2.0, to 2.8 ATA, to 7.46% O_2 at 2.8 ATA, and then back to air at ambient pressure. Narrow intervals between vertical dashed lines indicate times when O_2 partial pressure was changed.

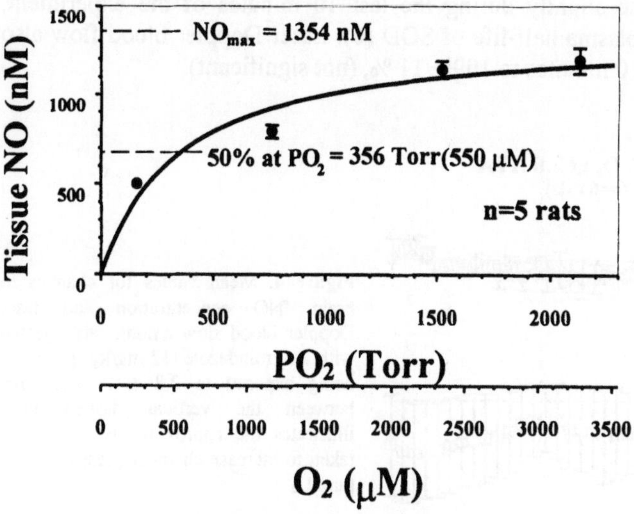

Figure 2. Mean ± SE brain $^{\bullet}$NO concentration as a function of brain tissue O_2 concentration measured in the same region. Rats were exposed to air, 1.0, 2.0 or 2.8 ATA O_2 after $^{\bullet}$NO and O_2 specific microelectrodes had been implanted. Calculations for fit to Michaelis-Menten kinetics were based on a baseline [$^{\bullet}$NO] of 500 nM.

The relationship between the elevation in brain $^{\bullet}$NO concentration and local blood flow measured by laser Doppler flowmeter at 2.8 ATA O_2 is shown in Figure 3. The $^{\bullet}$NO concentration changed immediately upon elevation of ambient O_2 tension and stabilized shortly after reaching 2.8 ATA. We found that blood flow vacillated during exposure to hyperbaric O_2 but was significantly elevated over control. The level on brain $^{\bullet}$NO and local blood flow decreased to the baseline level when rats were decompressed to ambient pressure and breathed air.

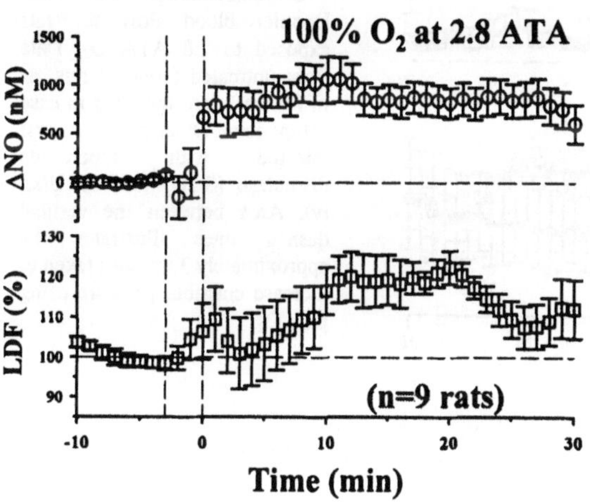

Figure 3. Mean values for changes in brain $^{\bullet}$NO concentration and laser Doppler blood flow among 9 rats exposed to 2.8 ATA O_2. Area between the vertical dashed lines illustrates the approximately 3 minutes taken to increase chamber pressure using pure O_2.

Pharmacological inhibitors of NOS activation in rats: Treatment with 7 nitroindazole (7-NI), a relatively specific inhibitor of nNOS, inhibited increases in brain $^{\bullet}$NO of rats exposed to 2.8 ATA O_2 and also blocked elevations in blood flow (Figure 4). To assess the possible role of oxidative stress in NOS responses, rats were infused with superoxide dismutase (SOD). SOD was an effective inhibitor, blocking elevation in brain $^{\bullet}$NO concentration by over 80% (Figure 5) for the first 20 minutes following infusion.

Brain •NO concentration rose slightly during the last 10 minutes of the experiment, presumably due to the short plasma half-life of SOD (6). Laser Doppler blood flow also rose slightly during the final 10 minutes to 109 ± 11 %, (not significant).

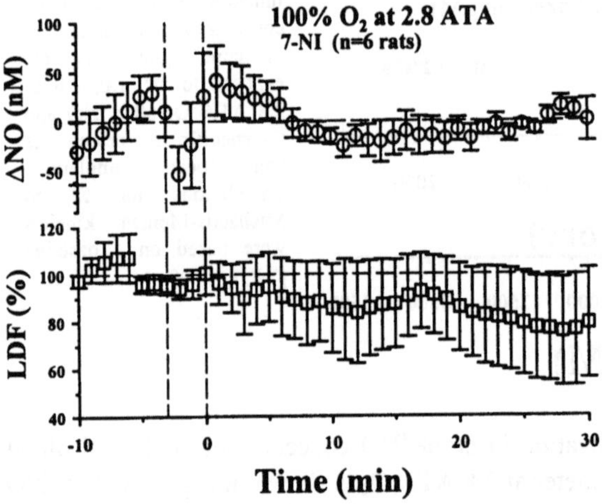

Figure 4. Mean values for changes in brain •NO concentration and laser Doppler blood flow among rats injected with 7-nitroindazole (12 mg/kg ip) before being exposed to 2.8 ATA O_2. Area between the vertical dashed lines illustrates the approximately 3 minutes taken to increase chamber pressure using pure O_2.

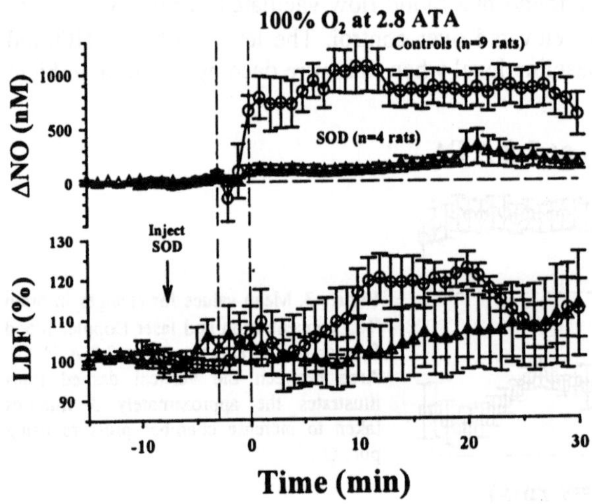

Figure 5. Mean changes in brain •NO concentration and laser Doppler blood flow for rats exposed to 2.8 ATA O_2. Data from untreated (control) rats, as in Figure 2, are included to ease comparison. Other rats were first injected with superoxide dismutase (SOD) (25,000 U/kg iv). Area between the vertical dashed lines illustrates the approximately 3 minutes taken to increase chamber pressure using pure O_2

4. DISCUSSION

Findings in this article are as follows: (1) hyperbaric oxygen causes an elevation in steady state •NO concentration in brains; (2) nNOS activity appears to be the major contributor to •NO elevation due to hyperoxia; (3) •NO elevation due to hyperoxia can be inhibited by infusion of SOD; (4) elevations in laser Doppler measurements of cortical blood flow occur with hyperoxia and this may be causally related to elevations of steady state •NO concentration.

Our data clearly demonstrate that elevated O_2 tensions will increase the production of •NO in brain. Contradictory conclusions in prior reports may have been due to measurements of break-down products of •NO (nitrite and nitrate) rather than •NO *per se* (3,4). The effect of SOD infusion suggests that NOS activation arises in response to oxidative stress. Further, because SOD cannot be expected to escape the circulation, NOS isoforms must be activated in response to perivascular changes mediated by one or more reactive species related to superoxide anion. Others have shown hyperoxia to reduce brain blood flow (3,4). The flow elevation observed in our study may be related to the laser Doppler device, which measures relative flow and not total blood flow. As O_2-induced vasoconstriction is heterogeneous, regional perfusion pressure to the cortex may have been elevated by hyperbaric O_2. This would be detected as increased flow by laser Doppler although perfusion of the whole brain may actually have been diminished. Others have also observed a small increase in cerebral flow during hyperbaric O_2 exposures using laser Doppler (7).

Some workers have suggested that elevated •NO synthesis associated with hyperbaric oxygen may be related to provision of increased concentration of substrate (1). The Michaelis constant (K_m) for O_2 of purified nNOS has been reported to be between 260 and 400 μM (1, 8). Oxygen tension is thought to influence NOS activity by mediating heme ferrous-to-ferric conversion at the active site (8,9). The physiological relevance of this process is unclear, however, as the propensity for oxidation in the intracellular milieu is likely to be limited. While our data can be fit to Michaelis-Menten kinetics, it appears that NOS activation is not simply a biochemical response to increased substrate, as SOD inhibits •NO elevation mediated by hyperoxia. Our results suggest that hyperoxia increases •NO production in brain via an oxidative stress response.

REFERENCES

1. I.M. Elayan, M.J. Axley, P.V. Prasad, S.T. Ahlers, and C.R. Auker, Effect of hyperbaric oxygen treatment on nitric oxide and oxygen free radicals in rat brain. *J. Neurophysiol.* **83,** 2022-2029 (2000).
2. W.J. Wang, X.P. Ho, Y.L. Yan, T.H. Yan, and C.L. Li, Intrasynaptosomal free calcium and nitric oxide metabolism in central nervous system oxygen toxicity. *Aviat. Space Environ. Med.* **69,** 551-555 (1998).
3. I.T. Demchenko, A.E. Boso, P.B. Bennett, A.R. Whorton, and C.A. Piantadosi, Hyperbaric oxygen reduces cerebral blood flow by inactivating nitric oxide. *Nitric Oxide* **4,** 597-608 (2000).
4. I.T. Demchenko, A.E. Boso, T.J. O'Neill, P.B. Bennett PB, and C.A. Piantadosi CA, Nitric oxide and cerebral blood flow responses to hyperbaric oxygen. *J. Appl. Physiol.* **88,** 1381-1389 (2000).
5. S.R. Thom, and M.E. Elbuken, Oxygen-dependent antagonism of lipid peroxidation. *Free Rad. Biol. Med.* **10,** 413-426 (1991).
6. J.F. Turrens J.D. Crapo, and B.A. Freeman, Protection against O_2 toxicity by intravenous injection of liposome-entrapped catalase and superoxide dismutase. *J. Clin. Invest.* **73,** 87-95 (1984).
7. J. Zhang, A.D. Sam, B. Klitzman , and C.A. Piantadosi, Inhibition of nitric oxide synthase on brain oxygenation in anesthetized rats exposed to hyperbaric oxygen. Undersea Hyperbaric Med **22,** 377-382 (1995).
8. H. M. Abu-Soud, D.L. Rousseau, and D.J. Stuehr. Nitric oxide binding to the heme of neuronal nitric-oxide synthase links its activity to changes in oxygen tension. *J. Biol. Chem.* **271,** 32515-32518 (1996).
9. A.R. Hurshman, and M.A. Marletta. Nitric oxide complexes of inducible nitric oxide synthase: spectral characterization and effect on catalytic activity. *Biochemistry* **34,** 5627-5634 (1995)

OXYGEN INDIRECTLY REGULATES NITRIC OXIDE AVAILABILITY

Philip E. James,[1*], Joan Parton, J.[2], Simon K. Jackson,[2], and Michael P. Frenneaux [1]

1. INTRODUCTION

Oxygen is a critical substrate for both nitric oxide (NO) and superoxide (O_2^-) biosynthesis. Previous studies have established that constitutive calcium-dependent NO synthase (NOS) and O_2^- producing NADPH-oxidase are affected at the level of gene expression and at a functional level. For example, hypoxia increases expression of the genes encoding both macrophage and endothelial NOS isoforms [1], and low oxygen tension attenuates NOS activity and NO-dependent vascular relaxation [2,3]. Both NOS and NADPH-oxidase participate in modulation of vascular tone [4] although how oxygen regulates endothelial response is unclear. We have previously demonstrated the dependence of macrophage NADPH-oxidase on the cellular distribution of oxygen [5], and both superoxide and NO production by endothelial cells [6].

The apparent Km for oxygen calculated from cells exposed to a range of oxygen concentrations depends on cell type, but is markedly lower for NOS (9-12 μM) compared to that of NADPH-oxidase (30-40 μM) [3,5,7]. We therefore hypothesized that at intermediate oxygen concentrations (i.e. between 10 and 30 μM), NO production by cells having substantial NOS and oxidase activity (such as endothelial cells) would be maximal whereas O_2^- production would be reduced. Since NO and O_2^- react at almost diffusion limited rates to produce peroxynitrite (ONOO-), reduced O_2^- might result in increased NO availability (without affecting NO production, per se). Thus oxygen may play an important yet indirect role in regulating vessel tone in vivo. To address this, we studied NO production by brain endothelial cells incubated at various oxygen concentrations.

---

* To whom correspondence should be addressed. [1]- Department of Cardiology, Wales Heart Research Institute, and [2]- Department of Microbiology, University of Wales College of Medicine, Heath Park, Cardiff, Wales, U.K.

2. METHODS

2.1. Cell Culture

A murine brain endothelial cell line (BEC) was a gift from Dr M. Gumbleton (School of Pharmacy, University of Cardiff). These were cultured in Dulbecco's minimal Essential Medium (supplemented with bovine serum albumin and antibiotic) at 37°C and air/CO_2 mix (95/5%, respectively). Twelve hours prior to the oxygen experiments, this medium was replaced by an equivalent volume that was lacking in phenol red, since this interferes with the NOx assay (below).

2.2. Experimental Protocol

We incubated BEC (10^6 cells/ml) in 12 well plates inside a chamber kept at 37°C and perfused with 10 L/min gas. Two chambers were always set up in parallel; one perfused with air (210 μM), the other with gas of the appropriate oxygen concentration (50, 20 or 10 μM). Air was mixed with oxygen-free nitrogen to produce the range of oxygen concentrations in the gas supply. Oxygen concentration was monitored continuously within the chambers using an oxygen electrode (World Precision Instruments). Parallel cell samples were treated with SOD (300 Units/ml). Cells were incubated for 30 min and then the culture medium collected via ports in the chamber. These samples were snap frozen in liquid nitrogen until batch analysis of NOx.

2.3. Nitrite and Nitrate (NOx) Measurement

Nitrite and nitrite (NOx) was measured in these studies using a fluorometric method[8] which has improved sensitivity over the conventional Griess detection. NADPH (10 μM) was added to each 50 μl sample of cell supernatant. Nitrite was measured from the fluorometric signal formed by the reaction between nitrite and naphthylethylene diamine (DAN; 0.62 M). Nitrate was measured by its conversion to nitrite by nitrate reductase and further reaction with DAN. 50 mM glucose-6-phosphate, 1.6 U/ml glucose-6-phosphate dehydrogenase and 0.8 U/ml nitrate reductase in sodium phosphate buffer 14 mM was added. For nitrite determination, a similar volume of sodium phosphate buffer was added. Samples were incubated for 1-1.5 hours at 37°C. Following incubation, DAN in 0.62M HCl was added with a further incubation for 10 minutes at room temperature protected from light. The reaction was halted by addition of 2.8N NaOH and the fluorescence measured after 10 minutes using a Perkin Elmer Fluorescent Spectrophotometer (emission= 450 nm, excitation=365 nm). Sodium nitrite and sodium nitrate were used as standards. The efficiency of conversion of nitrate to nitrite was >90% up to 150 μM nitrate added.

2.4. Electron Paramagnetic Resonance (EPR) Detection of NO

Spin trapping was used to detect NO produced by BEC. Carboxy-PTIO (10 μM; Alexis Corp.) was added to 2×10^7 cells/ml and the sample drawn into gas permeable teflon tubing (0.8mm i.d.) and placed into the cavity of the EPR spectrometer maintained at 37°C. C-PTIO has an EPR signal comprising 5 lines. NO detection is possible by monitoring conversion of this trap signal to that of the NO-cPTIO adduct (characterised by 7 apparent lines in the EPR spectrum). Following 2 min (for temperature equilibration within the cavity) spectra were recorded every 30 sec for 30 min on a Varian 104 EPR spectrometer linked to a microwave bridge operating at a frequency of 9.45 GHz. The spectrometer cavity was perfused with gas of the appropriate oxygen concentration, allowing equilibration across the teflon tubing and the cell sample (as previously described [5]).

3. RESULTS

NO production by BEC was measured as NOx accumulation in the culture medium and was found to be 8.2±0.9 μM/10^6 cells/ml/30 min, when the concentration of oxygen in the ambient gas was equal to that in air (210 μM). Addition of SOD to the culture medium resulted in a 139% increase in NOx detected ($p<0.05$). Levels of NOx were the same at 210 and 50 μM, but were reduced to around 64% at 10 μM (Figure 1). However, at 20 μM, NO levels were increased 118% ($p<0.05$ compared to 210 μM levels). Spin trapping studies were run in parallel to NOx measurements. NO production by BEC was assessed as the accumulation of NO-cPTIO adduct signal (Figure 2). The rate of NO-cPTIO formation was the same for Bend incubated at 210 and 50 μM oxygen. Similar to the NO studies above, NO-cPTIO formation was increased almost two-fold when the concentration of oxygen in the ambient gas was 20 μM.

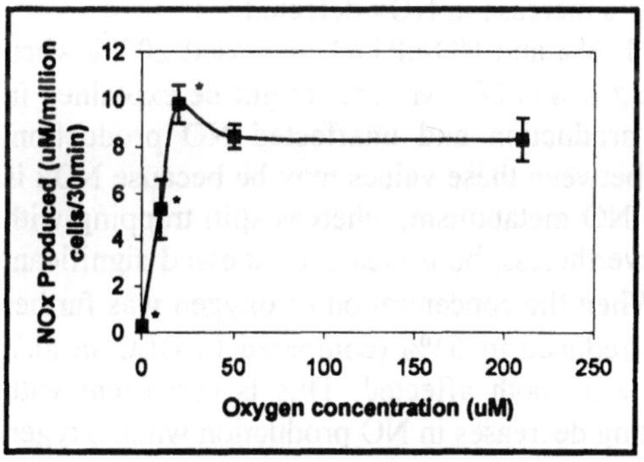

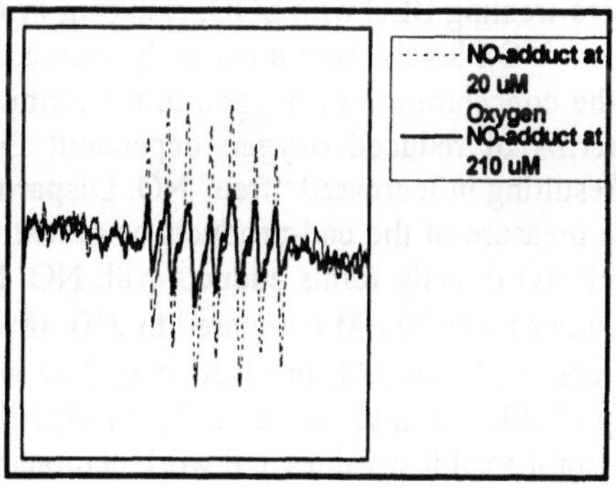

Figure 1. NOx production by BEC exposed to a range of oxygen concentrations in the ambient gas. *-denotes where mean was significantly different from 210 μM.
Figure 2. NO-cPTIO formation following 30 min incubation at 210 or 20 μM oxygen.

4. DISCUSSION

The steady state of NO is not merely determined by its rate of enzymatic formation by NOS, but also its rate of reaction with oxygen species [9], haemoglobin, and thiol-proteins in blood [10]. Among these, O_2^- is particularly efficient at inactivating NO since they react at an almost diffusion-controlled rate to form peroxynitrite [11]. Within the vasculature, O_2^- is synthesized by NADH/NADPH oxidases present in many cell types (especially endothelial cells and macrophages). Interestingly, NOS can also generate O_2^- when cofactors (such as tetrahydrobiopterin) are limiting [12]. Under normal conditions, anti-oxidants quickly inactivate O_2^- (for example, superoxide dismutase (SOD)) and in doing so help maintain NO levels and vessel dilatation [13]. However, dismutation of O_2^- occurs slower than reaction with NO ($2x10^{-9}$ versus $7x10^{-9}$ M sec-1) suggesting SOD can only incompletely compete for O_2^-. Furthermore, indirect control of available NO via SOD is unlikely to participate in immediate response of endothelial cells to rapid or transient changes in oxygen concentration.

The availability of substrate can limit production of NO and/or O_2^-, although for L-arginine (and also NADPH) under normal conditions, this seems unlikely. The Km for these substrates is believed to be <5 μM - a concentration well below that thought to exist in cells. Under pathological conditions, however, substrate and co-factor availability are undoubtedly important [14].

The Km for oxygen of NOS (9-12 μM) is markedly lower than that of O_2^- producing NADPH-oxidase (30-40 μM – depending on activation state). We hypothesized that at intermediate oxygen concentrations (i.e. between 10 and 30 μM), NO production would be maximal whereas O_2^- production would be reduced. This might result in increased NO availability without affecting NO production, per se.

We monitored NO production by brain endothelial cells exposed to a range of oxygen concentrations. NOx determination in the culture supernatant and spin trapping were used in parallel as measures of NO. We found that BEC incubated in air (210 μM) produced approximately 8 μM NO per million cells per 30 min. This reflects "free" NO and is the balance of NO that was not inactivated by interaction with O_2^-. This was confirmed by pre-treating BEC with SOD, resulting in 140% increase in NOx detected.

NOx levels were apparently increased 118% and NO-cPTIO increased 200% when the concentration of oxygen in the ambient gas was 20 μM. This might be explained in terms of reduced oxygen dependent O_2^- production and unaffected NO production, resulting in increased "free" NO. Disparity between these values may be because NOx is a measure of the end-products of oxidative NO metabolism, whereas spin trapping with cPTIO directly forms adducts with NO. Nevertheless, both measures showed significant increases at 20 μM compared to 210 μM. When the concentration of oxygen was further reduced to 10 μM, production of NO was reduced to 64% (compared to BEC in air), probably because NO and O_2^- production were both affected. This is consistent with established Km values and work demonstrating decreases in NO production when oxygen is reduced [2].

We cannot rule out the possibility that reduction to 20 μM oxygen does not have direct, but differential effects on NOS and NADPH-oxidase, in addition to that of limiting substrate. It is unlikely, however, that alterations in gene expression play an important role since cells were incubated for only 30 min. Furthermore, EPR spin-trapping showed an increased rate of NO-cPTIO adduct formation immediately following addition of the cPTIO and recording of spectra. Typically, hypoxia-induced changes in NOS gene expression relate to the inducible, non-calcium dependent NOS isoform, and increases are observed only following several hours (≈24), presumably time required for gene expression and protein production [1].

CONCLUSIONS

These results show brain endothelial cells incubated under conditions where oxygen is plentiful produce more NO than O_2^-. The level of available NO can be increased by dismutation of the O_2^-. Perhaps of greater importance, under conditions where oxygen dependent O_2^- production is reduced and NO production unaffected, available NO may be increased. This may have important implications for indirect oxygen sensing and control of vessel tone in vivo when oxygen delivery falls to a critical level.

ACKNOWLEDGEMENTS

This work was supported by a British Heart Foundation Project Grant.

REFERENCES

1. Liao, J.K., Zulueta, J.J., Yu., F.S., Peng, H.B., Cote, C.G., Hassoun, P.M. Regulation of bovine endothelial constitutive nitric oxide synthase by oxygen. *J Clin Invest* 96;2661-2666, 1995.
2. Whorton, A.R., Simmonds, D.B., and Piantadosi, C.A. Regulation of nitric oxide synthesis by oxygen in vascular endothelial cells. *Am J Physiol* 272; L1161-1166, 1997.
3. Rengasamy, A., and Johns, R.A. Determination of Km for oxygen of nitric oxide synthase isoforms. *J Pharmacol Exper Therap* 276; 30-33,1996.
4. Forstermann, U., Pollock, J.S., Nakane, M. Nitric oxide synthases in the cardiovascular system. *Trends Cardiovasc Med* 3; 104-110, 1993.
5. James, P.E., Grinberg, O.Y., and Swartz, H.M. Superoxide Production by Phagocytosing macrophages in Relation to the Intracellular Distribution of Oxygen. *J. Leuk. Biol.* 64:78-84, 1998.
6. Barchowsky, A., Klei, L.R., Dudek, E.J., Gesek, F.A., Swartz, H.M. and James, P.E. Stimulation of Reactive Oxygen, But Not Reactive Nitrogen Species, in Vascular Endothelial cells Exposed to Low Levels of Arsenite. *Free Radical Biology and Medicine* 27: 1405-1412, 1999.
7. Rossi, F. The O2- - forming NADPH-oxidase of the phagocytes; nature, mechanisms of activation and function. *Biochim. Biophys. Acta* 853, 65-89, 1986.
8. Misko TP, Schilling RJ, Salvemini D, Moore WM, Currie MG. A fluorometric assay for the measurement of nitrite in biological samples. Anal Biochem 1993; 214(1):11-16.
9. Wink, D.A., Mitchell, J.B. Chemical biology of nitric oxide: insights into the regulatory, cytotoxic, and cytoprotective mechanisms of nitric oxide. *Free Rad Biol Med* 25: 434-456, 1998.
10. Gow AJ, Luchsinger BP, Pawloski JR, Singel DJ, Stamler JS. The oxyhemoglobin reaction of nitric oxide [see comments]. Proc Natl Acad Sci U S A 1999; 96(16):9027-9032.
11. Huie, R.E.,, Padjama, S. The reaction of NO with superoxide. *Free Rad Res Commun* 18; 195-199, 1993.
12. Qury, T.D., Day, B.J., Crapo, J.D. Extracellular superoxide dismutase: a regulator of NO bioavailability. *Lab Invest* 75; 617-636, 1996.
13. Gryglewski, R.J., Palmer, R.M., Moncada, S. Superoxide anion is involved in the breakdown of endothelium-derived vascular relaxing factor. *Nature* 320; 454-456, 1986.
14. Xia, Y., Tsai, A.L., Berka, V., Zweier, J.L. Superoxide generation from endothelial nitric oxide synthase. *J Biol Chem* 273; 25804-25808, 1998

MODELING THE REGULATION OF OXYGEN CONSUMPTION BY NITRIC OXIDE

Kathleen Lamkin-Kennard, Dov Jaron and Donald G. Buerk*

1. INTRODUCTION

Nitric oxide (NO) is not only a potent gaseous vasodilator regulating blood flow, but also a reversible inhibitor of mitochondrial O_2 consumption. Mathematical models for NO biotransport were recently reviewed by Buerk[1], who pointed out that previous models for O_2 transport to tissue must be revised to include multiple effects of NO. To address this issue, we developed and solved coupled mass transport models including O_2-dependent properties of NO synthases, Michaelis-Menten kinetics, and known inhibitory effects of NO on mitochondrial O_2 consumption, recently reviewed by Brown[2].

2. LAYERED MODEL FOR NO BIOTRANSPORT

We performed preliminary model simulations for arterioles based on symmetrical cylindrical geometry to predict NO and PO_2 gradients from the vessel lumen out into the vascular wall and surrounding tissue. The model is based on 4 uniform layers: blood ($0 < r < R_1$), plasma ($R_1 < r < R_2$), endothelium ($R_2 < r < R_3$), and tissue ($R_3 < r < R_t$), and is expressed in terms of NO partial pressure (P_{NO}). PO_2 and P_{NO} are related to NO and O_2 concentrations by their respective solubilities, which may be different in each layer. As a first approximation, we neglected axial diffusion and convective transport in the blood.

2.1 NO Profile In Bloodstream

The steady state, one dimensional radial diffusion equation with first-order scavenging of NO by hemoglobin in red blood cells, using a pseudo-first-order rate for blood (λ_b) and

* Kathleen Lamkin-Kennard, Dov Jaron, School of Biomedical Engineering Science and Health Systems, Drexel University, Philadelphia, PA 19104, Donald G. Buerk, Departments of Physiology, Bioengineering, and The Institute for Environmental Medicine, University of Pennsylvania School of Medicine, Philadelphia, PA 19104

appropriate diffusion D_b and solubility S_b coefficients, is given by

$$\frac{D_b S_b}{r}\frac{d}{dr}\left(r\frac{dP_{NO}}{dr}\right) - \lambda_b S_b P_{NO} = 0 \tag{1}$$

The general solution for the radial P_{NO} gradient in the bloodstream[1] is

$$P_{NO}(r) = a_1\, I_0(B_b r) + a_2 K_0(B_b r) \tag{2}$$

where I_0 and K_0 are zero-order modified Bessel functions of the first and second kind, with a scaling factor defined as $B_b = (\lambda_b/D_b)^{0.5}$. The coefficients a_1 and a_2 will be determined by the boundary conditions as described below.

2.2 NO Profile In Plasma Layer

A thin plasma layer, possibly including a glycocalyx coating on the endothelium, was assumed to act as a simple diffusion resistance between the red blood cells and endothelium, with the differential equation

$$\frac{D_p S_p}{r}\frac{d}{dr}\left(r\frac{dP_{NO}}{dr}\right) = 0 \tag{3}$$

which has the general solution

$$P_{NO}(r) = a_3 + a_4 \log(r) \tag{4}$$

with coefficients a_3 and a_4.

2.3 NO Profile In Endothelium

A source term R_{NO} for NO production by endothelial cells is added to the radial diffusion equation

$$\frac{D_e S_e}{r}\frac{d}{dr}\left(r\frac{dP_{NO}}{dr}\right) + R_{NO} = 0 \tag{5}$$

which has the general solution

$$P_{NO}(r) = a_5 + a_6 \log(r) + \frac{R_{NO}}{4D_e S_e} r^2 \tag{6}$$

with coefficients a_5 and a_6. While R_{NO} will be treated as a uniform and constant production rate for the preliminary simulations in this paper, it can also be expressed as function of blood PO_2, shear stress, or other physiological variables.

2.4 NO Profile In Tissue

For the vascular wall and surrounding tissue, a uniform pseudo-first order rate constant is assumed for the scavenging of NO[1], with the resulting radial diffusion equation

$$\frac{D_t S_t}{r}\frac{d}{dr}(r\frac{dP_{NO}}{dr}) - \lambda_t S_t P_{NO} = 0 \quad (7)$$

which has the general solution

$$P_{NO}(r) = a_7\, I_0(B_t r) + a_8 K_0(B_t r) \quad (8)$$

with scaling factor $B_t = (\lambda_t/D_t)^{0.5}$ and coefficients a_7 and a_8.

2.5 Boundary Conditions

Coefficients (a_1 ... a_8) are determined from the following 8 boundary conditions: mass fluxes are zero at the centerline ($r = 0$) and the outermost boundary ($r = R_t$), with continuous P_{NO} and equal mass fluxes for NO at each interface ($r = R_1, R_2, R_3$). Additionally, the peak P_{NO} and its radial location in the endothelium can be found from the zero flux condition which must exist at the maximum of the P_{NO} profile.

3. RESULTS FOR NO PROFILES

NO profiles in blood and tissue were simulated for a 100 µm diameter arteriole with vessel hematocrit equal to systemic hematocrit (45%) and $\lambda_b = 117\ s^{-1}$ based on Carlson and Comroe[3], for a 1 µm thick plasma layer, a 2.5 µm thick endothelium, and zero flux boundary condition at $R_t = 100$ µm are shown in Fig. 1A. Case a is for $\lambda_t = 7.5\ s^{-1}$ from Thomas *et al.*[4] and case b with $\lambda_t = 0.01\ s^{-1}$ from Vaughn *et al.*[5] For simplicity, D and S for NO were assumed to be the same in each layer, and P_{NO} is converted to NO concentration. For this simulation, NO concentration at the endothelium/tissue interface ($r = R_3$) was specified to be 100 nM for each case. Detailed NO profiles near the endothelium, and peak values within the endothelium are shown in Fig. 1B. The majority of NO produced by the endothelium is scavenged by hemoglobin in the bloodstream, and the blood NO concentration at the center of the arteriole is negligible. With the lower λ_t, tissue NO is nearly constant (> 99 nM) throughout the tissue layer, but with the higher λ_t, falls below 20 nM at the outer radius.

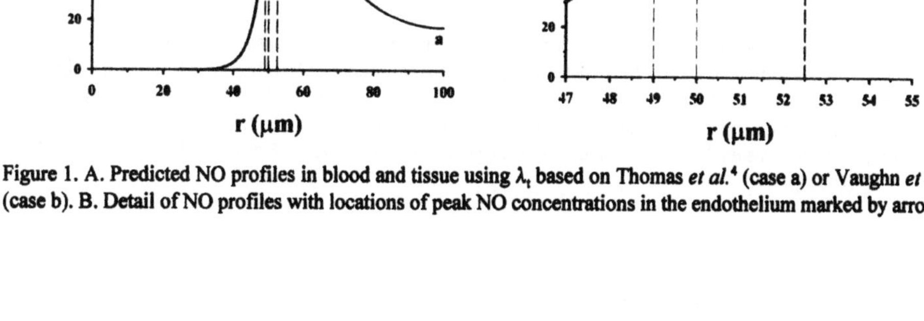

Figure 1. A. Predicted NO profiles in blood and tissue using λ_t based on Thomas *et al.*[4] (case a) or Vaughn *et al.*[5] (case b). B. Detail of NO profiles with locations of peak NO concentrations in the endothelium marked by arrows.

4. INHIBITION OF TISSUE O_2 CONSUMPTION BY NO

The steady state radial O_2 diffusion model with modified Michaelis-Menten kinetics

$$\frac{D_e S_e}{r}\frac{d}{dr}\left(r\frac{dPO_2}{dr}\right) - Q_{max}\frac{PO_2}{PO_2 + appK_m} = 0 \quad (9)$$

where D and S now represent the diffusion and solubility coefficients for O_2, was used to predict the inhibitory effects of NO, neglecting any O_2 loss due to NO autooxidation. The apparent K_m for O_2 consumption is not constant in this model. Instead, $appK_m$ increases linearly with NO concentration (see Buerk[1]). Equation (9) was solved numerically (MatLab®) to simulate PO_2 profiles for several different cases using the tissue NO profiles in Fig. 1, imposing a zero O_2 flux boundary condition at $R_t = 100$ μm.

5. RESULTS FOR PO_2 PROFILES WITH INHIBITION BY NO

The inhibitory effect of NO on O_2 consumption and resulting tissue PO_2 profiles around a 100 μm diameter arteriole with $PO_2 = 70$ Torr at the endothelium/tissue interface ($r = R_3$) are shown in Fig. 2 for several different cases. Case a is the limiting case where $appK_m$ is negligible and independent of NO (constant O_2 consumption), which leads to the well known Krogh-Erlang analytical solution for the radial PO_2 profile. Simulations in Fig. 2A are based on the findings of Brown and Cooper[6], who reported a low K_m of 2.6 Torr in the absence of NO (dashed line, case b). Cases c and d illustrate the inhibitory effect of NO for simulated NO profiles with $\lambda_t = 7.5\ s^{-1}$ and $\lambda_t = 0.01\ s^{-1}$ shown previously in Fig. 1. Simulations in Fig. 2B are based on the findings of Koivisto *et al.*[7], who reported a higher K_m of 16 Torr in the absence of NO (dashed line, case e). Cases f and g illustrate the inhibitory effect of NO respectively for NO profiles with $\lambda_t = 7.5\ s^{-1}$ and $\lambda_t = 0.01\ s^{-1}$ in Fig. 1.

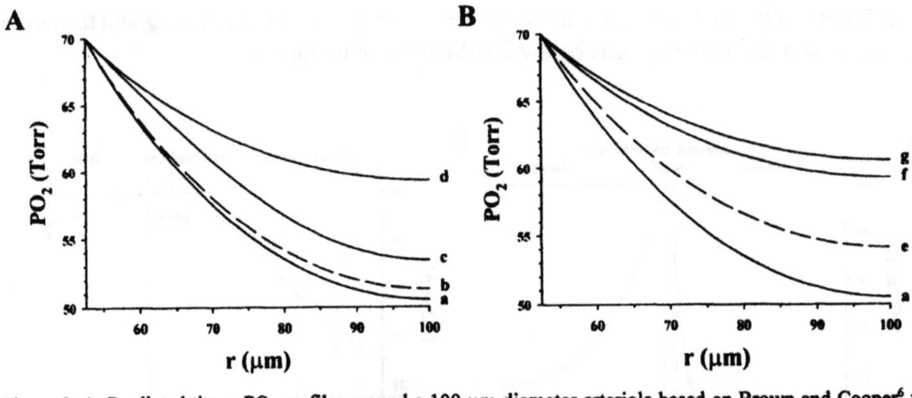

Figure 2. A. Predicted tissue PO_2 profiles around a 100 μm diameter arteriole based on Brown and Cooper[6] for constant O_2 consumption (case a), Michaelis-Menten kinetics in the absence of NO ($K_m = 2.6$ Torr, dashed line, case b), and inhibition based on NO profiles in Fig. 1 with $\lambda_t = 7.5\ s^{-1}$ (case c) and $\lambda_t = 0.01\ s^{-1}$ (case d). B. Predicted tissue PO_2 profiles based on Koivisto *et al.*[7] for constant O_2 consumption (case a), Michaelis-Menten kinetics in the absence of NO ($K_m = 16$ Torr, dashed line, case e), and inhibition based on NO profiles in Fig. 1 with $\lambda_t = 7.5\ s^{-1}$ (case f) and $\lambda_t = 0.01\ s^{-1}$ (case g).

6. DISCUSSION

Our preliminary modeling demonstrates that the inhibition of O_2 consumption by NO increases tissue PO_2 farther from the vessel, even with strong scavenging of NO in tissue. Furthermore, the effect is enhanced when tissue PO_2 is low, due to the functional dependence of appK_m on NO. Our results generally agree with the simplified planar model of Thomas *et al.*[4]. However, our model is based on much more realistic cylindrical geometry and includes greater information about blood and tissue properties.

In addition to the simulations shown above, we also modeled O_2 dependent NO production (see Buerk[1] for a summary of O_2 K_m values reported for NO synthases), which increases NO production as blood PO_2 increases. We also predict a sizeable reduction in NO scavenging by hemoglobin due to the Fåhraeus effect (lower tube hematocrit) in smaller vessels. This phenomenon was previously included in a compartmental O_2 transport model[8]. Our model predicts that blood NO concentrations are not negligible in smaller arterioles, capillaries and venules, and indicates that convective transport should not be neglected.

While we have made a number of simplifying assumptions that allowed analytical solutions to be obtained for steady state NO profiles in blood and tissue, we believe that this simple modeling approach can be very useful for understanding some basic principles of NO biotransport. We expect to further develop and refine our mathematical models, and anticipate that these modifications will require more complex numerical methods to solve. Analytical solutions can be used to test more elaborate models for certain limiting cases. Other aspects of NO biotransport, such as reaction with thiols, with storage and subsequent release of NO, need to be modeled. Other sources of NO production from neuronal sources, or from the immune system, can be incorporated. Experimental studies also need to be conducted to verify the relative contribution of different pathways and mechanisms for NO biotransport.

ACKNOWLEDGMENTS

Supported by EY09269 from NIH.

REFERENCES

1. Buerk, D.G., Can we model NO biotransport? A survey of mathematical models for a simple diatomic molecule with surprisingly complex biological activities, *Annu. Rev. Biomed. Eng.* **3**, 109-143 (2001).
2. Brown, G.C., Regulation of mitochondrial respiration by nitric oxide inhibition of cytochrome *c* oxidase, *Biochim. Biophys. Acta* **1504**, 46-57 (2001).
3. Carlsen, E. and J.H. Comroe, The rate of uptake of carbon monoxide and of nitric oxide by normal human erythrocytes and experimentally produced spherocytes, *J. Gen. Physiol.* **42**, 83-107 (1958).
4. Thomas, D.D., Liu, X., Kantrow, S.P. and J.R. Lancaster, Jr., The biological lifetime of nitric oxide: Implications for the perivascular dynamics of NO and O_2, *Proc. Natl. Acad. Sci. USA* **98**, 355-360 (2001).
5. Vaughn, M.W., Kuo, L., and J.C. Liao, Effective diffusion distance of nitric oxide in the microcirculation, *Am. J. Physiol. Heart Circ. Physiol.* **274**, H1705-H1714 (1998).
6. Brown, G.C., and C.E. Cooper, Nanomolar concentrations of nitric oxide reversibly inhibit synaptosomal respiration by competing with oxygen at cytochrome oxidase, *FEBS Lett.* **356**, 298-298 (1994).
7. Koivisto, A., Matthias, A., Bronnikov, G., and J. Nedergaard, Kinetics of the inhibition of mitochondrial respiration by NO, *FEBS Lett.* **417**, 75-80 (1997).
8. Ye, G-F., Park, J.W., Basude, R., Buerk, D.G., and D. Jaron, Incorporating O_2-Hb reaction kinetics and the Fåhraeus effect into a microcirculatory O_2-CO_2 transport model, *IEEE Trans. Biomed. Eng.* **45**, 26-35 (1998).

OXYGEN SENSOR DESIGN: ANALYSIS AND CORRECTION OF PROBLEMS ASSOCIATED WITH ZERO CURRENT, STABILITY AND STRUCTURE

Cameron J. Koch*

1. ABSTRACT

Membrane-covered polarographic sensors were designed by Leland Clark several decades ago. The membrane covering was a brilliant innovation because it provided an isolation of electrochemical elements from the external environment. These sensors have found a wide range of applications in many aspects of medicine, research and industry but they do not perform well at low oxygen levels because of high and unpredictable zero-current and drift characteristics. The key to overall improved zero-current and stability was a highly improved cathode seal design. Rather than heat-sealing wire into glass microtubing, a high temperature molding procedure was employed. Other improvements involved substantial modifications to the conventional methods of fixing the membrane in place, eliminating all crevices at the periphery of the membrane. This guards against potential contamination, entrapment of gas bubbles and carry over. The present sensor design can operate in gases and liquids with zero current less than 0.02% oxygen.

2. INTRODUCTION

The measurement of partial pressures and concentrations of oxygen is of great importance in many areas of biology, medicine, (bio)chemistry, (bio)physics and engineering. It can be accomplished with various degrees of accuracy and resolution using a variety of instrumentation and methodology[1]. For biological measurements, the most practical device is a membrane covered polarographic sensor of the type initially described by Clark[2]. In this type of sensor, oxygen is reduced by a noble metal cathode, whose potential is fixed with respect to a reference anode (often Ag-AgCl), both elements of the cell in an aqueous electrolyte and enclosed by a gas-permeable membrane. Unfortunately, many scientific studies involve reactions which are modulated by very small concentrations of oxygen and both the resolution and the response time of typical Clark sensors are often inadequate in monitoring such reactions[3].

Criteria of importance in determining the overall quality of a sensor have been described[4] but the nature of the problems of response time, zero current and noise is best

* Cameron J. Koch, University of Pennsylvania, Radiation Oncology Dep't, 195 John Morgan Bldg., Philadelphia, PA, 19104. (215) 898-0073 kochc@mail.med.upenn.edu

appreciated by measuring as a function of time the response of an oxygen sensor after its environment is rapidly changed from a high oxygen partial pressure (i.e. air) to zero. Such measurements, using the best responders of several commercial sensors, have shown an initial rapid decrease in current to 2 - 20% of that in air, but then a much slower decrease, taking many hours, to a minimum value of 0.2 - 10 % of that in air (Fig. 1). As stated, these were the "best responders", and more typical sensors had responses with much poorer overall dynamic range. Furthermore, it has been found that the magnitude of the zero current was not independent of the previous environment of the sensor[5], and this probably points to one source of the zero current, namely dissolved oxygen in the body, typically plastic, of the sensor itself. Such problems in response are perhaps to be expected considering that the cathode seal must be perhaps 100,000 times less than the engineering standard of a leakproof seal, and that the resistivity of all components of the connectors, cable and electrical elements of the sensor must be greater than 500,000 megohms in aggregate[4]. Such requirements are beyond the measured specifications of all components presently available.

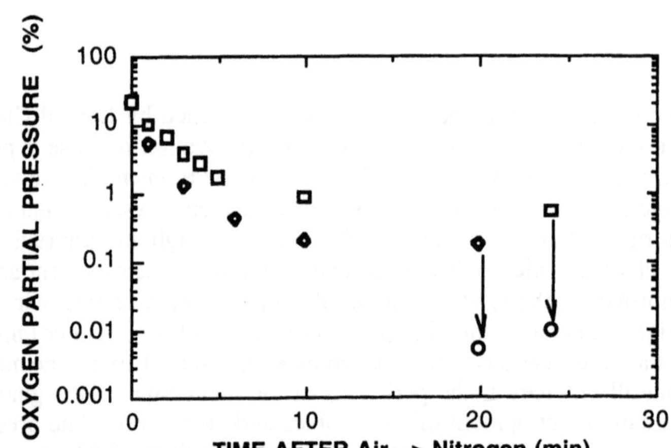

Figure 1. Sudden changes in the oxygen environment external to an oxygen sensor are characterized by various time constants and non-linearities in response. The data illustrate the best-responding of two sensors tested for their ability to track a change from ambient air (21% oxygen) to zero oxygen (open diamonds - Beckman 325814; open squares - Yellow Springs 5331). The zero current changes dramatically between the first few minutes (points as plotted) vs. final equilibrium at 16-24 hours (arrows). Short periods of re-exposure to high oxygen may allow more rapid equilibrium, suggesting oxygen contamination in various plastic components or other oxygen buffers. Many similar devices had dramatically poorer responses.

Another important aspect of the sensor's construction is the method of containment of the membrane and electrolyte. This subject has been the subject of several papers and patents (summarized in Fatt[1]). The electrolyte volume must be large enough to prevent significant changes in composition (i.e. pH and Cl^-) during the lifetime of operation between membrane changes and the membrane must be absolutely impermeable to water vapor to prevent desiccation of the electrolyte. Mechanical flaws may develop during membrane application or sensor use. Improved methods for membrane application have been described[5,6]. A significant practical problem is that before use the membrane must be trimmed around the membrane retainer (typically an 'O' ring) leaving a 'ruffled skirt' of residual membrane which traps air bubbles and makes the sensor tip difficult to clean.

3. METHODS

The fundamental design decision of the present work was to devise methods which would isolate as completely as possible the cathode, electrolyte and anode (reaction elements) from the influence of both internal (i.e. within the sensor) and external contaminants or chemical/physical modulations. The general method of sensor construction in the new design, as described below, consisted of several component sub-assemblies fastened together along matching tapers using high strength epoxy. The matching tapers allowed the extrusion of all but a thin film of adhesive, thus eliminating almost all plastics from the device. These subassemblies comprised cathode, anode, ceramic body, connector cap and electrical connector (Figure 2).

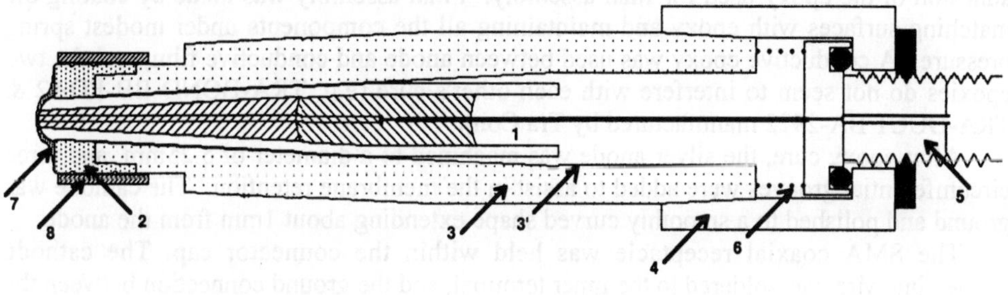

Figure 2. Cross-sectional view of component parts of complete sensor. The mating parts of all assemblies are connect by high-strength epoxies along matching tapers. The cathode seal extends for several mm along the inside of the cathode sub-assembly. The conductive thick film of the ceramic body assembly, and solid silver anode form a 100% electrical shield around the cathode.

The cathode sub-assembly was crucial for the formation of the molded glass seal, and will be described briefly (see Koch[4] for details). It was made by machining a ceramic tube, dimensions 4mm diameter, 33mm long, with tapered portion 2mm --> 4mm over a length of 20mm. The internal surface matched the outside surface, resulting in a wall thickness of about 0.5mm. At the ultimate cathode tip, two tiny holes (#80, 0.34mm) were drilled symmetrically to admit the ends of a 6 inch length of fine platinum wire (0.075mm diameter). Within the interior, these wires were surrounded by a tube of electrical sealing glass (Corning #8940). The entire part was suspended vertically by the ends of the platinum wire, with the proximal tip down, in a furnace @ 950° for three hours, then at 423° (annealing temperature) for 6 hr before return to room temperature. At 950° the glass viscosity (200 poises) was 50-fold lower than could be achieved with conventional glass-working (10,000 poises) - this allowed a very extensive and high quality seal to form along a distance of several mm. After the tip which contained the 0.5mm centering hole was ground away, the cathode wires were left exposed in roughly symmetrical position within the glass seal. The platinum wires were connected to a short length of teflon-coated multi-stranded wire and fixed into the distal end of the sub-assembly by epoxy.

The ceramic body sub-assembly was also made from a tube of ceramic with various tapers to accept the cathode internally, and the anode and connector cap externally. It also contained a standard (10:1) taper near its proximal end, extending from 8mm to 10mm over a distance of 20mm to allow fitting into standard-tapered glass joints (the smallest "standard" joint is 10mm to 7mm over 30mm length). The inside and both ends of the ceramic body were coated with a thick film composite made from precious metal and glass (e.g. Thick Film Systems Multifire® 3580). The film was fired in sealing contact with these surfaces (750°, 30 min.). Film resistance (end to end) was 1-3 ohm.

The anode was made from a rod of solid pure silver, with two internal tapers to match the outside tips of the cathode, extending through the ceramic body, and the body itself. After being carefully cleaned, the anode was etched in hot, 0.1N nitric acid to improve adhesion of the epoxy used for final assembly. Final assembly was made by coating the matching surfaces with epoxy and maintaining all the components under modest spring pressure. A conductive epoxy was used between anode and conductive film, and the two epoxies do not seem to interfere with each other's cure (e.g. TRA-BOND BB-FDA2 & TRA-DUCT BA-2922 manufactured by Tra Con).

After epoxy cure, the silver anode was machined to a diameter of 5.25mm, and three circumferential grooves were added to assist in the membrane retention. The cathode was ground and polished to a smoothly curved shape extending about 1mm from the anode.

The SMA coaxial receptacle was held within the connector cap. The cathode connecting wire was soldered to the inner terminal, and the ground connection between the connector and the anode, via the conductive film via a stainless-steel compression spring.

Before use of the sensor, the anode was carefully cleaned with a soft camel-hair brush or chamois cloth, using a polishing grade of alumina, and a weak ammonium hydroxide/detergent solution followed by thorough rinsing and final cleaning with ethanol. The composition of the electrolyte used was empirically determined, and typically contained 0.75% KCl or NaCl, with 1 mM DTPA (diethylenetriaminepentaacetic acid, a chelating agent) and 0.1% NP-40 (a non-ionic detergent) as a wetting agent. The pH was buffered to 8.5 with 25 mM borate.

A small amount of silicone grease was applied around the side of the anode and filling the circumferential grooves, and electrolyte was added to the tip of the sensor which was held vertically via a base with thread to match the SMA connector. The membrane was applied using an annular support as described previously[5]. The clamped membrane was then stretched over the tip of the sensor in such a manner that the membrane adhered tightly against the first two to three millimeters of the length of the anode (i.e. past the frontmost groove in the anode's periphery). While maintaining a downward pressure on the clamped membrane, a tight-fitting retaining sleeve, made from a cylinder of teflon about 6 mm long with a 0.5 mm thick wall, was then pushed over the anode so that its leading edge was just past the first groove of the anode. This groove caused the sleeve to be mechanically stable at this position. The membrane was released from its annular clamp and substantially trimmed. The retaining sleeve was then pushed further over the anode and its second and third grooves to cover completely both the anode and the trimmed peripheral edge of the membrane. This two-step procedure allowed no residual membrane edge portion to remain externally to the retaining sleeve

4. OPERATION

The completed sensor was connected to a sensitive current-measuring instrument, made by the author, at a polarizing voltage of 0.55 volts. The current sensor has a conventional polarogram and 0.55 V is well below the current-voltage plateau. However, there is a several-fold improvement in signal:zero current at the lower voltage, and operational stability does not appear to be affected[4,5]. A ceramic sleeve was glued to a leakproof aluminum chamber using Vac Torr® epoxy on the periphery, with its internal opening matching the taper of the sensor. The matching tapers of the outer sensor and inner sleeve were coated with a thin film of silicone grease and the chamber was connected to a gas/vacuum manifold[5]. When the sensor was switched from air (oxygen content 210,000 parts per million [ppm]) into a nitrogen atmosphere the current dropped exponentially with an initial response-time of 20 seconds to an absolute value corresponding to a value less than 100 ppm over the next 20 minutes (open squares - Figure 3). The zero current decreased to less than 50 ppm over 20 hr, and the noise was of the order of 2 ppm (data not shown). One can appreciate the importance of the present cathode seal design by considering the other curve shown (open diamonds - Fig 3). This response was for a sensor made in an identical fashion, but with an epoxy seal of the cathode wires. Interestingly, the change in response between air and nitrogen was similar for both designs. However, the leakage current, derived from plastic-dissolved oxygen and/or surface resistivity limitations, was higher than that from the sensor's expected operation.

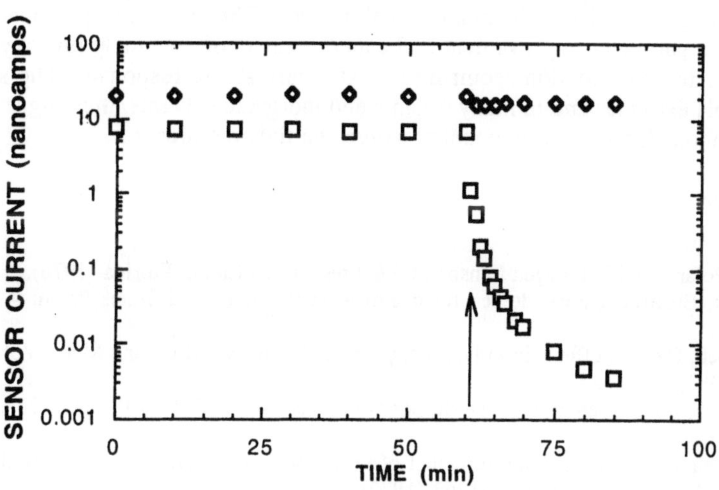

Figure 3. Transient response of current design sensor, after a switch from air to nitrogen in the gas phase extends for more than three decades with extremely stable zero-current and almost no noise (open squares). The importance of the extended glass cathode seal can be appreciated by comparison with an identical sensor but made using ultra high resistivity and strength epoxy instead of the glass (open diamonds).

The stability of the reading in continuous exposure to air at constant temperature was better than -2% per day after an initial operating time of 1 hour. In sealed containers with solutions held at constant temperature, the response of the sensor was constant within measurement limits (+/- 0.2%). This excellent stability was attributable to the sealing technique of the membrane. A measure of this seal was that the aerobic response of the sensor was not affected by prior continuous exposure to a high vacuum (less than 0.015 kPa) for 2 weeks (data not shown). Furthermore, the response at any given partial pressure of oxygen was unaffected by hydrostatic pressure. The response of the sensor, initially in air, was monitored after evacuating the chamber to quite a high vacuum (approximately 1 kPa), waiting for sensor equilibrium, and then refilling the chamber with pure nitrogen. The equilibrium response was the same in the partial vacuum as it was following the readmission of pure nitrogen (data not shown).

The linearity of the sensor was tested by filling the chamber with 66% oxygen and then every 10 minutes subjecting it to a 'gas change'. Each gas change was made by evacuating the chamber to 31.9 kPa pressure (as measured by a precision pressure gauge), then refilling the chamber to 101 kPa with pure nitrogen. As long as the vacuum gauge reading was reproducible, the oxygen partial pressure should decrease by the same proportion (roughly square root of 10) with each gas change, and therefore a plot of log sensor current vs. number of gas changes should be a straight line[7,8]. The response of the sensor was not different than the predicted exponential relationship over 4 decades of oxygen partial pressure if the absolute zero current (approximately 70 ppm in this experiment) was subtracted from all of the readings (data not shown).

In conclusion, the present sensor exhibits unprecedented accuracy, stability and utility. Many problems associated with the Clark-type sensor have been eliminated or greatly reduced and it now appears feasible to make additional improvements and to test various aspects of the design in an analytical manner. The operating specifications of the sensor offer a particular improvement in the measurement of extremely low oxygen partial pressures or in any situation requiring an extremely stable response. The sensor will operate at pH extremes and in many organic and inorganic solvents, thus suggesting many new applications for the measurement of oxygen partial pressures.

REFERENCES

1. I. Fatt, Polarographic Oxygen Sensors. CRC Press, Boca Raton, Florida (1976).
2. L. Clark, Electrochemical device for chemical analysis. United States Patent # 2,913,386, (1959).
3. C. J. Koch, Oxygen effects in radiobiology. In: Adv. in Exp. Biol. and Med. Vol. 157, 123-144 (1983).
4. C. J. Koch, Polarographic Oxygen Sensor with Glass Seal. Canada Patent, 1,304,449, June 30 (1992).
5. C. J. Koch, and J. Kruuv, Measurement of very low oxygen tensions in unstirred biological liquids. Anal. Chem. 44, 1258-1263 (1972).
6. J. A. Porter, and A. F. Dageforde, Membrane tensioning means and the use thereof. United States Patent # 3,887,194, (1975).
7. C. J. Koch, Measurement of very low oxygen tension in liquids: Does the extrapolation number for mammalian survival curves decrease after x-irradiation under hypoxic conditions? In: Proc. 6th L.H. Gray Conf. , The Institute of Physics, London (1974). 167-173 (1975).
8. L-W. Lo, C.J. Koch and D.F. Wilson. Calibration of oxygen-dependent quenching of the phosphorescence of Pd-meso-tetra (4-carboxyphenyl) porphine: a phosphor with general application for measuring oxygen concentration in biological systems. Analyt. Biochem. 236: 153-160 (1996).

COMPUTER SIMULATION OF OXYGEN MICROELECTRODE MEASUREMENTS IN TISSUES

Iuliana Toma-Daşu, Alexandru Daşu, Anthony Waites, and Juliana Denekamp*

1. INTRODUCTION

Several methods are available to investigate tissue oxygenation, a very important parameter in determining tissue response to radiation treatment. In spite of having different underlying principles, almost all these methods do some sort of averaging either in space or in time and the measured values might not correspond to the situation that has actually been measured. Therefore it is important to know the possible limitations of various methods and their implications on the measurements.

The purpose of this paper is to estimate the accuracy in determining the pO_2 distribution in tissues using an Eppendorf-like electrode. When the electrode measurements are correlated to treatment outcome, it is not very important if there is an identity between measurements and the real values. This becomes a very important issue only when the measurements are used to model the tissue response to new treatments.

By theoretically defining the pO_2 distribution in a certain tissue and then simulating the electrode measurement starting from these known values it is possible to investigate the efficiency of the method. Since the intention is to study the correlation between input values and measured ones, it is not critical to have the precise pO_2 distribution that might exist in a real tissue.

2. THE RESPONSE FUNCTION OF THE ELECTRODE

Several questions arise regarding the functioning of the Eppendorf electrode.

The electrode measures oxygen, but is it the oxygen in a point or in a whole volume? The electrode consumes the oxygen from the surrounding tissue and transforms it into an electric current. Depending on the measuring time, the electrode consumes the oxygen from a smaller or a larger volume. Therefore, the electrode measures the oxygen from a volume and not a point.

* Department of Radiation Sciences, Umeå University, 901 85 Umeå, Sweden

Oxygen Transport to Tissue XXIII, edited by D. Wilson *et al.*
Kluwer Academic/Plenum Publishers, 2003

But how big is this measuring volume? Given the fact that during the measurement the oxygen molecule moves towards the electrode, it means that the size of this volume depends on the measuring time. The actual movement is quite complex involving the diffusion of the molecules into the tissue. Therefore the probability of an oxygen molecule to reach the electrode is not the same throughout the measurement volume.

In order to simulate electrode measurements we have to know the probability that an oxygen molecule at a given distance will reach the electrode within the measuring time. This probability, multiplied by the local oxygen tension, gives the contribution to the electrode measurement of each point in the measurement volume. Several studies have investigated the absorption at the electrode and have found out a very complex function of variation with the distance to the electrode[1-5]. The equation that gives the probability that an oxygen molecule will reach the electrode, which we shall call the weighting factor, is:

$$w = (2/\pi)\sin^{-1}\left[2a/\left(\sqrt{(r-a)^2+z^2}+\sqrt{(r+a)^2+z^2}\right)\right] \tag{1}$$

where:

a is the radius of the electrode;

r is the radial distance (the distance to the central axis of the electrode in a plane parallel with the electrode surface);

and z is the axial distance (the distance along the central axis of the electrode).

The weighting factor given by Eq. (1) is represented in Figure 1 for an electrode having a diameter of 12 μm. On both axes are marked 12 μm intervals representing the average diameter of the cells in tissue. The function decreases very rapidly, being less than 20% after only two cell layers. After this distance, the asymmetry in the two directions becomes negligible, the function having then a spherical symmetry.

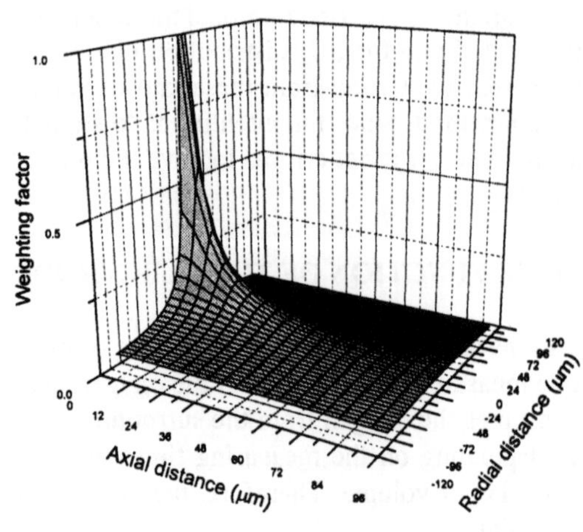

Figure 1. The response function of the electrode, represented as the weighting factor varying with the distance from the centre of the electrode.

3. CONDITIONS OF SIMULATIONS

In this study we have considered a two-dimensional matrix representing the tissue in which we are performing the simulated measurements. The matrix has regularly spaced, parallel blood vessels (capillaries) with the same oxygen tension. We are aware that this is a simplification compared to the non-parallel structure existing in a real tissue, but we consider that it is not critical for studying the accuracy of the measurement process. Furthermore, the simplicity of our system allows a better understanding of the averaging process during measurement.

To simulate the oxygen tension distribution that may exist around the blood vessels we have used Eq. (2) that takes into account the oxygen consumption due to cellular respiration[6].

$$p = p_0\left(R^2/R_0^{\,2}\right)\left[2\ln\left(R/r\right)-1+\left(r^2/R^2\right)\right] \qquad (2)$$

where:
p is the oxygen tension at radius r (mm Hg);
p_0 is oxygen tension in the blood vessel (mm Hg);
R is the diffusion radius (radial distance at which the oxygen concentration falls to zero) (μm);
and R_0 is a parameter given by Eq. (3).

$$R_0 = 60\sqrt{p_0 / Q_{O_1}} \qquad (3)$$

where Q_{O_1} is the rate of oxygen consumption (μl.O/mg dry weight/h^{-1}).

In the present simulation we have adopted the following values for the parameters: $Q_{O_1} = 5$ μl.O/mg dry weight/h^{-1} and the diameter of the electrode $D = 12$ μm.

Using Eq. (2) and (3), we have calculated the average pO_2 values in each cell around the blood vessels. This resulted in a discrete distribution with a limited number of values. However, the details of the input distribution are not critical for the comparison with the distribution representing the simulated measurements with the electrode.

According to the values of oxygen tension in each blood vessel and the distance between two blood vessels, the simulated tissue can have various degrees of oxygenation ranging from well oxygenated to hypoxic (poorly oxygenated). The distances between blood vessels have been chosen to correspond to those measured experimentally in real tissues by Konerding and his co-workers[7].

4. RESULTS AND DISCUSSIONS

Using the assumptions described above, we have simulated oxygen measurements in a range of tissues with various degrees of oxygenation. Figure 2 shows the real distribution of pO_2 values in various tissues and the results of the simulated electrode measurements. The input parameters for the three tissues were chosen to correspond to a well oxygenated tissue (Figure 2a) representing the normal tissue on the arteriolar side of the vasculature, a less oxygenated tissue (Figure 2b) representing the normal tissue on the venous side of the vasculature and a poorly oxygenated tissue (Figure 2c) representing the tumour.

As it can be seen in all the panels of Figure 2, the measured distributions of pO_2 values (the grey bars) do not correspond to the actual situation in tissues (the black bars). In spite of the relative simplicity of the input distribution, the simulations of measurements have resulted in very different distributions of values in all the cases taken into consideration. This can be seen most clearly for well oxygenated tissues (figure 2a and 2b) where the extreme values of the pO_2 were replaced by moderate values. Even for hypoxic tissues (figure 2c) characterised by low pO_2 values, the measurement yielded results quite different from the input values. This is due to the averaging effect of the measurement process at the electrode.

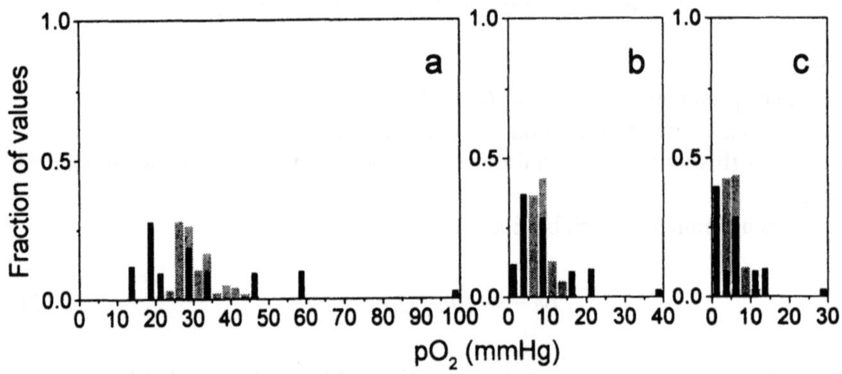

Figure 2. Histogram of the distribution of measurements for different theoretical tissues (gray bars represent the simulated measurements and black bars represent the input values):

a. Normal tissue with 6 cells between capillaries and $pO_2 = 100$ mmHg.

b. Normal tissue with 6 cells between capillaries and $pO_2 = 40$ mmHg.

c. Tumour with 6 cells between capillaries and $pO_2 = 30$ mmHg.

The results of our simulation have some very important implications for the practical use of the polarographic electrodes. The electrode measurements being quite different from the real distributions of pO_2 values in tissues cannot be used as input parameters for predictive purposes such as treatment modelling. We have shown elsewhere[8, 9] that even simplified parameters of the distributions such as the mean or the median values cannot be used to describe the real situation in tissues. It is therefore extremely important to distinguish between various areas of use for the electrode measurements.

For qualitative applications, such as empirical correlations between the Eppendorf measurements of tumour oxygenation and the outcome of treatment, simplified parameters of the measured distributions are generally enough for characterising the oxygenation status of the tissue. Therefore our results are not in contradiction with the numerous such correlations that have been derived clinically.

However, more attention should be paid when using electrode measurements for predictive purposes of a quantitative nature such as treatment simulations. For these applications it is very important to know the relationship between response and tissue oxygenation. The radiobiological response of the cells depends greatly on the oxygen tension and therefore extremely accurate distributions are needed in order to characterise the oxygenation of the tissue. Our simulation suggests that polarographic measurements cannot be used as input for theoretical modelling of radiation treatment. Inadequate input values could lead to results that may have disastrous effects if used for clinical decisions.

More work is in progress both on defining a broader input distribution of pO_2 values and in seeking a relationship between input and measured values.

CONCLUSIONS

Theoretical simulation of the measurements performed with polarographic electrodes having rather large diameters showed the great impact of the averaging process, the results of the measurement being quite different from the real distribution of oxygen in tissue. The extremely high and extremely low values are generally not detected. Therefore, such electrode measurements should not be used for modelling purposes since they do not reflect the actual distributions of oxygen tensions in the measured tissue.

ACKNOWLEDGEMENTS

This work has been supported by grants from the Cancer Research Foundation in Northern Sweden. We are grateful to Profs. Jack Fowler and Bo Littbrand for stimulating discussions.

REFERENCES

1. Y. Saito, A theoretical study of the diffusion current at the stationary electrodes of circular and narrow band types, *Rewieu of Polarography (Japan)* **15**, 177-187 (1968).
2. W. Grunewald, [Diffusion error and specific consumption of the Pt electrode during pO_2 measurements in the steady state], *Pflugers Arch.* **320**, 24-44 (1970).
3. I. Fatt, *Polarographic Oxygen Sensor* (CRC Press Inc, Cleveland, 1976).
4. K. Groebe, Relating measuring signals from PO_2 electrodes to tissue PO_2: a theoretical study, *Adv. Exp. Med. Biol.* **316**, 61-69 (1992).
5. K. Groebe and G. Thews, Basic mechanisms of diffusive and diffusion-related oxygen transport in biological systems: a review, *Adv. Exp. Med. Biol.* **317**, 21-33 (1992).
6. I. F. Tannock, Oxygen diffusion and the distribution of cellular radiosensitivity in tumours, *Br. J. Radiol.* **45**, 515-524 (1972).
7. M. A. Konerding, W. Malkusch, B. Klapthor, C. van Ackern, E. Fait, S. A. Hill, C. Parkins, D. J. Chaplin, M. Presta, and J. Denekamp, Evidence for characteristic vascular patterns in solid tumours: quantitative studies using corrosion casts, *Br. J. Cancer* **80**, 724-732 (1999).
8. I. Toma-Daşu, A. Waites, A. Daşu, and J. Denekamp, Theoretical simulation of oxygen tension measurement in tissues using a microelectrode: I. The response function of the electrode, *Physiol. Meas.* **22**, 713-725 (2001).
9. I. Toma-Daşu, A. Daşu, A. Waites, J. Denekamp, and J. F. Fowler, Theoretical simulation of oxygen tension measurement in tissues using a microelectrode: II. Simulated measurements in tissues, *Radiother. Oncol.* 2002 (submitted).

PHYSIOLOGICAL OXYGEN MEASUREMENTS USING OXYGEN ELECTRODES

David K. Harrison[1]

1. INTRODUCTION

It was the development by Clark[1] in 1956 of the membrane-covered "integral" electrode that opened the field for truly physiological *in vivo* - and indeed clinical - measurements of oxygen partial pressure in blood and tissue. Since then, much of our knowledge about oxygen transport to tissue has been gained by the use of polarographic oxygen (and hydrogen) electrodes – many of them of the Clark-type. It is not possible in this paper to examine all of these applications and the reader is referred to a comprehensive review by Lübbers[2].

The advantages of the Clark-type electrode are that it is an "integrated" electrode: the cathode and reference electrodes are contained in the same electrolyte space covered by an oxygen permeable membrane. It is protected from protein deposition and it is electrically insulated from the tissue. There can therefore be no interference to the electrode from changes in local ion concentrations and, on the other hand, the polarisation potential of the electrode does not cause ion gradients in tissue. Today, the main clinical application of the Clark-type electrode continues to be in blood gas analysis.

Worthy of mention here, however, is the transcutaneous pO_2 ($tcpO_2$) electrode technique[3], in which the blood flow at the site of measurement in skin is "arterialized" by local heating. Pulse oximetery has largely replaced this technique for neonatal, adult intra-operative and intensive care monitoring, but it continues to be used for some monitoring[4] and tissue viability[5] applications, usually combined with transcutaneous pCO_2 electrode measurements.

New oxygen measurement and imaging techniques (see this volume) have now supplanted the polarographic electrode to a large extent both for physiological and clinical applications. In order to interpret the measurements made using these new techniques, however, it is important to recognise the physiological principles of oxygen supply to tissue that were firmly established through the application of polarographic electrodes.

---

[1] David K. Harrison, Regional Medical Physics Department, Durham Unit, University Hospital of North Durham, North Road, Durham DH1 5TW, UK.

Oxygen Transport to Tissue XXIII, edited by D. Wilson *et al.*
Kluwer Academic/Plenum Publishers, 2003

One of the advantages of the electrode technique for the measurement of oxygen partial pressure in tissue is its resolution. It is possible to make measurements within volumes of a few cubic microns, which means that the real heterogeneity of oxygen supply can be quantified[6]. This, however, has also proved to be a disadvantage in that the question has always been asked: How applicable is a measurement in a small area of tissue to the situation in the whole organ?

This criticism was overcome to a great extent by the use of multiple measurements with multi-channel surface electrodes (see below) or stepped needle electrode measurements[7] to record pO_2 histograms. It is, therefore, the ability to measure the distribution of pO_2 in tissue that continues to make the oxygen electrode a useful tool both in understanding the physiology of oxygen supply and in measuring oxygen clinically in tissue.

The importance of the pO_2 histogram, both physiological and clinical, will therefore be reviewed in the light of past and recent findings.

Recent results of measurements using a combination of oxygen flux optodes and pO_2 needle electrodes that provide new insight into the mechanism of oxygen supply to and within the skin will also be presented.

2. THE PHYSIOLOGICAL PO_2 HISTOGRAM

The multiwire surface electrode, MDO[8,9], has been widely applied to physiological and sometimes clinical investigations of the oxygen supply to tissue. It has all the advantages of being a Clark-type electrode. It is also non-invasive, has a small catchment volume[10] and therefore has a high spatial resolution that allows intercapillary measurement of pO_2. The random orientation of the 8 measuring wires means that by rotating the electrode 13 times within a holder, a histogram of 104 measurement points can be accumulated within about 5 minutes. The separation between the wires is of the order of several hundred μm, so that although each wire measures intercapillary pO_2, the area covered encompasses several capillary supply units.

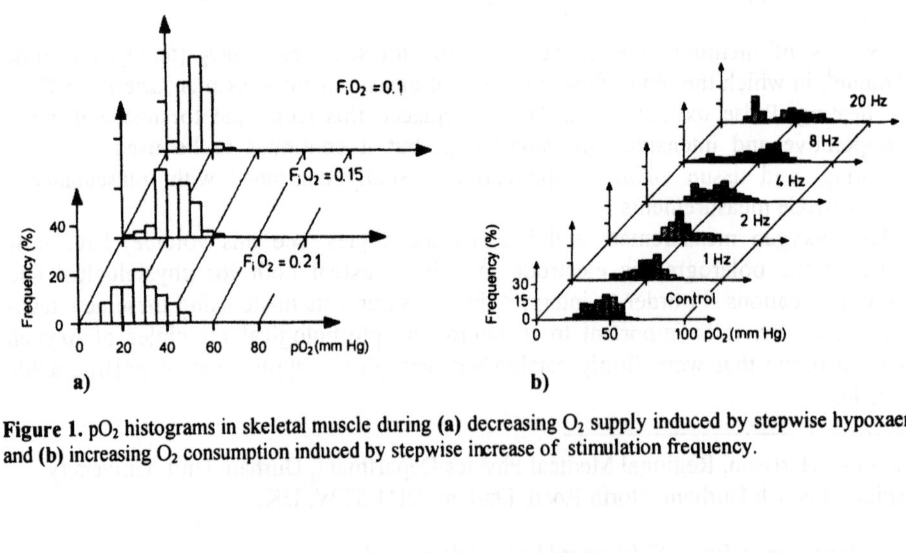

Figure 1. pO_2 histograms in skeletal muscle during **(a)** decreasing O_2 supply induced by stepwise hypoxaemia and **(b)** increasing O_2 consumption induced by stepwise increase of stimulation frequency.

The resultant histograms are therefore statistical histograms derived from several supply units. The lower left histograms in figure 1 (a) and (b) are mean histograms recorded in the sartorius muscle of anaesthetised dogs under resting, normoxic conditions. They are typical of those found in most organs, including the brain and liver[11]. However, do they represent the oxygen gradients to be found within a single capillary supply unit?

This question was recently investigated by Baumgärtl et al.[6], using needle micro-electrodes to measure the pO_2 in the renal cortex of anaesthetised dogs. Measurements were carried out, using an electrode with a spatial resolution of approximately 3 µm, at 10 µm steps from the surface of the kidney to depths of up to 11 mm. By comparing the differences between the pO_2 measured from one step to the next using 10, 50, 100 and 500 µm steps, the authors showed that heterogeneity of the differences between pO_2 values at adjacent sites increased for increasing step differences, reflecting the incursion into increasing numbers of capillary supply units. The resultant overall histograms, however, remained identical (figure 2).

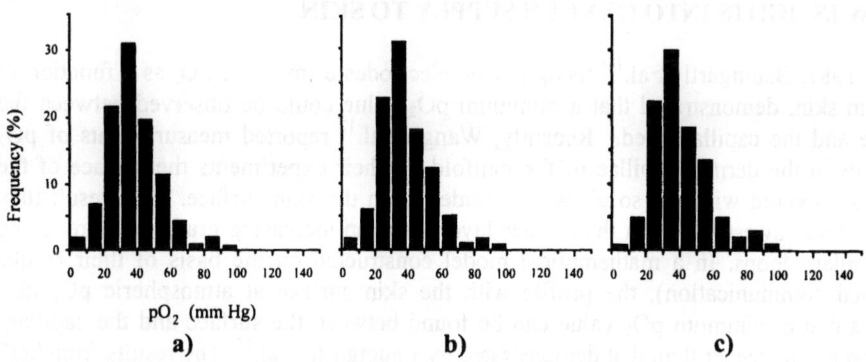

Figure 2. Histograms at increasing depths within the renal cortex measured at (a) 10, (b) 50 and (c) 500 µm steps. After Baumgärtl et al.[6].

Figure 1(a) shows mean histograms from 6 experiments during which the inspired oxygen fraction was reduced in two steps such that the final mean arterial pO_2 was reduced to 32 mmHg[12]. Figure 1(b) shows similar histograms recorded at increasing oxygen uptake rates induced by increased frequency of electrical stimulation of the muscle[13]. The constancy of the histograms is noteworthy, as is the fact that, despite a left shift of the histogram during hypoxia, no significant degree of tissue anoxia (defined as 5% of pO_2 less than 5 mmHg) is observed.

In summary, the organism has a remarkable capacity to maintain blood flow and cellular oxygen consumption in continual balance such that tissue anoxia is avoided – even for those cells at the end of the supply chain. The mechanisms by which this is achieved are discussed elsewhere[13].

3. THE PATHOLOGICAL PO_2 HISTOGRAM: CLINICAL APPLICATIONS

Under pathophysiological conditions, such as those existing in tumours or in ischaemia, the histogram will be shifted to the left and large percentages of values will appear in the 0-5 mmHg class. A significant proportion of the cells in the tissue represented in this class will be anoxic. The ensuing consequences are: anaerobic glycolysis, acidosis, cell membrane breakdown and necrosis. One area in which tissue hypoxia has been identified as an important parameter is in tumours, and the electrode measurement of histograms has become established (at least for clinical research purposes) as an indicator for survival and treatment[7]. This application, however, is dealt with in greater depth elsewhere in this volume.

Whilst not an electrode technique, we have used lightguide spectrophotometry to measure haemoglobin oxygen saturation (SO_2) histograms in the lower limbs of patients with critical limb ischaemia in order to predict amputation level viability[14]. A proportion of values greater than 15% in the class 0-10% SO_2 was found to be characteristic of necessity to above the knee. The routine clinical use of this parameter, along with consideration of the mean SO_2 at the proposed amputation flap site, has led, in our hands, to a 94% healing rate at a below knee to above knee amputation ratio of 82%. This represents a high healing rate for an exceptionally high ratio of below knee to above knee amputations[15].

4. NEW INSIGHTS INTO OXYGEN SUPPLY TO SKIN

In 1987, Baumgärtl et al.[16], using needle electrodes to measure pO_2 as a function of depth in skin, demonstrated that a minimum pO_2 value could be observed between the surface and the capillary bed. Recently, Wang et al.[17] reported measurements of pO_2 gradients in the dermal papillae of the nailfold. In their experiments the surface of the skin was covered with oil, so air was excluded from the skin surface. As a result they observed the lowest values in the surface layers with an increasing gradient approaching the capillary loops. In a mathematical model constructed on the basis of their results (personal communication), the profile with the skin surface at atmospheric pO_2 also predicts that a minimum pO_2 value can be found between the surface and the capillary bed, somewhat deeper than that demonstrated by Baumgärtl et al.[16]. The results from both of these studies using pO_2 electrodes indicate that there is an O_2 flux through the epidermis supplying the upper layers of the skin.

The recent development of a fluorescence-based O_2-fluxoptode for measuring oxygen flux[18], meant that this parameter could be measured directly. Recently, Stücker et al.[19] demonstrated, using simultaneous measurement of oxygen flux, $tcpO_2$ and laser Doppler perfusion that the epidermal oxygen uptake from the atmosphere is in balance with the blood-borne oxygen supply. In their most recent work[20] they carried out O_2 flux measurements in the forearm skin during suprasystolic brachial artery occlusion and observed less than a 10% increase in O_2 flux. They to simulated intracutaneous pO_2 profiles that indicated that the upper skin layers to a depth of 0.25 to 0.4 mm are almost exclusively supplied by atmospheric oxygen.

These optical measurements of oxygen flux combined with electrode measurements of pO_2 profiles indicate conclusively that capillary blood flow in skin at normal skin temperature contributes very little to the oxygen supply of the superficial layers. This has considerable implications for the treatment of diabetic, venous and ischaemic skin defects

ACKNOWLEDGEMENTS

I am most grateful to Horst Baumgärtl and Dietrich Lübbers (Max-Planck-Institute for Molecular Physiology, Dortmund, Germany), Markus Stücker (Department of

Dermatology, Ruhr-University, Bochum, Germany), Martina Günderoth (Helzel Medical Systems, Kaltenkirchen, Germany) and Wen Wang (Queen Mary College, University of London, UK) for supplying me with material for this review, and to Dawood Parker (Whitland Research Ltd, Whitland, Wales, UK) for financial support.

REFERENCES

1. L. C. Clark, Monitoring and control of blood and tissue oxygen, *Trans. Am. Soc. Artif. Intern. Organs* **2**:41-48 (1956).
2. D. W. Lübbers, Oxygen electrodes and optodes and their application *in vivo*, *Adv. Exp. Med. Biol.* **388**:13-34 (1996).
3. R. Huch, D. W. Lübbers and A. Huch, in: *Oxygen Measurements in Medicine and Biology*, edited by J. P. Payne and D. W. Hill (Butterworths, London, 1975) pp. 121-138.
4. R. Carter and S. Banham, Use of transcutaneous oxygen and carbon dioxide tensions for assessing indices of gas exchange during exercise testing, *Resp. Med.* **94**:350-355 (2000).
5. M. Mars, A. McKune and J. V. Robbs, A comparison of laser Doppler fluxmetry and transcutaneous oxygen pressure measurement in the dysvascular patient requiring amputation, *Eur. J. Vasc. Endovasc. Surg.*. **16**:53-58 (1998).
6. H. Baumgärtl, W. Zimelka and D. Lübbers, Evaluation of pO_2 profiles to describe the oxygen pressure field within the tissue, *Comp. Biochem. Physiol.* (In Press).
7. P. Vaupel, K. Schlenger, C. Knoop and M. Hockel, Oxygenation in human tumors: Evaluation of tissue oxygen distribution in breast cancers by computerized O_2 tension measurement, *Cancer Res.*, **51**:3316-3322 (1991).
8. M. Kessler and D. W. Lübbers, Aufbau und Anwendungsmöglichkeiten verschiedener PO_2-Elektroden, *Pflüg. Arch. ges. Physiol.* **291**:R82 (1966).
9. M. Kessler, D. K. Harrison and J. Höper, in: *Microcirculatory Technology*, edited by C. H. Baker and W. F. Nastuk (Academic Press, New York, 1986) pp. 391-425.
10. D. K. Harrison, Catchment depths of surface electrodes for H_2 and O_2 in tissue, *Adv. Exp. Med. Biol.* **411**:551-6 (1997).
11. M. Kessler, J. Höper, D. K. Harrison, K. Skolasinska, W. P. Klövekorn, F. Sebening, H. J. Volkholz, I. Beir, C. Kernbach, V. Rettig and H. Richter, Tissue O_2 supply under normal and pathological conditions, *Adv. Exp. Med. Biol.* **169**:69-80 (1984).
12. D. K. Harrison, M. Kessler and S. K. Knauf, Regulation of capillary blood flow and oxygen supply in skeletal muscle in dogs during hypoxaemia, *J. Physiol. (Lond.)* **420**: 431-446 (1989).
13. D. K. Harrison, M. Kessler, S. Birkenhake and S. K Knauf, Local oxygen supply and blood flow regulation in contracting muscle in dogs and rabbits. *J. Physiol. (Lond.)* **422**:227-243 (1990).
14. D. K. Harrison, P. T. McCollum, D. J. Newton, P. Hickman and A. S. Jain, Amputation level assessment using lightguide spectrophotometry, *Prosthet. Orthot. Int.* **19**:139-147 (1995).
15. P. T. McCollum, V. A. Spence and W. F. Walker, Amputation for peripheral vascular disease: the case for level selection, *Br. J. Surg.* **75**:1193-1195 (1988).
16. H. Baumgärtl, A. M. Ehrly, K. Saeger-Lorenz and D. Lübbers, in: *Clinical Oxygen Pressure Measurement*, edited by A. M. Ehrly, J. Hauss and R. Huch (Springer-Verlag, Berlin, 1987), pp. 121-128.
17. W. Wang, C. P. Winlove and C. C. Michel, Distribution of oxygen partial pressure (PO_2) in dermal papillae of human finger nailfolds, *J. Vasc. Res.* **38**:399 (2001).
18. G. Holst, T. Köster, E. Voges and D. W. Lübbers, FLOX – an oxygen-flux-measuring system using a phase-modulation method to evaluate the oxygen-dependent fluorescence lifetime, *Sens. and Actuators B* **29**:231-239 (1995).
19. M. Stücker, P. Altmeyer, A. Struk, K. Hoffmann, L. Schulze, A. Röchling and D. W. Lübbers, The transepidermal oxygen flux from the environment is in balance with the capillary oxygen supply, *J. Invest. Dermatol.* **114**:533-540 (2000).
20. M. Stücker, A. Struk, P. Altmeyer, M. Herde, H. Baumgärtl and D. W. Lübbers, The cutaneous uptake of atmospheric oxygen contributes significantly to the oxygen supply of human epidermis, *J. Physiol. (Lond.)* **538**: 985-994 (2002).

ACCURACY OF A POLAROGRAPHIC ELECTRODE AT HIGH OXYGEN CONCENTRATIONS

Mark D. Rollins, Miles B. Conrad, Thomas K. Hunt, Harriet W. Hopf *

1. INTRODUCTION

Oxygen plays a central role in wound healing. It is required for angiogenesis, collagen deposition, epithelization, and superoxide production for intracellular bacterial killing. Wound oxygen concentration depends upon the complex interactions of blood perfusion, intercapillary distance, local oxygen consumption, arterial oxygen concentration, and oxyhemoglobin dissociation conditions. Accurate determination of wound pO_2 during hyperbaric oxygen (HBO) treatment requires direct measurement. Transport of oxygen to the hypoxic center of a wound, or to a wound with impaired perfusion, is often only possible with HBO exposure[1-3]. Several studies have shown that HBO promotes wound healing, especially in hypoxic wounds[2-7].

In previous studies, human wound and subcutaneous tissue pO_2 levels during a single HBO treatment increased markedly (wound pO_2 values as high as 700 mmHg), but showed a high degree of variability[4, 8, 9]. Given the variation seen in the human case studies, it was first necessary to examine the accuracy of our current oxygen measuring system at high pO_2, prior to beginning any further animal or human studies.

2. METHODS

LICOX polarographic oxygen and temperature probe systems (Medical Systems Corp., Greenvale, NY) are designed for a two point calibration at 0% oxygen and 21% oxygen at 1 ATA pressure. The system assumes a linear relationship between pO_2 and current. Because the oxygen tensions anticipated were well out of this range, we examined the measurement error for gas mixtures ranging from 20% to 100% oxygen at 1 ATA pressure, to determine if error correction was necessary and possible.

Four LICOX polarographic oxygen probes were evaluated at both 37°C and 34°C, and allowed to equilibrate with bubbling oxygen-nitrogen gas mixtures containing 20, 40, 60, 70, 80, 90, and 100% oxygen. The electrode used was the "Microcatheter pO_2

* Mark D. Rollins, Bioengineering Graduate Group, University of California, San Francisco / Berkeley, CA. Miles B. Conrad, Dartmouth Medical School, Hanover, NH. Thomas K. Hunt, Department of Surgery, University of California, San Francisco, CA. Harriet W. Hopf, Departments of Anesthesia and Perioperative Care and Surgery, University of California, San Francisco, CA.

Oxygen Transport to Tissue XXIII, edited by D. Wilson *et al.*
Kluwer Academic/Plenum Publishers, 2003

probe", model C1 (OD=0.47mm, Length = 200 mm, integrating pO_2 sensitive area = 5 mm in length, sensitivity = 2.5×10^{-9} A / mmHg pO_2, gold cathode polarized to 795 mV). The electrode is housed in a weak polyethylene catheter with a thickness of approximately 70µm. Each electrode was initially calibrated in room air (atmospheric pressure = 748 mmHg, temp. = 22°C, and thus pO_2 = 152 mmHg) and had received prior factory calibration under nitrogen (pO_2 = 0). After room air calibration, the probe was inserted through the cover of a 500 mL container filled with normal saline, held at a constant temperature using a circulating water bath, and continuously bubbled with one of the oxygen-nitrogen gas mixtures. Following equilibration, as determined by less than a 1% change in measured pO_2 during a five minute period, the probe and cover were transferred to a second container being bubbled with another of the gas mixtures. This process was repeated in random order until all of the mixtures had been measured at a given temperature. Results from the four probes were combined and differences between actual pO_2 and measured pO_2 were examined as a function of measured pO_2, probe current, current normalized by initial calibration data, and current corrected for temperature. An empiric equation was determined for the relation showing the best correlation.

3. RESULTS

A total of 56 data points were collected from 4 probes in equilibrium with 7 gases containing 20-100% oxygen under room atmospheric conditions (barometric pressure = 748 mmHg) at both 34°C and 37°C. Actual pO_2 for each measured equilibrium point was determined from Eq. (1), where P_B is the barometric pressure, pvH_2O is the vapor pressure of water at the measured temperature (assuming 100% humidity), and FO_2 is the fraction of oxygen in the gas.

$$pO_2 = (P_B - pvH_2O) \times FO_2 \tag{1}$$

The absolute and percentage error for each equilibrium gas condition is displayed in Table 1. The relation between measured pO_2 and actual pO_2 is displayed in Figure 1. The calculation of measured pO_2 is shown in Eq. (2), (LICOX reference manual, Medical Systems Corp., Greenvale, NY).

$$\text{Measured } pO_2 = [(I - IcN_2)/(IcO_2 - IcN_2)](PcO_2)\exp[(TC - TT)(.01 \times TS)] \tag{2}$$

Where I = probe current, IcN_2 = probe calibration current under pure nitrogen, IcO_2 is the probe current during calibration in room air, PcO_2 is the calculated pO_2 in the calibration environment at the moment of IcO_2 determination, TC is the temperature of the calibration environment in °C, TT is temperature of the measured tissue or fluid, and TS is the temperature coefficient of the oxygen dependent fraction of probe current.

The differences between actual pO_2 and measured pO_2 were examined as a function of measured pO_2, the oxygen fraction of probe current ($I-IcN_2$), current normalized by initial calibration data [$(I-IcN_2)/(IcO_2-IcN_2)$], and current corrected for temperature ($(I-IcN_2)\exp[(TC-TT)(.01 \times TS)]$). The relation with the highest correlation for all the data obtained is displayed in Figure 2, showing an exponential relationship between absolute pO_2 error and probe oxygen current. The empiric equation is displayed in Figure 2 (r^2 = 0.84). All experimental equilibration data were corrected using this equation. The error from actual pO_2 after correction is also displayed in Table 1. The relation between measured pO_2 after error correction and actual pO_2 is displayed in Figure 3.

Table 1. The mean[a] absolute and percentage error for each equilibrium oxygen condition[b], displayed for both the original measurement and after correction using the empiric equation in Figure 2

		Original Measurement		After Empiric Correction	
Oxygen Fraction	Actual pO2 (mmHg)	pO2 Error (mmHg)	pO2 Error % Actual	pO2 Error (mmHg)	pO2 Error % Actual
0.20	141	2.7 ± 1.6	1.9 ± 1.2	3.8 ± 2.7	2.7 ± 1.9
0.40	282	8.0 ± 8.2	2.8 ± 2.9	7.7 ± 4.6	2.7 ± 1.7
0.60	423	27 ± 15	6.3 ± 3.6	12 ± 12	2.8 ± 2.9
0.70	494	48 ± 19	9.7 ± 3.9	10 ± 10	2.0 ± 2.0
0.80	564	73 ± 28	13 ± 5.0	20 ± 17	3.5 ± 2.9
0.90	635	108 ± 29	17 ± 4.6	22 ± 22	3.4 ± 3.4
1.00	705	153 ± 29	22 ± 4.1	32 ± 19	4.6 ± 2.7

[a] The means represent data taken from 4 probes at both 37°C and 34°C.

[b] The actual pO2 in the table represents an average between the two temperature conditions (35.5°C) and a barometric pressure = 748 mmHg.

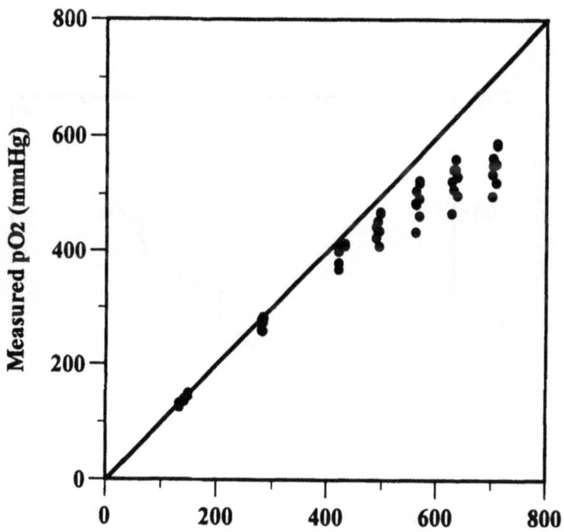

Figure 1. Relation between pO2 measured in vitro using a polarographic oxygen electrode measuring system vs actual pO2. Measurements obtained from 4 probes at both 37°C and 34°C. Diagonal line represents plot position if measured pO2 was equal to actual pO2.

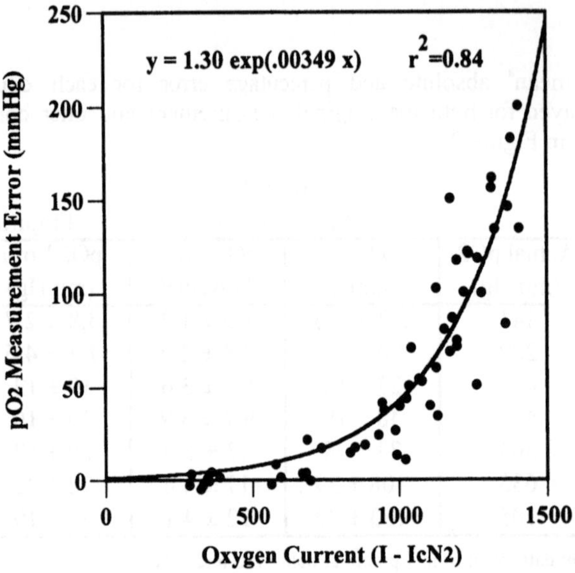

Figure 2. Relation between in vitro oxygen electrode measurement error (Actual pO2 - Measured pO2) vs probe oxygen current. Empiric equation using an exponential fit with correlation coefficient is displayed.

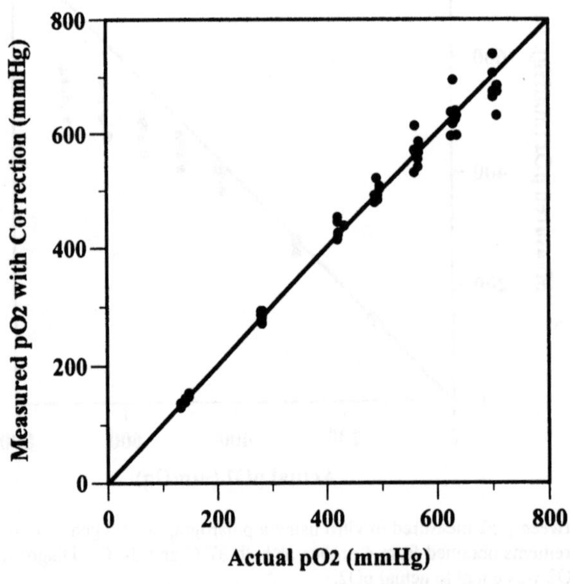

Figure 3. Relation between pO2 measured in vitro after correction using empiric equation from Figure 2 vs actual pO2. Diagonal line represents plot position if measured pO2 was equal to actual pO2.

4. DISCUSSION

The linear relationship between current and oxygen partial pressure, quantified using a calibration between 0 mmHg and room air at 1 ATA, is lost at a pO_2 greater than approximately 400 mmHg. The resulting deviation from linearity follows an exponential relationship with current (I) as shown in Figure 2. Correction of measured oxygen tensions based on this relationship results in a measured pO_2 error < 5% of the actual pO_2 over the entire range studied (140-700 mmHg). This creates a significant improvement over the 6% - 22% error seen in uncorrected measurements of pO_2 between 400 and 700 mmHg. This is extremely useful since individual calibration of each probe at higher pO_2 is not only time consuming, but shortens the useful lifetime of the probe.

The deviation from linearity most likely results from the high current generated at $pO_2 > 400$ mmHg, which causes the potential of the solution at the cathode to fall from its plateau value and the relationship to become nonlinear. This new dependence results in the deviation from linearity depicted in Figure 1. Another possible explanation for the deviation from linearity could be saturation of the current amplifier within the monitoring box. This is less likely since deviation from linearity due to amplifier saturation would be more abrupt than the gradual deviation noted. With correction applied to all measured oxygen tensions in which the current (I) was ≥ 650 (corresponding to a pO_2 approximately ≥ 400 mmHg, depending on the temperature and probe sensitivity), the measurement error is acceptable for HBO wound measurements.

CONCLUSION

HBO treatment greatly increases wound pO_2. With appropriate correction, the polarographic probe used in this study gives accurate measurement for oxygen tensions even as great as 700 mmHg, and is useful for future HBO wound studies.

REFERENCES

1. I. Silver, The measurement of oxygen tension in healing tissue, *Progress in respiration research.* **3**, 124 (1969).
2. J. Davis, The use of adjuvant hyperbaric oxygen in treatment of the diabetic foot, *Clin Podiatr Med Surg.* **4**, 429-437 (1987).
3. E. Faglia, F. Favales, A. Aldeghi, et al., Adjunctive systemic hyperbaric oxygen therapy in treatment of severe prevalently ischemic diabetic foot ulcer. A randomized study. *Diabetes Care.* **19**, 1338-1343 (1996).
4. P. Cianci, T. Hunt, Adjunctive hyperbaric oxygen therapy in the treatment of diabetic wounds of the foot. in: *The Diabetic foot.* 5th ed., edited by M. Levin, L. O'Neal, and J. Bowker (Mosby Year Book: St. Louis, 1993) xxiv, 635 , [612] of plates.
5. J. Niinikoski, Oxygen and wound healing, *Clin Plast Surg.* **4**, 361-374 (1997).
6. E. Uhl, A. Sirsjö, T. Haapaniemi, G. Nilsson, G. Nylander, Hyperbaric oxygen improves wound healing in normal and ischemic skin tissue, *Plastic and Reconstructive Surgery.* **93**, 835-841 (1994).
7. B. Smith, L. Desvigne, J. Slade, J. Dooley, D. Warren, Transcutaneous oxygen measurements predict healing of leg wounds with hyperbaric therapy, *Wound Repair and Regeneration.* **4**, 224-229 (1996).
8. P. Sheffield, Tissue Oxygen Measurements with Respect to Soft-Tissue Wound Healing with Normobaric and Hyperbaric Oxygen, *Hyperbaric Oxygen Review.* **6**, 18-31 (1985).
9. P. Sheffield, Tissue Oxygen Measurements, in: Davis JC, Hunt TK, editors. *Problem Wounds: The Role of Oxygen*, edited by T. Hunt and J. Davis (Elsevier, New York, 1988), pp. 17-51.

RECESSED OXYGEN ELECTRODES: GETTING MORE THAN PO_2

Donald G. Buerk*

1. INTRODUCTION

Several different experimental techniques have been developed using recessed gold cathode (Whalen/Nair-type)[1] O_2 microelectrodes to investigate O_2 transport to tissue. After relatively large O_2 cathodes were first used in tissue over 60 years ago, serious measurement errors were soon recognized, including electrochemical depletion of tissue O_2, sensitivity to blood flow or convective mixing (stirring artifact), signal averaging due to the tip size, and damage to cells or the microcirculation during penetration. To address these problems, O_2 electrodes were miniaturized, although this did not always eliminate PO_2 measurement errors.

2. THEORY FOR ACCURATE PO_2 MEASUREMENTS

The theoretical analysis by Schneiderman and Goldstick[2] identifies the major criteria that are critical for optimizing the design of recessed cathodes. They also briefly reviewed previous mathematical models in the literature, which were often based on bare or membrane covered needle-type microelectrodes. As shown by the lower curve in the left panel of Fig. 1, a bare electrode depletes O_2 from the surrounding medium due to the electrochemical reaction at the cathode and lowers O_2 tension in a relatively large region of tissue, even when the tip is very small. Mathematical models for recessed cathodes[2] predict that O_2 diffusion fields around a microelectrode tip are mostly confined within the recess (length, L) when it is relatively deep compared to the cathode diameter (d), as illustrated in the left panel of Fig. 1. When $L/d = 10$, 95% of the diffusion field is contained within the recess. Even a shallow recess ($L/d = 1$) has a substantial effect, with 67% of the diffusion field contained within the recess. Schneiderman and Goldstick[2] also examined how the recess affects current sensitivity and artifacts that are

* Donald G. Buerk, Departments of Physiology, Bioengineering, and the Institute for Environmental Medicine, University of Pennsylvania School of Medicine, Philadelphia, PA 19104

Oxygen Transport to Tissue XXIII, edited by D. Wilson *et al.*
Kluwer Academic/Plenum Publishers, 2003

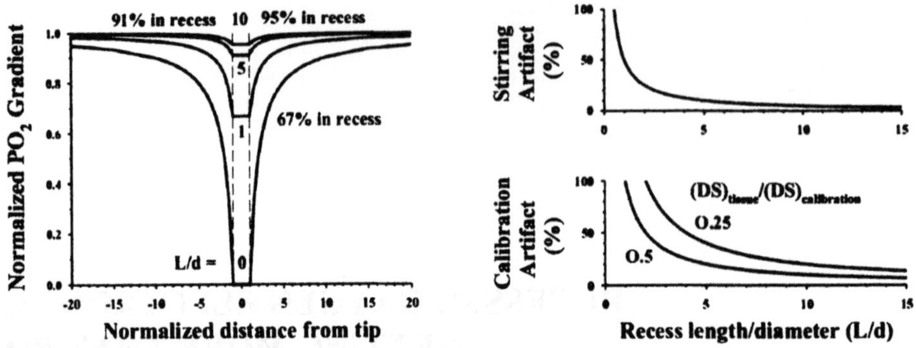

Figure 1. Left panel: Normalized PO_2 gradients in a homogenous medium with initially uniform PO_2 around a microelectrode tip as a function of recess length (L) to tip diameter (d) ratio ranging from 0 (no recess) to 10. Right panels: Increasing L/d reduces the maximum stirring artifact (top) due to convective mixing, and the calibration artifact (bottom) due to differences in diffusion (D) and solubility (S) coefficients in tissue and the calibration fluid.

commonly encountered with O_2 microelectrodes. These include stirring artifacts due to convective mixing and calibration artifacts due to differences between physical properties of tissue and the calibration fluid.

The maximum possible stirring artifact is shown in the top right panel of Fig. 1 as a function of increasing L/d. Tissue diffusion (D) and solubility (S) coefficients for O_2 are not well characterized, but the ratio probably falls within the range shown in the bottom right panel of Fig. 1. Increasing L/d reduces calibration errors, but it might be reasonable to use a calibration fluid with D and S similar to tissue. Grote *et al.*[3] used a ethylene glycol, borax, and phosphate buffer solution to calibrate recessed microelectrodes used in their rat brain studies. Crawford and Cole[4] filled their recesses with a hydrated polymer and found improved performance from tests with agar gels and measurements in canine femoral arteries. There is a practical trade off for increasing L/d or applying membranes since the current sensitivity will be lower and the time response will be slower. Modeling by Schneiderman and Goldstick[2] predicts that good current sensitivity, rapid time response, lack of stirring artifact, and minimal calibration artifacts can be achieved with tip diameters < 5 μm and L/d = 10.

3. PRACTICAL APPLICATIONS FOR RECESSED PO_2 MICROELECTRODES

Since it is not possible to comprehensively review the substantial body of biomedical research using O_2 electrodes in the past 60 years in this short paper, some relevant methods with which the author has direct experience will be discussed.

3.1 *In Vivo* Tissue PO_2 Distributions

Measurements in different organs from numerous laboratories clearly demonstrate that tissue PO_2 is heterogenous. However, it is difficult to compare findings, as noted from a comprehensive study of murine fibrosarcoma and normal subcutis tissue PO_2 measured by the Eppendorf® histograph method in 6 different laboratories[5]. They concluded that these tissue

PO_2 measurements were useful, despite differences in median PO_2 and frequency distributions between laboratories, but recommended caution in interpreting PO_2 data.

We reported very different results for Eppendorf® and recessed microelectrode PO_2 measurements in murine tumors[6]. Frequency distributions are shown in Fig. 2 for 5 untreated (left panel) and 5 VEGF transfected tumors (right panel), along with overall statistics for the 2 sets of recessed microelectrode measurements. Eppendorf® measurements (data not shown) were markedly lower, with 90% of PO_2 values < 2.5 Torr for the untreated group (7 tumors), and a median PO_2 of 2.5 Torr for the VEGF treated group (9 tumors). Since the Eppendorf® needle is much larger (~ 300 μm *vs.* < 5 μm diameter), we suspected that it caused much more damage to the tumor microcirculation.

Braun *et al.*[7] compared recessed PO_2 microelectrode and optical sensor (OxyLite®) measurements in tumors and normal tissues. The larger "measurement volume" of the optical sensor (~ 220 μm diameter) truncates upper and lower tissue PO_2 frequency distribution values compared to PO_2 measurements with microelectrodes. While there were not many statistical differences between mean or median PO_2 among different groups in this study, the optical method consistently produced a significantly higher fraction of hypoxic PO_2 values. This may reflect damage to the microcirculation rather than spatial averaging errors. Another laboratory compared OxyLite® and Eppendorf® PO_2 measurements in murine tumors[8], and found that the fraction of hypoxic values < 2.5 Torr were 50% and 69% respectively, with the Eppendorf® sensor generating significantly more hypoxic values.

It should be recognized that tissue PO_2 is not the whole story when trying to characterize O_2 transport to tissue. Blood flow, capillary density, tissue properties (D,S), O_2 consumption (Q), and Michaelis-Menten kinetics (K_m for O_2 dependant metabolism) are also key parameters that are usually not available to fully interpret the data. Other PO_2 microelectrode methods, summarized below, can be used to obtain additional information.

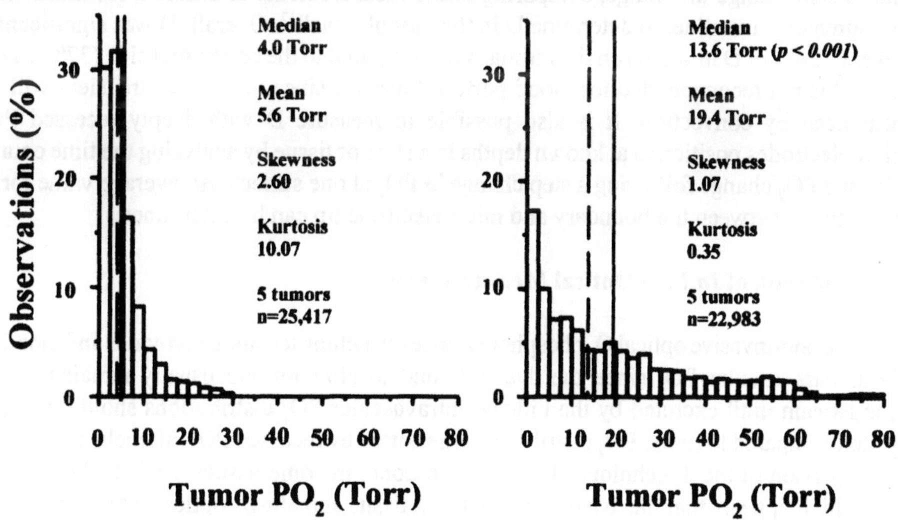

Figure 2. Frequency distribution of recessed PO_2 microelectrode measurements in untreated murine tumors (left panel) and tumors transfected with VEGF (right panel) reported by Polin et *al.*[6].

3.2 Analysis of Steady State PO_2 Gradients

When a one-dimensional tissue PO_2 gradient can be measured experimentally, an analysis of the PO_2 profile can be based on relatively simple diffusion and O_2 consumption models. *In vivo* examples include recessed PO_2 microelectrode measurements across avascular regions of the vascular wall[9] and the eye[10], which have been fit to analytical solutions for simple 1 or 2 layer models with constant Q/DS in each layer. Another *in vivo* example is the determination of O_2 fluxes out of retinal arterioles and into some venules (due to counter-current exchange) in the cat eye by analyzing PO_2 gradients measured in the *vitreous humor* immediately adjacent to small blood vessels[11]. *In vitro* examples include PO_2 profile measurements in brain and liver slices[12], which were fit to numerical solutions of the diffusion equation with Michaelis-Menten O_2 consumption kinetics[13] to determine Q/DS and K_m.

3.3 Determining Q and D From Transient Measurements

We have used recessed PO_2 microelectrodes to measure O_2 disappearance rates after stopping blood flow in gerbil brain[14] and cat carotid body[15] *in vivo* to determine Q and K_m. Bilateral carotid artery occlusion allows the method to be used in the gerbil brain since there are no collateral blood flow sources. In our study of the cortex and hippocampus of the aging gerbil brain[14], we found minimal changes in brain tissue PO_2 distributions with age, which can be explained by the reduction in Q with age that we quantified. We have also conducted numerous *in vitro* studies of neural activity, tissue PO_2, and Q in saline perfused cat and rat carotid bodies, and recently confirmed that nitric oxide inhibits carotid body Q[16].

Properties of bare or very shallowly recessed microelectrodes can be exploited to measure D with detailed spatial resolution in tissue. We analyzed current transients measured after a step change in voltage, comparing tissue measurements at different locations with measurements in saline, to determine D in the vascular wall[17]. Overall, D was significantly lower (~ 24% of D in water) on the intimal side compared to the adventitial side (37%). This method is not recommended for blood perfused tissues, since the current transient will be influenced by convection. It is also possible to measure D with deeply recessed PO_2 microelectrodes positioned at known depths in a slice of tissue by analyzing the time course of tissue PO_2 change following a step change in PO_2 at one surface. An average value for D in the tissue between the boundary and microelectrode tip can be determined.

3.4 Validation of *In Vivo* Optical Measurements

The non-invasive optical O_2 phosphorescence quenching technique provides information about intravascular PO_2 since the dye is bound to albumin and usually remains in the bloodstream until excreted by the kidney. Intravascular PO_2 distributions should be right shifted compared to tissue PO_2 distributions measured by recessed PO_2 microelectrodes, but a comparison of the 2 techniques has not been done. In some tissues, such as the hamster skinfold, it appears that the dye diffuses out into tissue and allows optical measurements of tissue PO_2 to be made. We have confirmed the accuracy of these measurements from recessed PO_2 measurements obtained in a narrow optical window around the tip[18].

In summary, with appropriate experimental methods and data analysis, much more than tissue PO_2 can be determined from recessed PO_2 microelectrode measurements.

ACKNOWLEDGMENTS

Supported by EY09269 from NIH.

REFERENCES

1. Whalen, W.J., Riley, J., and P. Nair, A microelectrode for measuring intracellular PO_2, *J. Appl. Physiol.* **23**, 798-801 (1967).
2. Schneiderman, G., and T.K. Goldstick, Oxygen electrode design criteria and performance characteristics: recessed cathode, *J. Appl. Physiol.* **45**, 145-154 (1978).
3. Grote, J., Laue, O., Eiring, P., and M. Wehler, Evaluation of brain tissue O_2 supply based on results of PO_2 measurements with needle and surface microelectrodes, *J. Autonom. Nerv. Sys.* **57**, 168-172 (1996).
4. Crawford, D.W., and M.A. Cole, Performance evaluation of recessed microcathodes: criteria for tissue PO_2 measurement, *J. Appl. Physiol.* **58**, 1400-1405 (1985).
5. Nozue, M., Lee, I., Yuan, F., Teicher, B.A., Brizel, D.M., Dewhirst, M.W., Milross, C.G., Milas, L., Song, C.W., Thomas, C.D., Guichard, M., Evans, S.M., Koch, C.J., Lord, E.M., Jain, R.K., and H.D. Suit, Interlaboratory variation in oxygen tension measurement by Eppendorf "Histograph" and comparison with hypoxic marker, *J. Surg. Oncol.* **66**, 30-38 (1997).
6. Polin, G.M., Bauer, T.W., Buerk, D.G., Hsi, A., Prabakaran, I., Menon, C., and D.L. Fraker, Eppendorf PO_2 histograph vs. recessed PO_2 microelectrode as methods of measuring in vivo oxygen tension in a murine tumor model, *Proc. Wangensteen Surgical Forum* **55**, 296-298 (2000).
7. Braun, R.D., Lanzen, J.L., Snyder, S.A., and M.W. Dewhirst, Comparison of tumor and normal tissue oxygen tension measurements using OxyLite or microelectrodes in rodents, *Am. J. Physiol. Heart Circ. Physiol.* **280**, H2533-H-2544 (2001).
8. Collingridge, D.R., Young, M.K., Vojnovic, B., Wardman, P., Lynch, E.M., Hill, S.A., and D.J. Chaplin, Measurement of tumor oxygenation: a comparison between polarographic needle electrodes and a time-resolved luminescence-based optical sensor, *Radiat. Res.* **147**, 329-334 (1997).
9. Buerk, D.G., and T.K. Goldstick, Arterial wall oxygen consumption rate varies spatially, *Am. J. Physiol. Heart Circ. Physiol.* **243**, H948-H958 (1982).
10. Linsenmeier, R.A., and L. Padnick-Silver, Metabolic dependence of photoreceptors on the choroid in the normal and detached retina, *Invest. Ophthalmol. Vis. Sci.* **41**, 3117-3123 (2000).
11. Buerk, D.G., Shonat, R.D. and C.E. Riva, O_2 gradients and countercurrent exchange in the cat vitreous humor near retinal arterioles and veins, *Microvas. Res.* **45**, 134-148 (1993).
12. Buerk, D.G. and I.S. Longmuir, Evidence for nonclassical respiratory activity from oxygen gradient measurements in tissue slices, *Microvas. Res.* **13**, 345-353 (1977).
13. Buerk, D.G. and G.M. Saidel, Local kinetics of oxygen metabolism in brain and liver tissues, *Microvas. Res.* **16**, 391-405 (1978).
14. Buerk, D.G. and P. Nair, P_tO_2 and $CMRO_2$ changes in cortex and hippocampus of aging gerbil brain, *J. Appl. Physiol.* **74**, 1723-1728 (1993).
15. Buerk, D.G., Nair, P.K. and W.J. Whalen, Evidence for second metabolic pathway for O_2 from P_tO_2 measurements in denervated cat carotid body, *J. Appl. Physiol.* **67**, 1578-1584 (1989).
16. Buerk, D.G. and S. Lahiri, Evidence that nitric oxide plays a role in O_2 sensing from tissue NO and PO_2 measurements in cat carotid body, In: Oxygen Sensing: Molecule to Man, Eds. S. Lahiri, N. Prabhakar, R.E. Forster II, Plenum Press, New York, N.Y., pp. 337-347 (2000).
17. Buerk, D.G. and T.K. Goldstick, Spatial variation of aortic wall oxygen diffusion coefficient from transient polarographic measurements, *Ann. Biomed. Eng.* **20**, 629-646 (1992).
18. Buerk, D.G., Tsai, A.G., Intaglietta, M., and P.C. Johnson, *In vivo* tissue pO_2 measurements in hamster skinfold by recessed pO_2 microelectrodes and phosphorescence quenching are in agreement, *Microcirc.* **5**, 219-225 (1998).

OXYGEN DISTRIBUTIONS IN TISSUE MEASURED BY PHOSPHORESCENCE QUENCHING

Sergei A. Vinogradov, Pavel Grosul, Vladimir Rozhkov, Isolde Dunphy, Lee Shuman, Benjamin W. Dugan, Sydney Evans, and David F. Wilson*

1. INTRODUCTION

Oxygen dependent quenching of phosphorescence is a powerful new method for measuring oxygen[1-4]. Although it can be adapted for almost any conditions, it is in measuring oxygen in biological systems, particularly tissue *in vivo*, that the advantages are most apparent. Some of the important features of the method are: high specificity, minimal invasiveness, absolute calibration with a linear relationship between the measured parameter and oxygen pressure (Stern-Volmer relationship), high accuracy in the physiological range of oxygen pressures, and, since phosphorescence lifetimes are measured, there is no interference by absorption or fluorescence of other tissue pigments. Phosphors are being synthesized that absorb and emit in the near infrared region of the spectrum and are dendrimer-encapsulated to make them non-toxic and soluble in the blood or other biological fluids. These new phosphors permit oxygen measurements throughout substantial thickness of tissue. New phosphorescence lifetime instruments are being tested that can determine the distributions of phosphorescence lifetimes within heterogeneous samples and therefore can measure the distribution of oxygen (oxygen histogram) within the samples[4].

The current paper is a "progress report" on the development of these new phosphors and instruments. It is designed to indicate the current state of the art and what can be expected to become available in the next year.

2. MATERIALS AND METHODS

Phosphorescence lifetime can be measured in either the time or frequency domains, but the frequency domain instruments have some advantages. When used with frequency lock techniques, they can be very stable and relatively insensitive to interference from ambient light. The phosphorescence lifetimes can be measured with high reproducibility

* Depts. of Biochem. & Biophys, and of Rad. Oncol., Univer. of Pennsylvania, Philadelphia, PA 19104

in individual samples (to better than one part in 10,000) and, because of the absolute calibration, there is no drift over time. The latter means that if the sample is chemically stable, the values measured over intervals of time are completely unchanged.

A new frequency domain phosphorometer has been built that can measure the distribution of oxygen in samples with heterogeneous oxygen levels, such as tissue[4]. Such samples give rise to phosphorescence signals that are the sum of signals with many different lifetimes. This instrument modulates the excitation in a waveform that is the sum of 200 or more frequencies ranging from 100 to 20,000 Hz. The collected phosphorescence signal is then analyzed to give the phase and amplitude for each of the frequencies used. Maximal entropy algorithm is used to deconvolute this data into the distribution of the phosphorescence lifetimes and amplitudes[5]. These recovered distributions of lifetimes and amplitudes are then converted into histograms of the oxygen pressures, indicating the oxygen pressures present and the fraction of the signal with that oxygen pressure. It will be shown that when measurements are made in a grid pattern, contour maps can be constructed showing the volume fraction of the sampled tissue with any selected range of oxygen pressures.

Oxyphor G2 is generation 2 polyglutamic dendrimer of Pd-tetracarboxyphenyl tetrabenzoporphyrin[6,7].

Subcutaneous tumors from 9L glioma cells were grown on the hind quarter of 250 g Fisher rats as previously described[8] and the animals were euthanized before the tumor volume reached 2 cm^3. Oxyphor G2 was injected i.v. as a 6 mg/ml solution in physiological saline to a final amount of 1.5 mg/kg body weight. All procedures were approved by the Institutional Animal Care and Use Committee.

3. RESULTS

3.1. Oxyphor G2, a Phosphor that Absorbs and Emits in the Near Infrared

Oxyphor G2 has a strong absorption band at 632 nm and a phosphorescence emission maximum near 800 nm. Oxyphor G2 is highly soluble in water and the presence of 16 carboxyl groups on the outer layer of the dendrimer gives it a net negative charge at neutral pH. It does, however, bind to albumin and this alters the quenching constant for oxygen. The effect of albumin is saturated when bovine serum albumin is added to a final concentration of 0.5% by weight. Oxygen dependent quenching follows the Stern-Volmer relationship:

$$T^o/T = 1 + k_Q * T^o * pO_2$$

where T^o and T are the lifetimes at zero oxygen and an oxygen pressure pO_2 while k_Q is the second order rate constant for quenching. Calibrations at 38 °C show that the quenching constant is 260 mm $Hg^{-1}sec^{-1}$ and the lifetime at zero oxygen is near 250 µsec. Although no systematic toxicity studies have been carried out on Oxyphor G2, in anesthetized animals there is no effect on the measured physiological parameters when the phosphor is injected or in the later experiment (blood pressure, heart rate, blood gases, blood pH etc.).

3.2. Histograms of Oxygenation in the Microvasculature of Tissue *in vivo*

Oxyphor G2 is fully water soluble and can be prepared in solution in physiological saline for i.v. injection. The amounts needed for oxygen measurements range from less than 1 mg/kg to 4 mg/kg body weight, depending on the sensitivity of the detector used. Figures 1A and 1B show the histograms of phosphorescence lifetime and oxygen measured in normal muscle tissue in the hind quarter of an anesthetized rat, both when breathing air and after the rat was euthanized. The skin was depilated in order to decrease scattering of the excitation light by fur prior to its entering the skin, which can greatly decrease the useful light intensity. The light was conducted to the tissue through a 4 mm diameter light guide with the tip placed in contact with the surface of the skin. The phosphorescence was collected with a similar light guide placed against the surface of the skin approximately 8 mm, unless otherwise indicated, from the excitation light guide. The lifetime histograms in normoxic muscle show a single symmetric peak at approximately 50 μsec and this shifted to 210 μsec after euthanasia. These lifetime histograms were converted to oxygen histograms simply using the Stern-Volmer relationship, without taking into account the intrinsic width of the lifetime and quenching constant distributions. The oxygen histograms in Figure 1B show that in normal tissue the peak oxygen pressure is near 40 mm Hg with a fairly symmetric distribution about the maximum. In general, the oxygen pressures are all above about 20 mm Hg and below 80 mm Hg. The highest oxygen values are the least well measured because, as the oxygen pressure increased, the phosphorescence lifetime and the signal strength decreased. After euthanasia the oxygen distribution rapidly fell to near zero. The calculated width of the peak is likely overestimated due to the intrinsic distribution of lifetimes for the phosphor, but this remains to be established.

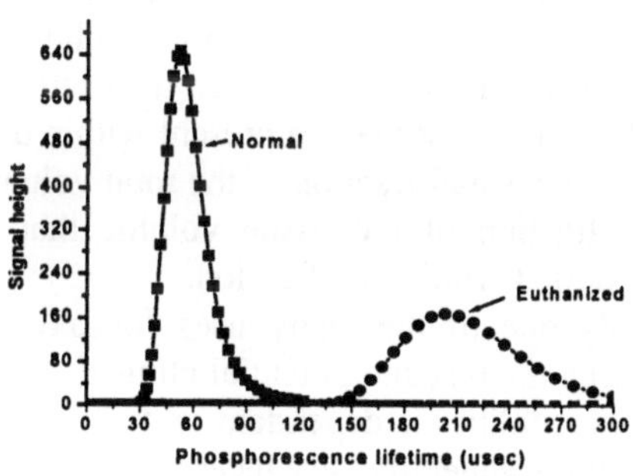

Figure 1A. Phosphorescence lifetimes.

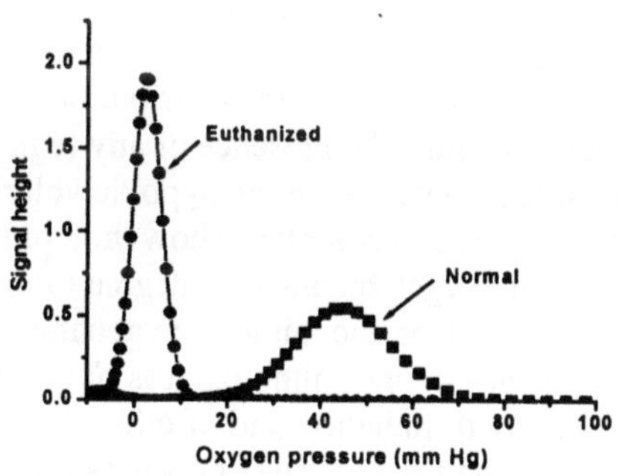

Figure 1B. Oxygen pressures.

Phosphorescence lifetime distribution measurements can provide a quantitative picture of the volume fraction of hypoxic tissue. We have examined a subcutaneous 9L glioma growing on the hind quarter of a rat. The skin over the tumor and surrounding tissue was depilated and a 5 x 6 grid was drawn on the surface. The oxygen histograms

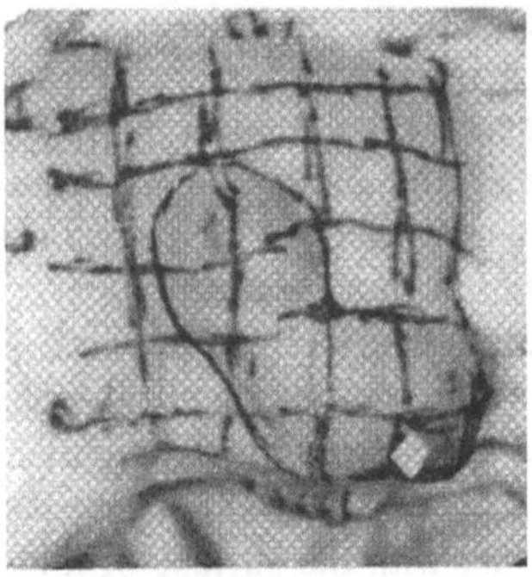

Figure 2A. Picture of rat tumor area.

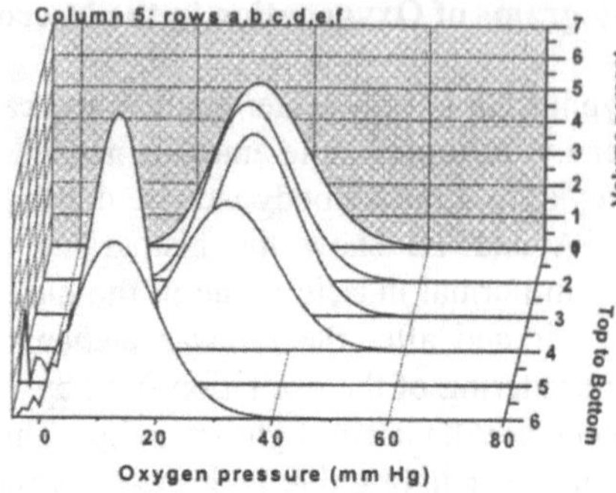

Figure 2B. Oxygen histograms.

were measured as described above. The histograms in regions of "normal" muscle tissue surrounding the tumor show very small fractions (< 3%) with oxygen pressures less than 15 mm Hg. In the tumor area the oxygen levels are markedly lower, and the fraction of the volume with less than 15 mm Hg of oxygen rises as high as 80%.

4. DISCUSSION

The availability of phosphors that both absorb and emit in the near infrared region of the spectrum greatly extends the applicability of phosphorescence quenching method for measuring oxygen. Near infrared light can penetrate a substantial thickness of living tissue, sampling all of the tissue that it passes through. This means that phosphorescence quenching can be used to determine oxygen pressure throughout the volume of a living tissue. When the phosphorescence emission of near infrared phosphors is measured using a multifrequency phosphorometer and analyzed using deconvolution algorithms, it is possible to reconstruct the distribution of oxygen within the tissue. Such distributions (histograms) show the presence of any regions of hypoxia that may be present within that tissue volume, even when the hypoxic volume is only a small fraction of the total volume measured. The histograms show not only the fraction of the tissue volume that is deficient in oxygen but also the degree of oxygen deficit within that fraction.

Evaluation of the tumor oxygenation is only one of the many uses for oxygen measurements *in vivo*, although it is of great research interest and potential clinical value. The efficacy of radiation and of other therapeutic modalities is dependent on the oxygen pressure in the tumor tissue. Oxygenation of tumors is generally very heterogeneous[9], so it is important that as much as possible of the tumor volume be sampled. In addition, the tumor prognosis is reported to correlate with tissue oxygenation, with metastatic potential increasing with increasing hypoxia[10]. Most of the currently available information on the role of oxygen in tumor physiology and therapy relies on oxygen measurements made with oxygen electrodes, particularly the Eppendorf electrode system. As a result, the fraction of tissue volume sampled was small, usually consisting of either a few individual measurements at locations within the tumor or, in the case of the Eppendorf system, a few tracks as the needle electrode was passed through the tumor. The oxygen electrode measurements do not distinguish between viable tissue and necrotic areas containing mostly dead cells. The necrotic areas can not contribute to the tumor growth or therapeutic response, and including them can lead to inappropriate evaluation. Binding of nitroimidazoles such as EF5[9], gives high resolution measurements of oxygen that do not include necrotic areas, but they are "one shot" measurements that require removal of tissue samples for measurement.

Phosphorescence quenching is unique among the currently available methods for measuring oxygen in being able to determine the histogram for each measurement. The measurements are "real time" and can be repeated as often as desired to provide temporal resolutions of a few seconds or less. Histograms provide an additional level of information on the oxygen in heterogeneous samples, information that is particularly valuable for evaluating tissue oxygenation. When the measurement sequences are designed to systematically sample most of the tissue volume, the histograms generated can be used to construct contour maps of the volume fraction of hypoxic tissue in each region of the tissue. Such contour maps are particularly useful when, for example, attempting to determine the fraction of a tumor that is radiologically hypoxic, i.e. has oxygen pressures less than that required for optimal radiation sensitivity. They also accurately depict the location and size of any regions of hypoxia as well as providing quantitative visualization of their degree of hypoxia. As shown here, the technology is now available for tumors grown in animals but clinical application awaits the development and testing of phosphors for human use.

ACKNOWLEDGEMENT. Supported by grants NS-31465 and CA-74062.

REFERENCES

1. Vanderkooi, J.M., Maniara, G., Green, T.J., and Wilson, D.F., An optical method for measurement of dioxygen concentration based on quenching of phosphorescence. *J. Biol. Chem.* 262: 5476-5482, 1987.
2. Wilson, D.F., Rumsey, W.L., Green, T.J. and Vanderkooi, J.M., The oxygen dependence of mitochondrial oxidative phosphorylation measured by a new optical method for measuring oxygen. *J. Biol. Chem.* 263: 2712-2718, 1988.
3. Lo, L.-W., Koch, C.J., and Wilson, D.F., Calibration of oxygen dependent quenching of the phosphorescence of Pd-meso-tetra-(4-carboxyphenyl) porphine: a phosphor with general application for measuring oxygen concentration in biological systems. *Analy. Biochem.* 236: 153-160, 1996.
4. Vinogradov, S.A., Fernandez-Seara, M.A. and Wilson, D.F., Frequency domain instrument for measuring phosphorescence lifetime distributions in heterogeneous samples. *Rev. Sci. Instr.*, 72: 3396-3406, 2001.
5. Vinogradov, S.A. and Wilson, D.F., Recursive maximum entropy algorithm and its application to the luminescence lifetime distribution recovery. *J. Appl. Spect.* 54: 849-855, 2000.
6. Vinogradov, S.A., and Wilson, D.F., Palladium Catalyzed Carbonylation of Br-substituted Porphyrins. *Tetr. Lett.*, 39: 8935-8938, 1998.
7. Vinogradov, S.A., Lo, L.-W. and Wilson, D.F., Dendritic polyglutamic porphyrins: probing porphyrin protection by oxygen dependent quenching of phosphorescence. *Chem. Eur. J.*, 5(4): 1338-1347, 1999.
8. Cerniglia, G.J., Wilson, D.F., Pawlowski, M., Vinogradov, S.A., and Biaglow, J., Intravascular oxygen distribution in subcutaneous 9L tumors and radiation sensitivity. *J. Appl. Physiol.* 82(6), 1939-1945, 1997.
9. Evans, S.M., Hahn, S., Pook, D.R., Jenkins, W.T., Chalian, A.A., Zhang, P., Stevens, C., Weber, R., Weinstein, G., Benjamin, I., Mirza, N., Morgan, M., Rubin, S., McKenna, W.G., Lord, E.M., and Koch, C.J., Detection of hypoxia in human squamous cell carcinoma by EF5 binding. *Cancer Res.* 60(7): 2018-2024, 2000.
10. Brizel, D.M., Dodge, R.K., Clough, R.W. and Dewhirst, M.W., Oxygenation of head and neck cancer: changes during radiotherapy and impact on treatment outcome. *Radiotherapy & Oncol.* 53(2): 113-117, 1999.

ADAPTING AND VALIDATING A NIRS DEVICE TO MEASURE *IN UTERO* FETAL SHEEP CEREBRAL OXYGENATION

James D. Reynolds, David W. Amory, and Paul B. Benni*

1. INTRODUCTION

Compared to the adult, the fetus exists in a relatively hypoxic environment, although, under normal conditions, there is excess oxygen available to meet its metabolic needs. Such adaptations are necessary because the developing central nervous system (CNS) is extremely sensitive to changes in oxygenation. Indeed, a reduction(s) in fetal blood oxygen levels can lead to *in utero* neurologic injury and subsequent life-long postnatal mental, behavioral, and motor deficits. Pathophysiologic insults that decrease delivery of oxygen to the fetus can be readily assumed to decrease the supply of oxygen to the fetal brain. In contrast, to hypothesize that maternal drug exposure specifically alters fetal cerebral oxygenation is somewhat equivocal. One reason for this uncertainty is the impracticability of measuring oxygenation in the intact fetus. To that end, we utilized a near infrared spectroscopy (NIRS) device of our own design and adapted it for *in utero* use. All developmental and experimental studies were conducted using near-term pregnant sheep. To validate our methodology we recorded fetal brain NIRS signals during periods of partial and full umbilical cord occlusion. We then monitored fetal cerebral oxygenation during maternal administration of two different drugs that human fetuses are frequently exposed to: ethanol and the anesthetic isoflurane.

---

* James D. Reynolds, Duke University Medical Center, Durham, NC, 27710. David W. Amory, Institute of Applied Physiology and Medicine, Seattle, WA, 98122. Paul. B. Benni, CAS Medical Inc., Branford, CT, 06405.

Oxygen Transport to Tissue XXIII, edited by D. Wilson *et al.*
Kluwer Academic/Plenum Publishers, 2003

2. JUSTIFICATION FOR USING FETAL SHEEP

There are several factors that make the pregnant sheep ideal for this line of research. In fetal sheep (similar to humans) the majority of brain development occurs prenatally[1], which allows one to examine the effects of drug exposure or pathophysiologic insult during the brain growth spurt. The fetal trunk and limbs, from gestational day 90 onwards, is of adequate size for surgical catheterization and instrumentation. Correspondingly, the fetal head is large enough to allow for implantation of NIRS fiberoptic probes onto the brain as well as providing ample spacing between the emitters and detector. Numerous studies have demonstrated that the pregnant sheep is sufficiently robust to survive such manipulations. With respect to cerebral circulation, the effects of vasoactive substances (such as nitric oxide donor compounds) on isolated fetal sheep and human fetal cerebral blood vessels are similar[2]. This suggests that the same vasoregulatory mechanisms are functional in the two species and that the fetal sheep is a reasonable model of human fetal cerebral circulation. Other pharmacologic similarities, such as the response of sheep spinal dorsal horn to IV opioids[3] and comparable minimum alveolar concentrations for volatile anesthetics[4] also exist between the two species. As such, the sheep has been identified for some time as a useful preparation for testing anesthetic agents[5]. Likewise, the pharmacokinetics and pharmacodynamics of ethanol in the maternal-fetal unit of the pregnant sheep have been well-characterized and are also similar to the human[6]. Collectively, the use of fetal sheep allows for examination of both direct (in this case cerebral oxygenation) and indirect (i.e. maternal, placental or other fetal sites) effects of maternal drug administration. By extension, experiments can be designed and executed that accurately mimic conditions of human fetal drug exposure.

3. SURGICAL INSTALLATION OF THE NIRS PROBES

On the day of surgery, the ewes are sedated by an IM injection of midazolam (0.07 mg/kg); subsequent surgical anesthesia is induced with IV sodium thiopental (500 mg) and maintained with 1-2 % inspired isoflurane in oxygen by mechanical ventilation. Maternal catheters are inserted into the descending aorta and inferior vena cava via the left femoral artery (high-pressure monitoring line) and vein (polyethylene PE 240 tubing), respectively. After this, the fetus is exteriorized and two catheters (PE 90) are introduced approximately 4 and 6 cm into the descending aorta via the left and right femoral arteries. For the hypoxia study (see below), an inflatable occluder is placed around the fetal umbilical cord. The upper body of the fetus is subsequently exteriorized through a separate uterine incision and the fetal head secured in a custom-designed stereotaxic frame. A midline incision is made to the fetal scalp and the skin is peeled back to reveal the skull. A stainless steel frame is secured directly to the skull with steel screws and then a burr hole is drilled through the skull (9.5 mm posterior to the coronal suture) with the frame acting as a guide. The fiberoptic bundle is affixed to the head so that the laser source is in direct contact with the dura. The detector is placed on the skull, approximately 15 mm anterior to the lasers, and the NIRS device activated. Once a strong signal is obtained, the detector is secured in place with dental cement. The fetus is returned to the uterus and the incisions closed. The fiber optic bundles are tunneled subcutaneously and exteriorized (along with the maternal and fetal catheters) through a small incision in the left flank of the ewe. The surgery takes approximately 4h to complete. There is at least a 3-day recovery period prior to experimentation during which post-operative analgesics and prophylactic antibiotics are administered as needed.

4. VALIDATION – NIRS MEASUREMENTS DURING FETAL HYPOXIA

We have used umbilical cord occlusion (UCO) as the definitive method to test the capabilities of our NIRS device because it is reproducible, reversible, and the occurrence of an hypoxic episode can be confirmed by procedures (e.g. fetal arterial blood sampling) independent of the NIRS measurements. Results from a representative animal depicting relative change in Hb, HbO_2, and totalHb, along with change in fetal blood oxygen saturation are presented in the Figure 1A. Basal data was collected for just over 1h, then the fetus was subjected to a 10-min period of partial and then full UCO by inflating the occluder that had been placed around the umbilical cord. As expected, partial UCO was associated with a small decrease in fetal blood oxygen saturation (from 30% to 23%) and then a recovery back to baseline following release of the occluder. Full occlusion produced a large decrease in oxygen saturation (nadir of 8%) followed by a rise upon occluder deflation.

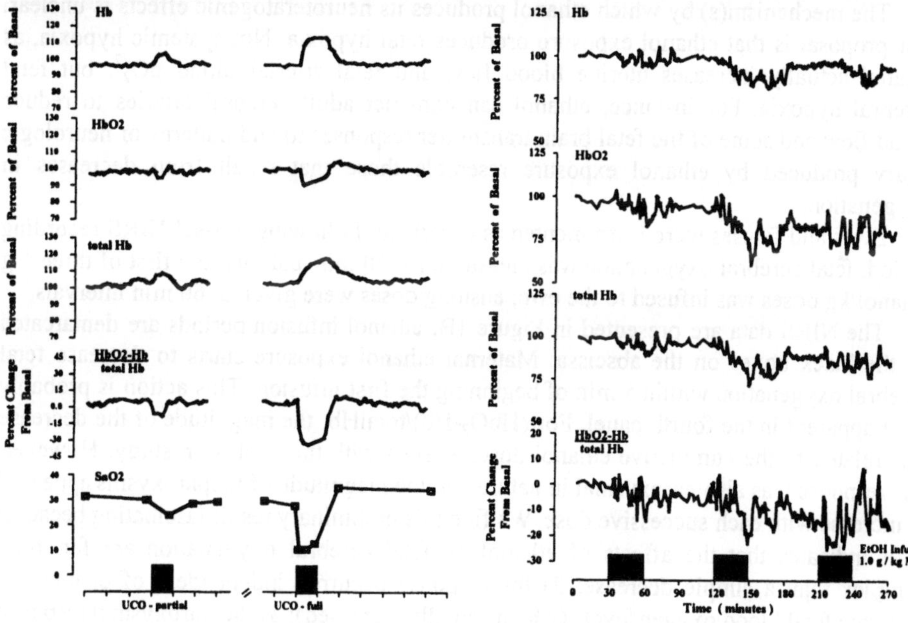

Figure 1A Results from a representative animal depicting relative change in fetal cerebral deoxygenated, oxygenated, and total hemoglobin (Hb, HbO2, and totalHb), along with change in fetal blood oxygen saturation before, during, and after 10-min periods of partial and then full umbilical cord occlusion (UCO). **Figure 1B**. Change in fetal cerebral Hb, HbO2, and totalHb during maternal ethanol exposure. Following a basal NIRS recording period, fetal cerebral oxygenation was measured and then the first of three 1 g ethanol/kg doses was infused to the ewe over 30 min; ensuing doses were given at 60-min intervals.

The NIRS measurements of changes in fetal cerebral oxygenation produced by partial and full UCO reflect the blood gas data. Inflation of the occluder was associated with a rapid increase in Hb and corresponding decrease in HbO_2. Also note the rise in totalHb that occurred during clamping, indicative of increased cerebral blood volume. (Centralization and increases of blood flow to the brain and other vital organs is a well-recognized fetal response to a decrease in blood oxygen content.) Overall change in fetal cerebral oxygenation for this procedure is best reflected by change in the calculated ratio of oxygenated/deoxygenated hemoglobin (the fourth panel). Similar to the blood oxygen measurements, the consequences of full occlusion on fetal cerebral oxygenation were magnifications of the effects of partial occlusion. Indeed, the data reflect the graded response of each of the five parameters to the increasing magnitude of insult. This result was replicated several times to confirm the ability of the NIRS device to measure small and large changes in fetal cerebral oxygenation. Collectively, the study demonstrates that we have developed methodology that enables us to continuously determine *in utero* fetal cerebral oxygenation.

5. DOES ETHANOL PRODUCE FETAL CEREBRAL HYPOXIA?

The mechanism(s) by which ethanol produces its neuroteratogenic effects is unclear. Our proposal is that ethanol exposure produces fetal hypoxia. Not systemic hypoxia, as ethanol actually increases uterine blood flow and fetal arterial blood pO_2[7], but fetal cerebral hypoxia. For instance, ethanol can constrict adult cerebral arteries to reduce blood flow and some of the fetal brain transmitter responses to and patterns of neurologic injury produced by ethanol exposure resemble those that result from decreases in oxygenation.

Ewes and fetuses were instrumented as described. Following a basal NIRS recording period, fetal cerebral oxygenation was measured for 30 min and then the first of three 1 g ethanol/kg doses was infused to the ewe; ensuing doses were given at 60 min intervals.

The NIRS data are presented in Figure 1B; ethanol infusion periods are demarcated by the black boxes on the abscissa. Maternal ethanol exposure starts to decrease fetal cerebral oxygenation within 5 min of beginning the first infusion. This action is probably most apparent in the fourth panel. For (HbO_2-Hb)/totalHb, the magnitude of the decrease was related to the cumulative ethanol dose, as seen with the occlusion study. However, this response was also paroxysmal in nature and the magnitude of theparoxysms appeared to increase with each successive dose. We find this preliminary result fascinating because: 1) it indicates that the affects of ethanol on fetal cerebral oxygenation are far more complex than a simple decrease; 2) the response occurred independent of changes in systemic fetal blood oxygen levels (which actually increased); 3) the paroxysmal response is not unfamiliar as it resembles the fluctuations in extracellular glutamate concentration in the fetal cerebral cortex that we have previously observed following maternal ethanol exposure[8]; and 4) fluctuations in cerebral blood flow in the developing brain have been associated with neuronal injury.

6. MATERNAL ANESTHESIA AND FETAL CEREBRAL OXYGENATION

The need for anesthetic medication can arise at any time during gestation. Amongst women of child-bearing age, the incidence of emergent non-obstetric related surgery is not altered by pregnancy status. This translates into 2% of pregnant women undergoing surgery at some point during gestation; more than 50,000 operations a year in the United States. In addition, the number of *in utero* surgeries conducted annually to correct fetal anomalies is increasing. Emergent cesarean section of both term and preterm parturients also occurs with significant frequency. In all situations, fetal organs, especially the brain, are or can be compromised with respect to the supply and exchange of oxygen. However, exact delineation of the fetal responses is complicated by the inability to separate anesthetic effect from the underlying pathology that necessitated its use. As a first step in such an elucidation, we monitored change in fetal cerebral oxygenation during maternal general anesthesia with isoflurane in the instrumented surgically recovered sheep.

Following a 60-min baseline recording period, the ewe underwent a rapid sequence induction-intubation, and then received 1.5% isoflurane in 100% oxygen. Fetal response to anesthesia is presented in figure 2. Almost immediately after intubation (time 0), fetal cerebral oxygenation started decreasing as seen by the rise in Hb and the decline in HbO_2. Isoflurane also appeared to decrease fetal cerebral blood volume as reflected by the decrease in totalHb. Note: these effects were observed independent of systemic fetal arterial blood oxygen levels: fetal pO_2 and SaO_2 did not change during the monitoring

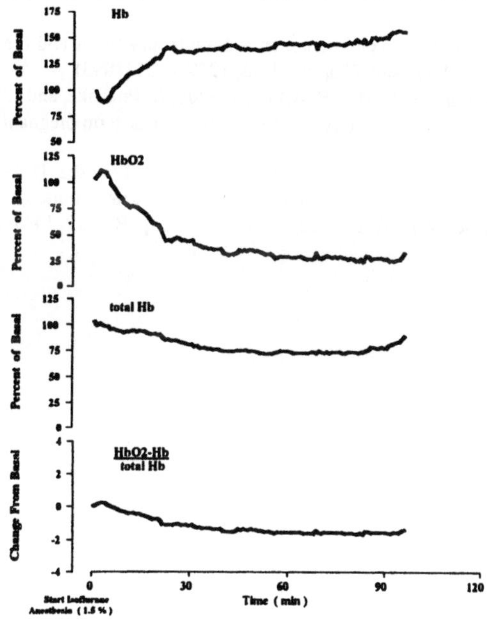

Figure 2. Change in fetal cerebral Hb, HbO2, and totalHb during maternal inspiration of 1.5 % isoflurane in oxygen. Hemoglobin values are expressed as percent basal, determined during a 60 min recording period immediately prior to induction at time 0.

period. One factor to point out is that the significance of the calculated Hb ratio (the fourth panel) is lessened by the reduction in totalHb. Nonetheless, a tentative conclusion from this preliminary work is that maternal inspiration of isoflurane decreases fetal cerebral oxygenation.

SUMMARY

NIRS has provided us with a reliable and sensitive method to monitor *in utero* fetal cerebral oxygenation in the surgically-recovered pregnant sheep. Our preliminary results indicate that monitoring fetal cerebral oxygenation may be important to understanding how maternal drug exposure can affect the fetal brain.

REFERENCES

1. J. Dobbing and J. Sands, Comparative aspects of the brain growth spurt, Early Hum.Dev. **3**, 79-83 (1973).
2. W. J. Pearce and L. D. Longo, Developmental aspects of endothelial function, Semin. Perinatol. **15**, 40-48 (1991).
3. H. Bouaziz, C. Tong, Y. Yoon, D. D. Hood, and J. C. Eisenach, Intravenous opioids stimulate norepinephrine and acetylcholine release from spinalcord dorsal horn: systematic studies in sheep and an observation in a human, Anesthesiology**84**, 143-154 (1996).
4. G. A. Gregory, J. G. Wade, D. R. Beihl, B. Y. Ong, and D. S. Sitar, Fetal anesthetic requirement (MAC) for halothane, Anesth. Analg. **62**, 9-14 (1983).
5. M. Lebeaux, Sheep: a model for testing spinal and epidural anesthetic agents, Lab. Anim. Sci. **25**, 629-633 (1975).
6. J. D. Reynolds and J. F. Brien, Ethanol neurobehavioural teratogenesis and the role of L-glutamate in the fetal hippocampus, Can. J. Physiol. Pharmacol. **73**, 1209-1223 (1995).
7. J. D. Reynolds, D. H. Penning, F. Dexter, B. Atkins, J. Hrdy, D. Poduska, and J. F. Brien, Ethanol increases uterine blood flow and fetal arterial oxygen tension in the near-term pregnant ewe, Alcohol **13**, 251-256 (1996).
8. J. D. Reynolds, D. H. Penning, F. Dexter, B. Atkins, J. Hrdy, D. Poduska, D. H. Chestnut, and J. F. Brien, Dose-dependent effects of acute in vivo ethanol exposure on extracellular glutamate concentration in the cerebral cortex of the near-term fetal sheep, AlcoholClin. Exp. Res. **19**, 1447-1453 (1995).

DETECTION OF THE FAST NEURONAL SIGNAL ON THE MOTOR CORTEX USING FUNCTIONAL FREQUENCY DOMAIN NEAR INFRARED SPECTROSCOPY

Martin Wolf, Ursula Wolf, Jee H. Choi, Rajarsi Gupta, Larisa P. Safonova, L. Adelina Paunescu, Antonios Michalos,, and Enrico Gratton*

1. ABSTRACT

Using non-invasive near infrared spectroscopy fast changes in the range of ms in the optical properties of neurons during brain activity have been described. Since the signal is small, the system to detect it has to be highly noise optimized.

We used a frequency-domain tissue oximeter, whose laser diodes were modulated at 110MHz and the amplitude (AC), mean intensity (DC) and phase (ϕ) of the modulated optical signal was measured at 96 Hz sample rate. In two volunteers, 36 and 37 years old, the probe consisting of 4 crossed source detector pairs was placed above the motor cortex (C3 position), contralateral to the hand performing the tapping exercise. The tapping frequency was set at 2.5 times the heart rate of the subject to avoid the influence of harmonics on the signal. An electronic device recorded the tapping movement. Control-data were obtained from a solid medium of approximately the same optical properties as the human head.

To reduce physiological noise the arterial pulsatility was removed using an adaptive filter, the data was detrended by a high pass filter and a cross correlation function between the optical data and the tapping signal was calculated.

The instrumental noise of the control data was very low (AC mean 0.0015% ± SD 0.00092%, DC 0.00037% ± 0.00023% and ϕ 0.00083° ± 0.00042°). On the head the noise level was AC 0.0042% ± 0.0031%, DC 0.0021% ± 0.0012% and ϕ 0.0020° ± 0.0017°. In 14 DC, 5 AC and 0 ϕ out of 30 locations a fast signal was detected, which was higher ($p < 0.001$) than the noise level. This signal disappeared during non-tapping periods. With the signal to noise ratio that we have achieved single subject measurements become feasible.

* Martin Wolf Laboratory for Fluorescence Dynamics, Department of Physics, University of Illinois at Urbana-Champaign, 1110 W. Green St., Urbana, IL 61801-3080, USA Tel: ++1 217 333 3525 Fax: ++1 217 244 7187

2. INTRODUCTION

Using non-invasive near-infrared spectroscopy (NIRS) fast changes in the range of milliseconds in the optical properties of cerebral tissue have been described, which are associated with brain activity[1]. These changes in light scattering are due to an alteration in the refractive index at neuronal membranes[2], which are simultaneous to electrical changes (action potential), cell swelling and increased heat production[3]. Thus the optical signal is directly related to neuronal activity, which is in contrast to conventional functional NIRS, fMRI (BOLD-signal) or PET, which only detect the slow hemodynamic response to neuronal activity.

The fast neuronal signal is difficult to detect, because the optical changes are small and other physiological signals such as the hemodynamic pulsatility due to systole and diastole are much larger. Therefore the system has to be highly noise optimized.

The aim of this study was to develop an instrumental low-noise system and data analysis algorithms to optimally detect the fast neuronal signal.

3. MATERIAL AND METHODS

We used a frequency-domain tissue oximeter (Oxy-Imager, ISS Champaign, IL, USA). For this instrument and application the largest source of noise is the shot noise, which results from the quantum nature of the light detection. To use higher light intensities is the only way to improve the signal to noise ratio concerning the shot noise. The light of four laser diodes at either 758nm or 830nm was combined to get a higher light intensity at each source location. The light intensity of the laser diodes was modulated at 110MHz and each quadruplet was synchronized to within 1° by specially tuned fiberoptics. The light was guided to the tissue by optical fibers of 400μm diameter. After passing through the tissue the light was collected by fibers of 3mm diameter. Four photomultiplier tubes were used as light detectors. The amplitude (AC), mean intensity (DC) and phase (ϕ) of the modulated optical signal were measured at 96 Hz sample rate.

The probe consisting of 4 crossed source detector pairs arranged in a circle of 3.2cm diameter was placed above the motor cortex (C3 position), contralateral to the hand performing the tapping exercise. The tapping frequency was set at 2.5 times the heart rate of the subject to avoid the influence of harmonics of the hemodynamic pulse due to systole and diastole on the fast neuronal signal, After 1min baseline measurement, the subject was tapping for 20s and resting for 20s and so on for 5min to 10min, followed again by 1min baseline. An electronic device recorded the tapping movement. The subject was seated in a quiet and dark room during the experiment. This study was approved by the Institutional Review Board of the University at Urbana-Champaign.

Control data were obtained from a solid medium of approximately the same optical properties as the human head.

4. SUBJECTS

Two volunteers 36 and 37 years old. Written informed consent was obtained.

5. DATA ANALYSIS

To reduce physiological noise the arterial pulsatility was removed using an adaptive filter (figure 1a), which is described in detail elsewhere[4]. In short it extracts a mean shape of the pulse by screening each trace separately for pulses, whose period is adjusted before the averaging. This mean shape corresponds to the best estimate of each pulse and is used to remove each pulse from the data. This again requires adjusting the period of the mean shape to the one of each pulse and by scaling the shape by a linear regression.

The data was detrended by a digital high pass filter with a variable cut-off frequency. The filter was very sharp (figure 1b). As a cut-off frequency we selected 2.2 times the mean heart rate, i.e. to further reduce the effect of the hemodynamic pulsatility, but still to be low enough not to affect the fast neuronal signal at 2.5 times the mean heart rate.

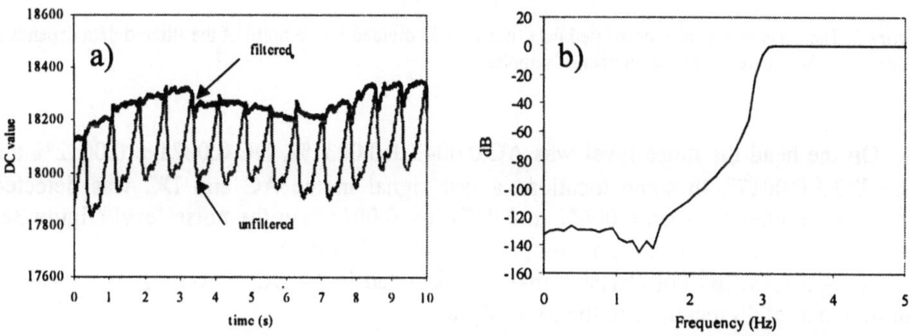

Figure 1. a) The effect of the adaptive filter on the signal caused by the arterial pulsatility: The grey trace corresponds to the data before the filtering, after filtering (black trace), the arterial pulsations are almost completely removed. **b)** The diagram shows the frequency response of the high pass filter. The filter is very sharp and effective.

A cross-correlation function (CCF) between the optical data (OD) and the tapping signal (TS) was calculated: $CCF(\tau)=\Sigma(OD(t)*TS(t+\tau))/\Sigma TS(t)$. The advantage of this equation is, that it removes the amplitude of the tapping signal and preserves the amplitude of the optical data.

The tapping signal was also cross-correlated to optical signals measured on a solid medium to test for instrumental noise. A noise analysis was carried out to determine the level of noise depending on the number of samples. Thus it was possible to identify fast neuronal signals, which are statistically above this level of noise.

6. RESULTS

The instrumental noise of the control data was very low (AC mean 0.0015 ± SD 0.00092%, DC 0.00037 ± 0.00023% and ϕ 0.00083 ± 0.00042°). Figure 2 shows the noise analysis on the solid medium. The data follows exactly the statistical laws.

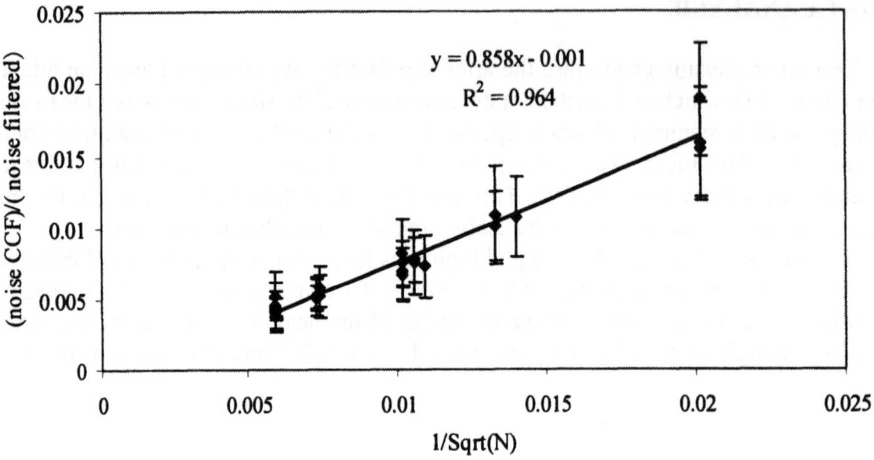

Figure 2. The noise of the cross-correlated data (noise CCF) divided by the noise of the filtered data depends as expected on the squareroot of the number of samples (N).

On the head the noise level was AC 0.0042 ± 0.0031%, DC 0.0021 ± 0.0012% and ϕ 0.0020 ± 0.0017°. In some locations a fast signal in the AC and DC was detected, which was higher between 0.006% to 0.01% ($p < 0.001$) than the noise level (figure 3a). This signal disappeared during non-tapping periods (figure 3b).

A significant ($p<0.001$) fast signal was detected in 14 DC, 5 AC and 0 ϕ locations out of a total of 30 locations in the two subjects.

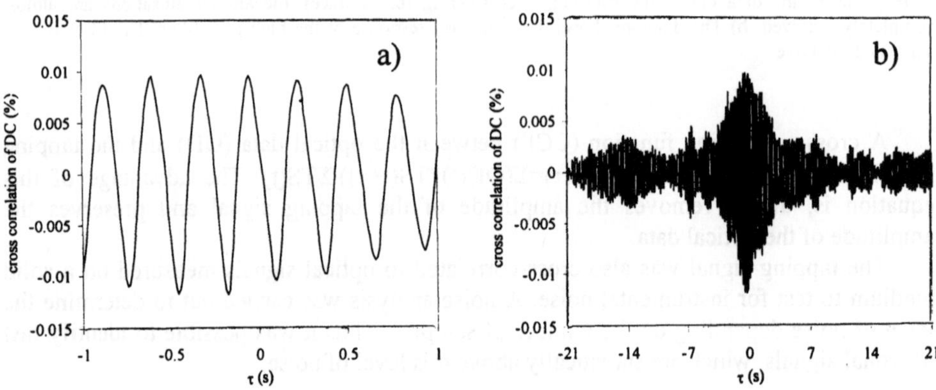

Figure 3. A cross correlation function showing a fast neuronal signal. The same is depicted in higher (a) and lower (b) resolution. τ denotes the period, by which the tapping signal was shifted with respect to the optical signal. As expected the signal is particularly strong for a τ between −1s and +1s, when the tapping and the optical signal are synchronous. At τ=20s the amplitude of the noise can be seen, because the subject was not tapping at this time. This amplitude is much smaller than the one of the fast neuronal signal.

7. DISCUSSION

Our current system and data analysis were studied in a noise analysis on a solid medium. The fast neuronal signal we detected in the two subjects clearly had statistically significant higher amplitudes than the noise.

A movement artifact, i.e. that the signal is caused by a movement of the sensor in relation to the finger tapping, cannot cause the detected fast neuronal signal, because otherwise the signal would appear in all locations.

The main aim of this work was to reduce the noise of every possible source in order to detect the fast neuronal signal optimally with a reasonable amount of repetitions (5min to 10min measurement time). This is crucial for possible clinical applications in the future. So far this signal was only detected using large averages[5] or by using very strong stimuli[6] (direct electrical neuronal stimulation). With the current system it becomes feasible to measure fast neuronal signal in a single subject by using non-invasive stimuli. This could open a new field in the application of functional investigations.

CONCLUSION

We have been able to considerably reduce the noise in the instrumental system and improve data analysis algorithms. With the signal to noise ratio that we have achieved, single subject measurements become feasible.

ACKNOWLEDGEMENTS

This research is supported by NIH grant PHS 2 RO1 CA57032 and RR 10966.

REFERENCES

1. G. Gratton, P. M. Corballis, E. Cho, M. Fabiani and D. C. Hood, Shades of gray matter: noninvasive optical images of human brain responses during visual stimulation, *Psychophysiology* **32**, 505-509 (1995).
2. A. Villringer and B. Chance, Non-invasive optical spectroscopy and imaging of human brain function, *Trends Neurosci.* **20**,435-42(1997).
3. I. Tasaki, Rapid structural changes in nerve fibers and cells associated with their excitation processes. *Jpn. J. Physiol.* **49**, 125-138 (1999).
4. G. Gratton and P. M. Corballis, Removing the heart from the brain: compensation for the pulse artifact in the photon migration signal, *Psychophysiology* **32**, 292-299 (1995).
5. G. Gratton, A. Sarno, E. Maclin, P. M. Corballis and M. Fabiani, Toward noninvasive 3-D imaging of the time course of cortical activity: investigation of the depth of the event-related optical signal, *Neuroimage* **11**, 491-504 (2000).
6. J. Steinbrink, M. Kohl, H. Obrig, G. Curio, F. Syre, F. Thomas, H. Wabnitz, H. Rinneberg and A. Villringer, Somatosensory evoked fast optical intensity changes detected non-invasively in the adult human head, *Neurosci. Lett.* **15**, 105-108 (2000).

Probing Rat Brain Oxygenation with Near-Infrared Spectroscopy (NIRS) and Magnetic Resonance Imaging (MRI)

Yu Chen[1,2], Xavier Intes[1], Dharmesh R. Tailor[2,3], Ravinder R. Regatte[3], HongYan Ma[1], Vasilis Ntziachristos[1], John S. Leigh[3], Ravinder Reddy[3] and Britton Chance[1*]

1. INTRODUCTION

Near-infrared spectroscopy (NIRS) has been a promising technique for non-invasively quantifying tissue chromophores concentrations. One of the potential applications is the monitoring of cerebral blood volume and blood oxygenation saturation. Theories and instruments have been developed to improve the measurement sensitivity and accuracy[1-6]. *In vivo* studies have been performed with calibration of NIRS saturation using the weighted average of global arterial and venous saturations[3,6]. Comparison of NIRS results with other non-invasive imaging modalities such as MRI can provide more information on the co-localized region, and has attracted a lot of interest recently[7,8]. Several studies have reported the temporal correlation between the fMRI BOLD (blood oxygenation level dependent) signal and deoxy-hemoglobin concentration measured by NIRS[8]. To our knowledge, the quantitative relation has not been reported. In this study, we show the correlation of deoxy-hemoglobin concentration as well as oxygen saturation measured with NIRS and the conventional T_2*-weighted BOLD MRI signal for an *in vivo* rat model during inhalation of gas with different concentrations of oxygen (FiO_2).

2. THEORY

Photon migration in biological tissues with high scattering and low absorption can be satisfactorily modeled by the diffusion equation. In monitoring brain function, we chose reflectance geometry, which can be approximated by the semi-infinite boundary condition. The analytical solution for homogeneous medium with this geometry was provided by Pogue *et. al.*[9]:

* [1]Department of Biochemistry and Biophysics, [2]Department of Bioengineering, [3]Department of Radiology, University of Pennsylvania, Philadelphia, PA 19104 USA

Oxygen Transport to Tissue XXIII, edited by D. Wilson *et al.*
Kluwer Academic/Plenum Publishers, 2003

$$R(\mu_a,\mu_s',v,\rho,f)=\frac{1}{2(2\pi)^{3/2}}\left[\frac{z_0(1+k\rho_0)}{\rho_0^3}\exp(-k\rho_0)+\frac{z_p(1+k\rho_p)}{\rho_p^3}\exp(-k\rho_p)\right] \quad (1)$$

where R is the number of photons crossing the tissue boundary per unit time per unit area at distance ρ, μ_a is the medium absorption coefficient (cm^{-1}), μ'_s is the reduced scattering coefficient (cm^{-1}), and v is the speed of light in the medium. Additionally, $k = [(\mu_a v + i\omega)/Dv]^{1/2}$ is the wave propagation factor for a source modulation frequency $f = \omega/2\pi$, $D = 1/(3\mu'_s)$ is the diffusion coefficient (cm), and $\rho_0^2 = \rho^2 + z_0^2$, $\rho_p^2 = \rho^2 + z_p^2$ with $z_0 = 1/\mu'_s$ and $z_p = z_0 + 4D/s$ ($s \sim 0.426$ for tissue-air boundary).

From the measured reflectance R (which is a complex number), we obtain the amplitude and phase of diffuse photon density wave. The amplitude and phase (A and θ) are functions of the medium optical properties, and also relate to the initial amplitude and phase of the instrument. To account for the instrument parameters, calibration of the instrument on a model with known μ_a and μ'_s is performed. By fitting A(medium)/A(model) and [θ(medium) – θ(model)], we can estimate the μ_a and μ'_s of the investigated medium.

The concentration of deoxy- and oxy-hemoglobin can be calculated from the absorption coefficients at multiple wavelengths using the Beer-Lambert law:

$$\begin{bmatrix}\mu_a^{\lambda 1}\\ \mu_a^{\lambda 2}\\ \vdots \\ \mu_a^{\lambda n}\end{bmatrix}=\begin{bmatrix}\varepsilon_{Hb}^{\lambda 1} & \varepsilon_{HbO_2}^{\lambda 1}\\ \varepsilon_{Hb}^{\lambda 2} & \varepsilon_{HbO_2}^{\lambda 2}\\ \vdots & \vdots \\ \varepsilon_{Hb}^{\lambda n} & \varepsilon_{HbO_2}^{\lambda n}\end{bmatrix}\begin{bmatrix}C_{Hb}\\ C_{HbO_2}\end{bmatrix}+\begin{bmatrix}\mu_a^{B,\lambda 1}\\ \mu_a^{B,\lambda 2}\\ \vdots \\ \mu_a^{B,\lambda n}\end{bmatrix} \quad (2)$$

where $\mu_a^{\lambda n}$ is the absorption coefficient at wavelength λ_n ($n \geq 2$), $\varepsilon_x^{\lambda n}$ are the extinction coefficients, C_x are hemoglobin concentrations and $\mu_a^{B,\lambda n}$ is the background absorption coefficient, which takes into account the absorption due to other factors except the hemoglobin, such as lipid, water, cytochrome c oxidase, etc. We use a Nelder_Mead simplex search algorithm to minimize a least square function as maximum likelihood estimator.

3. EXPERIMENTAL DESIGN AND METHODS

3.1. Instrumentation

A frequency-domain multi-wavelength instrument was used in monitoring rat brain oxygenation (Figure 1). The light sources were 690 nm, 750 nm, 780 nm, 810 nm and 830 nm laser diodes, which were modulated at 140 MHz, time-multiplexed and coupled to a 1 mm diameter fiber guide. The light was collected by a 3 mm fiber bundle, and a photomultiplier tube (Hamamatsu, R928) was used to detect the light signal. After amplifying and filtering, the signal from the PMT is mixed with the reference channel through the IQ demodulator, and converted into in-phase (I) and

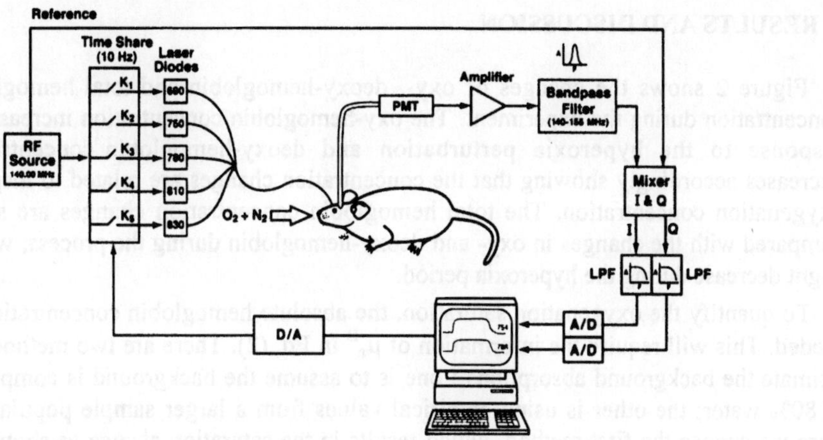

Figure 1. Instrument diagram and Experimental setup for detecting the oxygenation changes in rat brain (PMT: photomultiplier tube; LPF: low pass filter; I & Q: in-phase and quadrature-phase).

quadrature-phase (Q) signals which carry the information of amplitude and phase[10]. The source and detector fibers were placed over the frontoparietal region of the rat brain and separated by 1.5 cm.

3.2. Experiment Protocol

Three rats (Female Sprague-Dawley) were anesthetized with Nembutol® (50mg/kg), and kept with normal body temperature using a waterbed throughout the experiment. A mixture of oxygen and nitrogen was delivered to them through a gas flow meter and a cylindrical mask. The inspired oxygen concentration (FiO_2) was incremented evenly from 20% to 100% by 20% steps, and returned back to baseline (20%) after each perturbation. The stimulation period lasted 1 minute and the resting period was 2 minutes.

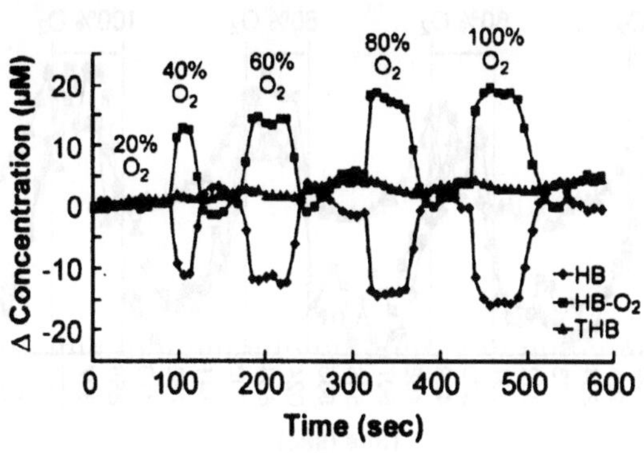

Figure 2. Changes of deoxy-hemoglobin (HB), oxy-hemoglobin (HB-O2) and total hemoglobin (THB) concentrations during the inhalation of gas with different FiO_2.

4. RESULTS AND DISCUSSION

Figure 2 shows the changes of oxy-, deoxy-hemoglobin and total hemoglobin concentration during the experiment. The oxy-hemoglobin concentration increases in response to the hyperoxia perturbation and deoxy-hemoglobin concentration decreases accordingly showing that the concentration changes are related to inspired oxygenation concentration. The total hemoglobin concentration changes are small compared with the changes in oxy- and deoxy-hemoglobin during the process, with a slight decrease during the hyperoxia period.

To quantify the oxygenation saturation, the absolute hemoglobin concentration is needed. This will require the information of μ_a^B in Eq. (2). There are two methods to estimate the background absorptions[3]: one is to assume the background is composed of 80% water; the other is using empirical values from a larger sample population. Here we choose the first method, which results in the saturation change as shown in Figure 3, from which we can see that in the normal state, the oxygen saturation is

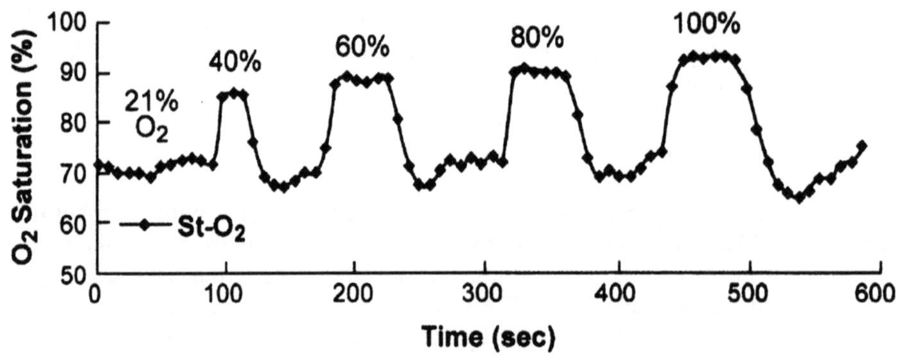

Figure 3. Change of oxygen saturation (St-O2) during the inhalation of gas with different FiO_2. (Rat #1)

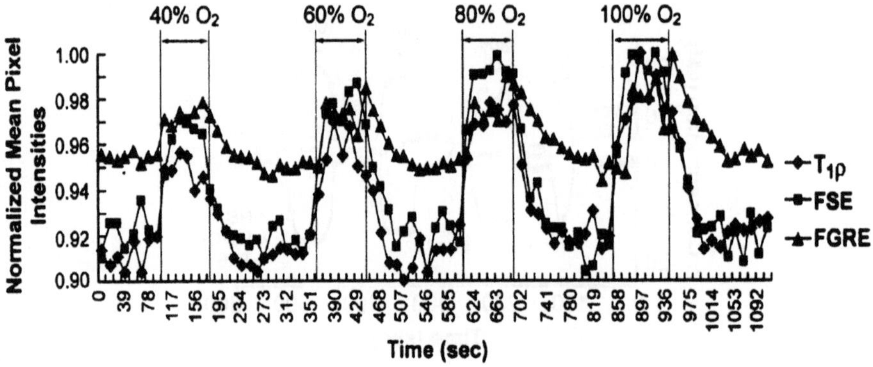

Figure 4. The magnetic resonance signals (T_1 - ρ, FSE and FGRE) for rat brain mapping during the inhalation of gas with different FiO_2.

around 70%. After inhalation of gas with a higher oxygen concentration than the normal state (21%), the saturation increases to above 80% and the higher the FiO_2, the slightly higher the oxygen saturation. This trend is repeated on the other two rats we had tested. Although there exist individual differences, the changes of oxygen saturation in those three rats are similar. The increment is large from the normal physiological state to hyperoxia, and the increases are much slower when further increasing the FiO_2.

MRI study was performed on a different set of 3 rats under the same experimental paradigm as described in a companion paper[11]. Figure 4 shows representative normalized mean pixel intensities for the ROI consisting of bilateral frontoparietal cortex in a single coronal brain slice in $T_1\rho$-weighted, FSE (fast spin echo) and FGRE (fast gradient recall echo) images under different FiO_2. All three methods show the physiological changes brought on by variations in FiO_2.

To study the quantitative relations between BOLD MRI signals and the NIRS signal, we normalized the changes for both types of signals. For MRI signal, we choose the conventional T_2*-weighted BOLD signal (FGRE). For NIRS signal, we select the deoxy-hemoglobin concentration and oxygen saturation to calculate the correlation. Figure 5 shows the correlation between the NIRS signals (oxygen saturation and deoxy-hemoglobin concentration) and MRI signal with high correlation coefficients ($c \geq 0.98$), which means those two kinds of signals are linearly related.

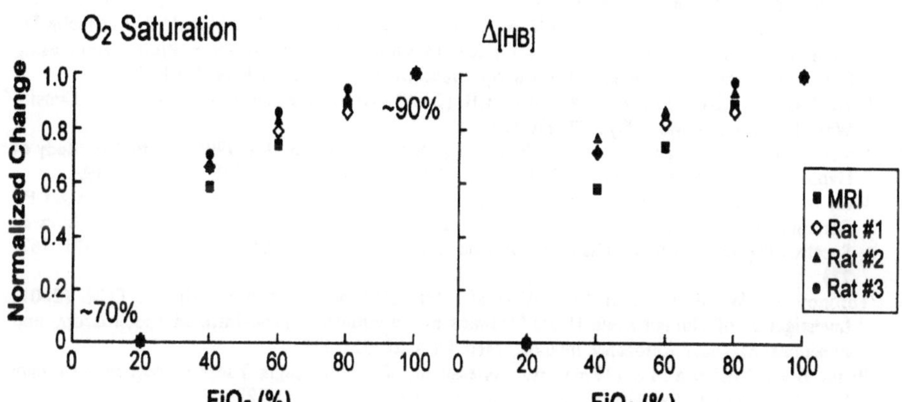

Figure 5. Correlation between MRI signal and averaged NIRS (A) oxygen saturation ($c = 0.99$) and (B) deoxy-hemoglobin concentration ($c = 0.98$).

Because the BOLD signal is proportional to the changes in deoxy-hemoglobin concentration[8], the results in Figure 5 (B) show reasonable correlation. While the data do not show the exact same curve, which could be due to the error in our simplified model, or the difference between two sets of experimental subjects, note that the oxygen saturation also correlated very well with the BOLD signal, which can be explained by the small changes in total hemoglobin concentration.

In conclusion, we have observed significant NIRS signal changes from normoxia to hyperoxia, from which we can calculate the changes in blood oxygenation saturation and hemoglobin concentration. And the changes in deoxy-hemoglobin correlate well with the BOLD fMRI signal changes sampled in the co-localized region. This correlation encourages further investigation, which might help to probe the tissue saturation and hemoglobin concentration more accurately for these two modalities.

ACKNOWLEDGMENTS

We would like to thank Mr. T. Durduran for relevant discussions. This study is supported in part by NIH grant CO97065.

REFERENCES

1. Villringer A., Chance B. (1997) Non-invasive Optical Spectroscopy and Imaging of Human Brain Function. Trends Neurosci. 20: 435-442.
2. Sevick E.M., Chance B., Leigh J., Nioka S., Maris M. (1991) Quantitation of Time- and Frequency-Resolved Optical Spectra for the determination of Tissue Oxygenation. Anal. Biochem. 195: 330-351.
3. Hueber D.M., Franceschini M.A., Ma H.Y., Zhang Q., Ballesteros J.R., Fantini S., Wallace D., Ntziachristos V., Chance B. (2001) Non-invasive and Quantitative Near-Infrared Haemoglobin Spectrometry in the Piglet Brain during Hypoxic Stress, using a Frequency-Domain Multidistance Instrument. Phys. Med. Biol. 46: 41-62.
4. Fantini S., Hueber D.M., Franceschini M.A., Gratton E., Resonfild W., Stubblefield P.G., Maulik D., Stankovic M.R. (1999) Non-invasive Optical Monitoring of the Newborn Piglet Brain using Continuous Wave and Frequency Domain Spectroscopy. Phys. Med. Biol. 44: 1543-1563.
5. Ntziachristos V., Kohl M., Ma H.Y., Chance B. (2000) Oximetry based on Diffuse Photon Density Wave Differentials. Med. Phys. 27: 410-421.
6. Ma H.Y., Xu Q., Ballesteros J.R., Ntziachristos V., Zhang Q., Chance B. (1999) Quantitative Study of Hypoxia Stress in Piglet Brain by IQ Phase Modulation Oximetry. Proc. SPIE 3597: 642-649.
7. Wolf M., Weber O., Keel M., Golay X., Scheidegger M., Bucher H.U., Kollias S., Boesiger P., Banziger O. (1999) Comparison of Cerebral Blood Volume measured by Near Infrared Spectroscopy and Contrast Enhanced Magnetic Resonance Imaging. Adv. Exp. Med. Bio. 471: 767-773.
8. Toronov V., Webb A., Choi J.H., Wolf M., Michalos A., Gratton E., Hueber D.M. (2001) Investigation of Human Brain Hemodynamics by Simultaneous Near-Infrared Spectroscopy and Functional Magnetic Resonance Imaging. Med. Phys. 28: 521-527.
9. Pogue B.W., Patterson M.S. (1996) Error Assessment of a Wavelength Tunable Frequency Domain System for Noninvasive Tissue Spectroscopy. J. Biomed. Opt. 1: 311-323.
10. Yang Y., Liu H., Li X., Chance B. (1997) Low Cost Frequency-Domain Photon Migration Instrument for Tissue Spectroscopy, Oximetry and Imaging. Opt. Eng. 36: 1562-1569.
11. Tailor D.R., Regatte R.R., Chen Y., Intes X., Chance B., Leigh J., Reddy R., this volume.

VARIATION IN BLOOD OXYGENATION AND CEREBRAL pO_2 IN A MOUSE MODEL MEASURED BY EPR SPECTROMETRY

Matthew P. Thomas[1], Simon K. Jackson[2] and Philip E. James[3]

1. INTRODUCTION

The partial pressure of oxygen (pO_2) is one of the most important parameters in many physiological and pathological processes. As a highly aerobic organ, the brain is very sensitive to changes in oxygen availability [1-4]. The importance of a sufficient oxygen supply is reflected in the brain through many complex mechanisms that ensure adequate blood flow and oxygen delivery. Methods for investigating brain oxygenation are therefore of importance especially in pathological conditions. In the past, one of the primary methods for measuring pO_2 in organs such as the brain, was to use small oxygen-sensitive electrodes. These had several disadvantages including; mechanical damage to the tissue, consumption of oxygen by the electrode itself and the inability to make repeated measurements from the same site [5, 6]. Over the past 10 years, oxygen measurements made in the brain using near infrared absorption or phosphorescence quenching have become more common. These methods have the advantages of being largely non invasive compared to using oxygen electrodes but can be limited to making measurements in tissue to a certain depth, or do not directly measure pO_2. EPR oximetry has some advantages over other oximetric methods including; the ability to directly measure pO_2, the ability to obtain repeatable measurements at well resolved sites, and a high degree of accuracy and sensitivity across a range of oxygen levels [7].

Electron Paramagnetic Resonance (EPR) oximetry was employed in this investigation to measure mouse brain oxygenation [8]. It centres on the introduction of specific paramagnetic materials to the tissue of interest, where their interaction with molecular oxygen can be observed and quantified by EPR spectroscopy [9]. Significant

[1] To whom correspondence should be addressed; Department of Cardiology, Wales Heart Research Institute, University of Wales College of Medicine, Heath Park. Cardiff, UK CF14 4XN.
[2] Department of Medical Microbiology, Cardiff, UK.
[3] Department of Cardiology, Cardiff, UK.

advances in EPR oximetry have been made possible by the development and characterisation of new oxygen sensitive paramagnetic probes, and low frequency EPR spectrometers (around 1 GHz) which are more suitable for use with living tissue. Its major disadvantages are the small degree of invasiveness required, such as the implantation of Lithium Phthalocyanine (LiPc) crystals into the mouse brain, the high cost of the instruments, and its current limitation to small animals.

The aim of this study was to develop a model system to investigate the relationship between blood oxygenation and cerebral tissue pO_2 in a mouse model. LiPc was the probe of choice to report on cerebral tissue pO_2 as it has a high spin density providing excellent signal : noise. It also has good sensitivity to oxygen in terms of the magnitude of changes of line widths per unit of oxygen across a wide range [8]. Measurements of blood pO_2 were made using the soluble free radical Oxo 63 (or Trityl). As a soluble material it has the advantage of not requiring surgery for administration of the probe to the site of interest. Trityl diffuses freely into the blood providing a uniform sampling environment. It provides very strong Lorentzian line shapes that broaden modestly, but reproducibly in the presence of oxygen.

2. MATERIALS & METHODS

2.1 Materials & Oxygen Calibration

LiPc was obtained as a gift from Oleg Y. Grinberg and Harold M. Swartz of Dartmouth Medical School, Hanover, NH, USA. LiPc oxygen calibrations were conducted on a Varian E-104 X-Band spectrometer, linked to a microwave bridge operating at 9.45 GHz. Calibrations were performed by placing a single crystal in a NaCl solution, in oxygen permeable teflon tubing. It was then exposed to various oxygen concentrations by mixing N_2 gas and an air pump (ranging from 0-21% O_2 at 37°C). The modulation amplitude was set at less than one third of the EPR line width at 40-dB to avoid over modulation and power saturation effects, which also affect line widths.

Trityl was obtained as a gift from Nycomed-Amersham. Oxygen calibrations were made using a Varian E-104 X-Band and a Bruker 1.1 GHz L-Band EPR spectrometer. A trityl solution was calibrated in oxygen permeable teflon tubing at a range of oxygen concentrations (0-21% at 37°C) using the X-Band spectrometer. Modulation amplitude was set to one third of the nitrogen line width (0.04 G) with a power of 30 dB. The calibration curve was found to be linear from 0-21% O_2.

Trityl oxygen calibrations were also conducted *in-vivo* through an I.P. injection to a mouse (400μl I.P. - 15mg/200μl diluted ¼ in NaCl). After approximately 40 minutes, the mouse was sacrificed by cervical dislocation. Spectra were recorded using the L-Band EPR spectrometer until a minimal line with was achieved, which represented the zero oxygen point. This value was confirmed by a comparison with the previous calibration at X-Band. The blood from another mouse (given the same trityl solution), was collected after 40 minutes, and plasma extracted. Spectra from the plasma were recorded at 37°C in air to give the 158 mmHg oxygen point. Both paramagnetic probes were found to have line widths that are a linear function of the pO_2 from 0 to 158 mmHg O_2.

In order to use Trityl to measure blood pO_2 we must consider the following:

- The line widths of the spectra could be converted to pO_2 if the calibration was performed in mouse blood.
- The spectrum produced arose from blood in the cerebral circulation and not the cerebral tissue itself.
- The trityl signal will reflect the average blood pO_2 from all over the cerebral circulation (arterial and venous).

2.2 Animal Preparation

Female Balb/c mice (obtained from Biomedical Services, UWCM, UK) weighing 20-23g were used in these studies. Three to five days prior to EPR oximetry, mice to be implanted with LiPc crystals were anaesthetized using halothane at 3% and maintained under anaesthesia at 1.5% halothane, in a carrying gas of 95% O_2. Body temperature was maintained at 37°C during the operation using a heated bed. LiPc crystals were implanted into the right cerebral hemisphere (1 – 1.5 mm from the midline, via a scalp incision at the level of the Bregma). This was achieved by drilling a small hole in the cranium by gently rotating a 23 gauge needle. A 22 gauge needle containing the crystals was then positioned in the hole. Crystals were then deposited by pushing a stylus through the barrel of the needle to a previously specified depth.

2.3 Trityl Validation

Trityl solubility in aqueous and lipid solutions were also investigated. Two trityl solutions, one containing 100μl (15mg/200μl NaCl, diluted 1/1000) in 2 ml of H_2O, the second containing the same amount of trityl in 1ml of H_2O and 1ml of olive oil. Both solutions were vortex mixed and given 24 hours to settle. By transferring a sample of each separated component into a quartz flat cell, the degree of trityl solubility was estimated by comparing the concentration of trityl in each of the samples using EPR spectroscopy.

Female Balb/c mice weighing 20-23g were also used to investigate the proportion of trityl signal arising from blood. Mice were injected with trityl as described in section 2.1, and brought under anaesthesia with 3% isoflourane and maintained at 1.5% isoflourane, in a carrying gas of 24% O_2. Spectra were recorded at L-Band until maximal signal amplitude was achieved, representing maximal trityl absorbance to blood. Mice were then exaunguinated by cardiac puncture to remove as much blood as possible. By comparing signals recorded prior to and after exangunation, the proportion of trityl signal arising from the blood was estimated.

2.4 Electron Paramagnetic Resonance & Oximetry

All in-vivo oximetry measurements were made on a Bruker 1.1 GHz L-Band EPR spectrometer with a 34mm birdcage resonator (active length 20 mm). It is of low Q design, making it suitable for samples with strong loading characteristics, such as animal tissue. The resonator has built in field modulation coils that require water cooling when gradients are applied. As gradients were not applied in the oximetry studies, this feature was used to heat the birdcage resonator to 37°C using an external heating water circulator to maintain body temperature, also monitored for the duration of the experiment. Mice were maintained under anaesthesia (1.5% isoflourane, in a carrying gas of 24% O_2) for the duration of the study. Ten 1 minute spectra were accumulated and averaged to maximise the line width accuracy. Modulation amplitude was always kept to one third of the line width.

3. RESULTS

Partition of the trityl between aqueous and lipid phases showed it to be highly lipophobic. We found no trityl spectra in the oil once the solution was permitted to equilibrate, with all the trityl being recovered in the aqueous compartment. Animal bleed experiments showed most of the trityl signal was lost from the brain, following cardiac puncture. Measurements taken before and after bleeding the animals (n=3) show 61.9% ±2.8 of the signal was lost. It was impossible to remove all the blood from the cerebral circulation by cardiac puncture, and undoubtedly some trityl remained in the smaller vessels of cerebral circulation.

The average pO_2 from cerebral tissue and blood over the 200 minute investigating period is shown in figure 1. The pO_2 in both locations was found to slowly decrease, with a small difference between blood and cerebral tissue pO_2 in the mouse brain (<15.9 mmHg, mean difference 8.8 ±3.7). Statistical analysis found the average of both lines to be significantly different at the 95% confidence level, however only 30, 40, 60, 180 and 190 minute time points were significantly different from each other. Cerebral pO_2 was stable for 130 minutes before the decrease in pO_2 was found to be significant (compared

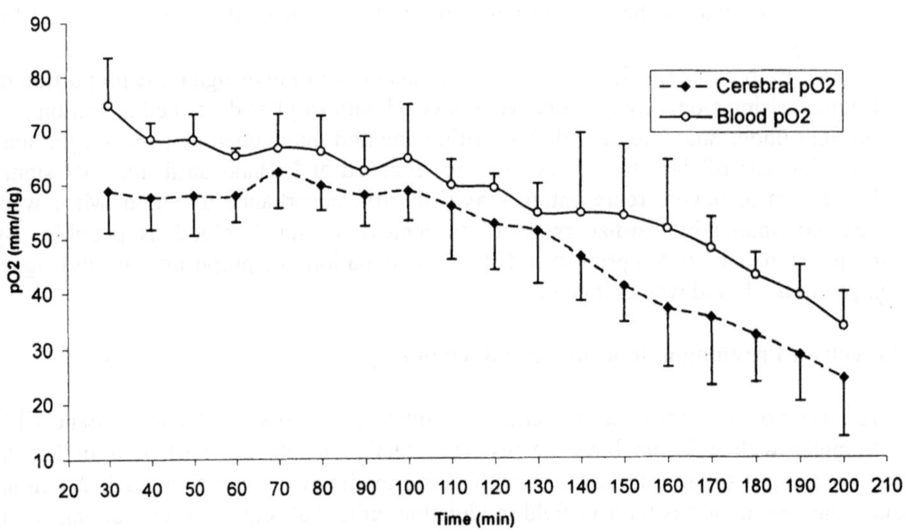

Figure 1. LiPc Vs Trityl, the average cerebral and blood pO_2 in a mouse brain over time .

to pO_2 at 30 minutes). By comparison, blood pO_2 was stable for 80 minutes. The cerebral blood shows a slightly greater total decrease in pO_2 compared to cerebral tissue. Cerebral pO_2 fell from 58.8 mmHg (±7.7) at 30 minutes to 24.4 mmHg (±10.5) at 200 minutes, an average fall of 34.4 mmHg over the experiment (n=6). Cerebral blood pO_2 fell from 74.7 mmHg (±8.8) at 30 minutes to 33.7 mmHg (±6.4) at 200 minutes, an average decrease of 41 mmHg (n=5).

The systolic blood pressure measured using a tail cuff over the same investigating period (n=6) is shown in figure 2. It fell from an average of 114 mmHg (± 8.3) to 72.3 mmHg (± 4.4), a change of 41.7 mmHg, similar to the fall in blood pO_2.

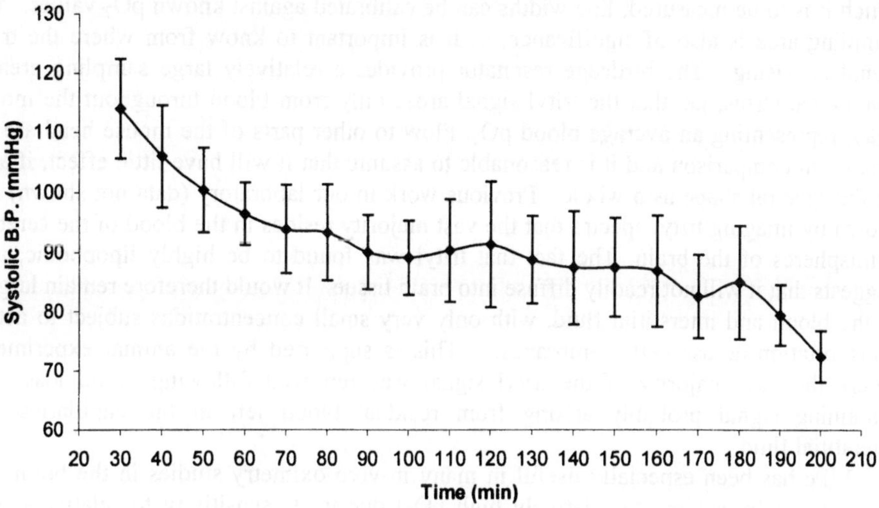

Figure 2. The average systolic blood pressure over time.

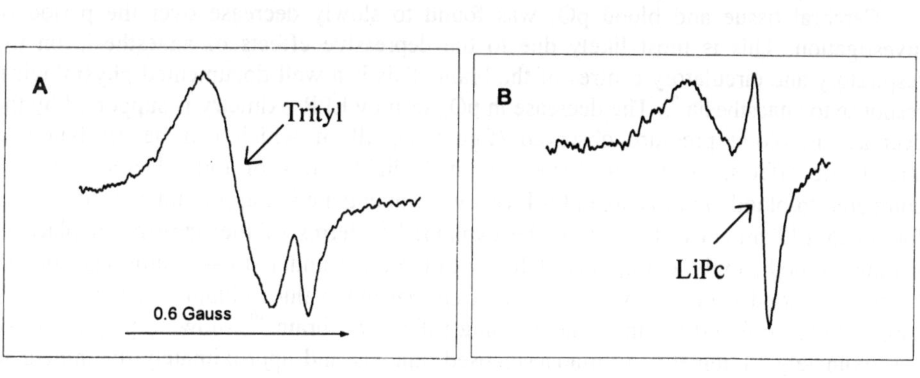

Figure 3. The spectra showing differentiated signals produced by trityl and LiPc. This occurs as trityl is subject to 'washout' from the circulation (A) and at very low pO_2 levels where the LiPc line width decreases and signal amplitude increases (B).

Under certain conditions (i.e. low trityl concentration & low pO_2), both trityl (in blood) and LiPc (cerebral tissue) sites could be differentiated and monitored simultaneously, see figure 3.

4. DISCUSSION

Measuring pO_2 with trityl is inherently more complicated than using particulate paramagnetic materials, because trityl line widths reflect the collision rate with molecular O_2 (a product of O_2 solubility and O_2 diffusion in the media being investigated). Thus it does not directly measure pO_2. However, as long as trityl is calibrated in the medium in which it is to be measured, line widths can be calibrated against known pO_2 values. The sampling area is also of significance, as it is important to know from where the trityl signal is arising. The birdcage resonator provides a relatively large sampling area in anatomical terms, i.e. that the trityl signal arose only from blood throughout the mouse head, representing an average blood pO_2. Flow to other parts of the mouse head are far smaller in comparison and it is reasonable to assume that it will have little effect, if any, on the spectral shape as a whole. Previous work in our laboratory (data not shown) has shown by imaging trityl spectra that the vast majority resides in the blood of the cerebral hemispheres of the brain. The fact that trityl was found to be highly lipophobic, also suggests that it will not readily diffuse into brain tissue. It would therefore remain largely in the blood and interstitial fluid, with only very small concentrations subject to active transportation across cell membranes. This is supported by the animal experiments where the vast majority of the trityl signal was removed following blood loss. The remaining signal probably arising from residual blood left in the capillaries and interstitial fluid.

LiPc has been especially useful in many in-vivo oximetry studies in the brain, (an organ normally subject to relatively high pO_2) due to its sensitivity to relatively high concentrations of oxygen. pO_2 measurements with LiPc are highly localised and are ideal for reporting on site specific tissue pO_2. The resultant spectra are the average pO_2 of the tissue in contact with the particle.

Cerebral tissue and blood pO_2 was found to slowly decrease over the period of investigation. This is most likely due to the depressive effects of anaesthesia on the respiratory and circulatory centres of the brain. This is a well documented physiological response to anaesthesia [6]. The decrease in pO_2 seen by EPR oximetry is supported by the decrease in blood pressure observed (figure 2), all of which can be attributed to anaesthetic effects. Cerebral tissue exhibited slightly less of total decrease in pO_2 compared to blood, and average pO_2 levels were maintained at approximately 55 mmHg for up to 130 minutes. This might be explained in terms of mechanisms in place to maintain blood supply and oxygen delivery to the brain under stress conditions, such as 'shunting' blood from less vital organs. Cerebral pO_2 values obtained using LiPc are similar to those found in other studies using LiPc in rat brain [10], showing pO_2 values of approximately 51 mmHg in un-anaesthetised animals and approximately 49 mmHg in those anaesthetised with Ketaset.

Simultaneous measurements of blood and cerebral pO_2 in the mouse brain using both probes were possible under certain conditions. Initial experiments found that the g-values of the two probes were very close (g-values of 384.621 and 384.781 gauss for trityl and LiPc, respectively), making it difficult to differentiate between the two spectra, with the trityl spectra obscuring the LiPc at high pO_2 levels, and high trityl concentrations. As these conditions were reversed, the two spectra could be differentiated, and potentially be used to simultaneously report on blood and tissue pO_2.

CONCLUSION

This provides a model with which to study brain oxygenation for relatively long periods under physiological conditions. Using this method cerebral tissue and blood pO_2 can be selectively measured, and in certain conditions can be monitored simultaneously. The ability to measure and differentiate between tissue and blood pO_2 can aid our understanding of oxygen's role in normal and pathological states.

REFERENCES

1. C. Iadecola, D. A. Pelligrino, M. A. Moskowitz, and N. A. Lassen. Nitric oxide synthase inhibition and cerebrovascular regulation. *J.Cereb.Blood Flow Metab* 14 (2):175-192, 1994.
2. D. A. Pelligrino, V. L. Baughman, and H. M. Koenig. Nitric oxide and the brain. *Int.Anesthesiol.Clin.* 34 (4):113-132, 1996.
3. D. A. Pelligrino, H. M. Koenig, and R. F. Albrecht. Nitric oxide synthesis and regional cerebral blood flow responses to hypercapnia and hypoxia in the rat. *J Cereb.Blood Flow Metab* 13 (1):80-87, 1993.
4. M. Fabricius, I. Rubin, M. Bundgaard, and M. Lauritzen. NOS activity in brain and endothelium: relation to hypercapnic rise of cerebral blood flow in rats. *Am.J Physiol* 271 (5 Pt 2):H2035-H2044, 1996.
5. F. Goda, J. A. O'Hara, K. J. Liu, E. S. Rhodes, J. F. Dunn, and H. M. Swartz. Comparisons of measurements of pO2 in tissue in vivo by EPR oximetry and microelectrodes. *Adv.Exp.Med.Biol.* 411:543-549, 1997
6. K. J. Liu, G. Bacic, P. J. Hoopes, J. Jiang, H. Du, L. C. Ou, J. F. Dunn, and H. M. Swartz. Assessment of cerebral pO2 by EPR oximetry in rodents: effects of anesthesia, ischemia, and breathing gas. *Brain Res.* 685 (1-2):91-98, 1995.
7. H. M. Swartz, G. Bacic, B. Friedman, F. Goda, O. Grinberg, P. J. Hoopes, J. Jiang, K. J. Liu, T. Nakashima, J. O'Hara, and . Measurements of pO2 in vivo, including human subjects, by electron paramagnetic resonance. *Adv.Exp.Med.Biol.* 361:119-128, 1994.
8. H. M. Swartz and R. B. Clarkson. The measurement of oxygen in vivo using EPR techniques. *Phys.Med.Biol.* 43 (7):1957-1975, 1998.
9. H. M. Swartz, J. Dunn, O. Grinberg, J. O'Hara, and T. Walczak. What does EPR oximetry with solid particles measure--and how does this relate to other measures of PO2? *Adv.Exp.Med.Biol.* 428:663-670, 1997.
10. K. J. Liu, P. Gast, M. Moussavi, S. W. Norby, N. Vahidi, T. Walczak, M. Wu, and H. M. Swartz. Lithium phthalocyanine: a probe for electron paramagnetic resonance oximetry in viable biological systems. *Proc.Natl.Acad.Sci.U.S.A* 90 (12):5438-5442, 1993.

FUNCTIONAL OXYGEN-17 MAGNETIC RESONANCE IMAGING AND LOCALIZED SPECTROSCOPY

Gheorghe D. Mateescu*

ABSTRACT

Functional oxygen-17 magnetic resonance (MR) imaging and localized spectroscopy is defined as the ensemble of MR measurements aiming at *in vivo*, noninvasive characterization of oxygen transport and utilization. After a brief description of the present status of *in vivo* ^{17}O-MR, preliminary results are reported on oxygen delivery and consumption in cell suspensions of *Saccharomyces cerevisiae*. It is shown that parallel ^{31}P-MR at high magnetic fields has an important corroborative value.

INTRODUCTION

In vivo and *in vitro* ^{17}O-magnetic resonance imaging and localized spectroscopy in tissues and organs were reported as early as 1987.[1, 2] First described by Mateescu et al.[3] in 1989, metabolic water was measured *in vitro* by Arai et al.[4] in 1990. The first *in vivo* measurements of the rate of formation of "nascent" mitochondrial water in animals breathing ^{17}O-enriched air were reported independently by Mateescu et al.[5, 6] and by McLaughlin et al.[7] in 1991. The effect of temperature on mitochondrial metabolism (Q_{10} -factor) was reported at the ISOTT-XVIII meeting.[8]

Subsequently it has been shown that interleaved ^{17}O-^{31}P magnetic resonance imaging and spectroscopy is a powerful means for the noninvasive investigation of the effects of uncouplers and inhibitors on oxidative phosphorylation.[9] The results indicated that chloroform poisoning affects the *complex V* of the mitochondrial respiratory chain. This method has particular potential in the discrimination between degenerative diseases resulting from DNA mutations in the cell nucleus and in the mitochondrion.[10]
Preliminary reports indicate that ^{17}O MR will be an important means for functional imaging in humans (brain, heart, kidneys, etc).[9, 11] Hence, the acronym FOX-17. It is

*Gheorghe D. Mateescu, Department of Chemistry, Case Western Reserve University, Cleveland, OH 44106

estimated that in brain function mapping FOX-17 will be 4 to 5 times more sensitive than the current proton detection *via* BOLD (Blood Oxygen Level-Dependent) contrast. Currently, the prohibitive cost of the oxygen-17 isotope is the principal factor delaying experiments on humans. Should the isotope be readily available, questions regarding blood recirculation effects[12] would most likely be favorably answered. In the meantime it has been demonstrated that ^{17}O-MR is sufficiently sensitive to allow *in vivo* metabolic water measurements in very small organisms such as larvae of Tenebrio molitor.[8,9] This opens the possibility of conducting studies on cell cultures.

Preliminary results of such attempts with cells of *Saccharomyces cerevisiae* are reported here. They provide evidence for the feasibility of direct, noninvasive investigations of oxygen delivery and consumption in these cells. Phosphate metabolism being intimately connected with oxygen consumption, combining ^{17}O-MR measurements with ^{31}P-MR proves to be most beneficial for the interpretation of mitochondrial metabolism.

MATERIALS AND METHODS

Magnetic resonance was performed at two field strengths: 4.7 Tesla (Varian Gemini-200) and 14.1 T (Varian INOVA-600). Both instruments yielded similar results, but with the expected difference in sensitivity and dispersion. The 14.1 T work is described here.

Dry, gluten-free yeast (*Fleischmann's Active Dry Yeast,* Burns Philp Food, Inc., Fenton, MO 63026) was suspended in water (0.3 g yeast / 2.5 ml water) at 40°C in a 10 mm NMR tube capped with a two-hole rubber stopper. First, the water was partially deoxygenated by bubbling nitrogen gas for 10 min at room temperature. The inlet gas was delivered via 1.7 mm ID polyethylene tubing at the bottom of the tube. The excess gas was eliminated at the top through tubing of same diameter.

When the suspension was settled, the NMR tube was placed in the magnet of the spectrometer and ^{17}O-spectra were taken for approximately one minute at 3.2-second intervals (32 scans, acquisition time 0.1 s without pulse delay). A baseline of natural abundance ^{17}O-water (20.3 mM) was thus obtained, as shown in Figure 1 from –60 to 0 s on the time scale. At this point, the sample tube was removed from the magnet and 20 ml $^{17}O_2$-gas (40 atom % enrichment, Cambridge Isotope Laboratories) was slowly injected from a syringe through the bottom of the suspension. The displaced gas was suctioned in another syringe from the top in such a way as to partially replace the N_2 with $^{17}O_2$. This four-minute operation is indicated by the rectangle marked 0-240 s on the time scale in Figure 1. Acquisition was then resumed and 145 spectra were taken (from 240 to 704 s). Of these only 50 are shown in the Figure. Since oxygen consumption was no longer observed in this period, the acquisition was stopped and the sample was removed from the magnet. After vigorous stirring for ~3 min it was reintroduced in the magnet and acquisition was resumed 240 seconds after the interruption (704-944 s on the scale). At 1056 s the sample was again removed and 0.5 g sucrose (Domino sugar) was dissolved in the suspension by vigorous stirring. After addition of a new $^{17}O_2$ supply, the measurement was continued from 1296 to 1405 s.

P-31 MR spectra were taken in parallel experiments following the same protocols as above, with acquisition time 0.4 s, pulse delay 1 s, number of scans 185. Total time per spectrum 259 s. Spectra taken before and after sugar addition are shown in Figure 2.

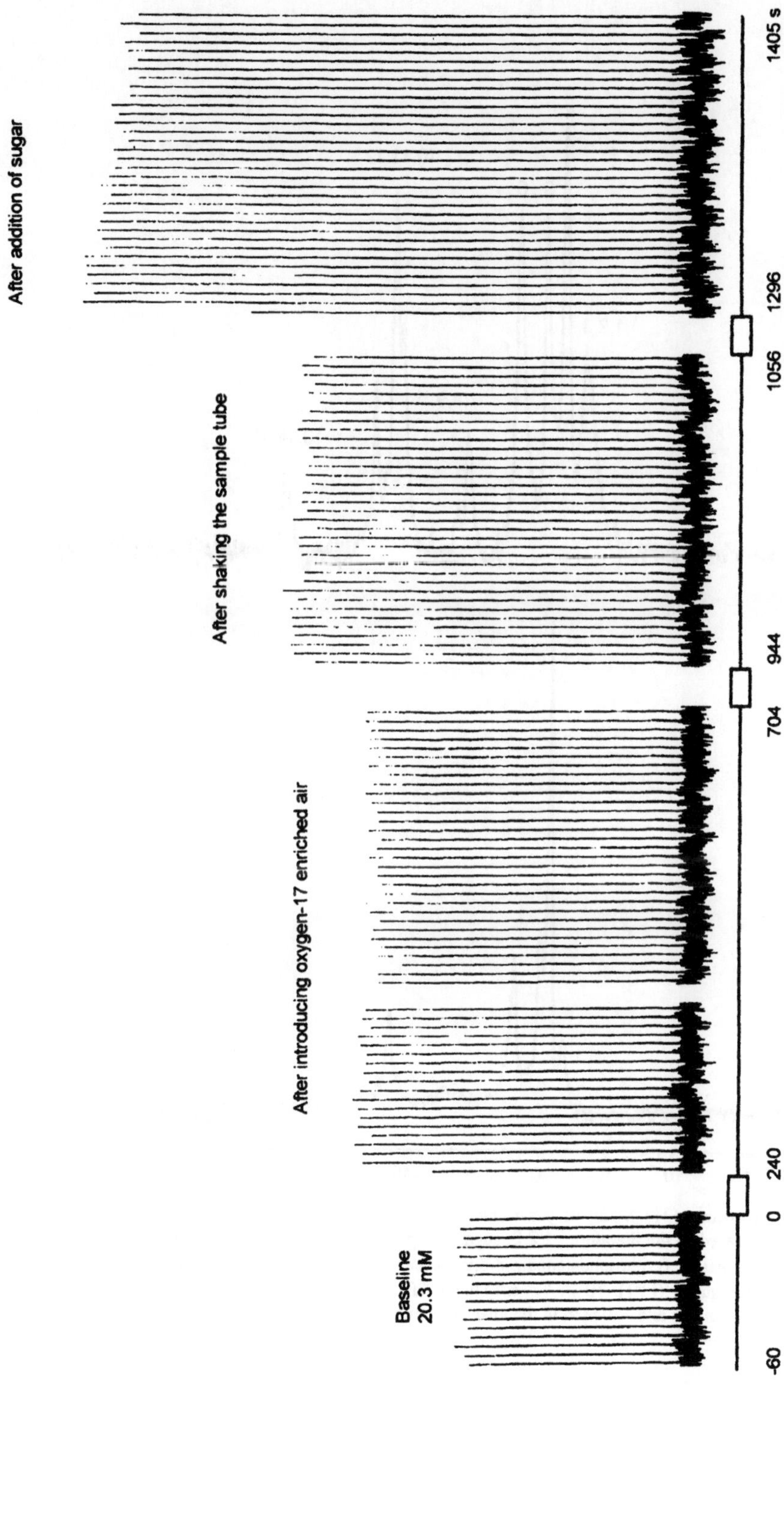

Figure 1. O-17 spectra of yeast cells. From -60 to 0 s. baseline (natural abundance); 0-240 s. 704-944 s. 1056-1296 s. respiration with O-17 enriched air

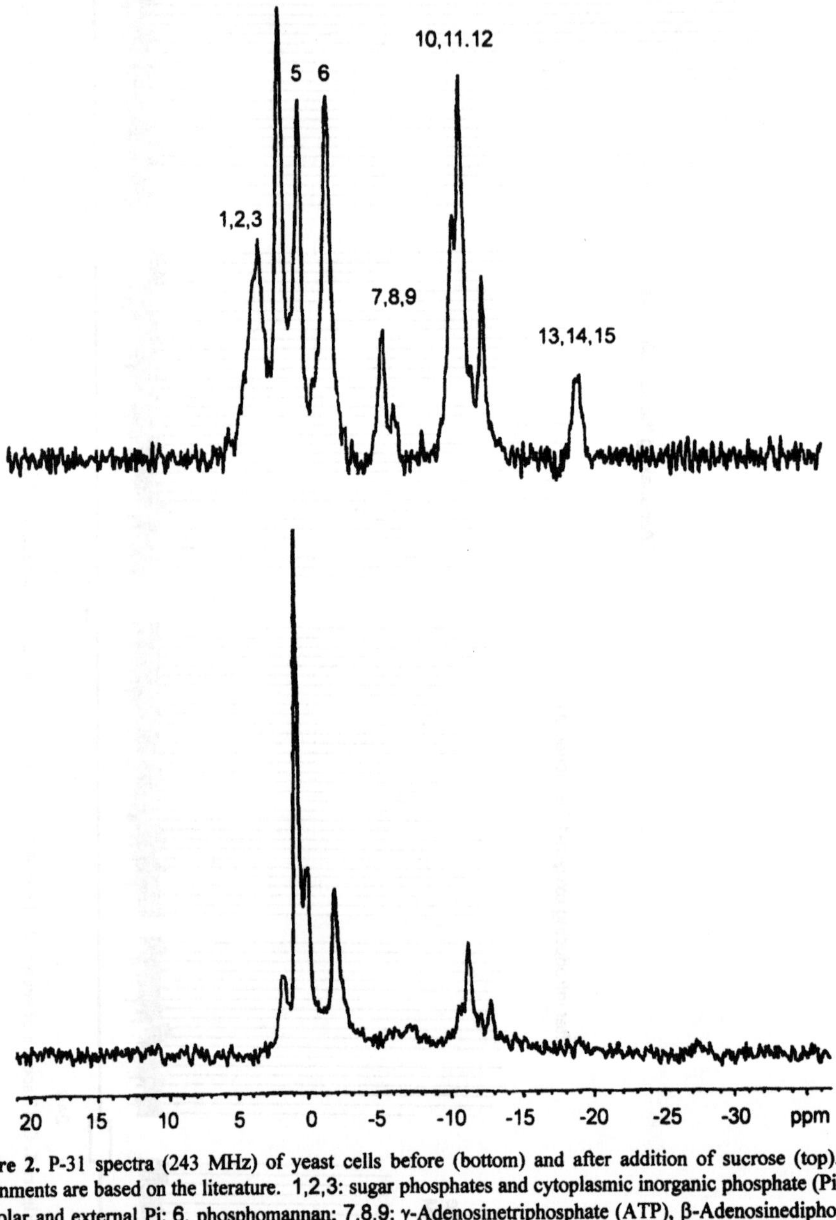

Figure 2. P-31 spectra (243 MHz) of yeast cells before (bottom) and after addition of sucrose (top). Peak assignments are based on the literature. 1,2,3: sugar phosphates and cytoplasmic inorganic phosphate (Pi); 4,5: vacuolar and external Pi; 6, phosphomannan; 7,8,9: γ-Adenosinetriphosphate (ATP), β-Adenosinediphosphate (ADP), polyphosphate-P1 (PP1); 10,11,12: αATP, αADP, NAD(P)(H); 13,14,15: PP2, β-ATP, PP3. PP4,5 are observed between –25 and –30 ppm in the lower spectrum.

RESULTS AND DISCUSSION

Figure 1 reveals changes in oxygen consumption in active yeast cells caused by three consecutive events. During the first event (0-240 s) the cell suspension was aerated with $^{17}O_2$-enriched air. A 50% increase in the $H_2{}^{17}O$ peak intensity and a rapid leveling off indicate that all available oxygen was readily consumed within the four-minute period while the sample was out of the magnet. Shaking the sample again (second event, 704-944 s) renewed the oxygen supply and another 30% intensity increase again followed by leveling off demonstrates that oxygen delivery is diffusion limited. In the absence of agitation the cells agglutinate and remain anaerobic. Addition of sugar (third event, 1056-1296 s) caused an almost 3-fold intensity increase that also occurred within a four-minute period.

Parallel ^{31}P-MR measurements (Figure 2) also reveal the dramatic (well-known) changes in "glucose-repressed" yeast cells. A vast ^{31}P-MR literature[13-21] exists on this subject. By showing the spectra in Figure 2 we only want to emphasize the important advantage brought by better dispersion at higher magnetic fields. Indeed, better peak separation will greatly facilitate much needed quantitative measurements. An important observation is that the oxygen signal intensity decreases significantly after the initial enhancement caused by addition of sugar. This is a strong indication that the metabolic water readily participates in ATP hydrolysis and the label is incorporated in the departing phosphate. Indeed, in experiments with *Tenebrio molitor* larvae we identified ^{17}O-labeled phosphate ions.[9]

But the most important element of this novel tool for cell metabolism investigation is the possibility of direct determination of oxygen delivery and consumption. This will greatly enhance the validity of the ^{31}P/pH-scale as aerobic-anaerobic marker and the quantitative correlation with other metabolites. Of course, our results also show that there is an imperative need to use an adequate NMR bioreactor in order to determine the time course of oxygen consumption that in the present experiment could not be seen in detail. Construction of a bioreactor based on published designs[17, 20] is in progress.

CONCLUSION

The complete genome of *Saccharomyces cerevisiae* having been elucidated, great efforts are invested in the understanding of its metabolism in order to achieve the goal of "designing and genetically engineering biochemical pathways of cell factories."[20] Our preliminary results demonstrate the feasibility of combined $^{17}O/^{31}P$ spectroscopy in studies of yeast metabolism. At high magnetic fields, ^{2}H, ^{13}C, and ^{15}N nuclei may be added for interleave measurements. This will contribute to the reduction of the great variability in data reported in the literature.[20]

ACKNOWLEDGMENT

Dr. Dale Ray is gratefully acknowledged for his assistance with NMR measurements.

REFERENCES

1. G. D. Mateescu, T. Dular and G. Yvars, Oxygen-17 magnetic resonance imaging, *Proc. Soc. Magn. Reson. Med.*, **6**, 929 (1987).
2. G. D. Mateescu, G. M. Yvars and T. Dular, Water, ions, and O-17/proton magnetic resonance imaging, in: *Proc. 4th Int. Conf. on Water and Ions in Biological Systems*, edited by P. Läuger, L. Packer and V. Vasilescu (Birkhauser, Boston, 1988), pp. 239-250.
3. G. D. Mateescu, G.Yvars, D.I. Pazara, N.A. Alldridge, J.C. LaManna, D.W. Lust, M. Mattingly and W. Kuhn, Combined ^{17}O-^{1}H magnetic resonance imaging in plants, animals, and materials, in: *Synthesis and Application of Isotopically Labeled Compounds*, edited by T. A. Baillie and J. R. Jones (Elsevier, Amsterdam, 1989) pp. 499-508.
4. T. Arai, S. Nakao, K. Mori, K. Ishimori, I. Morishima, T. Miyazawa, and B. Fritz-Zieroth, Cerebral oxygen utilization analyzed by the use of O-17 and its MR, *Biochem. Biophys. Res. Comm.*, **169**, 153-158 (1990).
5. G. D. Mateescu, From materials testing to brain-function testing, *Spectroscopy International* **3**(3), 14-18 (1991).
6. G. D. Mateescu, J. LaManna, D. Lust, L. Mars, and J. Tseng, Oxygen-17 magnetic resonance: *in vivo* detection of nascent mitochondrial water in animals breathing ^{17}O-enriched air, *Proc. Soc. Magn. Reson. Med.*, **10**, 1031, 1991.
7. J. Pekar, L. Ligetti, Z. Ruttner, R. Lyon, T. Sinnwell, P. van Gelderen, D. Fiat, C. T. Moonen and A. McLaughlin, *In vivo* measurement of cerebral oxygen consumption and blood flow using ^{17}O magnetic resonance imaging, *Magn. Reson. Med.*, **21**, 313-319 (1991).
8. G. D. Mateescu and M.E. Cabrera, *In vivo* ^{17}O magnetic resonance spectroscopy: determination of temperature effects on metabolic rates (Q_{10} factor), *Adv. Exp. Med. Biol.*, **411**, 585-590 (1997).
9. G. D. Mateescu, M. Cabrera and D. Fercu, ^{17}O and ^{31}P magnetic resonance imaging and spectroscopy: *In vivo* investigations of cell bioenergetics, in: *Spatially Resolved Magnetic Resonance*, Ed. by P. Bluemler, B. Bluemich, R. Botto and E. Fukushima (Wiley-VCH, Weinheim, New York, 1998), pp. 421-429.
10. D. C. Wallace, Mitochondrial Diseases in Man and Mouse, *Science*, **283**, 1482-1488 (1999).
11. X-H Zhu, R. Tian, K. Ugurbil and W. Chen, Feasibility studies of ^{17}O MRS imaging for fast and repeatable measurements of 3D $CMRO_2$ image in the rat brain, *Proc. Int. Soc. Magn. Reson. Med.*, **9**, 649 (2001).
12. J. Pekar, T. Sinnwell, L. Ligeti, A.S. Chesnick, J.A. Frank and A. C. McLaughlin, Simultaneous Measurements of Cerebral Oxygen Consumption and Blood Flow Using ^{17}O and ^{19}F Magnetic Resonance Imaging, *J. Cerebr. Blood Flow and Metab.*, **15**:312-320 (1994).
13. G. Navon, R. G. Shulman, R. G. Yamane, T. R. Eccleshall, K. B. Lam, J. J. Barnofski, and J. Marmur, P-31 nuclear magnetic resonance studies of wild-type and glycolytic pathway mutants of *Saccharomyces cerevisiae Biochemistry*, **18**, 4487-4499 (1986).
14. J. A. den Hollander, K. Ugurbil, T. R. Brown and R. G. Shulman, P-31 nuclear magnetic resonance studies of the effect of oxygen upon glycolysis in yeast, *Biochemistry*, **20**, 5871-5880 (1981).
15. J. R. Alger, J. A. den Hollander, and R. G. Shulman, In vivo P-31 nuclear magnetic resonance saturation transfer studies of adenosinetriphosphatase kinetics in *Saccharomyces cerevisiae*, *Biochemistry*, **21**, 2957-2963 (1982).
16. D. Reibstein, J. A. den Hollander, S. J. Pilkis, and R. G. Shulman, Studies on the regulation of yeast phosphofructo-1-kinase: its role in aerobic and anaerobic glycolysis, *Biochemistry*, **25**, 219-227 (1986).
17. B. K. Melvin, and J. V. Shanks. Influence of aeration on cytoplasmic pH of yeast in an NMR airlift bioreactor, *Biotechnol. Tech.*, **12**, 257-265 (1996)
18. A. Teleman, P. Richard, M. Toivari, and M. Pentillä, Identification and quantitation of phosphorus metabolites in yeast neutral pH extracts by NMR spectroscopy, *Analytical Biochemistry*, **272**, 71-79 (1999).
19. S. Kim, K. D. Barrow, and P. Rogers, Application of NMR to analysis of ethanol fermentation kinetics in yeasts and bacteria, *Biotechnology Letters*, **21**, 839-848 (1999).
20. B. Gonzalez, A. de Graaf, M. Renaud and H. Sahm, Dynamic *in vivo* ^{31}P nuclear magnetic resonance study of *Saccharomyces cerevisiae* in glucose-limited chemostat culture during the aerobic-anaerobic shift, *Yeast*, **16**, 483-497 (2000).
21. R. Lagunas, C. Gancedo, Role of phosphate in the regulation of the Pasteur effect in *Saccharomyces Cerevisiae, Eur. J. Biochem.*, **137**, 479-483 (1983)

DEVELOPMENT OF A RETRACTOR FOR MONITORING THE OXYGENATION STATE OF BRAIN TISSUE

Yoichi Onodera, Yukio Yamada, Kaoru Sakatani, and Tomokatsu Hori *

1. INTRODUCTION

In the majority of neurosurgical operations, brain retraction is one of the essential techniques. When the stress of retraction becomes too high, because of the occlusion of the blood vessel, saturation of oxygen (SO_2) in the retracted tissue would be decreased. Every neurosurgeon knows the safe limits of retraction from one's experience. However, the clinical technique to measure the oxygenation state of retracted tissue has not been developed yet. We have developed a new optical probe (retractor-type probe) that enables quantitative measurement of SO_2 in the retracted tissue. From the experimental use of the retractor-type probe, the feasibility of the SO_2 measurement during brain retraction was confirmed. Additionally, the relationship between stress and SO_2 in the brain tissue is discussed.

2. RETRACTOR-TYPE PROBE

The retractor-type probe consisted of a tapered flexible retractor with tip width of 6 mm and with optical fibers built in. The fiber built in the retractor 5 mm from the tip acted as source, and the other fibers acted as detectors 10, 20, 30 mm apart from the source (Figure 1).

3. MEASURING SYSTEM

The retractor-type probe was connected to the oxygenation monitor (OM-200, Shimadzu) and lights of three wavelengths in the near infrared (NIR) wavelength range were supplied. The incident light was scattered and absorbed in the tissue sample. The light was detected continuously at the two different points with a photomultiplier tube (Figure 2). Measuring the intensity of the light at the wavelength of 780, 805, 830 nm, SO_2 was calculated by means of spatially resolved spectroscopy[1,2].

* Yoichi Onodera, Biomedical Sensing and Imaging Group, Institute for Human Science and Biological Engineering, National Institute of Advanced Industrial Science and Technology, Tsukuba, Ibaraki 305-8564, Japan. Yukio Yamada, University of Electro-Communications. Kaoru Sakatani, China-Japan Friendship Hospital. Tomokatsu Hori, Tokyo Women's Medical University.

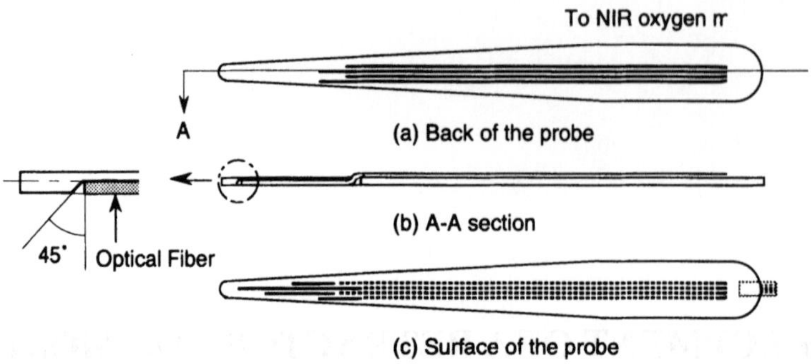

Figure 1. A retractor-type probe.

4. MEASUREMENT OF SO_2 CHANGE IN TISSUE-LIKE PHANTOM

4.1 Experimental Setup

To confirm the feasibility of SO_2 measurement with the retractor-type probe, SO_2 of the tissue like phantom was measured. The phantom was made from Intralipid as scatterer and red blood cell of human blood as absorber (Figure 3). The reduced scattering coefficient of the phantom was made 1.0 mm^{-1}. The volume concentration of the red blood cell in the phantom is about 2%. By changing the oxygenation state of the tissue-like phantom by the bubbling of nitrogen gas, SO_2 was measured using the retractor and a blood gas analyzer.

4.2 Result

The line in Figure 4 shows SO_2 measured by the retractor-type probe connected with the NIR tissue oxygen monitor, and the dots show SO_2 measured with a blood gas analyzer. A good agreement between the measured temporal profiles using those two devices was observed. We think that the differences after 6000 s are due to the temporal change of the scattering coefficient caused by the destruction of the red blood cells in the tissue-like phantom.

5. CHANGE OF SO_2 DURING THE RETRACTION OF PIGLET BRAIN

5.1 Experimental Setup

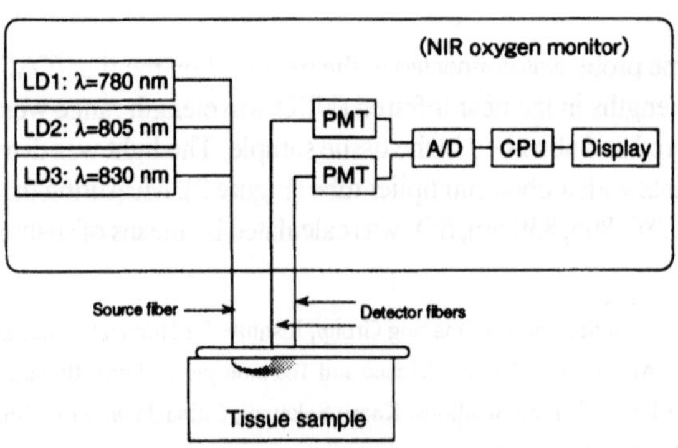

Figure 2. Measurement system.

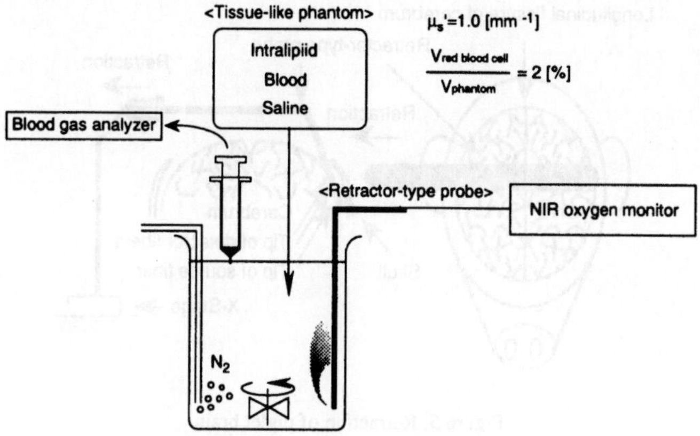

Figure 3. Tissue-like phantom.

We made a retraction experiment of a piglet brain to confirm the capability of the retractor-type probe. After anesthetizing a piglet, the skull and dura mater were removed. The tip of the retractor-type probe was inserted between the skull and cerebrum, and was applied directly to the brain tissue. The retractor-type probe was held with a retainer attached to the X-stage (Figure 5). Then, SO_2 change during brain retraction was measured.

5.2 Results

The temporal change of SO_2 during the retraction of piglet brain is shown in Figure 6. The SO_2 change observed in 0 to 800 s could not be explained at present. Decrease of SO_2 caused by the retraction and rapid increase of SO_2 after release of retraction were observed.

6. CHANGE OF SO_2 DURING THE INDENTATION OF PIGLET BRAIN

The tissue deformation is caused by the brain retraction. The tissue deformation will cause the occlusion of blood flow, and hypoxia of the retracted tissue will be caused. We have examined the relationship between the stress caused in the brain and the change in SO_2.

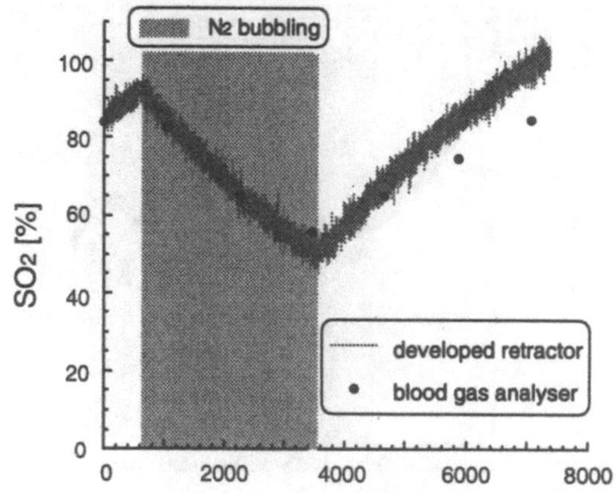

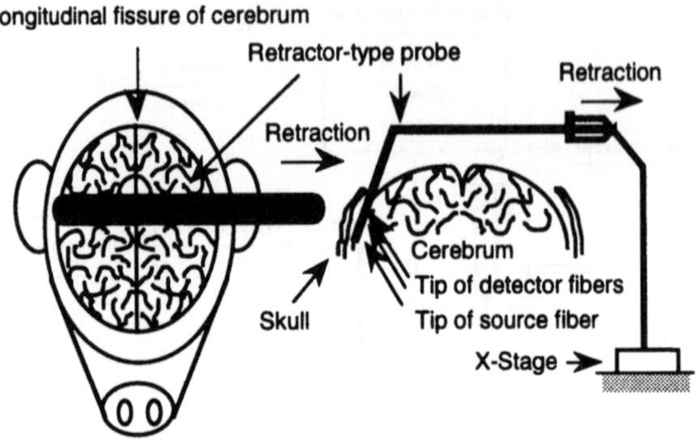

Figure 5. Retraction of piglet brain.

6.1 Experimental Setup

We have prepared an indentor to measure the stress caused in the brain and the change in SO_2. The source and detector fibers were built in a black painted block made of Plexiglas. The distances from source fibers to detector fibers were 10 and 20 mm. After anesthetizing the piglet, the skull and dura mater were removed. The indentor was directly attached to the piglet brain and the force from the brain was measured with a load cell (Figure 7). The stress in the brain was obtained dividing the force by the cross-sectional area of the indentor.

6.2 Results

The temporal changes of stress, SO_2 and total hemoglobin concentration of the piglet brain given by the NIR oxygen monitor are shown in Figure 8. The indentor began to push in at a constant speed (0.1 mm/s) at time A. The stress in the brain tissue increased while SO_2 decreased. The total hemoglobin concentration increased to compensate the hypoxia of tissue. The indentor stopped and maintained the position at time B. The stress was relaxed and SO_2 was recovered. The indentor began to be pulled out at constant speed at time C, and the indentor was fully retracted at time D. The change in the stress and SO_2 during indentation is plotted in Figure 9. The data were

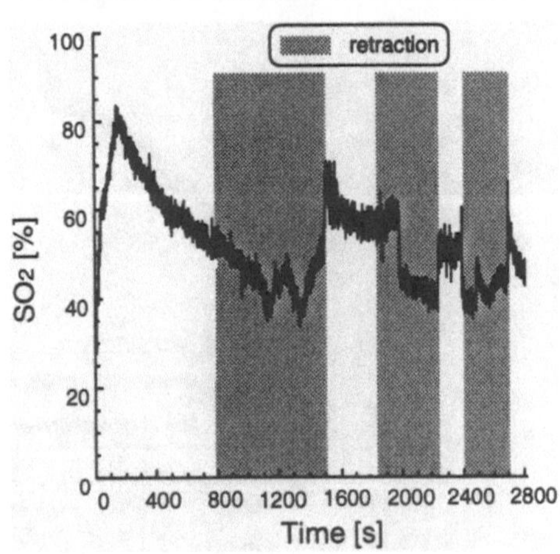

Figure 6. SO2 change in piglet brain during retraction.

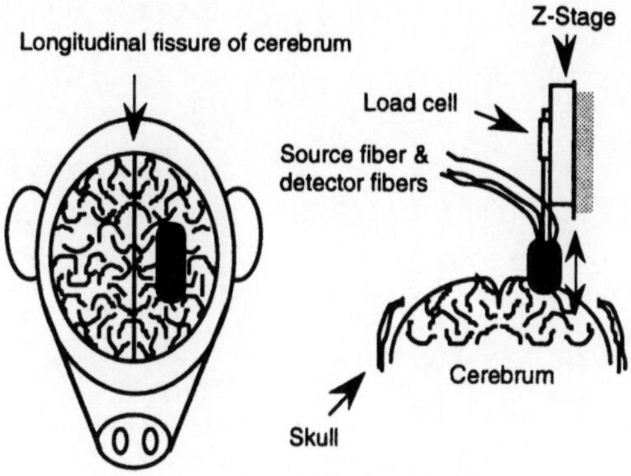

Figure 7. Indentation of piglet brain.

all from one piglet. We have found the curve of stress and SO_2 shows hysteresis. The experiment of indentation depth of 5 mm was done in the same position. The curve showed the same tendency as the result of the latter part of 3mm indentation. From this result, we deduce that the brain tissue adapted to the stress.

CONCLUSIONS

A retractor for continuous oxygenation monitoring of brain tissue was developed to prevent damage during the brain retraction. Using the developed probe, measurement of SO_2 change during the retraction of piglet brain was demonstrated and the capability of the developed probe was confirmed. The relationship between stress and SO_2 of brain tissue was obtained through the indentation experiment of piglet brain.

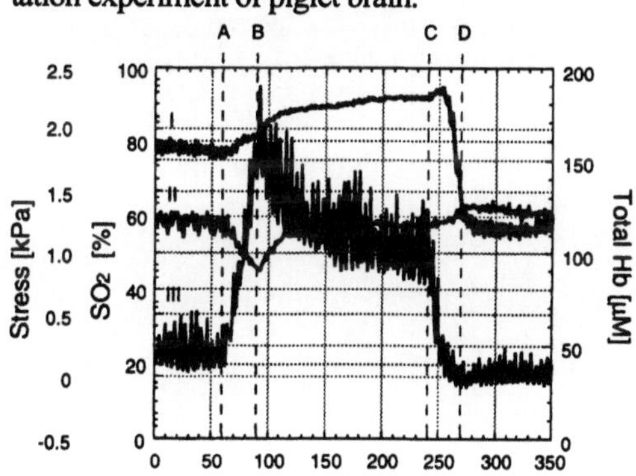

I. Total Hb [μM], II. SO_2 [%], III. Stress [kPa]

Figure 8. Temporal change of stress, SO_2 and total hemoglobin concentration in piglet brain.

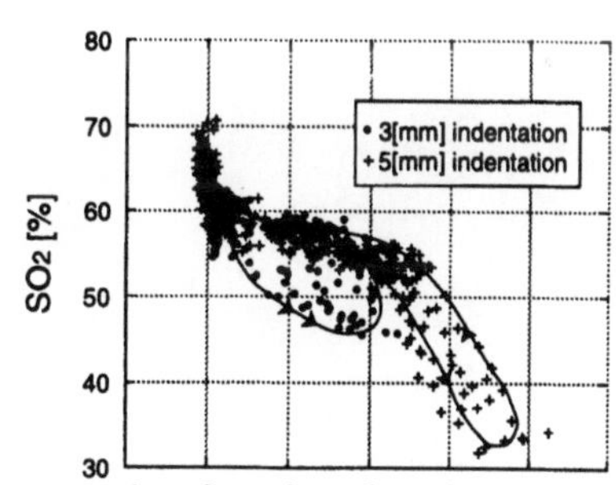

Figure 9. A relationship between stress and SO_2

References

1. T. J. Farrel, M. S. Patterson and B. Wilson, A diffusion theory model of spatially resolved, steady-state diffuse reflectance for the noninvasive determination of tissue optical properties *in vivo*, Med. Phys., 19(4), 879-888 (1992)

2. I. Konishi, S. Takeuchi, Y. Oikawa, Y. Wada, N. Sakauchi, Y. Ito, I. Oda and Y. Tsunazawa, Development of OMM-2000 optical multichannel monitor, SHIMADZU REVIEW, 57(1,2), 141-151 (2000)

MAPPING OF HEMODYNAMICS ON THE HUMAN CALF WITH NEAR INFRARED SPECTROSCOPY AND THE INFLUENCE OF THE ADIPOSE TISSUE THICKNESS

Ursula Wolf, Martin Wolf, Jee H. Choi, L.Adelina Paunescu, Larisa P. Safonova, Antonios Michalos, and Enrico Gratton*

1. ABSTRACT

We investigated the influence of the adipose tissue thickness (ATT) on near-infrared spectroscopy (NIRS) measurements of the absorption coefficient (μa), the reduced scattering coefficient (μs') and changes in concentrations of oxyhemoglobin ([O_2Hb]) and deoxyhemoglobin ([HHb]). We used a frequency domain spectrometer and a special probe to generate maps of these parameters on the human calf during venous occlusion. For ATT below 6 mm μa remained constant, whereas for ATT between 6 and 14 mm μa decreased quickly and became almost constant again for ATT larger than 14 mm. μs' was not significantly altered by the ATT but the values showed a high variability between subjects. We found significantly different changes in both the [O_2Hb] and the [HHb] between the proximal and distal locations of measurement. Although ATT influences the recovery of the optical properties of the underlying tissue, these differences depending on the location cannot be sufficiently explained by the ATT for the following reasons. The ATT varied little within one subject (mean difference 0.88 ± 1.80 mm). The inter-subject variability was 5 times higher. For a given ATT within one subject we observed different values for changes in [O_2Hb] and [HHb] depending on the measurement location. Moreover for a smaller ATT the difference between the values of Δ[O_2Hb] and Δ[HHb] proximal versus distal were more distinct. The thinner the overlying tissue (ATT) the higher is the proportion of muscle tissue in the probed tissue volume. Therefore these differences are most likely coming from the muscle tissue rather than the ATT. This indicates that although the ATT has an evident influence on the measurement of optical

* Ursula Wolf, Laboratory for Fluorescence Dynamics, Department of Physics, University of Illinois at Urbana-Champaign, 1110 W. Green St., Urbana, IL 61801-3080, USA Tel: ++1 217 333 3525 Fax: ++1 217 244 7187

parameters and hemodynamics and should therefore be recognized when performing NIRS measurements, other factors will have to be considered as well.

1. INTRODUCTION

In this study we have investigated the influence of the adipose tissue thickness (ATT) on non-invasive near infrared spectroscopy (NIRS) measurements. We generated maps of optical parameters, the absorption coefficient (μa) and reduced scattering coefficient ($\mu s'$) and changes in concentrations of oxyhemoglobin ($\Delta[O_2Hb]$) and deoxyhemoglobin ($\Delta[HHb]$) on the human calf during venous occlusion measuring at different locations simultaneously. Preliminary data were previously reported[1]. We found that $\Delta[O_2Hb]$ and $\Delta[HHb]$ varied depending on the measurement location. A high variability of changes in $[O_2Hb]$, [HHb], oxygen consumption, blood flow and of optical density (OD) was described by several authors[2-4]. Their studies were performed as single location measurements and they thought the ATT to be the reason for the high intersubject variability. However, our results revealed the differences in the changes in $[O_2Hb]$ and [HHb] not only between but also within subjects even though the ATT varied little within individuals. Our goal was to study the relationship between the optical and hemodynamic parameters and the ATT.

2. MATERIAL AND METHODS

We used a frequency domain spectrometer (Oxy-Imager, ISS, Champaign, IL). The principle of the frequency-domain spectrometer is described in detail elsewhere[5]. Our instrument operates at 758nm and 830nm. The light generated by 32 laser diodes (16 per wavelength) is intensity modulated at a frequency of 110 MHz. The light from the instrument to the tissue and back to the instrument is guided through optical fibers. The incoming light is collected in 4 photomultiplier tubes (PMT)detectors, demodulated and its mean intensity (DC), modulation amplitude (AC) and phase (Φ) are determined. The output signals from the PMTs are sent to a computer for data processing.

The probe covers an area of 18.5x6 cm and its geometry is shown in figure 1.

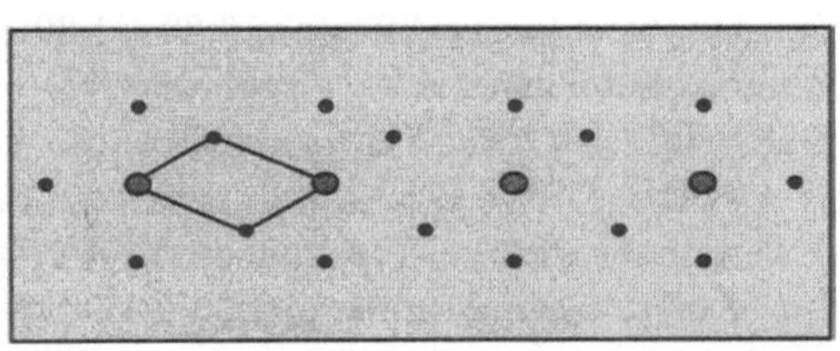

Figure 1. Sketch of the sensor: Large circles are detector locations, black dots are source locations. The 4 detector fibers are located on the middle line of the sensor. The source fibers are arranged around the detector fibers. The interoptode distance to the outer 10 source fibers is 3.0 cm. In the center of the probe are two different interoptode distances, 2.4 cm and 3.5 cm in a symmetrical parallel arrangement.

Maps over the entire region were generated simultaneously at a sample rate of 6 Hz. The pneumatic cuff pressure curve was recorded (Cole Parmer digital manometer). ATT was measured at 9 representative locations with a skinfold caliper (Lange).

Subjects were studied in a supine position with the legs suspended just above the level of the heart to provide a quick venous drainage. Subjects rested for 15 minutes prior to the measurements. The probe was placed over the lateral part (lateral gastrocnemius muscle) of the subject's calf. A pneumatic cuff was wrapped around the subject's thigh. Venous occlusion was achieved by inflating the cuff within 2 seconds to a pressure of 60 mmHg. Venous occlusion was held for one minute, after which the pressure was quickly released. The rest period between the venous occlusions was 2 minutes. 5 cycles of venous occlusion-rest period were performed. Preceding and following these cycles a two-minute baseline measurement was recorded. The protocol was approved by the Institutional Review Board of our university.

The symmetrical arrangement of the two different source detector distances in the middle part of the sensor allows the calculation of μa and μs' of the tissue[6] at three locations in the inner part of the probe. From μa and μs' we calculated the differential pathlength factor (DPF)[7] for three areas. Quantitative values of $\Delta[O_2Hb]$ and $\Delta[HHb]$ in the different locations were calculated from the attenuation changes by the DPF method[8]. To test for the statistical significance we used the Mann-Whitney U test for independent and the Wilcoxon test for paired data.

Fifteen subjects, 9 female, 6 male with an age range of 26-37 years were included in the study. Both calves of each individual were measured (30 legs). Written informed consent was obtained from all subjects prior to the experiments.

3. RESULTS

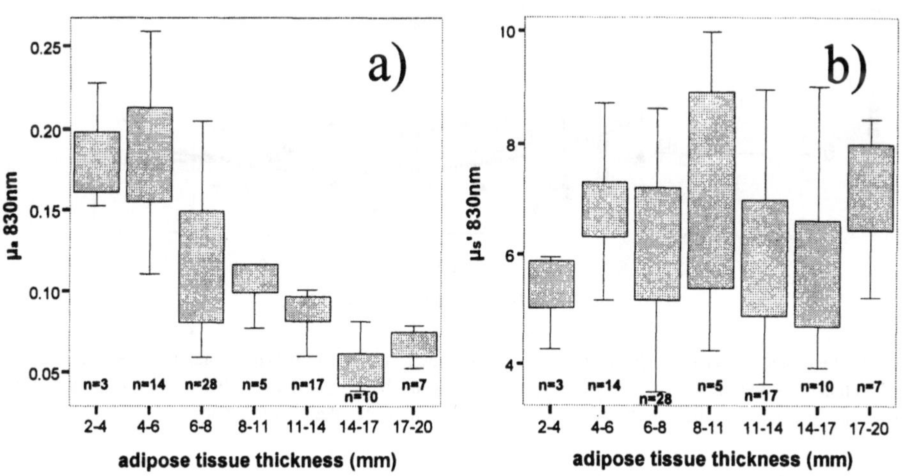

Figure 2. The diagrams show μa (a) and μs' (b) of the tissue depending on the adipose tissue thickness. For an ATT lower than 6 mm no change in μa was observed. For an ATT higher than 6 mm μa decreased constantly, whereby the most distinct changes were observed in the ATT range between 6 and 14 mm. The decrease stopped above 14 mm ATT. μs' is almost constant over the range of the ATT. The variability of the μs' values is relatively large for a given ATT.

Figure 2a shows the dependence of the μa at 830 nm on the ATT. In figure 2b the relationship between μs' at 830 nm and the ATT can be seen. A similar pattern was also found for 758 nm.

The distribution of the ATT over the subjects' calves showed slightly higher values in the distal compared to proximal area. The difference was significant. The inter-individual differences in the ATT were about 5 times larger than the intra-individual differences. The mean ATT for the 2 areas over 28 legs is given in table 1.

Changes in [O_2Hb] during venous occlusion were significantly larger in the proximal area compared to the distal area ($p< 0.0001$). Also the changes in [HHb] proximal versus distal showed significantly different values ($p< 0.0001$). The mean of the changes in [O_2Hb] and [HHb] over all legs is given in table 1.

Table 1. Shows the mean over all legs of [O_2Hb], [HHb] and ATT over 30 legs at the proximal and distal area of the calf (*p = 0.02, **p < 0.0005).

	Proximal	Distal
Δ[O_2Hb] ± SD	7.30 ± 3.79**	5.05 ± 2.86**
Δ[HHb] ± SD	2.31 ± 1.07**	1.26 ± 1.11**
ATT ± SD	9.7 ± 4.38*	10.6 ± 4.56*

In figure 3 the correlation between the mean Δ[O_2Hb] of each subject at the proximal and distal locations and the ATT for all subjects is presented.

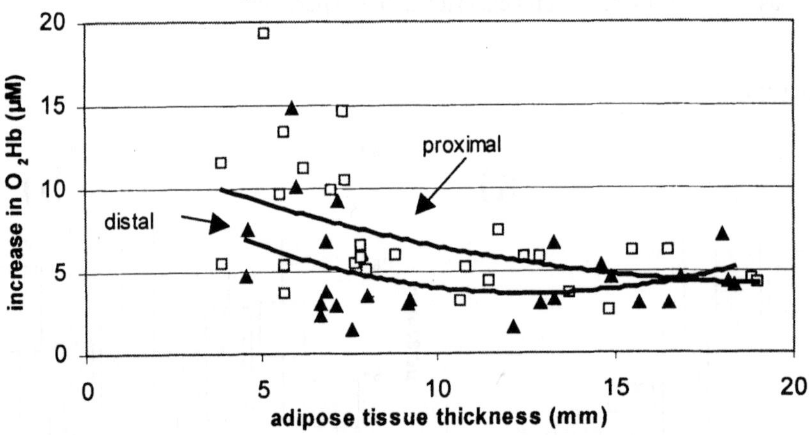

Figure 3. Relationship between the mean Δ[O_2Hb] of each subject at the proximal (squares) and distal (triangles) locations and the ATT for all subjects. The values for the Δ[O_2Hb] are higher in the proximal than in the distal part. It can clearly be seen that for a given ATT the values for the Δ[O_2Hb] vary.

4. DISCUSSION

We performed maps of hemodynamics during venous occlusion where we observed different values for Δ[O_2Hb] and Δ[HHb] depending on the point of measurement. To address the question whether these differences are truly generated by local variations in hemodynamics or by the overlying tissue we investigated the influence of the ATT on μa and μs'. From our measurement we found a clear influence of the ATT on the μa at both wavelengths, 758 and 830 nm. With increasing ATT μa decreased revealing a sharp decline for ATT between 6 and 14 mm. For ATT larger than 14 mm μa was almost the same.

For superficial layers below 6 mm we obtained steady values for μa. Our findings are in accordance with results from Franceschini[9] whose study on a two-layer phantom model is relevant to our data because they used the same optical properties, approximately the same source detector distances and the multi-distance approach as in our study. They showed that superficial layers up to 6 mm do not influence the recovery of the optical properties of the underlying tissue. For superficial layers of more than 16 mm they found that only the optical properties of these layers could be detected. The μs' values are virtually independent of the ATT. The variability of the μs' values is relatively large, which may hide a possible difference between muscle tissue (for ATT smaller than 6 mm) and adipose tissue (for ATT larger than 14 mm). Our data are in agreement with the literature concerning the size of the values as well as their variability[10].

Homma[4] did studies on the human lower leg using a continuous wave instrument and could therefore not separate μa and μs'. They determined the optical density (OD) and found a negative correlation between ATT and OD. Since with increasing ATT we observed a constant μs' and decreasing μa, which corresponds to a decreasing OD, our findings are comparable and coincide with their results. Niwayama[11], measuring the human forearm, described with increasing ATT a decrease in mean optical pathlength in the muscle, while it increased in the adipose tissue layer, thus being more sensitive to the latter. This again confirms our results, since the adipose tissue has a lower absorption than the muscle tissue. Homma[4] and Niwayama[11] attribute the difference in absorption to the lower concentration of the main chromophores O_2Hb and HHb in the adipose tissue.

Binzoni[12] investigated the hemodynamics in the human calf and found that with increasing ATT absolute values for [O_2Hb] decreased while [HHb] did not change much. They commented that with increasing ATT the volume, which is probed contains less muscle and more adipose tissue. A decreasing Δ[O_2Hb] with increasing ATT would therefore indicate that the adipose tissue contains less [O_2Hb] than the muscle tissue.

All mentioned studies were performed by single location measurements over the tissue of interest. Our measurements were probing several locations simultaneously and thus we obtained maps of hemodynamic changes in the proximal and distal part of the lateral human calf. Although we observed on the one hand an influence of the ATT on μa, we measured on the other hand for a given ATT within one subject different values for ΔO_2Hb and ΔHHb depending on the measurement location. Moreover for a smaller ATT the difference between the values of Δ[O_2Hb] and Δ[HHb] proximal versus Δ[O_2Hb] and Δ[HHb] distal were more distinct. The smaller the overlying tissue the greater the proportion of muscle tissue in the probed tissue volume. Therefore these differences are most likely to come from the muscle tissue. This indicates that although the ATT has an evident influence on the measurement of hemodynamics and should therefore be taken into consideration when performing measurements not all the effects described in this study can be explained by it. Other factors in the human tissue such as structural and functional parameters may play an important role.

CONCLUSION

The ATT has an influence on μa and therefore on Δ[O_2Hb] and Δ[HHb]. The ATT does not influence μs'. The ATT on the human calf varied little within a subject while the variations in Δ[O_2Hb] and Δ[HHb] with respect to the location were significant. To explain these effects other parameters (anatomical and physiological) than ATT should be considered.

ACKNOWLEDGEMENT

This research is supported by NIH grant PHS 2 RO1 CA57032 and RR 10966.

REFERENCES

1. U. Wolf, M. Wolf, J. H. Choi, L. A. Paunescu, A. Michalos, and E. Gratton, Mapping of oxyhemoglobin and deoxyhemoglobin changes of the human calf during venous occlusion, *Abstract ISOTT* **2000**,Thursday p. 5 (2000).
2. W. N. Colier, I. B. Meeuwsen, H. Degens, and B. Oeseburg, Determination of oxygen consumption in muscle during exercise using near infrared spectroscopy, *Act. Anaesthesiol. Scand. Suppl.* **107**,151-155 (1995).
3. R. Krageli, T. Jarm, and D. Miklavcic, Reproducibility of parameters of postocclusive reactive hyperemia measured by near infrared spectroscopy and transcutaneous oximetry, *Annals of Biomedical Engineering* **28**,168-173 (2000).
4. S. Homma, T. Fukunaga, and A. Kagaya, Influence of adipose tissue thickness on near infrared spectroscopic signals in the measurement of human muscle, *J. Biomed. Optics.* **1**,418-424 (1996).
5. S. Fantini, M. A. Franceschini, J. S. Maier, S. A. Walker, B. Barbieri, and E. Gratton, Frequency-domain multichannel optical detector for non-invasive tissue spectroscopy and oximetry, *Opt. Eng.* **34**,32-42 (1995).
6. D. Hueber, S. Fantini, A. E. Cerussi, and B. Barbieri, New optical probe design for absolute (self-calibrating) NIR tissue hemoglobin measurements, *SPIE* **3597**,618-631 (1999).
7. S. Fantini, D. Hueber, M. A. Franceschini, E. Gratton, W. Rosenfeld, P. Stubblefield, D. Maulik, and M. R. Stankovic, Non-invasive optical monitoring of the newborn pigletbrain using continuous-wave and frequency-domain spectroscopy, *Phys. Med. Biol.* **44**,1543-1563 (1999).
8. D. T. Delpy, M. Cope, P. van der Zee, S. R. Arridge, S. Wray, and J. S. Wyatt, Estimation of optical pathlength through tissue from direct time of flight measurements, *Phys. Med. Biol.* **33**,1433-1442 (1998).
9. M. A. Franceschini, S. Fantini, L. A. Paunescu, J. M. Maier, and E. Gratton, Influence of a superficial layer in the quantitative spectroscopic study of strongly scattering media, *Applied. Optics.* **37**,7447-7458 (1998).
10. V. G. Peters, D. R. Wyman, M. S. Patterson, and G. L. Frank, Optical properties of normal and diseased human breast tissues in the visible and near infrared, *Phys. Med. Biol.* **35**,1317-1334 (1990).
11. M. Niwayama, L. Lin, J. Shao, T. Shiga, N. Kudo, and K. Yamamoto, Quantitative measurement of muscle oxygenation by NIRS: Analysis of the influence of a subcutaneous fat layer and skin, *SPIE* **3597**, 291-299 (1999).
12. T. Binzoni, V. Quaresima, G. Baratelli, E. Hiltbrand, L. Guerke, F. Terrier, P. Cerretelli, and M. Ferrari, Energy metabolism and interstitial fluid displacement in human gastrocnemius during short ischemic cycles, *J. Appl. Physiol.* **85**,1244-1250 (1998).

EPR IMAGING OF VASCULAR CHANGES IN OXYGEN IN RESPONSE TO CARBOGEN BREATHING

William Stein, S. Subramanian, James B. Mitchell, Murali C. Krishna*

1. INTRODUCTION

EPR (Electron Paramagnetic Resonance) spectroscopy is being used for in vivo imaging with the recent availability of non-toxic paramagnetic spin probes.[1] The EPR spectral line widths of these agents are directly dependent on oxygen levels.[2] The narrow line width and the long in vivo half-lives make these agents suitable to obtain high resolution spatial images. In this study, the effects of changes in the venous and arterial oxygen in response to changes in oxygen content in the breathing atmosphere is evaluated.

2. MATERIALS AND METHODS

2.1. Chemicals

The spin probe, Oxo 63, a triaryl methyl structure, used for the studies was a gift from Nycomed Innovations, Malmo, Sweden. Gas cylinders of various mixtures of oxygen, nitrogen and carbon dioxide were obtained from Roberts Oxygen, Rockville, Maryland.

2.2. EPR Spectroscopy and Imaging Measurements

The Continuous Wave (CW)EPR spectroscopic and imaging measurements were performed using an RF spectrometer operating at 300 MHz.[3] The gradient values were in the range of 1.5 – 2.5 G/cm.

*Radiation Biology Branch, National Cancer Institute, Bethesda, Maryland 20892, USA. Email: murali@helix.nih.gov

Oxygen Transport to Tissue XXIII, edited by D. Wilson *et al.*
Kluwer Academic/Plenum Publishers, 2003

2.3. Mice

Female C3H mice (weight: 25±5 g, APA, Frederick) were used. The experiments were approved by the NCI animal care review board and conducted in accordance with the principles outlined in the *Guide for Care and Use of Laboratory Animals,* prepared by the Institute of Laboratory Animal Research, National Research Council.

2.4. Animal Preparation and Administration of the Spin Probe

The mice were anesthetized with a solution of 100 mg/kg ketamine and 10 mg/kg of xylazine (J.A. Webster, Sterling, MA). Once fully anesthetized, the mice were placed on a heated blanket (40°C) and positioned inside the magnet/gradient assembly housing a 10 mm diameter resonator.[4] The tail of the mouse was held vertical through the resonator.[4] The spin probe (50 µl of 100 mM Oxo63) was injected through the cannula placed into the lateral tail vein positioned outside the resonator. To eliminate any artifacts, the cannula was inserted through the distal aspect of the tail outside the resonator.

2.4. Oxygen Measurements with Fluorescence Quenching

For localized oxygen measurements, the fiber optic oxygen sensor (OxyLite System, Oxford Optronix, London) was inserted through a Abbocath T 26-gauge X 3/4" radiopague FEP teflon IV catheter (Abbot Hospitals Inc., North Chicago, IL) into the right hamstring muscles.

The oxygen sensor consisted of a fiber optic cable coated with a ruthenium complex. When excited with blue light, the ruthenium complex enters into an excited state. When oxygen interacts with the ruthenium exciplex, its lifetime is altered. By measuring the time required for the complex to return to its ground state and by knowing the temperature of the animal, i.e. the temperature of the muscle in which the probe has been inserted, the oxygen tension at the probe tip can be determined by using the Stern-Volmer relationship. Oxygen in the environment of the ruthenium complex interacts with, but does not react with, the ruthenium exciplex. Thus the sensor provides a means to measure the amount of oxygen at a given point *in vivo* without consuming the oxygen in the environment. The Oxylite system was connected to a Macintosh Powerbook 3400 running MacChart. Data were collected every second for the duration of the experiment.

3. RESULTS AND DISCUSSION

3.1. Phantom Imaging for Oxygen Induced Changes in Image Intensities

The schematic of the phantom used for imaging is shown in Figure 1a. In brief, four glass capillaries were positioned in a polymeric mold inside the resonator. Inside each tube was placed a segment of gas permeable tubing (ID 0.08 mm) filled with approximately 25 µl of 7 mM Oxo63 in phosphate buffered saline (PBS) solution. The oxygen level in the solution was changed by flushing the glass capillaries with the desired gas mixture. Once the solutions were equilibrated, imaging sequence was initiated. The intensities from the 2-dimensional EPR images were calculated. When all the four tubes containing the spin probe were exposed to air, the signal intensity was almost similar (Figure 1b). However, when two tubes were maintained in nitrogen atmosphere (left and right) and the two tubes (in air), the image intensities from the tubes experiencing nitrogen atmosphere had a higher intensity than those equilibrated with air (Figure 1c). When the atmosphere was changed to carbogen (95% oxygen, 5% carbon dioxide), the image intensity from these tubes was significantly reduced as a result of oxygen-induced line broadening (Figure 1d). This behavior, i.e., the image intensity dependence on the inverse level of oxygen may therefore provide qualitative information on oxygen in vivo.

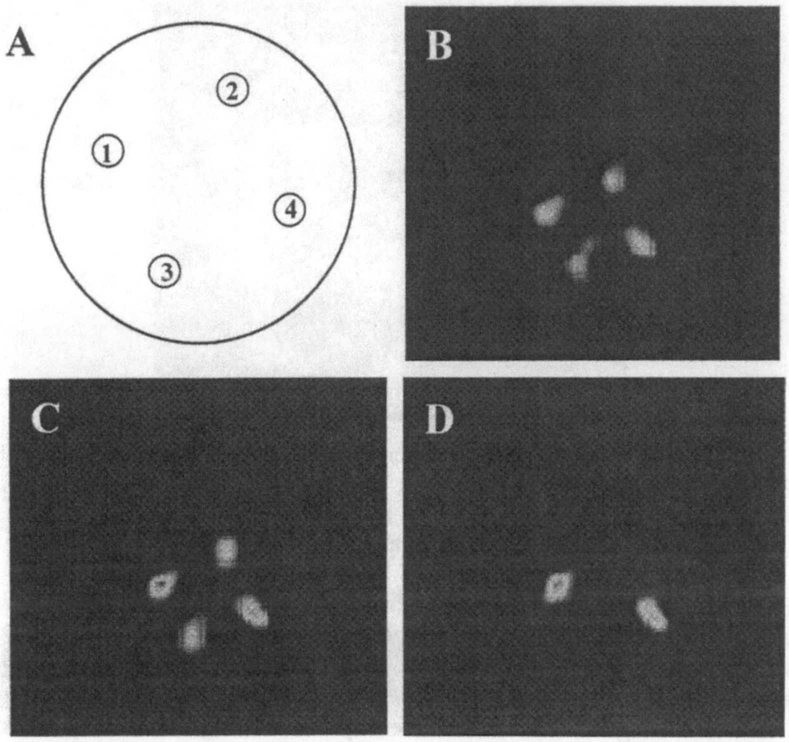

Figure 1. (A) Schematic of a four tube phantom used for EPR imaging. The four circles represent the capillary tubes containing 25 μl of the spin probe Oxo63 (7 mM). (B) 2D EPR image of the four tube phantom when solutions in all the capillary tubes are air-saturated. (C) 2D EPR image of the four tube phantom when the solutions in the capillary tubes (2, 3) are air saturated and tubes (1, 4) are nitrogen saturated. (D) 2D EPR images of the four tube capillary when the solutions in tubes (2, 3) are air saturated and tubes (1, 4) are saturated with 95% oxygen.

3.2. EPR Imaging of Oxygen Changes in a Mouse Tail

A C3H mouse's tail imaged with the CW-EPRI is shown in Figure 2. The asterisk denotes the site where the cannula was positioned in the tail. The three pairs of intense signals correspond to the anatomic vein-artery pairs, which supply the blood to the tail.

The central region of a mouse's tail is avascular cartilage, which accounts for the absence of spin probe in the center of the EPR image. The results from the imaging experiment show that it is possible to resolve three of the four vein-artery pairs in this image. The fourth pair of vessels was not visualized because it was occluded by the cannula.

Given that the mouse's tail, as shown in Figure 2a, is approximately 4 mm in diameter, the separation between the pairs of vessels in the EPR image corresponds to the distances between the vessels of a histological preparation.

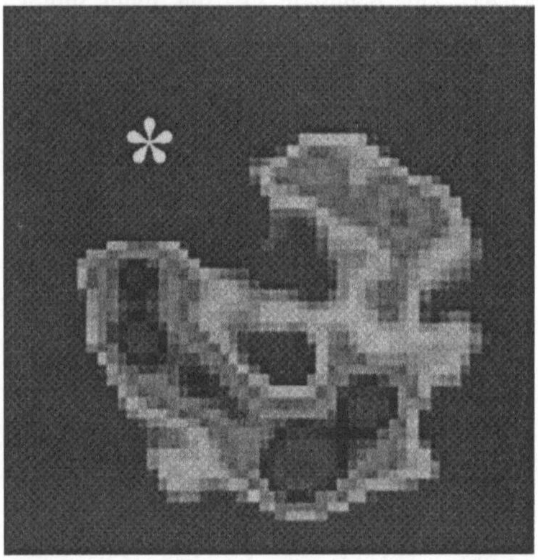

Figure 2. 2D EPR image of the spin probe distribution in the tail veins of a mouse after administration of the spin probe (50 μl of 100 mM Oxo63).

3.3. EPR Imaging of Oxygen Changes in a Mouse Tail

Figure 3A is a trace of the oxygen tension in the right hamstring muscle while the mouse is breathing different levels of oxygen. The time represented by region 1 is the time when the mouse was exposed to 20% O_2 / 5% CO_2 / 75% N_2. The image of the mouse's tail while breathing the 20% O_2 / 5% CO_2 / 75% N_2 is shown in Figure 3B. There is a toroid region of increased intensity surrounding a relatively weak central region. The central region corresponds to the anatomic avascular region in the mouse's tail. Within the toroid of increased signal are smaller regions of maximum signal that correlate to the vessels of the tail. In the lower left quadrant of the mouse's tail is a region of low intensity signal. This low intensity signal corresponds to the location of the cannula in the tail.

In region 2 of Figure 3A the mouse was allowed to breath 7% O_2 / 5% CO_2 / 88% N_2. There was approximately an 85% decrease in oxygen tension when compared to the oxygen tension recorded when the mouse was breathing 20% O_2 / 5% CO_2 / 75% N_2. The two dimensional image taken during region 2 is shown in Figure 3B2. The image has the same basic structure as seen during 20% O_2 / 5% CO_2 / 75% N_2 breathing but the points demonstrate increased signal.

In region 3 the mouse was allowed to breath 100% O_2. The oxygen tension in the muscle increased by 20% when compared to that of an animal breathing 20% O_2 / 5%

CO_2 / 75% N_2. The image taken while the animal breathed 100% O_2 (Figure 3B-3) shows lower intensity signal than the image taken while breathing 20% O_2 / 5% CO_2 / 75% N_2.

In region 4, the mouse was allowed to breath room air (20% O_2 and no CO_2). The oxygen tension in the leg decreased approximately 35% when compared to the oxygen tension determined during 20% O_2 / 5% CO_2 / 75% N_2 breathing. The image of the tail taken while the animal breathed room air (Figure 3B-4) shows at least six regions of increased intensity.

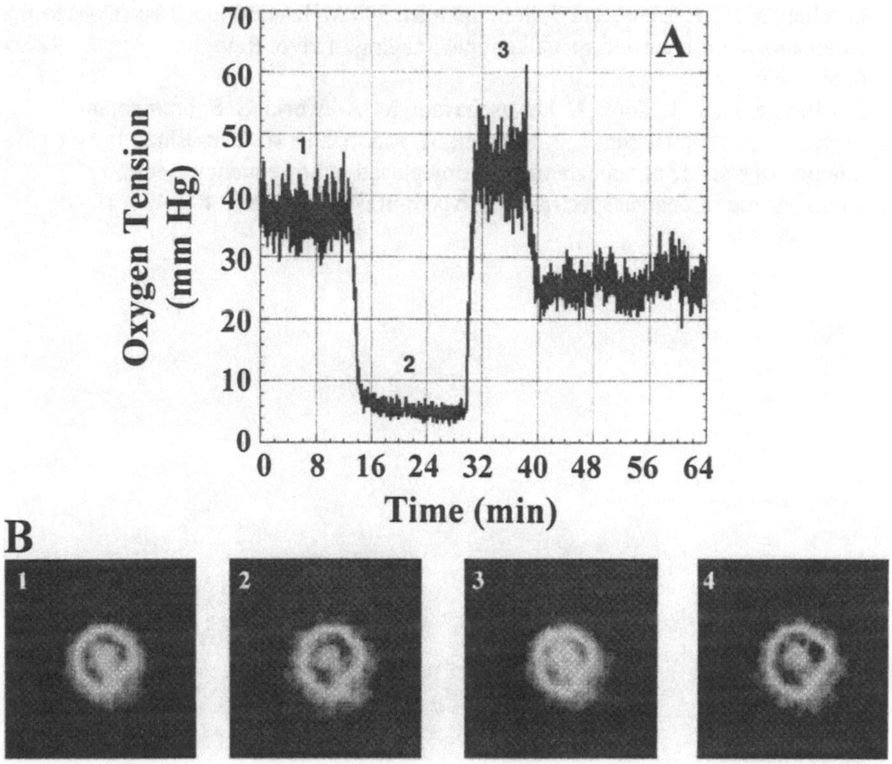

Figure 3. (A) Trace of pO_2 in the muscle as a function of time and the corresponding EPR images (B, C, and D). Region 1, 2, 3, and 4 (A) correspond to conditions when the inspired oxygen content of the mouse was maintained at 20, 7, 100, and 20%, respectively.

CONCLUSIONS

The present investigation provides data supporting the use of EPR imaging with narrow line spin probes. Using EPR imaging it is possible to obtain well-resolved oxygen changes in localized regions of anatomy. Future studies will be directed toward determining oxygen concentrations in rodent tumors before and after radiation treatment.

REFERENCES

1. K. Golman, I. Leunbach, J. H. A. Larsen, G. J. Ehnholm, L. G. Wistrand, J. S. Petersson, A. Jarvi, and S. Vahasalo. Acta Radiologic. Overhauser Magnetic Resonance Imaging. 39, 10-17, 1998.
2. J. H. A. Larsen, I. Leurson, I. Leunbach, G. Ehnholm, L. G. Wistrand, J. S. Petersson, and K. Golman. EPR and DNP properties of certain novel single electron contrast agents intended for oximetric imaging. J. Magn. Reson. 133, 1-12, 1998.
3. J. Koscielniak, N. Devasahayam, M. S. Moni, P. Kuppusamy, K. Yamada, J. B. Mitchell, M. C. Krishna, and S. Subramanian. 300 MHz continuous wave electron paramagnetic resonance for small animal imaging in vivo. Rev. Sci. Instr. 71, 4273-4280
4. R. Murugesan, J. A. Cook, N. Devasahayam, M. Afeworki, S. Subramanian, R. Tschudin, J. H. A. Larsen, J. B. Mitchell, A. Russo, and M. C. Krishna. In vivo imaging of a stable paramagnetic probe by pulsed radiofrequency electron paramagnetic resonance spectroscopy. Magn. Reson. Med. 38, 409-414, 1997.

MOLECULAR IMAGING BY PET

[1]Martin Reivich

1. INTRODUCTION

Positron emission tomography (PET) allows quantitative measurements of regional physiological and biochemical processes to be made in vivo in humans. PET determines the three-dimensional distribution of activity in the organ being studied after the subject is administered a tracer labeled with a positron emitting radionuclide. Quantitation of the parameter under study is accomplished by the use of a mathematical model that describes the biochemical or physiological process being examined.

The positron is a particle with a positive charge and a mass equal to that of an electron. It is emitted from a radioactive nuclide and interacts with an electron in the tissue. The two particles undergo the process of matter anti-matter annihilation in which their masses are converted into electromagnetic radiation in the form of two photons, each with an energy of 511 keV, which travel in directions nearly 180° apart. These photons are detected by the PET scanner to produce an image of the distribution of the positron activity in the tissue.

The detection and collimation of the photons in PET devices is achieved by having an array of radiation detectors in which two detectors on opposite sides of the array are electronically connected by a coincidence circuit. An event is registered only if both detectors sense the annihilation photons virtually simultaneously. These events are then reconstructed into an image of the distribution of these annihilation events within the organ of interest using algorithms similar to those developed for computerized tomography (CT) scanners.

The most frequently used positron-emitting radionuclides are isotopes of oxygen, nitrogen and carbon (common constituents of organic molecules), as well as fluorine. The specific isotopes used are oxygen-15, nitrogen-13, carbon-11, and fluorine-18 with half-lives of 2, 10, 20 and 110 min, respectively. These short half-lives mean that these radionuclides must be produced close to where they will be used, requiring a cyclotron or other accelerator on site.

---

[1] Cerebrovascular Research Center, Department of Neurology, University of Pennsylvania, Philadelphia, PA 19104.

Oxygen Transport to Tissue XXIII, edited by D. Wilson *et al.*
Kluwer Academic/Plenum Publishers, 2003

In addition to cyclotron-produced positron emitters, many potentially useful positron emitter generator systems have been developed. These systems offer convenient access to a choice of positron emitters with a wide range of half-lives.

2. PARKINSONISM

Parkinsonism is a feature of a number of neurodegenerative diseases, including Parkinson's disease, multiple system atrophy, and progressive supranuclear palsy. Separating these various parkinsonian syndromes using clinical criteria alone can be difficult. Discrimination of these various disorders is important in view of differences in prognosis and therapy. These syndromes can be differentiated partially on the basis of the involvement of different components of the dopaminergic system.

Dopamine neurons reside predominantly in the brainstem, and project mainly to the striatum (presynaptic nigrostriatal dopaminergic projection). ^{18}F-DOPA can be used to examine this system. The uptake of ^{18}F-DOPA is determined by the transfer of DOPA across the blood-brain barrier, its decarboxylation to fluorodopamine, and its retention in nerve terminals.

Another approach to the visualisation of dopaminergic neurons is the use of radiotracers for the dopamine transporter. Dopamine is synthesised in dopaminergic neurons where it is stored in vesicles that protect it from oxidation by monamine oxidase. In response to an action potential dopamine is released into the synaptic cleft and interacts with dopamine receptors. To terminate the interaction with these receptors, the extracellular dopamine is actively pumped back into the dopaminergic terminal by the dopamine transporter. Following re-uptake, the dopamine molecules present in the neuronal cytosol are either catabolized or actively transported via the vesicular monoamine transporter back into the vesicles to be stored.

2.1 Parkinson's Disease

The major cause of parkinsonism is Parkinson's disease (PD), accounting for 60-85% of all cases. Nigral dopamine projections to the striatum are targeted in PD, especially those to the putamen, while those to the caudate are relatively spared. Degeneration of the nigrostriatal dopaminergic projection leads to loss of dopamine transporters and dopamine in the striatum.

^{18}F-DOPA studies show a more pronounced reduction of striatal uptake in the putamen than in the caudate nucleus that is usually asymmetric corresponding with the usually asymmetric symptomatology in these patients and is correlated with disease severity. Morrish et al.[1] reported a decline in ^{18}F-DOPA uptake in early PD. ^{18}F-DOPA studies appear to be sensitive enough to detect preclinical PD. Rakshi and coworkers[2] reported bilateral alterations in ^{18}F-DOPA uptake in hemiparkinsonian patients.

Studies of the dopamine transporter also show a more pronounced reduction of striatal binding in the putamen than in the caudate, asymmetry, and correlation with disease severity. Even in patients with mild PD, PET studies have shown a marked loss of striatal dopamine transporters[3].

Post-mortem studies have shown small increases or no change in striatal dopamine D_2 receptors in PD[4]. In agreement with these studies, D_2 receptor imaging (using ^{11}C-raclopride) has shown normal or increased D_2 receptor binding[5].

2.2 Multiple System Atrophy

Multiple system atrophy (MSA) accounts for up to 10% of patients with parkinsonism and is poorly responsive to dopaminergic medication. Neuropathologically multiple system atrophy is characterised by neuronal degeneration in the caudate, putamen, globus pallidus, brain stem, cerebellum and spinal cord.

Some ^{18}F-DOPA studies have shown relatively more impairment of the nigrostriatal dopaminergic projections to the caudate in patients with MSA than in patients with Parkinson's disease[6]. However, other studies have shown comparable changes in patients with multiple system atrophy and Parkinson's disease[7,8]. PET studies also have shown a reduced post-synaptic dopamine D_2 receptor density in patients with multiple system atrophy[8].

2.3 Progressive Supranuclear Palsy

In agreement with post-mortem findings, ^{18}F-DOPA uptake is reduced in patients with progressive supranuclear palsy (PSP)[9], putamen and caudate uptake being similarly affected. The pattern of loss of striatal dopamine transporters in patients with PSP is comparable to that in PD patients. Striatal D2 receptor binding also is reduced[9].

In summary PD patients can be distinguished from MSA and PSP patients with PET studies of the dopamine system. In PD, pre-synaptic function (^{18}F-DOPA uptake, dopamine transporter and vesicular transporter binding) is impaired while post-synaptic function (D_2 receptor binding) is normal or increased; while in MSA and PSP both pre- and post-synaptic function are impaired.

3. GENE THERAPY

Several approaches to gene therapy are being tested using a variety of vectors and therapeutic genes. All rely on the same rationale: genetically engineering a target tissue to produce a new therapeutic protein. This requires the selective introduction of genes into a target tissue using a vector. Such vectors include adenoviruses, retroviruses, DNA-protein complexes, and liposomes. The vector has several roles. It delivers the gene across the cell membrane into the nucleus and drives expression of the gene to produce the therapeutic protein.

Reliable assessment of gene transfer remains a major challenge. Gene expression is typically assessed by analyzing random tissue samples. This has several limitations.

1. Gene expression within the target tissue may be unevenly distributed which may render tissue samples unrepresentative.
2. Tissue biopsy cannot assess the extent and overall distribution of gene transfer within the tissue.
3. Only a limited number of biopsies can be obtained in any one patient.
4. A biopsy may not be able to be performed because of the tumor location.
5. The biopsy is invasive and not without risk.

Therefore, a non–invasive method such as PET would be of great value in evaluating the efficacy of the transduction and in determining its relationship to clinical outcome.

3.1 Enzyme/Prodrug Systems

One major approach in cancer gene therapy that has been extensively investigated is the use of enzyme/prodrug systems, such as the herpes simplex virus thymidine kinase/ganciclovir (HSVtk/GCV) system, which introduce a drug-activating enzyme. Phosphorylation by the viral kinase triggers the therapeutic effect of GCV. The viral kinase activates the drug to a monophosphate. Subsequent phosphorylation of the monophosphate by cellular kinases leads to inhibition of DNA synthesis and cancer cell death. The extent of gene expression can be monitored by imaging the accumulation of phosphorylated derivatives of radiolabeled HSVtk substrates (e.g. ^{18}F-GCV or ^{18}F-FHPG (fluoro-hydroxy-propoxy methyl guanine). In 1998, Gambhir and coworkers[10], in murine *in vivo* micro-PET studies, demonstrated the feasibility of HSV1-tk as a reporter gene and ^{18}F-GCV as an imaging reporter probe. Mice were injected i.v. with either Ad-HSVtk or a control virus, 48 hrs. later they were administered ^{18}F-GCV i.v., and images were then obtained. The liver, which is the primary site of adenovirus uptake after iv administration, was visualized only in the mice receiving Ad-HSVtk. They found that there was a significant correlation between the HSVtk enzyme activity in the liver and the ^{18}F-GCV activity in the liver[11]. This suggests that the uptake of ^{18}F- GCV can be used as a marker for the distribution of HSVtk in tumors. Such data obtained early in the course of GCV therapy in patients might influence the dose or duration of GCV to be administered to achieve optimal therapeutic outcome.

Another tracer that has been studied is ^{18}F-FHPG. When this tracer is injected into mice 48 hr. after the i.v. administration of a recombinant adenovirus carrying the HSVtk gene, only liver and bladder activity was detected by PET imaging[12]. No hepatic uptake was seen in mice injected with a control adenovirus vector. Hustinx and coworkers[13] also showed that ^{18}F-FHPG could visualize intra-tumor gene delivery. PET studies were performed in rats bearing II-45 tumors on each flank. The one tumor was injected with Ad.HSVtk, and the other with a control vector 48 hrs prior to ^{18}F-FHPG administraion i.v. and imaging. Only the AdHSVtk injected tumor was visualized.

3.2 Receptor/Ligand Systems

A second reporter gene technique is based on the expression of a transmembrane receptor as the reporter gene and its labeled receptor ligand as the marker. This method was developed by the UCLA group using the dopamine D_2 receptor and ^{18}F-fluoroethyl-spiperone as the ligand[14,15]. The vectors used in these studies were either an adenovirus carrying the D_2R reporter gene (Ad-D_2R) or a control adenovirus expressing β-galactosidase (Ad-βGal). Studies in whole cell homogenates of C6 cells infected with Ad-D_2R or Ad-βGal showed spiperone binding only in the Ad-D_2R infected cells. Studies were also performed in mice administered either Ad-D_2R or Ad-βGal i.v. and 2 days later given FESP and imaged. Only those animals given Ad-D_2R showed significant activity in the liver. The ^{18}F-activity in the liver was found to be proportional to the amount of hepatic FESP present and to the levels of hepatic D_2R-dependent ^{3}H-spiperone binding levels.

The monitoring with these techniques of patients undergoing gene therapy may be improved greatly from the ability to determine repetitively and non-invasively where the transferred DNA is being expressed, and to what extent expression occurs over time.

REFERENCES

1. Morrish P.K., Sawle G.V., and Brooks D.J. 1995, Clinical and [^{18}F]dopa PET findings in early Parkinson's disease. *J Neurol Neurosurg Psychiatry* **59**:597-600.
2. Rakshi J.S., Uema T., Ito K., et al. 1996, Statistical parametric mapping of three dimensional ^{18}F-dopa PET in early and advanced Parkinson's disease [abstract]. *Mov Disord* **11**:147.
3. Frost J.J., Rosier A.J., Reich S.G., et al. 1993, Positron emission tomographic imaging of dopamine transporter with ^{11}C-WIN 35,428 reveals marked declines in mild Parkinson's disease. *Ann Neurol* **34**:423-431.
4. Joyce J.N. and Murray A. 1994, Distribution of D_1- and D_2-like dopamine receptors in human brain, in: *Dopamine receptors and transporters: pharmacology, structure, and function*, H.B. Niznik, ed., Marcel Dekker, New York pp. 345-383.
5. Antonini A., Schwarz J., Oertel W.H., et al. 1994, ^{11}C-raclopride positron emission tomography in previous untreated patients with Parkinson's disease: influence of L-dopa and lisuride therapy on striatal dopamine D2 receptors. *Neurology* **44**:1325-1329.
6. Brooks D.J., Ibanez V., Sawle G.V., et al. 1990, Differing patterns of striatal ^{18}F-dopa uptake in Parkinson's disease, multiple system atrophy, and progressive supranuclear palsy. *Ann Neurol* **28**:547-555.
7. Burn D.J., Sawle G.V., and Brooks D.J., 1994, Differential diagnosis of Parkinson's disease, multiple system atrophy, and Steele-Richardson-Olszewski syndrome: discriminant analysis of striatal ^{18}F-dopa PET data. *J Neurol Neurosurg Psychiatry* **57**;278-284.
8. Antonini A., Leenders KL., Vontobel P., et al. 1997, Complementary PET studies of striatal neuronal function in the differential diagnosis between multiple system atrophy and Parkinson's disease. *Brain* **120**:2187-2195.
9. Brooks D.J., Ibanez V., Sawle G.V., et al. 1992, Striatal D_2 receptors in patients with Parkinson's disease, Striatonigral degeneration, and progressive supranuclear palsy measured with ^{11}C-raclopride and positron emission tomography. *Ann Neurol* **31**:184-192.
10. Gambhir S.S., Barrio J., Wu L., et al. 1998 Imaging of adenoviral directed herpes simplex virus type 1 thymidine kinase gene expression in mice with ganciclovir. *J Nucl Med* **39**:2003-2011.
11. Gambhir S.S., Barrio J.R., Iyer M., et al. 1999, In vivo validation of PET reporter gene/reporter probe assay for herpes simplex virus type 1 thymidine kinase with 8-[F-18]-fluoropenciclovir [abstract]. *J Nucl Med* **40**:25P-26P.
12. Hustinx R., Eck S.L., and Alavi A. 1999, Potential applications of PET imaging in developing novel cancer therapies. *J Nucl Med* **40**:995-1002.
13. Hustinx R., Shiue C.,Y., Alavi A., et al. 2001, Imaging in vivo herpes simplex virus thymidine kinase gene transfer to tumour-bearing rodents using positron emission tomography and [^{18}F]FHPG. *Eur J Nucl Med* **28**:5-12.
14. MacLaren D.C., Gambhir S.S., Satyamurthy N., et al. 1999, Repetitive, non-invasive imaging of the dopamine D_2receptor as a reporter gene in living animals. *Gene Therapy* **6**:785-791.
15. MacLaren D.C., Toyokuni T., Cherry S.R., et al. 2000, PET imaging of transgene expression. *Biol Psychiatry* **48**:337-348.

FREQUENCY DOMAIN IMAGING OF OXYGEN TENSION IN THE MOUSE RETINA

Preliminary Instrumentation Development

Ross D. Shonat and Amanda C. Kight*

1. INTRODUCTION

Insufficient oxygen delivery and retinal hypoxia have been implicated as causal for the development of many devastating diseases of the eye, including diabetic retinopathy, glaucoma, and retinopathy of prematurity. The imaging of oxygen tension (PO_2) using the phosphorescence lifetime measuring technique has received considerable attention in the literature[1-5] and its potential for advancing our understanding of the oxygen dynamics in the retina is well recognized.[3] It is also recognized that, as the number of transgenic and knockout mouse models displaying the characteristics of human retinal diseases rapidly increases, an ability to image PO_2 in these very small eyes will likely be of great benefit. In this study, we describe and test new instrumentation for generating maps of PO_2 in the mouse retina. These maps are generated from the phosphorescence lifetimes of injected palladium-porphyrin probes using a frequency-domain measurement.

2. MATERIALS AND METHODS

2.1. Frequency-Domain Imaging of Oxygen Tension

Methodology for the frequency-domain imaging of PO_2 has been described in detail elsewhere.[4] Briefly, a frequency-resolved, phase-sensitive spectroscopy technique was employed in the mouse retina, using a palladium (Pd)-porphyrin derivative, Pd *meso*-tetra [4-carboxyphenyl] porphine, as the oxygen-sensitive vascular probe for phosphorescence lifetime measurements.[5] Modulation of the excitation light was accomplished with an optical chopper and a generation-IV image intensifier was used to modulate the

---

* Ross D. Shonat and Amanda C. Kight, Department of Biomedical Engineering, Worcester Polytechnic Institute, Worcester, Massachusetts, 01609.

Oxygen Transport to Tissue XXIII, edited by D. Wilson *et al.*
Kluwer Academic/Plenum Publishers, 2003

sensitivity of the camera detector. The probe was excited by light modulated by the optical chopper at a circular frequency ω. Phosphorescence resulting from this excitation was emitted at the same frequency ω, but delayed in phase by an angle θ. The camera intensifier was also modulated at ω, but delayed in phase relative to the excitation by an angle θ_D. Under this acquisition scheme, the intensity reaching the camera is

$$I(\theta_D) = k[Pd]\{1 + 1/2\, m_D m \cos(\theta - \theta_D)\}, \quad (1)$$

where k is a constant, [Pd] is the concentration of the probe, and m_D is the modulation of the camera intensifier.[6] θ is related to the apparent phase lifetime τ by

$$\tan\theta = \omega\,\tau \quad (2)$$

and τ is converted to PO_2 using the Stern-Volmer equation,

$$\tau_o/\tau = 1 + k_Q\tau_o PO_2, \quad (3)$$

where τ_o (μs) is the lifetime in the absence of O_2 and k_Q ($mmHg^{-1}\mu s^{-1}$) is the bimolecular rate, or quenching, constant. Phase shift (θ) maps were generated by the method of least-squares from $I(\theta_D)$ measurements taken at a number of different θ_D values, using a linearized form of Eq. (1).[6] Equations (2) and (3) were used to convert phase shift maps to PO_2 maps.

2.2. Instrumentation

A research-grade fluorescence microscope (model E600, Nikon) equipped with a long-working distance, high NA 4X objective (CFL_{60} series, Nikon) and standard accessories for fluorescence imaging was used. This is schematized in Figure 1. The specific narrow-band excitation wavelength for PO_2 mapping was generated by a

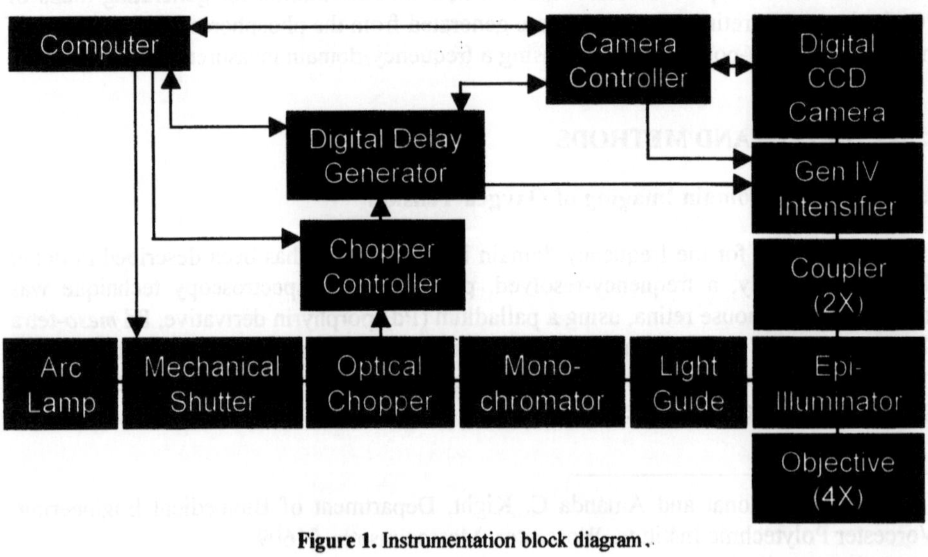

Figure 1. Instrumentation block diagram.

computer-controlled monochromator (0.2-meter design, Photon Technologies, Inc) coupled to the excitation port of the microscope via a liquid light guide (Photon Technologies, Inc). A generation IV intensified digital CCD camera (model I-Max, Thompson 512 X 512 CCD chip, Roper Scientific) was placed in the imaging path of the microscope for acquiring the raw phosphorescence images. The spinning of a digital optical chopper disk (model OC-4000D, Photon Technologies, Inc.), mounted at the entrance-slit of the monochromator, and gain modulation of the camera intensifier was used for the phase-sensitive detection of phosphorescence lifetimes. The precise phase relationship between the optical chopper and the camera intensifier was controlled by a digital delay generator (model DG535, Stanford Research Systems), with TTL signals from the optical chopper triggering the digital delay generator.

2.3. Animal Preparation and Experimental Procedures

Female C57BL/6J mice were anesthetized with 2.5% Avertin (0.015 ml/g, IP) and maintained under anesthesia with additional doses of Avertin as needed. The head was immobilized and the retina made visible for optical microscopy by placing ophthalmic ointment (Goniosoft) and a cover glass over the eye to negate the refractive power of the cornea. The oxygen probe was injected (10 mg/kg) into the femoral vein by venous puncture and excited by modulated light (wavelength 524 ± 10 nm, linear frequency =

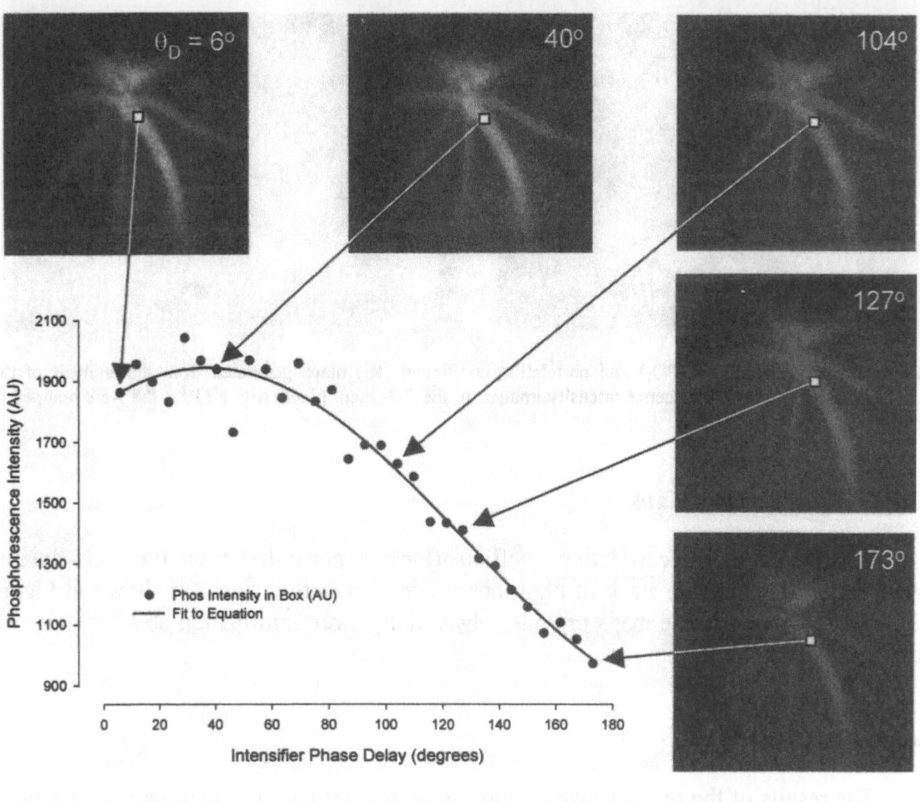

Figure 2. Phosphorescence intensity images at five different instrument phase delays. Intensity values within the venous region demarcated by the white box and the best fit of this data to Eq. (1) are also plotted.

1600 Hz). Exposure time for each phosphorescence intensity image was 1 second. Thirty images were acquired and transferred to the computer as θ_D was varied from 0 to 180°. All animal protocols were approved by the Institutional Animal Care and Use Committee at Worcester Polytechnic Institute.

3. RESULTS

3.1. Phosphorescence Intensity Images

Phosphorescence intensity images in one mouse retina acquired at five different phase delays (in degrees) are shown in Figure 2. For the venous region demarcated by the square box, phosphorescence intensity values are also plotted as a function of phase delay for the 30 images acquired. A best-fit curve to Eq. (1) is also plotted. The calculated phase shift is approximately 40 degrees, corresponding to a phosphorescence lifetime of 83 μs and a PO_2 of 28 mmHg.

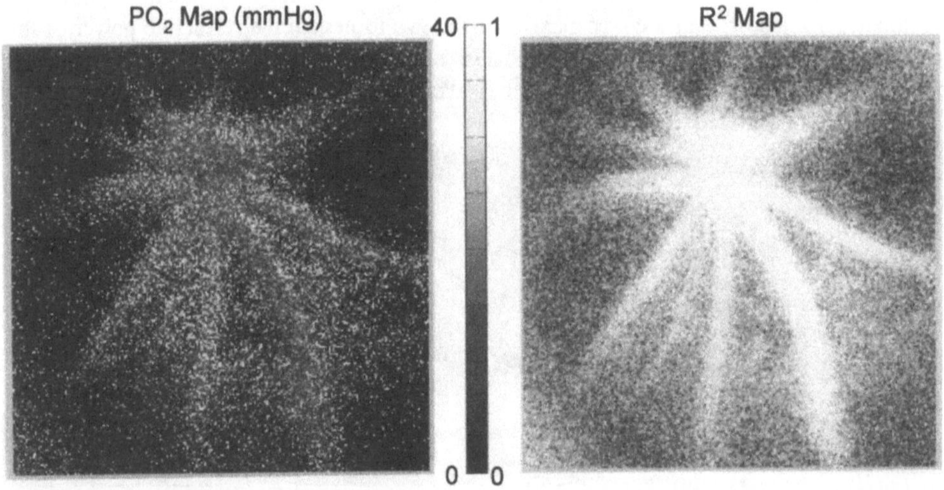

Figure 3. Oxygen tension (PO_2) and correlation coefficient (R^2) maps generated from the analysis of the complete stack of phosphorescence intensity images. In the PO_2 map, pixels with an $R^2 < 0.8$ were mapped to black.

3.2. Oxygen Tension Maps

The PO_2 and R^2 (correlation coefficient) maps generated from the least-squares analysis of the complete stack of phosphorescence intensity images are shown in Figure 3. When colorized, these maps provide substantially more information about the oxygen dynamics in the eye.

4. DISCUSSION

The results of the present investigation show that retinal PO_2 maps can be generated in the very small mouse eye. Ongoing investigations with this instrumentation are directed at a more complete characterization of the instrumentation for mouse retinal studies, including detailed calibration and optimization studies. When fully tested and characterized, this instrumentation will be used to investigate how changes in retinal PO_2 correlate with ocular disease progression in cases where abnormalities in the delivery and consumption of oxygen are thought to be contributing factors, such as diabetic retinopathy.

ACKNOWLEDGMENTS

This research was supported in part by a pilot and feasibility grant from the Diabetes and Endocrinology Resource Center (DERC) at the University of Massachusetts Medical School, Worcester, MA (Center Grant DK32520) and a research grant from the Whitaker Foundation, Rosslyn, VA.

REFERENCES

1. J.M. Vanderkooi, G. Maniara, T.J. Green, and D.F. Wilson, An optical method for measurement of dioxygen concentration based upon quenching of phosphorescence, *J. Biol. Chem.* **252**:5476-5482 (1987).
2. D.F. Wilson, A. Pastuszko, J.E. DiGiacomo, M. Pawlowski, R. Schneiderman, and M. Delivoria-Papadopoulos, Effect of hyperventilation on oxygenation of the brain cortex of newborn piglets, *J. Appl. Physiol.* **70**:2691-2696 (1991).
3. R.D. Shonat, D.F. Wilson, C.E. Riva, and M. Pawlowski, Oxygen distribution in the retinal and choroidal vessels of the cat as measured by a new phosphorescence imaging method, *Applied Optics* **31**:3711-3718 (1992).
4. R.D. Shonat, E.S. Wachman, W-H. Niu, A.P. Koretsky, and D.L. Farkas, Near-simultaneous hemoglobin saturation and oxygen tension maps in the mouse brain using an AOTF microscope, *Biophys. J.* **73**:1223-1231 (1997).
5. L. Lo, C.J. Koch, and D.F. Wilson, Calibration of oxygen-dependent quenching of the phosphorescence of Pd-meso-tetra (4-carboxyphenyl) porphine: a phosphor with general application for measuring oxygen concentration in biological systems, *Anal. Biochem.* **236**:153-160 (1996).
6. J.R. Lakowicz, H. Szmacinski, K. Nowaczyk, K.W. Berndt, and M. Johnson, Fluorescence lifetime imaging, *Anal. Biochem.* **202**:316-330 (1992).

OXYGEN DELIVERY TO THE RETINA AND RELATED VISUAL PATHOLOGY

Brief Review

Ross D. Shonat*

1. INTRODUCTION

The survival of the retina and optic nerve is dependent on an adequate supply of oxygen and nutrients to fuel the metabolic processes that are vital for tissue homeostasis and primary visual function. Ocular diseases that affect the retina, including diabetic retinopathy, retinopathy of prematurity, glaucoma, and numerous proliferative retinopathies, are devastating diseases leading to major vision loss in humans. While the etiologies of these diseases are different, abnormalities in the delivery and consumption of oxygen are thought to be significant in all of them. Despite this common factor, quantitative studies that adequately correlate disease progression with changes in retinal oxygenation are lacking. In part, this is due to the almost complete absence of instrumentation for monitoring oxygenation in humans and the relatively few animal models that exhibit the full pathophysiology of these different retinal diseases.

This brief review will begin with a consideration of how oxygen is delivered and utilized in the retina, followed by a consideration of the possible role that oxygen might play in three different visual pathologies, and finally to a brief consideration of developing and advanced technologies for understanding, diagnosing, and treating human eye disease.

2. OXYGEN DELIVERY AND UTILIZATION IN THE RETINA

The retina of most mammals is supplied with oxygen and nutrients from two separate and distinct vasculatures: the retinal vasculature distributed within the inner layers of the

* Ross D. Shonat, Department of Biomedical Engineering, Worcester Polytechnic Institute, Worcester, Massachusetts, 01609.

retina and the choroidal vasculature situated behind the retina. The light absorbing photoreceptors (rod and cone layer), which have an extremely high oxygen demand, are located within an avascular retinal layer sitting between these two vasculatures. PO_2 values in the choroidal vasculature are very high (> 60 mmHg),[1,2] a physiologic state hypothesized as necessary for maintaining an adequate diffusive flux of oxygen to the highly-oxidative photoreceptors.[3] By contrast, PO_2 values in the retinal vasculature are similar to the values observed in most other vascular beds (15-30 mmHg).[1,2] PO_2 measurements in the retina are complicated by the presence of two distinct vasculatures with different oxygen dynamics.

2.1. PO_2 profiles through the retina

Most of what we know about oxygen levels in the retina has come from invasive oxygen microelectrode studies in animal models and during some surgical procedures in humans. These polarographic electrodes are inserted into the vitreous through a hole in the side of the eye and have been used to measure transretinal PO_2 during oxygen breathing,[2,4] light and dark adaptation,[3] elevated intraocular pressure,[5,6] retinal vascular occlusion,[7] after panretinal photocoagulation therapy,[8,9] and in long-term diabetic cats.[10] PO_2 values in the avascular outer retina have been shown to be extremely low, reflecting the high oxidative demands of the photoreceptors.[1,11] This is especially true in the dark-adapted retina, where the metabolic consumption of oxygen by the photoreceptors is approximately twice that under light-adapted conditions.[1] It is thought that the retina is vulnerable to the complications associated with suboptimal oxygen delivery and vascular disease because of the high oxidative demand for oxygen and the fact that the retinal vasculature must be sparsely distributed above and away from the photoreceptors.

3. POSSIBLE ROLE OF OXYGEN IN VISUAL PATHOLOGIES

Our understanding of oxygen's role in many human retinal diseases is very incomplete and experiments and technologies to understand, diagnose, and treat these human diseases where oxygen is implicated are sorely needed. Due to space constraints, we will consider only three in this review: diabetic retinopathy, glaucoma, and retinal detachment.

3.1. Diabetic Retinopathy

Diabetic retinopathy, a vascular complication of both type 1 (IDDM) and type 2 (NIDDM) diabetes, is the leading cause of new blindness for people between the ages of 20 and 74 years.[12] Vision loss from diabetic retinopathy is associated primarily with long-term changes in the retinal vasculature, including basement membrane thickening, pericyte loss, vascular leaks leading to macular edema, capillary nonperfusion, and ultimately to the proliferative growth of new vessels which can lead to tractional retinal detachment.[13,14] Because diabetic retinopathy is primarily a disease of the vasculature, insufficient oxygen delivery and developing retinal hypoxia have been implicated in many of the complications of this retinopathy.

Because oxygen is implicated so strongly in the proliferative stages of diabetic retinopathy, it is natural to ask whether subtle changes in oxygenation appear before clinically visible symptoms of retinopathy appear. There has been some recent progress in this area[10] and application of advanced non-invasive imaging technologies will likely lead to a greater understanding of oxygen's role in the early phases of diabetic retinopathy.[15]

3.2. Glaucoma

Glaucoma is phenomenologically characterized by the appearance of optic nerve atrophy and visual field defects. Because a chronically elevated intraocular pressure (IOP) is often associated with clinically diagnosed glaucoma, impaired blood flow and oxygen delivery are thought by some to play a role in the pathogenesis of this disease.[15,16] Using a phosphorescence oxygen imaging technique, the effects of acute increases in IOP on vascular PO_2 in the optic nerve head has been studied in the cat[18] and, recently, in glaucomatous monkeys with chronically elevated IOP.[19] A reduction in ONH PO_2 with elevation of IOP was not observed in cats,[18] but preliminary evidence suggests that it may be reduced in monkeys.[19] Clearly, more studies are needed in this important area.

3.3. Retinal Detachment

Retinal detachment is typically defined by the presence of subretinal fluid. As a result, the distance from the choroid to the inner segments of the photoreceptors increases. Photoreceptors in the detached retina lose outer segments and then undergo apoptosis.[20] Recently, mathematical modeling was used to explore the spatial relationship between the choroid and the photoreceptors in the normal and detached retina.[3] Results of this modeling show that detachment of the retina severely reduces oxygen delivery from the choroid to the photoreceptors. The modeling also shows that the choroid is not normally overperfused, since the diffusive flux of oxygen to the photoreceptors is barely adequate to maintain a normal metabolic function. Based on this modeling, hyperoxia is predicted to have a clinical benefit to patients with detached retinas prior to reattachment.[3]

4. ADVANCED TECHNOLOGIES FOR UNDERSTANDING, DIAGNOSING, AND TREATING HUMAN EYE DISEASE

A major challenge to researchers working in the field of eye research is an understanding of the disease state in humans. This is complicated by the paucity of appropriate animal models and the relatively few non-invasive technologies that can be employed to measure oxygenation in the human eye. New technologies for creating animal models and advances in non-invasive imaging technologies are likely to lead to major advances in eye research and are considered briefly below.

4.1. New animal models of human ocular diseases

Advances made in molecular biology and in the manipulation of the mouse genome have generated an explosion of transgenic and knockout mouse models. While available in ever-increasing numbers, directed physiologic studies on these mice have not kept pace with production. It is, therefore, reasonable to expect that many of these mice might prove to be excellent animal models of human ocular diseases. For example, within the last few years, there has been an explosion of new transgenic and knockout mouse models of both type 1 and type 2 diabetes. Various substrates and enzyme molecules in the insulin signaling pathway have been removed, including the glucose transporters GLUT2[21] and GLUT4,[22] the β cell glucokinase (GK) enzyme,[23] and the insulin receptor substrates 1 (IRS-1)[24] and 2 (IRS-2).[25] These mice were made to have the characteristics of non-insulin dependent type 2 diabetes mellitus, considered a polygenic disorder in which insulin resistance and insulin secretory defects are both present.[25] For glaucoma research, directed mutations in the myocilin gene can cause primary and juvenile-onset open angle glaucoma and may be a useful model for studying the molecular pathophysiology of glaucoma.[26]

4.2. Non-invasive oxygenation imaging technologies

Three important technologies that that might some day be used to measure oxygenation in the human eye include: 1) the phosphorescence lifetime imaging of oxygen tension,[15] 2) oxygen-sensitive magnetic resonance imaging (MRI),[27] and 3) spectral imaging of oxygenation, including hemoglobin saturation.[28,29] Because the eye provides an optically clear window to the neural tissues of the retina, non-invasive optical imaging technologies have the potential to provide retinal oxygenation maps in the diseased human eye with high spatial resolution and might be used to diagnose or monitor disease onset and progression in cases where oxygen is implicated.

CONCLUSIONS

While our understanding of oxygen dynamics in the retina has advanced significantly over the last 30 years, an ability to diagnose and treat human ocular diseases, where abnormalities in retinal tissue oxygen delivery and consumption are thought to be significant, remains an elusive goal. New non-invasive oxygenation imaging technologies, including nuclear magnetic resonance imaging, phosphorescence lifetime imaging, and spectral imaging hold great promise and are currently being developed and applied to the study of different visual pathologies. At the same time, advances in genomics are rapidly yielding new animal models of human ocular disease. It is certainly reasonable to expect that these new technologies might ultimately be used to understand, diagnose, and treat vascular diseases in the eye where oxygen is implicated.

ACKNOWLEDGMENTS

The author thanks Robert Linsenmeier, Bruce Berkowitz, Charles Riva, and Stephane Chamot for their assistance in the preparation of this review.

REFERENCES

1. R.A. Linsenmeier, Effect of light and darkness on oxygen distribution and consumption in the cat retina, *J. Gen. Physiol.* **88**:521-542, (1986).
2. D.-Y. Yu, S.J. Cringle, V.A. Alder, and E.-N. Su, Intraretinal oxygen distribution in the rat with graded systemic Hyperoxia and hypercapnia, *Invest. Ophthalmol. Vis. Sci.* **40**:2082-2087 (1999).
3. R.A. Linsenmeier and L. Padnick-Silver, Metabolic dependence of photoreceptors on the choroid in the normal and detached retina, *Invest. Ophthalmol. Vis. Sci.* **41**:3117-3123 (2000).
4. R.A. Linsenmeier and C.M. Yancey, Effects of Hyperoxia on the oxygen distribution in the intact cat retina, *Invest. Ophthalmol. Vis. Sci.* **30**:612-618 (1989).
5. A. Alm and A. Bill, The oxygen supply to the retina I: Effects of changes in intraocular and arterial blood pressures, and in arterial PO_2 and PCO_2 on the oxygen tension in the vitreous body of the cat, *Acta Physiol. Scand.* **84**:261-274 (1972).
6. C.M. Yancey and R.A. Linsenmeier, Oxygen distribution and consumption in the cat retina at elevated intraocular pressure, *Invest. Ophthalmol. Vis. Sci.* **30**:600-611 (1989).
7. R.D. Braun, R.A. Linsenmeier, and T.K. Goldstick, Oxygen consumption in the inner and outer retina of the cat, *Invest. Ophthalmol. Vis. Sci.* **36**:542-554 (1995).
8. I. Molnar, S. Poitry, M. Tsacopoulos, N. Gilodi, and P.M. Leuenberger, Effect of laser photocoagulation on oxygenation in the retina of miniature pigs, *Invest. Ophthalmol. Vis. Sci.* **26**:1410-1414 (1985).
9. E. Stefansson, D.L. Hatchell, B.L. Fisher, F.S. Sutherland, and R. Machemer, Panretinal photocoagulation and retinal oxygenation in normal and diabetic cats, *Am. J. Ophthalmol.* **101**:657-664 (1986).
10. R.A. Linsenmeier, R.D. Braun, M.A. McRipley, L.B. Patnick, J. Ahmed, D.L. Hatchell, D.S. McLeod, and G.A. Lutty, Retinal hypoxia in long-term diabetic cats, *Invest. Ophthalmol. Vis. Sci.* **39**:1647-1657 (1998).
11. V.A. Alder, S.J. Cringle, and I.J. Constable, The retinal oxygen profile in cats, *Invest. Ophthalmol. Vis. Sci.* **24**:30-36 (1983).
12. L.P. Aiello, T.W. Gardner, G.L. King, G. Blankenship, J.D. Cavallerano, F.L. Ferris, and R. Klein, Diabetic retinopathy, *Diabetes Care* **21**:143-156 (1998).
13. V.A. Alder, E.-N. Su, D.-Y. Yu, S.J. Cringle, and P.K. Yu, Diabetic retinopathy : Early functional changes, *Clin. Exp. Pharmacol. Physiol.* **24**:785-788 (1997).
14. R.N. Frank, On the pathogenesis of diabetic retinopathy : A 1990 update, *Ophthalmology* **98**:586-593 (1991).
15. R.D. Shonat, D.F. Wilson, C.E. Riva, and M. Pawlowski, Oxygen distribution in the retinal and choroidal vessels of the cat as measured by a new phosphorescence imaging method, *Applied Optics* **31**:3711-3718 (1992).
16. S.S. Hayreh, The pathogenesis of optic nerve lesions in glaucoma, *Trans. Am. Acad. Ophthalmol. Otolaryngol.* **81**:OP-197-OP-213 (1976).
17. B. Schwartz, J.C. Rieser, and S.L. Fishbein, Fluorescein angiographic defects of the optic disk in glaucoma, *Arch. Ophthalmol.* **95**:1962-1974 (1977).
18. R.D. Shonat, D.F. Wilson, C.E. Riva, and S.D. Cranstoun, Effect of acute increases in intraocular pressure on intravascular optic nerve head oxygen tension in cats, *Invest. Ophthalmol. Vis. Sci.* **33**:3174-3180 (1992).
19. S. Chamot and C.E. Riva, Unpublished observations.
20. P.A. Erickson, S.F. Fisher, D.H. Anderson, W.H. Stern, and G.A. Borgula, Retinal detachment in the cat : The outer nuclear and outer plexiform layers, *Invest. Ophthalmol. Vis. Sci.* **24**:927-941 (1983).
21. M.-T. Guillam, E. Hummler, E. Schaerer, J.-Y. Wu, J. Birnbaum, F. Beermann, A. Schmidt, N. Deriaz, and B. Thorens, Early diabetes and abnormal postnatal pancreatic islet development in mice lacking Glut-2, *Nature Genetics* **17**:327-330 (1997).
22. A.E. Stenbit, T.-S. Tsao, J. Lu, R. Burcelin, D.L. Geenen, S.M. Factor, K. Houseknecht, E. B. Katz, and M.J. Charron, GLUT4 heterozygous knockout mice develop muscle insulin resistance and diabetes, *Nature Medicine* **3**:1096-1101 (1997).
23. Y. Terauchi, H. Sakura, K. Yasuda, K. Iwamoto, N. Takahashi, K. Ito, H. Kasai, H. Suzuki, O. Ueda, and N. Kamada, Pancreatic-β cell specific targeted disruption of glucokinase gene: Diabetes mellitus due to defective insulin secretion to glucose, *J. Biol. Chem.* **268**:30253-30256 (1995).
24. E. Araki, M.A. Lipes, M.-E. Patti, J.C. Bruning, B. Haag, III, R.S. Johnson, and C.R. Kahn, Alternative pathway of insulin signaling in mice with targeted disruption of the *IRS-1* gene, *Nature* **372**:186-190 (1994).

25. D.J. Withers, J.S. Gutierrez, H. Towery, D.J. Burks, J.-M. Ren, S. Previs, Y. Zhang, D. Bernal, S. Pons, G.I. Shulman, S. Bonner-Weir, and M.F. White, Disruption of the IRS-2 causes type 2 diabetes in mice, *Nature* **391**:900-904 (1998).
26. S.I. Tomarev, E.R. Tamm, and B. Chang, Characterization of the mouse Myoc/Tigr gene, *Biochem. Biophys. Res. Commun.* **28**:887-893 (1998).
27. B.A. Berkowitz, C.A. Wilson, Quantitative mapping of ocular oxygenation using magnetic resonance imaging, *Magn. Reson. Med.* **33**:579-581 (1995).
28. M. Hammer, D. Schweitzer, L. Leistritz, M. Scibor, K.-H. Donnerhacker, and J. Strobel, Imaging spectroscopy of the human ocular fundus in vivo, *J. Biomed. Optics* **2**:418-425 (1997).
29. R.D. Shonat, E.S. Wachman, W-H. Niu, A.P. Koretsky, and D.L. Farkas, Near-simultaneous hemoglobin saturation and oxygen tension maps in mouse brain using an AOTF microscope, *Biophysical J.* **73**:1223-1231 (1997).

ACTIVATION OF THE VISUAL CORTEX IN NEWBORN INFANTS UNDER NATURAL SLEEP USING MULTICHANNEL NEAR-INFRARED SPECTROSCOPY

Takashi Kusaka*, Kensuke Okubo, Keiko Nagano, Saneyuki Yasuda, Kou Kawada*, Tadashi Imai, Kenichi Isobe, Susumu Itoh

1. Introduction

During development, the brain undergoes sequential anatomical, functional, and organizational changes necessary to support the complex adaptive behavior of a mature normal individual. Delineation of developmental changes occurring in different brain regions would provide a means of relating various behavioral phenomena to maturation specific brain structures, thereby enhancing our understanding of structure-function relationships in both normal and disease states. Positron emission tomography (PET) and functional magnetic resonance imaging (fMRI) have been used widely for detection and determination of the location of cerebral functional activation. However these technologies can not be used for investigations in infants without sedation. Near-infrared spectroscopy (NIRS) is a noninvasive method for detecting changes in the concentrations of oxyhemogobin (oxyHb), deoxyhemoglobin (deoxyHb) and total hemoglobin (totalHb) at the bedside. In recent years, NIRS has been used to study functional activations of various areas of the brain. This is based on the assumption that an increase in recorded oxyHb concentration represents an increase in blood flow, which in turn reflects neural activation. Most studies in which NIRS was used to assess cortical function have been performed on adult subjects. In newborn infants, NIRS has been used to assess the activities of the visual cortex,[1] frontal cortex[2], olfactory cortex[3] and sensorimotor cortex[4] after repeated light simulation, music stimulation, odor stimulation and passive motor stimulation, respectively. In this study, we used NIRS to monitor the activities of the visual cortex as mirrored by hemodynamic responses in newborn infants subjected to photostimulation during natural sleep, and we compared the activities with normal adult response patterns.

*Maternal and Children's Medical Center and Department of Pediatrics, Kagawa Medical University, 1750-1 Mikicho, Kitagun, Kagawa , 761-0793 Japan.

Oxygen Transport to Tissue XXIII, edited by D. Wilson *et al.*
Kluwer Academic/Plenum Publishers, 2003

2. Methods

2.1. Near-infrared Optical Imaging System

We used multichannel NIRS (Near-infrared optical imaging system, OMM-2000, Shimadzu Corp., Japan), in which three laser diodes, 776, 804 and 828 nm, are used as the light source.[5, 6] Three wavelength light fluxes are put into a source fiber, constituting one transmission channel. Sixteen such source fibers are provided in the system, and the light fluxes are sent through the fibers to probe-pads attached to the head of the infant. Diffusely reflected light fluxes from the infant are caught by another 16 fibers and respectively guided to photomultipliers for final detection. From the possible combination of 256 (16 x 16) source-detector signals, 64 ch data are picked up according to an operator-definable table and are used for imaging or time-course graphics. In this study, we used 8 sources and 8 detectors, and the resulting 24 source-detector signals were picked up for time-course graphics for every 85 msec. We measured the changes in concentrations of [oxyHb], [deoxyHb] and [totalHb] from prebaseline values.

2.2. Subjects and Methods

We examined five infants, aged 9 d to 16 wk (34 – 43 wk's postconceptional age) and three healthy adult volunteers using multichannel NIRS. All five infants had undergone clinical MRI examinations for a variety of pathologic conditions unrelated to visual function, which was normal in routine clinical examinations. Informed consent in writing was obtained from the parents of the infants and from the adult volunteers, and the protocol was approved by the local ethics committee. A probe consisting of 16 optical fibers, 8 for transmission and 8 for detection, was placed over the bilateral occipital region, with the center of probe at the level of the calcarine sulcus. The interoptode distance was 3 cm for adults and 2 cm for infants. Measurements in the infants were performed in the spine or prone position with the occipital region of the head touching the probe, and measurements in the adults were performed in a comfortable sitting position in a dark, quiet room. The adult subjects were asked to keep their eyes closed throughout the examination. The subjects were exposed to stroboscopic white flashing light at 8 Hz projected on the eyelids during the stimulation period (15 seconds) and to non-flashing light during the rest period (45 seconds). The stimulation cycle was repeated 11 to 30 times. All infants were monitored by pulse oximetry during the examination. The infants were clinically in quiet sleep, although the sleep state was not systematically assessed by electroencephalogram.

Means and standard deviations of changes in each parameter were calculated from 8 to 10 trials in which continuous stable measurements were possible. Changes in parameters were determined from means of values measured during a 5-sec period from the start of measurement, with the start and end points of measurement (15sec before onset stimulus and 30 sec after end of stimulus, respectively) set to zero. From the 24 source-detector signals, the signal with the greatest change in [oxyHb] in each subject was selected for statistical analysis. Wilcoxon`s signed rank test was used for statistical analysis of changes in parameters from 5 sec before the start of measurement (10 sec before the onset of stimulus), and a p value of less than 0.05 was considered significant.

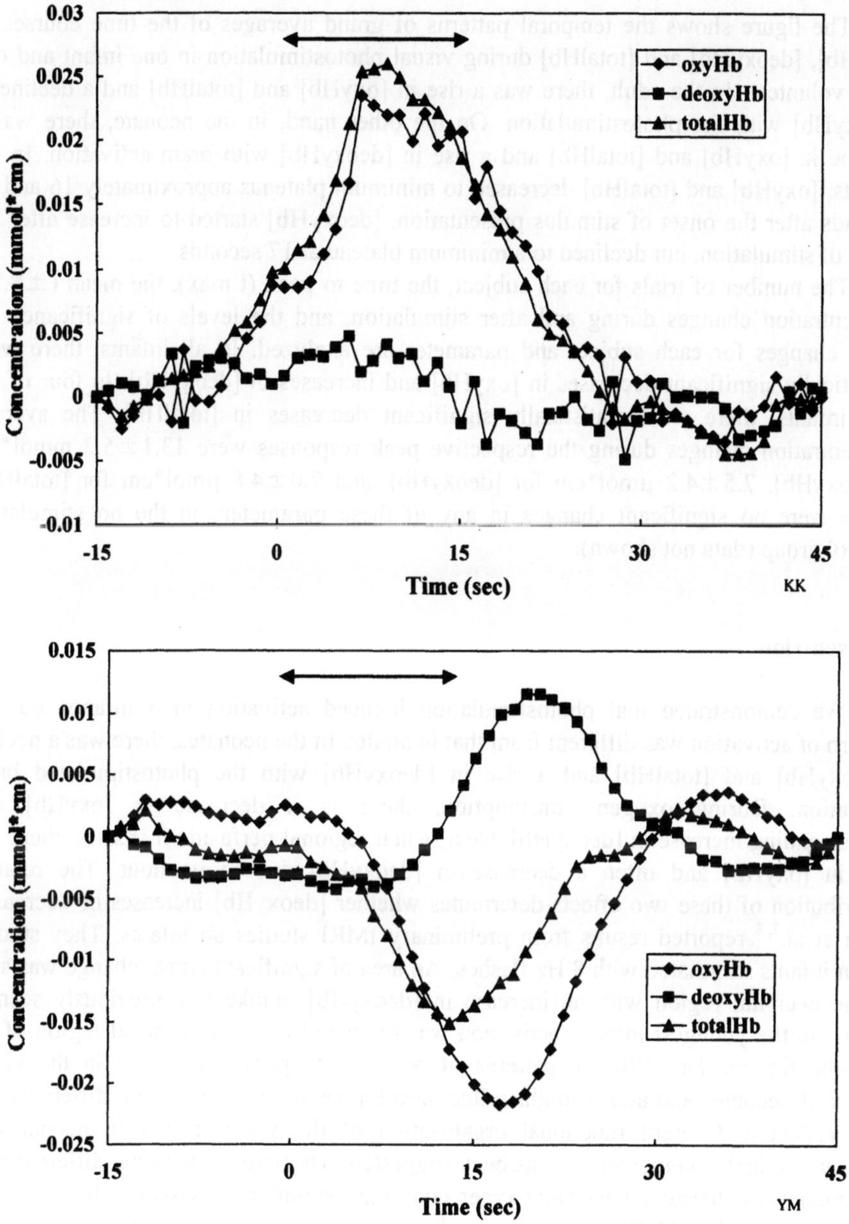

Figure 1. The grand average of concentration changes for one subject as a function of time. (Top) Average from 10 trials in an adult volunteer. (Bottom) Average from 10 trials in an infant aged 16 wk (43wk's postconceptional age). The arrow indicates the 15-second visual stimulation period. Changes in oxyHb, deoxyHb and totalHb are given in mmol*cm.

3. Results

The figure shows the temporal patterns of grand averages of the time courses of [oxyHb], [deoxyHb] and [totalHb] during visual photostimulation in one infant and one adult volunteer. In the adult, there was a rise in [oxyHb] and [totalHb] and a decline in [deoxyHb] with the photostimulation. On the other hand, in the neonate, there was a decline in [oxyHb] and [totalHb] and a rise in [deoxyHb] with brain activation. In the infants, [oxyHb] and [totalHb] decreased to minimum plateaus approximately 16 and 13 seconds after the onset of stimulus presentation. [deoxyHb] started to increase after the onset of stimulation, but declined to a minimum plateau at 17 seconds.

The number of trials for each subject, the time to peak (t max), the mean (±S.D.) concentration changes during and after stimulation, and the levels of significance for these changes for each subject and parameter are analyzed. In all infants, there were statistically significant decreases in [oxyHb] and increases in [deoxyHb]. In four of the five infants, there were statistically significant decreases in [totalHb]. The average concentration changes during the respective peak responses were 13.1±5.2 mmol*cm for [oxyHb], 7.5±4.2 μmol*cm for [deoxyHb], and 7.0±4.8 μmol*cm for [totalHb]. There were no significant changes in any of these parameters in the no stimulation control group (data not shown).

4. Discussion

We demonstrated that photostimulation induced activation in 5 infants, but the pattern of activation was different from that in adults. In the neonates, there was a decline in [oxyHb] and [totalHb] and a rise in [deoxyHb] with the photostimulated brain activation. During oxygen consumption, there is a decrease in [oxyHb] and corresponding increase in [deoxyHb]. Next, when regional perfusion increases, there is a rise in [oxyHb] and often a decrease in [deoxyHb] due to washout. The relative contribution of these two effects determines whether [deoxyHb] increases or decreases. Born et al.[7, 8] reported results from preliminary fMRI studies on infants. They studied seven infants stimulated with 8 Hz flashes. An area of significant signal change was seen in the occipital region with an increase in [deoxyHb], unlike that previously seen in adults. In the youngest infants, activation was restricted to the anatomical region of the calcaine fissure. The different patterns of response to photostimulation in the visual cortex of neonates and adults might reflect developmental and behavioral differences. It may reflect a different functional organization of the visual cortex in neonates or on-going retinal development. It has been suggested that there are regional differences in the amounts of change in CBF and oxygen consumption within the visual cortex.

Meek et al.[1]used NIRS in a study in which ten infants aged 3 d to 14 wk in an awake state viewed a checkerboard with a 5 Hz pattern reversal. All of the infants showed an increase in [totalHb] in the occipital region with the stimulation, Nine of the ten infants showed an increase in [oxyHb] and 9 showed an increase in [deoxyHb] related to the stimulus. These results are different to our results showing that [oxyHb] and [totalHb] decreased with photostimulation in neonate. This discrepancy in results may be due to the difference in the sleeping stages or pattern of photostimulation. Further studies are needed. The use of NIRS is expected to enable study of the physiology of the developing brain and the brain`s response to damage.

Conclusions

We have reported NIRS measurements of functional hemodynamic responses in infants during natural sleep. The NIRS signal response patterns in neonates and adults are different. This difference may reflect differences in the behavioral or developmental state.

Acknowledgements

This research was supported by grants-in-aid for scientific research (C) no. 13671140 and 12671065, and encouragement of young scientists no. 13770619, from the Ministry of Education Science and Culture of Japan.

References

1. J. H. Meek, M. Firbank, C. E. Elwell, J. Atkinson, O. Braddick, J. S. Wyatt, Regional hemodynamic responses to visual stimulation in awake infants, *Pediatr. Res.* **43**(6), 840-843(1998).
2. K. Sakatani, S. Chen, W. Lichty, H. Zuo, Y. P. Wang, Cerebral blood oxygenation changes induced by auditory stimulation in newborn infants measured by near infrared spectroscopy, *Early Hum. Dev.* **55**(3), 229-236(1999).
3. M. Bartocci, J. Winberg, C. Ruggiero, L. L. Bergqvist, G. Serra, H. Lagercrantz, Activation of olfactory cortex in newborn infants after odor stimulation: a functional near-infrared spectroscopy study, *Pediatr. Res.* **48**(1), 18-23(2000).
4. K. Isobe, T. Kusaka, K. Nagano, K. Okubo, S. Yasuda, M. Kondo, S. Itoh, S. Onishi, Functional imaging of the brain in sedated newborn infants using near infrared topography during passive knee movement, *Neurosci. Lett.* **299**(3), 221-224(2001).
5. T. Kusaka, K. Isobe, K. Nagano, K. Okubo, S. Yasuda, K. Kawada, S. Itoh, S. Onishi, I. Oda, Y. Wada, I. Konishi, T. Tsunazawa, Estimation of regional cerebral blood flow distribution in infants by multichannel near-infrared spectroscopy with indocyanine green, *Proc. SPIE* **4250**, 301-305(2001).
6. I. Oda, Y. Wada, S. Takeuchi, Y. Oikawa, N. Sakauchi, Y. Ito, I. Konishi, Y. Tsunazawa, T. Kusaka, K .Isobe, S. Itoh, S. Onishi, Near infrared optical imager for cerebral blood flow and oxygenation detection, *Proc. SPIE* **4250**, 371-379(2001).
7. P. Born, H. Leth, M. J. Miranda, E. Rostrup, A. Stensgaard, B. Peitersen, H. B. Larsson, H. C. Lou, Visual activation in infants and young children studied by functional magnetic resonance imaging, *Pediatr Res.* **44**(4), 578-83(1998).
8. P. Born, E. Rostrup, H. Leth, B. Peitersen, H. C. Lou, Change of visually induced cortical activation patterns during development, *Lancet.* **347** 543(1996).

EFFECT OF BLOOD FLOW ON PvO_2-VO_2 RELATION IN CONTRACTING *IN SITU* SKELETAL MUSCLE

Hisaharu Kohzuki, Hiromi Misawa, Susumu Sakata, and Miyako Takaki*

1. INTRODUCTION

It has been reported that high O_2 affinity of blood reduced O_2 uptake (VO_2) of maximally and submaximally stimulated muscle[1, 2], beating heart[3] and brain[4] with a decrease in venous effluent PO_2 (PvO_2). These studies suggest that VO_2 in organs of high metabolic activity is determined by the interaction of the convective O_2 delivery and the subsequent diffusion of O_2 from the erythrocytes to the mitochondria, rather than by O_2 delivery alone (blood flow multiplied by arterial O_2 content), and PvO_2 reflects the diffusive O_2 supply.

In previous studies using maximally and submaximally stimulated canine gastrocnemius during hypoxic hypoxia[5], high O_2 affinity hypoxia[1, 2] and anemic hypoxia[6], we found a linear relationship between VO_2 and PvO_2 with VO_2-axis intercept, suggesting that PvO_2 is a PO_2 value of mixed effluent blood from both a region supplied by a blood flow (Q_{diff}) to tissue of which VO_2 is limited by O_2 diffusion and a region supplied by a blood flow (Q_{conv}) to tissue of which VO_2 is limited by O_2 delivery. We hypothesized that VO_2 consists of PvO_2-dependent VO_2 (O_2 diffusion-limited VO_2) and PvO_2-independent VO_2 (O_2 delivery-limited VO_2, $VO_{2,conv}$)[7]. It indicates $VO_2 = D \times PvO_2 + VO_{2,conv}$, where D is apparent diffusion capacity. One of aims of the present study is to clarify the effect of blood flow on the PvO_2-VO_2 relation or whether blood flow directly changes the extent of the VO_2-axis intercept of the PvO_2-VO_2 relation in anesthetized canine gastrocnemius muscle. The other aim is to estimate the fraction of Q_{conv} to total blood flow ($Q_{conv} + Q_{diff}$) using the PvO_2-VO_2 relation according to an additive two-compartment model. We evaluated the effect of blood flow on the fraction of Q_{conv}, and obtained the result that an increase in blood flow increased the fraction of Q_{conv} under the moderate change of perfusion pressure.

* Department of Physiology II, Nara Medical University, Kashihara, 634-8521, Japan

Oxygen Transport to Tissue XXIII, edited by D. Wilson *et al.*
Kluwer Academic/Plenum Publishers, 2003

2. METHODS

Seven mongrel dogs (weight 12.7±1.8 kg) were anesthetized with pentobarbital sodium (30 mg/kg body weight). The in situ isolation method of gastrocnemius-plantaris muscle group and the perfusion method were similar to that described previously[5]. Repetitive isometric, tetanic contraction was induced by supramaximal stimuli (4 V, 0.2 msec duration, 50 Hz in frequency) in 1 Hz trains of 200 msec duration. Each experiment (n=7) consisted of four 3 min contraction trials with 20 min rest interval. To construct the PvO_2-VO_2 relation, least square method was used for pooled data for high flow (n=14), and that for low flow (n=14). Lactate concentration was measured enzymatically (Lactate test, Boehringer Mannheim). All samples had their SO_2 measured with an OSM-2 (Radiometer, Copenhagen, Denmark) and their pH, PCO_2, and PO_2 with a blood gas analyzer (Radiometer ABL 330). VO_2 was calculated by flow × arterio-venous O_2 content difference.

To determine the fraction of Q_{conv} from the observed PvO_2-VO_2 relation, D was calculated in Q_{conv} ranging from 40 to 90 $ml \cdot min^{-1} \cdot 100g^{-1}$ for high flow, from 28 to 70 $ml \cdot min^{-1} \cdot 100g^{-1}$ for low flow. O_2 extraction of the diffusion-limited compartment is set at 0.8. Arterial O_2 content was reduced by steps of 1 $ml \cdot dl^{-1}$ from 20 to 1 $ml \cdot dl^{-1}$ to simulate hypoxic hypoxia. VO_2-axis intercepts determined by the PvO_2-VO_2 relation for high and low flows were used as $VO_{2,conv}$. In each step, we calculated PvO_2 as a value of mixed venous effluent from both compartments according to the oxygen dissociation curve.

Statistical comparisons were made using repeated measures analysis of variance (ANOVA) or paired t-test. Significance was considered to be at $P < 0.05$. All data are presented as mean ± SEM (unless otherwise noted). An analysis of covariance (ANCOVA) was applied to compare the regression lines of the PvO_2-VO_2 relation for high and low flows.

3. RESULTS AND DISCUSSION

Table 1 shows O_2 supply and VO_2 parameters in in situ canine gastrocnemius muscle under 1 Hz tetanic contracting condition. The VO_2, O_2 delivery and O_2 extraction were significantly different from each other ($P < 0.05$). The VO_2/PvO_2 ratio for hypoxia was significantly higher than that for normoxia ($P < 0.05$). PvO_2 for high flow was significantly higher than that for low flow in normoxia ($P < 0.05$). Figure 1 A and B show the relationship between VO_2 and O_2 delivery, and the relationship between VO_2 and PvO_2, respectively. The calculated VO_2-axis intercept for high flow was significantly different from that for low flow ($P < 0.05$, ANCOVA). The slope of the linear regression line was not significantly different from each other ($P > 0.05$, ANCOVA).

Figure 2 shows the relationship between the distributed blood flow to O_2 delivery-limited compartment and the slope of the PvO_2-VO_2 relationship. Increase in Q_{conv} reduced the apparent O_2 diffusion capacity. For low flow of 80 (= $Q_{conv} + Q_{diff}$), 0.25 of the slope of the PvO_2-VO_2 relation is Q_{conv}=30 or 100×30/80=37.5 (%) of the fraction of Q_{conv}. For high flow of 110 (= $Q_{conv} + Q_{diff}$), 0.31 of the slope is Q_{conv}=55 or 100×55/110=50 (%) of the fraction of Q_{conv}. Increase in blood flow increased the fraction of Q_{conv}.

Table 1. O_2 supply parameters

	High flow		Low flow	
	Normoxia	Hypoxia	Normoxia	Hypoxia
pH_a[1)]	7.37±0.02	7.41±0.03#	7.35±0.05	7.42±0.02#
P_aCO_2[2)]	38.3±0.7	35.1±1.1	39.7±1.4	34.6±1.4#
P_aO_2[3)]	84.1±6.3	29.0±2.2#	80.7±5.8	28.2±1.8#
C_aO_2[4)]	19.1±1.1	10.9±1.5#	19.4±1.1	10.8±1.4#
Blood flow[5)]	111±5*	112±6*	75±3	77±4
Perfusion pressure[6)]	132±9*	113±3*#	92±6	84±3
$Lactate_a$[7)]	2.2±0.3	2.2±0.3	2.0±0.3	2.0±0.1
pH_v[8)]	7.28±0.02	7.32±0.03	7.26±0.02	7.28±0.02
P_vCO_2[9)]	56.8±2.4	49.5±2.3#	61.0±2.5	54.1±2.2
P_vO_2[10)]	25.3±1.3*	11.7±0.9#	21.3±1.6	9.3±0.7#
O_2 delivery[11)]	21.1±1.4*	12.0±1.6*#	14.4±0.8	8.2±0.9#
VO_2[12)]	13.5±0.6*	8.8±1.0*#	10.4±0.3	6.8±0.7#
O_2 extraction[13)]	65±3*	75±4*#	73±4	84±2#
VO_2/P_vO_2 ratio[14)]	0.55±0.04	0.76±0.07#	0.50±0.04	0.76±0.10#
Force[15)]	16.6±2.3	13.3±1.5	14.3±2.0	11.4±1.6
iForce[16)]	27.6±4.6	23.1±6.0	27.2±5.2	26.1±5.7
Lactate release[17)]	43±29	130±48	41±16	153±34

Values are mean±SEM (n=7). 1), arterial pH; 2), arterial PCO_2, Torr; 3), arterial PO_2, Torr; 4), arterial O_2 content, ml/dl; 5), $ml{\cdot}min^{-1}{\cdot}100g^{-1}$; 6), mmHg; 7), lactate concentration of arterial blood, mM; 8), venous pH, 9), venous PCO_2, Torr; 10), venous PO_2, Torr; 11), $ml{\cdot}min^{-1}{\cdot}100g^{-1}$; 12), $ml{\cdot}min^{-1}{\cdot}100g^{-1}$; 13), 100 × VO_2/O_2 delivery; 14), $ml{\cdot}min^{-1}{\cdot}100g^{-1}{\cdot}Torr^{-1}$; 15), kg/g muscle; 16), initial force, kg/g muscle; 17), $\mu mole{\cdot}min^{-1}{\cdot}100g^{-1}$.

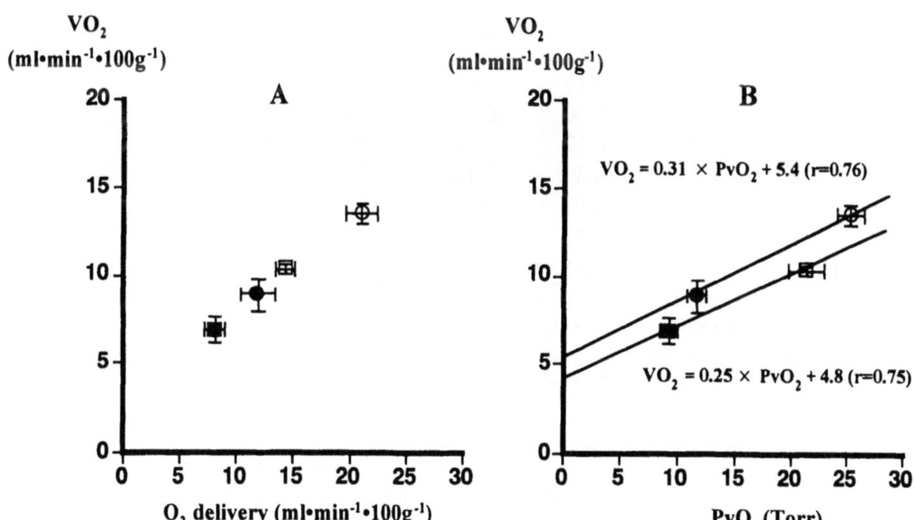

Figure 1. Effect of blood flow on the O_2 delivery-VO_2 relation (A) and the PvO_2-VO_2 relation (B). Circle indicates high flow, and square indicates low flow. Open symbol indicates normoxia, and closed symbol indicates hypoxia.

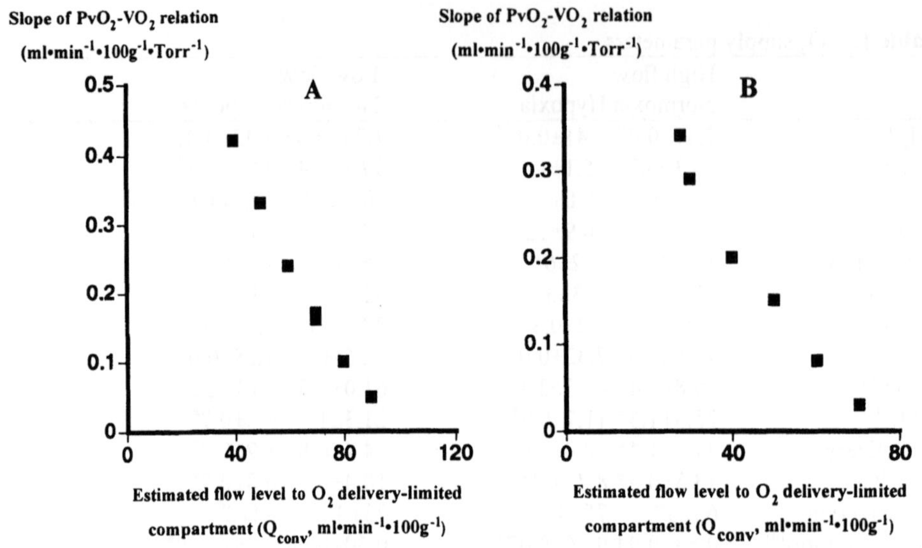

Figure 2. Relationship between distributed blood flow passing through the O_2 delivery-limited compartment (Q_{conv}) and the slope of the PvO_2-VO_2 relationship to simulate hypoxic hypoxia at high flow (A) and low flow (B).

Table 2. Slope and VO_2-axis intercept of the PvO_2-VO_2 relation in seven preparations of in situ gastrocnemius muscle

Experimental number	High flow		Low flow	
	D[1)]	$VO_{2,conv}$ [2)]	D	$VO_{2,conv}$
1	0.15	10.4	0.24	5.88
2	0.16	9.05	0.1	7.74
3	0.56	0	0.23	3.97
4	0.17	5.74	0.75	0
5	0.3	5.16	0.11	7.63
6	0.91	0	0.56	0
7	0.42	4.5	0.49	1.47
mean ± SD	0.38±0.28	4.98±4.01	0.35±0.25	3.81±3.39*

1), slope (D, ml•min^{-1}•100g^{-1}•Torr^{-1}); 2), VO_2-axis intercept ($VO_{2,conv}$, ml•min^{-1}•100g^{-1}); PvO_2-VO_2 relation was calculated by the following equation, $VO_2 = D \times PvO_2 + VO_{2,conv}$.
If the calculated $VO_{2,conv}$ is within SD of VO_2 for pooled data, $VO_{2,conv}$ is assumed to be zero. * Significant difference between high and low flows ($P < 0.05$).

The most important result of the present study is that we successfully obtained the significantly different calculated VO_2-axis intercept of the PvO_2-VO_2 relation between high and low blood flows. According to our additive two-compartment model, an increase in blood flow induced the increased fraction of blood flow supplied to the O_2

delivery-limited compartment of tissue in in situ contracting canine gastrocnemius muscle. The increase in blood flow did not seem to affect the apparent O_2 diffusion capacity.

Hepple et al.[8], from their pioneer work on O_2 diffusion limitation to peak VO_2 of contracting skeletal muscle, discussed that one of the interpretations of the linear relationship, passed through the origin, usually observed between peak VO_2 and mean capillary PO_2 is that maximal mitochondria capacity is not limiting peak muscle VO_2. Under these conditions, the slope of this relationship reflects muscle O_2 diffusion capacity. Conversely, deviation from linearity suggests that factors other than diffusion capacity are constraining peak VO_2, and thus the slope of the line is blunted by some region in the absence of O_2 diffusion limitation. Our hypothesis of two additive VO_2 compartments is not inconsistent with their thought of the deviation from linearity due to the limitation of maximal mitochondria capacity. The slope of the PvO_2-VO_2 relation with the VO_2-axis intercept could reflect apparent O_2 diffusion capacity because VO_2 consists of both additive O_2 diffusion- and O_2 delivery-limited VO_2.

We observed the value of the slope of the PvO_2-VO_2 relation to range from 0.1 to 0.91 $ml \cdot min^{-1} \cdot 100g^{-1} \cdot Torr^{-1}$ and the value of $VO_{2,conv}$ from 0 to 10.4 $ml \cdot min^{-1} \cdot 100g^{-1}$ (Table 2), suggesting a wide variation of the PvO_2-VO_2 relation due to the experimental protocol used or individual variation among preparations: e. g. difference in myoglobin concentration[9]. Experiment number 6 indicated $VO_{2,conv}=0$ and VO_2 for high flow and normoxia of this muscle is 14.78 $ml \cdot min^{-1} \cdot 100g^{-1}$ at blood flow of 115 $ml \cdot min^{-1} \cdot 100g^{-1}$ under the perfusion pressure of 100 mmHg. Although, in the present experiment, we used pooled data to compare the effect of blood flow on the PvO_2-VO_2 relation, it is better to compare the PvO_2-VO_2 relation in the same contracting period in each preparation to evaluate the effect of single factor.

REFERENCES

1. H. Kohzuki, Y. Enoki, K. Matsumura, S. Sakata, and S. Shimizu, Flow-dependent influence of high-O_2-affinity erythrocytes on peak VO_2 in exercising muscle in situ. *J. Appl. Physiol.* **80**(3), 832-838 (1996).
2. H. Kohzuki, Y. Enoki, S. Sakata, S. Shimizu, and Y. Ohga, High affinity of blood for oxygen reduces oxygen uptake in contracting canine gracilis muscle. *Exp. Physiol.* **79**(1), 71-80 (1994).
3. C. S. Apstein, R. C. Dennis, L. Briggs, W. M. Vogel, J. Frazer, and C. R. Valeri, Effect of erythrocyte storage and oxyhemoglobin affinity changes on cardiac function. *Am. J. Physiol.* **248**(17), H508-H515 (1985).
4. R. D. Woodson, J. H. Fitzpatrick, Jr., D. J. Costello, and D. D. Gilboe, Increased blood oxygen affinity decreases canine brain oxygen consumption.. *J. Lab. Clin. Med.* **100**(3), 411-424 (1982).
5. H. Kohzuki, H. Misawa, T. Kishi, Y. Ohga, S. Sakata, and M. Takaki, Muscle venous PO_2 and VO_2 are linearly related in repetitive tetanic contractions of canine muscle during hypoxic hypoxia. *Clin. Exp. Pharmacol. Physiol.* **26**(8), 639-644 (1999).
6. H. Kohzuki, Y. Enoki, Y. Ohga, S. Sakata, S. Shimizu, T. Morimoto, T. Kishi, and M. Takaki, Effect of blood flow and haematocrit on relationship between muscle venous PO_2 and oxygen uptake in maximally contracting dog gastrocnemius in situ. *Clin. Exp. Pharmacol. Physiol.* **24**(2), 182-187 (1997).
7. H. Kohzuki, Y. Enoki, S. Sakata, S. Shimizu, and T. Kishi,. PvO_2-VO_2 diagram permits differentiation of O_2 delivery- and diffusion-limited VO_2 components. *Jpn. J. Physiol.* **44**, Suppl 1, s114 (1994)
8. R. T. Hepple, M. C. Hogan, C. Stary, D. E. Bebout, O. Mathieu-Costello, and P. D. Wagner, Structural basis of muscle O_2 diffusion capacity: evidence from muscle function in situ. *J. Appl. Physiol.* 88(2), 560-566 (2000)
9. H. Kohzuki, M. Takaki, S. Sakata, Y. Ohga, H. Misawa. Origin of individual deviation of metabolic activity for oxygen in contracting in situ skeletal muscle. *Ono Sports Science* **8**, 15-23 (2000).

OXYGEN TRANSPORT AND INTRACELLULAR BIOENERGETICS ON STIMULATED CAT SKELETAL MUSCLE

S. Nioka, K. McCully, G. McClellan, Jane Park, and B. Chance[1]

1. ABSTRACT

A unique multiparameter recording of skeletal muscle bioenergetics, biochemistry and biomechanics has permitted determination of novel relationships among hemodynamics, cellular high-energy metabolites and mitochondrial bioenergetics in feline skeletal muscle. The study utilizes ^{31}P NMR, NIR, and NADH fluorescence spectrophotometry, biochemical assays and muscle performance. Seven cats were anesthetized and mechanically ventilated. Calf muscles were stimulated through sciatic nerve electrical stimulation and tension was monitored by a strain gauge connected to the Achilles tendon. We stimulated the muscle to produce several workloads up to V_{max}. We also changed FiO_2 from normoxia to hypoxia for each $\%V_{max}$. From these results, the most sensitive indicators of cellular hypoxia leading to a reduction in muscle performance can be determined. Hemoglobin deoxygenation generally does not correlate with cellular hypoxia, although when the HbO_2 drops below 30% saturation there is an increased incidence of cellular hypoxia. The [ADP], which is known to regulate mitochondrial function, has a close relation to the work, not to the hypoxia. On the other hand, the mitochondrial NADH does respond to cellular PO_2. The degree of oxidation (NADH decrease) due to the ATP flux shifts with oxygen availability in mild to moderate hypoxia (at FiO_2 down to 9%). As cellular hypoxia causes decreases in muscle performance (moderate to severe hypoxia), NADH is being reduced rather than oxidized with increasing workloads.

---

[1] Shoko Nioka, George McClellan and Britton Chance, Department of Biochemistry and Biophysics, School of Medicine, University of Pennsylvania, Philadelphia, PA. Kevin McCully, University of Georgia, Athens, GA. Jane Park, University of Vanderbilt.TN.

2. INTRODUCTION

One of our interests was to determine if oxygen is a limiting factor for any competitive sports event. Many strategies to enhance muscle performance relate to oxygen delivery, extraction and the capacity of oxidative metabolism for greater ATP production. We wanted to know in what situation the oxygen limitation is reached, and how it can be detected. We designed a study to determine the conditions of oxygen limitation in a hypoxia model using cat. Cat muscle is similar to human in the mixed populations of Type I and Type II muscle cells. In a design to determine the role of oxidative muscle, we used only the oxidative muscle fiber population. Therefore, in order to have only oxidative muscle in use, we stimulated submaximally for a long period of time (about 45 min) to make sure the muscle performance reached steady state. Only oxidative muscle can perform steady state metabolism. We have used NADH changes, tissue hemoglobin/myoglobin changes, cellular high-energy biochemistry and other physiological measurements to correlate muscle performance and degree of hypoxia. We have found that NADH changes are the best indicator of the hypoxic responses. We will discuss the potential for using these measurements to evaluate hypoxic stress.

3. MATERIALS AND METHODS

3.1 Preparation

The left gastrocnemius-soleus muscle complex was studied in vivo under sciatic nerve stimulation in seven cats. The animals were anesthetized with Ketamine hydrochloride and ventilated. For force measurements, the Achilles tendon was connected to a strain gauge transducer and the knee joint was fixed to a mechanical ground. A 1.5 cm surface coil was placed on the skin over the muscle for the MRS. Electrodes were placed under the sciatic nerve and pulses of approximately 10V, 0.5 msec, were used for stimulation. Workload was varied by changing rates of submaximal stimulation from 0.25 Hz to approximately 5 Hz which produced maximum workload. Calculation of tension time integral (TTI) was made by taking area of tension curve (integral of tension time) and frequency (1/sec) yielding units of g x sec/ sec.

NADH fluorescence spectroscopy was used to determine %NADH change in the muscle tissue. Excitation light from a mercury lamp at 366 nm with two filters illuminated the surface of a skinned gastrocnemius muscle through a quartz fiber whose diameter was 3 mm and located in the middle of the MRS surface coil. This yielded the emission fluorescence light of NADH at 450 nm. The NADH signal was calculated by subtracting the reflectance at 366 nm from the intensity at 450 nm (1). The fluorometer was calibrated from resting to anoxia (maximal reduction).

A NIRS spectrometer was used for detecting blood volume and percent saturation change of myoglobin and hemoglobin (Hb/Mb) at 750, 780 and 850 nm. The gastrocnemius muscle was transilluminated through two glass fibers (diameters of 5 mm) coupled to source and detector and separated by 1.5 cm. The percent saturation of Hb/Mb in gastrocnemius was calculated by taking 0% at death and 100% at highest possible value after recovery from exercise with 100% FiO_2.

For analysis of high-energy phosphate concentration, ^{31}P MRS (2.1 Tesla, 30 cm bore) was used. Control cat gastrocnemius muscles (n=5) were analyzed biochemically to determine cytosolic [ATP] (5.9 mM) and cytosolic total [creatine] (35.2 mM) (R. Veech personal communication). These values were used for quantitation of MRS data. Detailed analysis of MRS data has been described elsewhere (2). Calculated [ADP] was obtained using the creatine kinase equilibrium constant with the assumption that the creatine kinase reaction is in equilibrium (3).

3.2 Experimental Procedure

Prior to the main protocol, the calf muscle was stimulated for 45 min submaximally until the glycolytic muscle contribution was minimized. FiO_2 was decreased stepwise (33, 15, 13, 11, 9, 8, 7 and to 4%) to give graded hypoxia. Four frequencies of nerve stimulation were used to vary workload.

3.3 Data Analysis

In order to determine the significance in $\%HbO_2$ and %NADH in the stages of hypoxia and workloads, One way ANOVA was used for means. In [ADP] and TTI, the significant difference in the hypoxic stages was examined by regression analysis for sensitivity ([ADP]/TTI). Probability of less than 5% was used for significance of null hypothesis.

4. RESULTS

The prior muscle stimulation worked well to suppress the glycolytic metabolism of the Type II muscle, as seen by lactate elevation and decreased pH_i. There were no large pH_i shifts at the maximum level of work, nor during the extremely severe hypoxia (lowest pH_i was 6.73). In the following section, each candidate of hypoxic indicators is evaluated in the working muscle.

4.1. Hemoglobin saturation and cellular hypoxic states

Since oxygen is carried to the tissue by the hemoglobin, the cellular response to hypoxia will ultimately depend on the oxygen carrying capacity and unloading characteristics of the hemoglobin. The amount of oxygen released to the tissue by the hemoglobin was estimated by the percent saturation of the hemoglobin ($\%HbO_2$) and was measured through the muscle by NIRS. The results are shown in Figure 1a. It is evident that mild hypoxia results in a significant decrease in the percent saturation of the hemoglobin. Both in normoxia (25%FiO_2) and mild hypoxia (12% FiO_2), the hemoglobin unloads about 30% of its oxygen as it passes through the muscle. That this amount of oxygen is adequate to support the maximum workload is shown in Figure 1b. Imposing mild hypoxia has no effect on the ability of the muscle to perform work. As the hypoxia is increased to moderate (9%FiO_2), there is no significant change in the %HbO_2 and the hemoglobin continues to release about 30% of its oxygen as the workload approaches its maximum. However, the total amount of oxygen being released to the tissue is much less and the muscle is not able to sustain the workload that was seen in normoxia and mild hypoxia. Even in severe hypoxia (3-4% FiO_2), the %HbO_2 in the resting muscle is not depressed much below the values seen in the mild and moderate states. This does not hold for working muscle. When the muscles are stimulated, there is an immediate decline in %HbO_2 and the workload supported by the muscle is also very small.

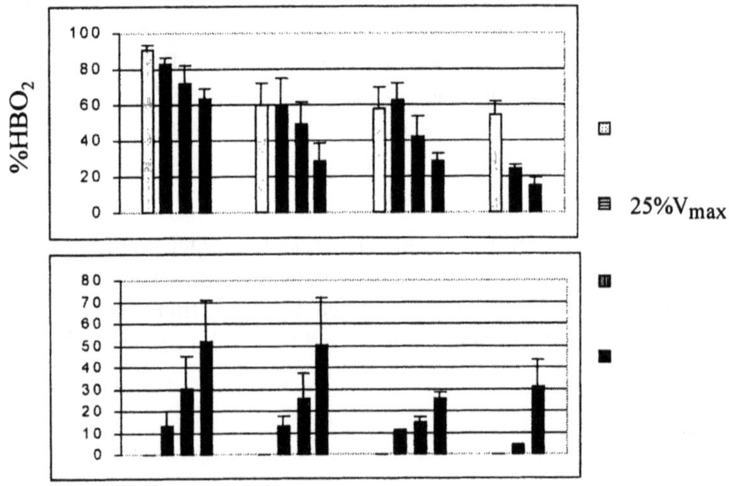

Figure 1. The muscle oxygenation (%HbO_2, upper histogram) is plotted for four hypoxic treatments. The %HbO_2 values are expressed as means and standard errors. For each stage of hypoxia, there is %HbO_2 under four workload conditions. The lower histogram shows the muscle work plotted for the four hypoxic conditions as indicated in the abscissa. TTI values are expressed as means and standard errors. In each hypoxic stage, TTI under the four workloads is shown.

4.2. The role of ADP during the hypoxia in the working muscles

ADP is a primary regulator of oxidative metabolism and is very sensitive to muscle performance as shown in Figure 2. In the resting muscle, the ADP concentration is maintained at around 16 μM and does not change significantly as the PO_2 is reduced into the severely hypoxic range.

There is no difference in the rate of ADP increase in normoxia and mild hypoxia. In both conditions, the free [ADP] rises about three times from resting to maximum work level. In this, the relationship between [ADP] and workload appears to be more closely related to the TTI as shown in Figure 1b. The TTI is stable while the %HbO2 does change significantly from normal to mild hypoxia as shown in Figure 1a. When the PO_2 is reduced further to moderate hypoxia, there is a significant increase in rate of ADP

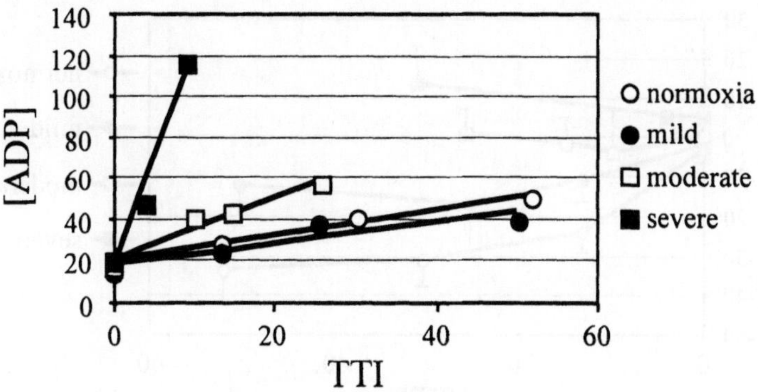

Figure 2. The relationship between workload (TTI) and [ADP] is shown with four levels of hypoxia. The [ADP] increased significantly with the TTI. In this figure, regression line slopes for the four oxygen states are positive. With severity of hypoxia, the [ADP] concentration increased with the amount of workload. There was no difference in the slopes of normal and mild hypoxia but the moderate and severe hypoxic states did differ significantly from each other and the other two.

accumulation in the muscle. As the workload increases from rest to the maximum sustainable, the free [ADP] increases to around 60 μM. Moreover, the maximum sustained workload is around 50% of the workload in normoxia and mild hypoxia. When the hypoxia becomes severe, going into the lethal range, the rate of ADP production is higher by an order of magnitude than in the normoxic state. This strongly suggests that there is an uncontrollable depression of mitochondrial oxidative activity.

4.3. Response of NADH to cellular hypoxia

The sensitivity of the NADH concentration to both workload and hypoxia is shown in Figure 3. In response to workload, muscle cells increase their oxidative metabolism to increase the production of ATP. One of the mechanisms used to produce ATP is the oxidation of NADH to NAD. As the rate of oxidative metabolism is increased, the concentration of NADH decreases and NAD rises. This change is clearly seen in Figure 3. In the case of the normoxic muscle, the NADH concentration decreases as a log function by around 35% when the workload is increased from resting to maximum. As the animal becomes more hypoxic, the rate of NADH oxidation is reduced. In mild hypoxia, the decrease in NADH is half that seen in the normoxic animal, and in moderate hypoxia, the concentration does not change significantly from resting. As the hypoxia becomes severe, a significant increases in NADH concentration results. Since the oxidation of NADH requires the presence of oxygen in the mitochondria, the slowing of the rate of NADH oxidation and its associated production of ATP represents a very sensitive indicator of hypoxia at the cellular and mitochondrial level.

5. SUMMARY AND DISCUSSION

This extensive study of bioenergetics of cat skeletal muscle demonstrate that the mitochondrial redox state is the most sensitive indicator of the adequacy of O_2 delivery to

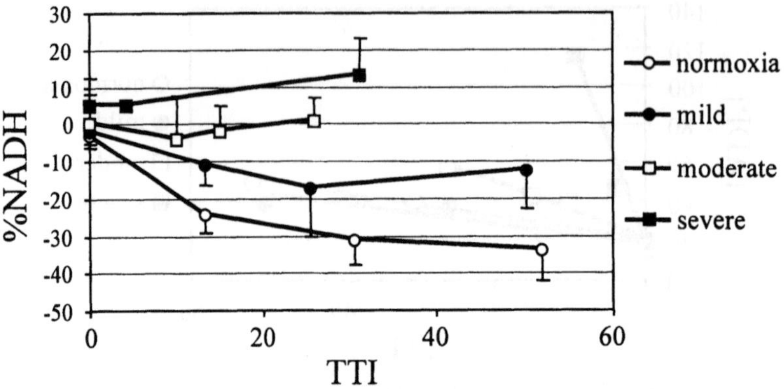

Figure 3. NADH and work relationship in the cat gastrocnemius muscle. Note that an NADH increase can be seen in extreme hypoxia. On the other hand, in normal and mild hypoxia, workload results in oxidation of NADH (NAD).

the myocytes and their mitochondria. Although the O_2 extraction in mild hypoxia is sufficient to support a workload comparable to that seen in the normoxic muscle, the decreased oxidation of NADH strongly suggests that there is a significant deficit of O_2 at the level of the mitochondria. The workload in mild hypoxia is probably maintained by compensatory effects of ADP and the influence it exerts on the mitochondrial production of ATP. On the other hand, in both moderate and severe hypoxia, there is a progressive decrease in work capacity that is correlated with the very large increase in [ADP] and the failure to maintain NADH oxidation.

ACKNOWLEDGEMENTS

This research is funded by NIH HL44125 and Medical Diagnostics Research Foundation. We thank Dr. R. Veech for the biochemical assays of the cat gastrocnemius-soleus muscles, and to Ms. D. Coleman for preparation of the manuscript.

REFERENCES

1. Mayevsky, A. J. Brain oxidative metabolism of the newborn dog: correlation between ^{31}PNMR spectroscopy and pyridine nucleotide redox state. J.Cereb. Blood Flow Metab. 8:201-207 (1988).
2. Nioka, S., DS. Smith, B. Chance, HV. Subramanian, S. Butler, M.Katzenberg. Oxidative phosphorylation system during steady-state hypoxia in the dog brain. J.Appl.Physiol.68: 2527-35, 1990.
3. Veech, R.L., Lawson, J.W.R., Cornell, N.W. and Krebs, H.A. Cytosolic phosphorylation potential. J. Biol. Chem. 254:6538-6547 (1979).

MUSCLE OXYGENATION BY NEAR INFRARED SPECTROSCOPY AND LACTATE THRESHOLDS IN ENDURANCE TRAINED AND RECREATIONALLY ACTIVE CYCLISTS

Melissa Ladewig, Robert Robertson, Edwin M. Nemoto[1*]

1. INTRODUCTION

The development of noninvasive near infrared spectroscopy (NIRS) to monitor tissue oxyghemoglobin/myoglobin saturation provides a means of continuously and noninvasively monitoring muscle oxygenation during exercise (1-3). The lactate threshold has been used to gauge endurance training (4,5). It would be of interest to determine whether the noninvasive monitoring of muscle oxygenation can be used as a surrogate marker of lactate threshold or as a gauge of conditioning during endurance training. The ability to identify the lactate threshold by a noninvasive, continuous monitor such as NIRS would greatly aid athletes in assessing their level of performance during training and to guide that training.

Thus, we hypothesized that NIRS measurement of muscle oxygenation can differentiate the lactate threshold, i.e. the inflection point of the increase in plasma lactate levels during exercise, in endurance trained and recreationally active athletes. The objective of this study is to test this hypothesis by continuous monitoring of muscle oxygenation during graded exercise in endurance trained and recreationally active males with venous blood lactate measures.

[1] Department of Health Physical Recreation and Education, School of Education and Department of Neurological Surgery*, School of Medicine University of Pittsburgh, Pittsburgh, PA 15213

Oxygen Transport to Tissue XXIII, edited by D. Wilson *et al.*
Kluwer Academic/Plenum Publishers, 2003

2. METHODS

2.1 Study Design

The study is a cross-sectional experimental design with a single cycle exercise bout administered to each subject according to an approved protocol by the Institutional Review Board. All subjects signed informed consents. Trained and untrained subjects were tested in a counterbalanced order. All subjects reported to the laboratory in a three-hour post-absorption state.

2.2 Screening Procedures

Health and medical history questionnaire to exclude any possible participants with chronic dysrhythmias, asthma or exercise induced asthma and/or any recent chronic physical injury, which may be aggravated by the exercise test. Excluding these individuals greatly reduces the risk of possible complications during exercise. Training and physical activity questionnaire allows the investigators to place subjects in the correct cohort. Endurance Trained subjects trained 100 miles a week or more whereas recreationally active subjects cycled 25 miles per week. In the event an individual is identified as having one of the previous conditions listed above, the individual was excluded from participation.

2.3 Experimental Paradigm

A total of 21 males 18 to 36 years old were studied in two groups. Ten were recreationally active (RA) and eleven were endurance trained (ET) cyclists. Trained endurance cyclists are currently participating in competitions, have a $VO_{2peak} \geq 55$ ml·kg^{-1}·min^{-1}, and have at least 4 years of competitive cycling experience. RA subjects are those that cycle at least 2 times per week. All subjects were exposed to a single exercise bout on a Monark cycle ergometer. A 20 gauge Intracath was inserted into the antecubital vein of the forearm for blood lactate analyses. The catheter and stopcock were secured to the forearm for blood sampling (1 ml) during exercise.

NIRS sensors (INVOS4100, Somanetics Corp, Troy, MI) were placed on the forehead for continuous monitoring of cerebral hemoglobin oxygen saturation ($CrSO_2$) and relative blood volume (CRBVI) and a second placed on the lateral vastus muscle to monitor muscle hemoglobin/myoglobin oxygen saturation (MrSO2) and relative blood volume (MRBVI). The INVOS 4100 is a spatially resolved, continuous wave device using two wavelengths between 730 and 810 nm with detectors at 3 and 4 cm from the light emitting diode. (Oxygen consumption (VO_2) was continuously monitored (Truemax 2400 Metabolic Measurement System, Paramedics, Salt Lake City, Utah). The Polar Heart rate monitor was used to monitor heart rate (Polar Electro Oy, Finland) The subjects began pedaling at 0 Watts and a cadence of 50 rpm. Every 3 minutes, power output was increased by 25 Watts until exhaustion. Blood lactate measurements were measured at each exercise level at the 3 min mark, using a YSI 2700 Select Biochemistry Analyzer (Yellow Springs, Ohio).

2.4 Data Analysis

Muscle oxygenation along with other variables except for blood lactate, were continuously monitored throughout the study. The lactate threshold was determined from the blood lactate versus power output (watts) curve extrapolated to intersect power output (watts) at zero blood lactate. Muscle oxygenation (MrSO2) at VO_2 max was recorded at exhaustion. The data were analyzed by parametric statistics including one-way ANOVA and Tukey's multiple comparisons and the Student-Newman Keuls tests for posthoc analysis. Significance was set at $P<0.05$.

3. RESULTS

Endurance trained (ET) athletes were distinguished from the RA males by their two-fold greater power output in watts at the lactate threshold which was also reflected by a significantly greater oxygen consumption (VO_2) (Table 1). VO_2max was greater by about 25% in the ET athletes. VO_2 at the lactate threshold was about 77% higher in the ET athletes. However, muscle hemoglobin/myoglobin oxygen saturation was similar between ET and RA athletes at both the lactate threshold and at VO_2 max.

There was a tendency for the RA males to desaturate faster and greater during exercise and especially at maximum exercise (Fig. 1) but there were individuals that departed from others in the group. The rate and magnitude of MrSO2 desaturation in the ET athletes (Fig. 2) was generally slower and less than in the RA males. Whether the two subjects in particular showing the greater decrease in muscle oxygenation is due to a poorer conditioning status remains to be determined. In the RA group (Fig.1), on the other hand, there were those that showed slower and less marked muscle hemoglobin/myoglobin desaturation but whether these were also due to differences in conditioning status remains to be determined.

Table 1. Physiological variables related to maximal exercise and lactate thresholds

Group	Power LA	Blood LA	VO2LA	VO2max	VO2LA	MrSO2 SatVO2max	MrSO2 SatLA
(n)	Watts	mM	L/min	L/min	% max	%	%
RA (10)	101 ± 45	1.56 ± 0.62	1.65 ± 0.66	3.27 ± 0.45	50.35 ± 17.5	48.40 ± 17.10	70.40 ± 8.53
ET (11)	214 ± 27*	1.72 ± 0.51	2.92 ± 0.39*	4.08 ± 0.20	71.37 ± 7.73*	51.50 ± 17.55	68.20 ± 9.64

* = P<0.05
Power LA = power output at the lactate threshold
Blood LA = Blood lactate value at the lactate threshold.
VO2LA = Oxygen consumption at the lactate threshold
VO2max = Oxygen consumption at the maximum exercise
VO2LA (% max) = VO2 at the lactate threshold as a percent of VO2 max.
MrSO2 SatVO2max = Muscle hemoglobin/myoglobin saturation at peak exercise and VO2 max.
MrSO2 SatLA ≐ Muscle saturation at the lactate threshold.

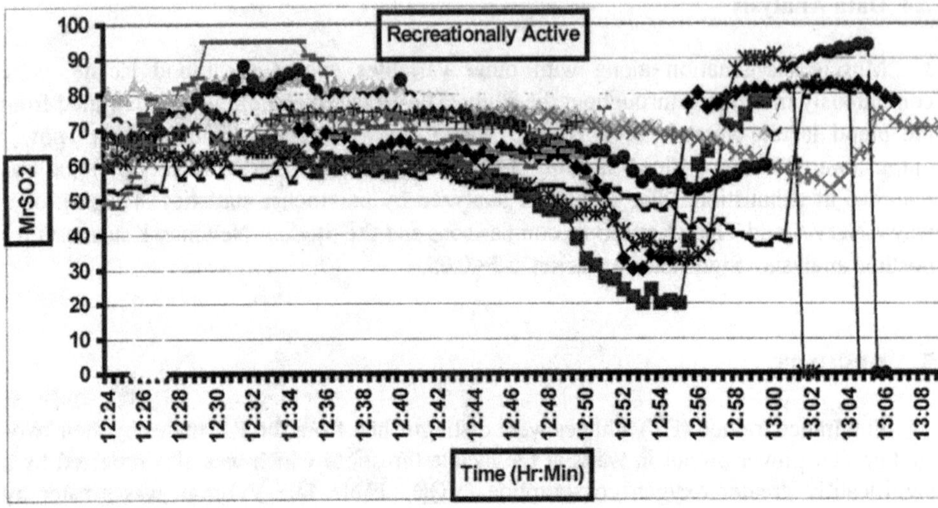

Figure 1. Muscle (vastus lateralis) hemoglobin oxygen saturation (MrSO2) in recreationally active males (n=10) by NIRS during cycling exercise with increasing power output in 25 watt increments every 3 minutes. Note the two subjects who showed marked desaturation and an earlier cessation at maximal exercise than the other subjects. MrSO2=muscle hemoglobin oxygen saturation.

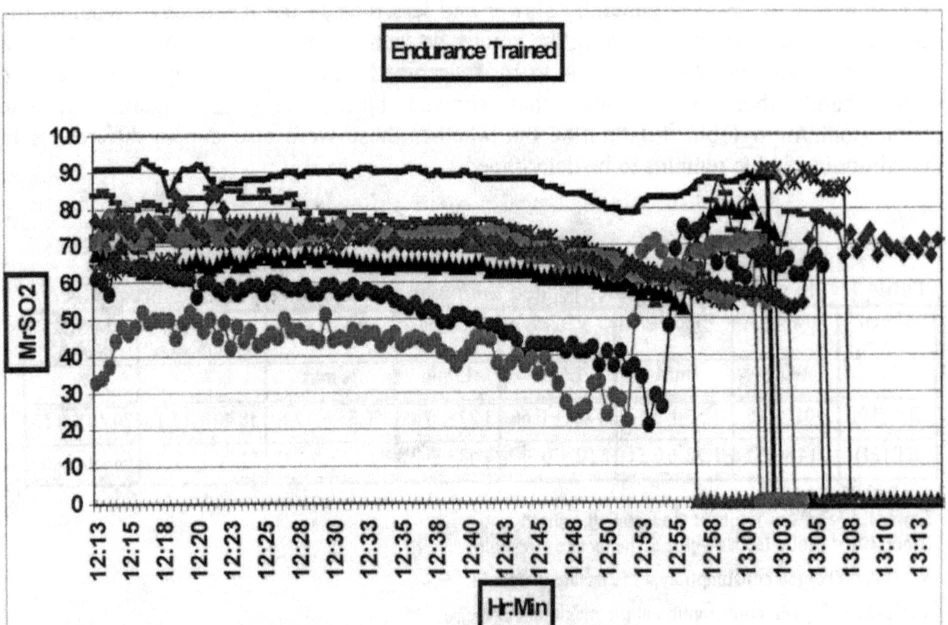

Figure 2. Muscle oxyhemoglobin saturation (MrSO2) in 11 endurance trained athletes during cycling exercise with increasing power output in 25 watt increments every 3 minutes. Note the wide range of rates of desaturation between the subjects.

4. DISCUSSION

The differences between the endurance trained and the recreationally active athletes could be seen in terms of their power output at the lactate threshold, VO_2 at the lactate threshold, and their VO_2max. However, muscle oxygenation at the lactate threshold and at maximum exercise did not differentiate these athletes. However, the lack of a difference between these groups may be simply attributable to a wide range of different levels of conditioning in both ET and RA groups.

The wide range of desaturation profiles in the RA group compared to the ET group may be due to differences in conditioning. This was particularly evident in the ET group where two of the 10 subjects showed an accelerated and marked decrease in MrSO2 compared to the rest of the group. The RA group on the other hand showed for the most part, a pattern consistent with the 2 subjects that showed a rapid and early desaturation of the muscle during exercise. A higher anaerobic threshold is associated with a greater endurance performance capacity and is a well-documented response to over-distance training. Therefore, the rate of oxygen desaturation appears to be affected by endurance training status and may be a rate limiting mechanism for peripheral extraction and utilization of oxygen during prolonged sub-maximal exercise.

Endurance exercise training produces a number of physiological adaptations that facilitate oxygen availability to exercising muscle tissue. These include an increase in size of slow twitch fibers, an increase in capillary density, a slight increase in muscle myoglobin content, an increase in the number and size of mitochondria and an increase in the activity of enzymes that play a crucial role in the oxidative breakdown of fuels (i.e. succinate dehydrogenase and citrate synthase)(6). There is also a significant difference between endurance trained and untrained individuals in terms of energy sources used. For example, endurance trained athletes are more efficient in conserving glycogen and utilizing their fat stores (6,7).

The endurance-trained athlete's body has a comparatively greater capacity to take up, transport and use oxygen during dynamic exercise. This allows the trained athlete to perform at a higher intensity and longer duration compared to the untrained individual before anaerobic glycolysis is initiated, as indicated by the lactate threshold. The lactate threshold is defined as the highest VO_2 that can be attained during incremental exercise before blood lactate increases above the baseline concentration of $1 mM \cdot L^{-1}$. The lactate threshold is identified as an abrupt and non-linear increase in blood lactate concentration above $1 mM \cdot L^{-1}$ in the presence of increasing exercise intensity (5). As a result of endurance training, the lactate threshold occurs at a higher percentage of VO_{2max} as compared to the non-trained individual (5,6). Exercise at an intensity equivalent to the lactate threshold has been shown to be an effective training stimulus to increase endurance performance.

5. CONCLUSIONS

We were unable to show a difference in muscle hemoglobin desaturation during maximal exercise at the lactate threshold and at VO_2 max between the endurance trained and the recreationally active athlete. This may not be due to a lack of difference but rather due to the mix of endurance trained and recreationally active athletes. For example, some of the ET group were not truly conditioned as one might expect whereas some of those in the recreationally active group may have been in better condition that they supposed. A future study that might determine whether a NIRS monitor can distinguish between a conditioned and unconditioned athlete may depend upon longitudinal studies that would follow the improvement in one athlete with time in an exercise program.

ACKNOWLEDGEMENTS

The authors gratefully acknowledge the assistance of Ms. Debby Stroop, RN, BSN, CCRN, who inserted the peripheral venous catheters in all subjects and withdrew the blood samples for lactate analyses.

The loan of the equipment and the sensors used for the study by Somanetics, Inc., Troy, MI is gratefully acknowledged.

REFERENCES

1. De Blasi RA. Quaglia E. Ferrari M. Skeletal muscle oxygenation monitoring by near infrared spectroscopy. *Biochem Int* 25(2), 241-248 (1991).
2. De Blasi RA. Alviggi I. Cope M. Elwell C. Ferrari M. Noninvasive measurement of forearm oxygen consumption during exercise by near infrared spectroscopy. *Adv Exp Med Bio*. 345:685-692 (1994).
3. Sahlin K. Non-invasive measurements of O2 availability in human skeletal muscle with near-infrared spectroscopy. *Int J Sports Med.* 13 (Suppl 1), S157-S160 (1992).
4. Yoshida T. Effect of exercise duration during incremental exercise on the determination of anaerobic threshold and the onset of blood lactate accumulation. *Eur. J. Appl. Physiol.* 53,196-199 (1984).
5. Weltman A., Snead D., Seip R., Schurrer R., Rutt R., Weltman J: Reliability and validity of a continuous incremental treadmill protocol for the determination of lactate threshold, fixed blood lactate concentrations, and VO_{2max}. *Int. J. Sports Med.* 11,26-32 (1990).
6. Wilmore J., Costill D., 1994, *Physiology of Sport and Exercise*. Human Kinetics, Champaign, IL, pp.149-158.
7. Bassett D., Howley E.: Limiting factors for maximum oxygen uptake and determinants of endurance performance. *Med. Sci. Sports Exerc.* 32(1), 70-84 (2000).

CARDIAC OUTPUT, OXYGEN CONSUMPTION AND MUSCLE OXYGEN DELIVERY IN SUBMAXIMAL EXERCISE

Normal and low O_2 states

Christopher B Wolff *

1. ABSTRACT

Cardiac output (Q) changes linearly with oxygen consumption (VO_2) in normal subjects undertaking submaximal exercise ($Q = A + B.VO_2$ where A is the y intercept and B the slope). If (hypothesis 1) the increase in cardiac output above the resting state represents the blood flow to exercising muscle (qm) and the increase in VO_2 represents the oxygen consumption of exercising muscle (VO_{2m}) then, where C_aO_2 is the arterial oxygen content, oxygen extraction, $Em = 1/(B \times C_aO_2)$. Secondly, exercising muscle venous oxygen content, $C_vO_{2m} = C_aO_2 - 1/B$. Limiting the hypothesis just to the calculation of VO_{2m} (hypothesis 2) allows calculation of qm if C_aO_2 and C_vO_{2m} are available. From Koskolov et al. (Am. J. Physiol.: Heart and Circ. Physiol. 273, H1787-H1793, 1997), exercising muscle blood flow (qm) is equal to the increment in cardiac output when C_aO_2 is normal but exceeds it when C_aO_2 is low. Muscle Oxygen extraction (Em) is found to be 68% in submaximal exercise. Hence, muscle oxygen delivery (D_aO_{2m}) for a given metabolic rate is sustained in low O_2 states (at 1.48 ml D_aO_{2m} per ml VO_{2m}), confirmed by analysis of Roach et al. (Am. J. Physiol.: Heart and Circ. Physiol. 276, H438-H445, 1999).

2. INTRODUCTION

Oxygen delivery (D_aO_2) in submaximal exercise must exceed oxygen consumption (VO_2). The question arises as to what exactly is the appropriate increase in cardiac output to provide this oxygen delivery. The usual account suggests that mixed venous blood at rest acts as an oxygen reserve for exercise (Astrand and Rodahl[1]; McArdle, Katch, and Katch[2].

However, oxygen uptake at the tissues is a rate of oxygen consumption not a finite quantity. A continuous, sufficient rate of oxygen delivery is therefore required.

* Clinical Pharmacology, St Bartholomew's and the Royal London School of Medicine and Dentistry, Charterhouse Square, London, EC1M 6BQ, UK.

Oxygen Transport to Tissue XXIII, edited by D. Wilson *et al.*
Kluwer Academic/Plenum Publishers, 2003

3. ANALYTICAL METHODS

The submaximal exercise cardiac output (Q) oxygen consumption (VO_2) relationship is remarkably constant in slope and intercept (figure 1, left).

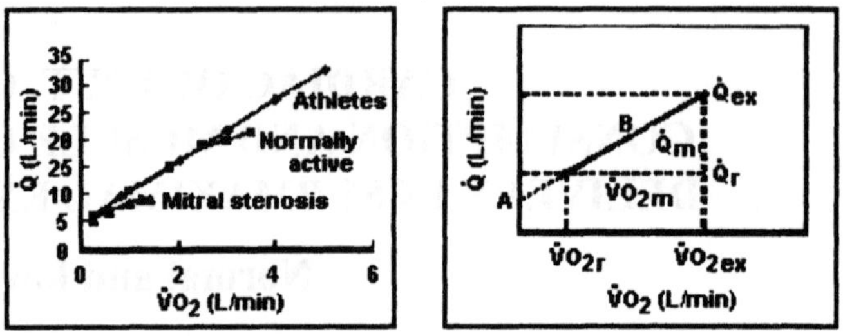

Figure 1. The chart on the left (after Rowell[3, 4]) illustrates the way cardiac output varies with oxygen consumption. It is strikingly linear in the submaximal exercise range in normal subjects and is represented diagrammatically on the right. Q_r and VO_{2r} represent resting cardiac output and oxygen consumption; Q_{ex} and VO_{2ex} values in exercise. Hypothesis 1 would mean that muscle blood flow, $Qm = Q_{ex} - Q_r$; muscle oxygen consumption, $VO_{2m} = VO_{2ex} - VO_{2r}$.

An initial hypothesis (hypothesis 1) is that the excess oxygen consumption and cardiac output above rest values represent muscle oxygen consumption (VO_{2m}) and blood flow (Qm) for the submaximal, linear, range of exercise (figure 1). The slope, B, is then Qm/VO_{2m}. Exercising muscle oxygen extraction, $Em = VO_{2m} /(Qm \times C_aO_2)$, where C_aO_2 is the arterial oxygen content. Hence, $Em = 1 /(B \times C_aO_2)$. Since the muscle a-v difference is VO_{2m} /Qm, or 1/B, exercising muscle venous oxygen content, $C_vO_{2m} = C_aO_2 - 1/B$.

A simpler hypothesis (hypothesis 2), that exercising muscle oxygen consumption equals the increment in VO_2 (amount by which submaximal exercise VO_2 exceeds the resting value), allows calculation of muscle blood flow if C_aO_2 and exercising muscle C_vO_2 are known. Two such studies will be analysed.

4. RESULTS

The two studies, in which C_aO_2 and exercising muscle C_vO_2 are given, are those of Koskolou, Roach, Calbet, Rådegran and Saltin[5] and Roach, Koskolou, Calbet and Saltin[6]. Values from Koskolou et al.[5] are listed in Table 1 with calculated muscle blood flow (qm) for normal and temporarily anaemic (but normovolaemic) subjects (using hypothesis 2). Values are available for 30 watts and half maximal exercise.

The difference in VO_2 between rest and each submaximal exercise state has been taken to be exercising muscle VO_2 (hypothesis 2). Muscle blood flow (qm) is obtained from the Fick equation. Flow to non-exercising tissue (qo) is then Q – qm. As an example: for normal C_aO_2 at 30 watts exercise VO_{2m} = 0.801 - 0.349 L O_2 min^{-1} (= 0.452): the a-v difference is 0.1905-0.0647 = 0.1258 vol/vol. Hence, qm (muscle blood

Table 1. Values of cardiac output and oxygen consumption from Koskolou et al.[5]

	VO_2 (L min^{-1})	C_aO_2 (vol/vol)	C_vO_{2m}	Q (L min^{-1})	qm	qo	D_aO_{2m} (L min^{-1})	Em %
Normal								
Rest	0.349	0.1901	0.0965	6.3				49.7
30 watts	0.801	0.1905	0.0647	10.3	3.4	6.7	0.684	66.0
73 watts	1.425	0.191	0.0565	14.7	8.0	6.7	1.528	70.4
Low Hb								
Rest	0.411	0.1511	0.0666	7.7				55.9
30 watts	0.89	0.1531	0.0502	12.1	4.7	7.4	0.713	67.2
55 watts	1.188	0.1538	0.0489	13.7	7.4	6.3	1.139	68.2

flow) is 0.452/ 0.1258 = 3.593 L min^{-1}. Oxygen delivery (D_aO_{2m}) is therefore 3.593 x 0.1905 = 0.684 L O_2 min^{-1} and extraction, Em = 0.452/ 0.684 or 66%.

The values for cardiac output and for blood flow to non-exercising tissues (see table) have been plotted against whole body oxygen consumption in figure 2 (left). For normal subjects blood flows to non-exercising tissues are very similar to whole body rest values (consistent with hypothesis 1).

However, blood flow to exercising muscle is higher in the anaemic than in the normal subjects at the expense of non-exercising tissues (no longer consistent with hypothesis 1). Figure 2 (right) shows that muscle oxygen delivery (D_aO_{2m}) is sustained in anaemia at the normal level for the work rate. The increased flow in the anaemic subjects has therefore compensated for their low C_aO_2.

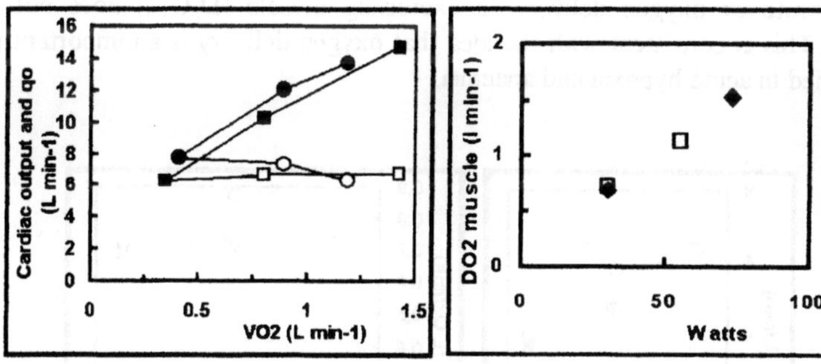

Figure 2. In the left hand panel normal subjects had undertaken exercise at 30 watts and half maximal exercise either when normal or when acutely anaemic (but normovolaemic). Filled rectangles represent normal cardiac output (normal oxygenation), filled circles represent cardiac output in anaemic subjects. The open rectangles and circles represent flows to non-exercising tissues. Flows to non-exercising tissues remain at the resting level in the normally oxygenated subjects but are reduced in the anaemic subjects. In the right hand panel values for arterial oxygen delivery to exercising muscle plotted against watts fall on the same line (D_aO_{2m} = 0.019 x watts + 0.121 L O_2 min^{-1}) whether the subjects were anaemic (filled diamonds) or normal (open squares). The increased muscle blood flow in the anaemic subjects therefore sustains muscle D_aO_2.

Table 2. Values are given here for the study of Roach et al.[6]

	VO_2	C_aO_2	C_vO_{2m}	Q	qm	qo	D_aO_{2m}	Em
	(L min^{-1})	(vol/vol)		(L min^{-1})			(L min^{-1})	%
CONTROL rest	0.35	0.1903	0.0939					
30 watts	0.84	0.1903	0.0626	10.2	3.8	6.4	0.730	67.1
ANEMIA rest	0.4	0.1507	0.0654					
30 watts	0.88	0.1513	0.0485	12.1	4. 7	7.4	0.706	67.9
HYPOXIA rest	0.3	0.1624	0.0783					
30 watts	0.88	0.1507	0.0501	11.2	5.8	5.4	0.869	66.8
AN&HYPOX rest	0.38	0.131	0.0592					
30 watts	0.91	0.1147	0.031	12.7	6.3	6.4	0.726	73.0

The finding that oxygen delivery obeys the same relationship to metabolic rate in normally oxygenated and anaemic subjects suggests that muscle D_aO_2 is a major priority in sub maximal exercise.

The unique rate of oxygen delivery for a particular metabolic rate is confirmed by the study of Roach et al.[6] for exercise at 30 watts. Oxygenation was varied from normal (C_aO_2 0.1901 vol O_2 per vol blood) to low (hypoxia, normovolaemic anaemia and hypoxia and anaemia combined). Table 2 shows values for muscle blood flow, oxygen delivery and percentage oxygen extraction calculated from the data of Roach et al.[6]

Values of calculated muscle blood flow (qm) and oxygen delivery for 30 watts exercise have been plotted against arterial oxygen content in figure 3 (left, qm and right, D_aO_{2m}) both from the study of Roach et al.[6] and that of Koskolou et al.[5].

Muscle blood flow can be seen to increase as arterial oxygen content decreases whereas the rate of oxygen delivery to exercising muscle (D_aO_{2m}) does not change significantly. This is consistent with the idea that oxygen delivery is an important priority and is sustained in acute hypoxia and anaemia.

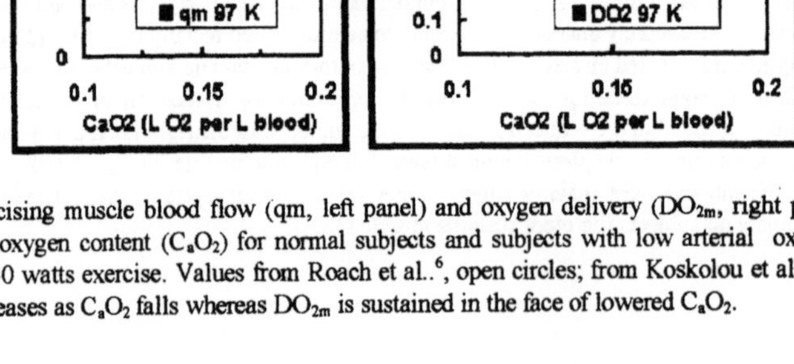

Figure 3. Exercising muscle blood flow (qm, left panel) and oxygen delivery (DO_{2m}, right panel) have been plotted against oxygen content (C_aO_2) for normal subjects and subjects with low arterial oxygen content. All points are for 30 watts exercise. Values from Roach et al..[6], open circles; from Koskolou et al.[5], closed squares. Blood flow increases as C_aO_2 falls whereas DO_{2m} is sustained in the face of lowered C_aO_2.

5. DISCUSSION

The present study explores the implications of hypotheses regarding cardiac output and exercising muscle blood flow utilising data from papers by Koskolou et al.[5] and Roach et al.[6] Hypothesis 1 is that both muscle blood flow and oxygen consumption are provided by the excess cardiac output and VO_2 above rest. Hypothesis 2 retains the assumption that the increase in VO_2 represents exercising muscle oxygen consumption but assumes nothing about blood flow. Hypothesis 2 has been applied to the studies of Koskolov et al.[5] and Roach et al.[6] where both arterial and muscle venous O_2 content were available, allowing calculation of exercising muscle blood flow. Hypothesis 1 then fits data for normally oxygenated subjects: the excess cardiac output above rest provides exercising muscle blood flow and the two equations derived from hypothesis 1 are therefore applicable to the normally oxygenated subjects. In contrast, where C_aO_2 is low supplementary blood flow is diverted to the exercising muscle from non-exercising tissues. There is also constant muscle oxygen extraction (Em = 67.6 $\pm$ 0.53 %, $\pm$SEM) in submaximal exercise. This means that for exercising muscle 1.48 ml O_2 delivery (1/Em) is required per ml of oxygen consumption (similar to cardiac muscle). Andersen and Saltin[7] have shown that use of quadriceps muscle alone for exercise, is accompanied by oxygen extraction similar to heart muscle.

Koskolou et al.[5] and Roach et al.[6] measured exercising muscle blood flow in their studies using the constant infusion thermo-dilution method of Andersen and Saltin[7]. The measured flows are higher than those calculated here (around 8% for the highest flows and almost 40% for the lowest ones). The difference could either be from the assumption, made here being faulty (that the extra VO_2 above rest is that used by the exercising muscle) or an over estimation of exercising muscle blood flow in the studies of Koskolou et al.[5] and Roach et al.[6] from, for example, heat losses when using thermodilution. Nevertheless, Koskolou et al.[5] and Roach et al.[6] found, as in the present paper, that D_aO_2 is sustained when C_aO_2 is low. Roach et al. (1999) noted that similar observations were made by Jones, Traystman, Simmons and Molteni[8] for cerebral D_aO_2 in the lamb and commented that "...the close coupling of C_aO_2 to regulation of blood flow to exercising skeletal muscle ..... may be a widespread mechanism of O_2 sensing in other vascular beds". The coupling of cerebral blood flow to C_aO_2 is supported for man by the analyses of Wolff[9, 10] of the high altitude data of Severinghaus, Chiodi, Eger, Brandstater and Hornbein[11]. Sea level cerebral oxygen delivery was sustained at high altitude.

Constant oxygen extraction appears to be a prominent feature of heart, skeletal muscle and brain. Oxygen delivery appears to be their major priority. It would be interesting to see whether the specific rates of oxygen delivery to skeletal muscle are upset when there is endothelial dysfunction. If the description of skeletal muscle blood flow and oxygen delivery in submaximal exercise presented here is a correct one, the question remains as to why cardiac output does not increase enough in anaemia (and possibly in hypoxia) to supply all the extra blood flow required by exercising muscle.

REFERENCES

1. P. Astrand and K. Rodahl, *Textbook of Work Physiology: Physiological Bases of Exercise*, 3rd edition (McGraw-Hill International Editions, New York, 1986).
2. W. D. McArdle, F. I. Katch and V. L. Katch, *Exercise Physiology: Energy, Nutrition and Human Performance*, Fourth Edition (Williams and Wilkins, Baltimore, 1996).
3. L. B. Rowell, *Human Circulation: Regulation During Physical Stress* (OUP *Oxford*, (1986).
4. L. B. Rowell, *Human Cardiovascular Control* (OUP *Oxford*. 1993).
5. M. D. Koskolou, R. C. Roach, J. A., Calbet, G. Rådegran, and B. Saltin, Cardiovascular responses to dynamic exercise with acute anemia in humans. Am. J. Physiol. (Heart and Circulatory Physiology) **273**: H1787-H1793 (1997).
6. R. C. Roach, M. D. Koskolou, J. A. L. Calbet, and B. Saltin, Arterial O_2 content and tension in regulation of cardiac output and leg blood flow during exercise in humans. Am. J. Physiol. (Heart and Circulatory Physiology) **276**: H438-H445 (1999).
7. P. Andersen, and B. Saltin, Maximal perfusion of skeletal muscle in man. J. Physiol. **366**: 233-249 (1985).
8. M. D. R. Jones, R.J. Traystman, M. A. Simmons, and R. A. Molteni, Effects of changes in arterial O_2 content on cerebral blood flow in the lamb. Am. J. Physiol. (Heart and Circulatory Physiology) **240**: H209-H215 (1981).
9. C. B. Wolff, Cerebral blood flow and oxygen delivery at high altitude. High Altitude Medicine and Biology, **1**: 33-38 (2000).
10. C. B. Wolff, P. Barry, D. J. Collier Cardiovascular and respiratory adjustments at altitude sustain cerebral oxygen delivery - Severinghaus revisited. Comparative Biochemistry and Physiology **132**: 221-229 (2002).
11. J. W. Severinghaus, H. Chiodi, E. I. Eger, Brandstater, B. and T. F. Hornbein, Cerebral blood flow in man at high altitude. Circulation Research, **19**: 274-282 (1966).

NON-INVASIVE PET AND SPECT IMAGING OF TISSUE HYPOXIA USING ISOTOPICALLY LABELED 2-NITROIMIDAZOLES[+]

Cameron J. Koch and Sydney M. Evans*

1. ABSTRACT

The measurement of pathologically low levels of tissue pO_2 is an important diagnostic goal for determining the prognosis of many clinically important diseases including cardiovascular insufficiency, stroke and cancer. A class of bioreductively activated drugs, typified by the 2-nitroimidazoles, has excellent potential for application to this goal. Such drugs bind to cells at a rate which is maximal under conditions of severe hypoxia (e.g. less than 0.05% oxygen) and is inhibited, with Michaelis-Menten kinetics, as a function of increasing oxygen concentration. A number of detection possibilities exist for the drug adducts, including invasive assays which can measure drug adducts in tissue sections at cell-to-cell resolution. Use of such agents in non-invasive assays is important and, to this end, a number of drugs have been conjugated with radioactive isotopes suitable for detection by Nuclear Medicine techniques. In contrast with the invasive assays, resolution and contrast is much more limited with the non-invasive assays. Thus, there are many factors contributing to the balance of pros and cons for the non-invasive vs. invasive use of 2-nitroimidazole drugs as hypoxia detectors. These factors will be summarized in this review, with emphasis on compounds suitable for clinical use. PET (positron emission tomography) imaging with ^{18}F-labeled EF5 (a drug in current clinical trials using invasive assays) will be described.

2. INTRODUCTION

Over the past decade, it has been clearly shown that invasive measurements of tumor hypoxia using needle electrodes provide important prognostic information. For several tumor types and treatment modalities, tumor hypoxia impacts negatively on patient survival and tumor control[1-4]. It is also well-established that hypoxia is associated with genetic changes which may allow selection of more aggressive tumor phenotypes[5-7].

+ Work support by RO1 CA87645 from the NCI, National Institutes of Health, U.S.A.
* Cameron J. Koch and Sydney M. Evans, Radiation Oncology, University of Pennsylvania, Philadelphia, PA, 19104-6072. Phone (215) 898-0073 or 898-0074. email: kochc@mail.med.upenn.edu or sydevans@mail.med.upenn.edu

Oxygen Transport to Tissue XXIII, edited by D. Wilson *et al.*
Kluwer Academic/Plenum Publishers, 2003

One of the more versatile assays for tissue hypoxia involves use of 2-nitroimidazole drugs (see Adams[8] for review). Such drugs are reduced by accepting electrons from cellular bioreductive pathways. At the stage of first electron addition, the resulting nitro radical anion can be further reduced, leading to nitroso or hydroxylamine derivatives which bind to cellular macromolecules (referred to as 'binding' in this review) or oxidized back to the original compound through electron transfer to oxygen. The kinetics of this process allow an assessment of cell or tissue oxygenation as an inverse function of 2-nitroimidazole binding. This assessment may be possible on a very quantitative basis with suitable understanding of the oxygen-dependence of binding[9-11]. Binding of 2-nitroimidazoles to hypoxic cells was first demonstrated in 1976[12] and was subsequently developed extensively by Chapman and colleagues[10,13]. Detection techniques for the bound 2-nitroimidazole adducts have evolved steadily. In early studies, uptake of radioactive (^{3}H- or ^{14}C-) drug was followed by tissue-average (liquid scintillation counting) or histology (autoradiography) based methods[13,14]. More recently, immunohistochemical methods allowing detection of adducts at cell-to-cell resolution have been optimized[15-17]. Several clinical trials are now in progress whose specific purpose is to assess the prognostic importance of 2-nitroimidazole binding as a surrogate for hypoxia[18-21]. Since needle electrodes and 2-nitroimidazole binding measure different qualitative aspects of hypoxia[21-23] it is presently unclear whether the previous success of Eppendorf measurements in predicting cancer-treatment failure will be surpassed or even equaled by the binding methods.

With suitable choice of radioactive isotope it was possible to apply the binding method to non-invasive imaging of tumor hypoxia by Nuclear Medicine techniques. Non-invasive imaging was considered essential for extension of hypoxia detection methods to non-neoplastic pathological processes such as cardiac or brain infarcts, because biosy-based or invasive assays would be contraindicated. Since there was existing experience with misonidazole in humans[24], bromine- and iodine-substituted misonidazole were the first drugs tested as non-invasive imaging agents[25,26]. The radioactive halogen linkages were unstable *in vivo* (rodent studies) and these compounds did not progress to clinical trials. More recent approaches have used either a fluorine-18 labeled misonidazole derivative (^{18}F-FMISO), or azomycin-nucleoside conjugates, wherein iodine labeling of the sugar was much more stable than the prior labeling of the 2-nitroimidazole ring (^{123}I-azomycin arabinoside[27]. Both of these compounds have been tested in humans, with a trend towards drug binding being correlated with outcome[27,28]. The lack of a more definitive correlation with outcome, similar to the Eppendorf data, is perhaps surprising since these and several other compounds have undergone extensive animal testing. This raises the dilemma that there could be some inherent problem with the binding methods or at least the non-invasive assays. It is the purpose of this review to consider possible causes of this dilemma and further, to consider possible approaches to improvements in future assays.

2.1 Basic Requirements for Non-Invasive Imaging

All imaging methods rely upon differentiation of the desired signal in the tissue of interest (in the present case, binding to hypoxic tumor tissue) from the signal present due to all other causes. Ideally, signal from non-target normal tissue will be low. So another way to phrase the above-mentioned dilemma is: "Why have the hypoxic cells in human tumors (as determined by needle electrode studies) remained relatively 'invisible' using non-invasive hypoxic-cell imaging agents?".

We suggest the following possibilities, emphasizing 2-nitroimidazole drugs:

i) Rodent pilot studies are not parallelled by clinical results.
ii) Currently employed detection agents have significant non-oxygen-dependent variations in metabolism.
iii) The drug structures are unstabie *in vivo.*
iv) The isotope linkages are unstable *in vivo.*
v) Drug elimination by clearance pathways can obliterate imaging in gut and other normal tissues (e.g. thyroid, salivary gland, etc.).
vi) The use of tracer concentrations of labeled drug may impact negatively on drug biodistribution.
vii) Imaging resolution may be inadequate to detect therapy-relevant hypoxia.

2.1.i The first possibility listed above arises from considerations of possible differences between the extent and degree of 'hypoxia' in rodent vs. human tumors. The present information is, on the surface, somewhat baffling. For rodent tumor models, the importance of hypoxia in limiting treatment response has been very well characterized[29,30] but measurement techniques, with only three exceptions[23,31,32], have been almost universally unsuccessful at predicting individual rodent tumor outcome. Conversely, for human tumors, the importance of hypoxia in limiting therapeutic outcome in patients had been widely disputed, but predicting outcome based on Eppendorf needle electrode measurements has been very successful.

Transplanted rodent tumors are often characterized as having substantial fractions of severely hypoxic cells[29]. Additionally, rapidly-growing murine tumors or xenografts often contain significant central regions of necrosis. Presumably these regions are also severely hypoxic. Therefore, we suggest a two part mechanism responsible for the failure of Eppendorf measurements to predict outcome in murine tumors: first, geometrical factors including the presence of central necrosis have caused the overemphasis of severe hypoxia in Eppendorf measurements of murine tumors[22], further complicated by our demonstration that severely hypoxic cells may be less viable, and hence less important for therapeutic outcome, than those at more modest levels of hypoxia[31,33].

In contrast, human tumors are not characterized by large fractional necrotic volumes and their median pO_2 is substantially higher than that seen for rodent tumors[34]. These dual features may improve the predictive accuracy of Eppendorf measurements in human tumors. In support of this concept, work using EF5 has clearly shown that human tumors are prone to contain only moderately hypoxic cells[20,21]. EF5 was one of the first 2-nitroimidazoles specifically designed as a hypoxia imaging agent, and methods have been developed to carefully quantify the relationship between EF5 binding and oxygen concentration[11]. Additional effort has been made to validate the relationship between binding *in vivo* and *in vitro*. The inhibition of 2-nitroimidazole binding by oxygen is very efficient at low pO_2 levels. For example, 1% oxygen may decrease binding by 10-20 fold, compared with severe hypoxia. Thus, if human tumors have higher oxygen levels than their rodent counterparts we must expect a much smaller 'signal' from the non-invasive assays.

2.1.ii-v Additional concern stems from flaws in the metabolic characteristics of some 2-nitroimidazoles (possibilities ii-v, above). It is crucial to remember that a non-invasive imaging camera can only describe the distribution of decays of the isotope under consideration. Thus, if the imaging agent is biochemically unstable, or undergoes metabolism (e.g. liver, kidney, gut) which is unrelated to the oxygen-dependence of 2-nitroimidazole reduction and binding, a component of the resulting images will be comprised of these non-oxygen-dependent pathways. Similarly, if the isotope does not stay associated with the drug (as indicated above for iodine and bromine linkages to misonidazole) then a background signal develops which is characteristic of the isotope itself or whatever molecules may complex it. Common examples of this problem are exemplified by free iodide being concentrated by thyroid, or fluoride by bone.

Under ideal circumstances, free drug would clear rapidly from the organism being imaged, leaving only the retained biochemical image of hypoxia. In practice however, the clearance half-time of drugs in humans is typically several hours. This means that the organs of clearance, especially kidney, bladder and gut will represent the dominant regions of image density. For drugs with this problem, the entire abdomen contains high signal, severely constraining the ability to image true hypoxia in a major portion of the body.

Although the above problems have been described at length in the literature[35-37] there has not been a consensus as to their solutions. Discussions often center around characteristics of specific isotopes or drugs (see Table I). In recent reviews, the requirement for very polar molecules characterized by rapid renal clearance has been emphasized[35,36]. Rapid renal clearance will enhance the contrast between bound *vs.* free drug and this concept is consistent with the development of fluoromisonidazole and recently fluoroetanidazole[38]. Nevertheless, in a recent paper, Rasey and colleagues reported that only 3% of ^{18}F-Fluoromisonidazole's radioactivity was recovered in the urine of patients and that the majority of this radioactivity was not Fluoromisonidazole[39,40]. Such findings likely apply to many of the other drugs, but corresponding data have not been published. A similar problem was found for the Tc chelate, HL91. Despite its high polarity, the signal from gut clearance eliminated the possibility of imaging in all but peripheral limbs and head[41]. In humans, tumor-labeling was often found to be small compared with binding in normal tissue (data communicated at 'International Symposium on Oxygen and Oncotherapy.' Egmond aan Zee, The Netherlands, November, 1998). One of the problems with HL91, recently dropped from clinical trials, is almost certainly caused by dissociation of isotope from drug. However, this hypothesis could not be tested since the concentration of drug administered to patients or animals was in the picomolar range, eliminating the possibility of chemical analysis. Thus, drug polarity is certainly no guarantee of drug stability *in vivo*, and may cause tremendous heterogeneity in initial biodistribution (polar compounds are usually excluded from brain and nervous system) .

2.1.vi We have been unable to find literature examples which have specifically addressed the problem posed by the typical use of very low drug concentration. This is despite the general concern regarding non-specific binding by serum and tissue proteins as well as non-specific metabolism by organs such as liver, gut and kidney[36,37]. Clearly, as the amount of drug administered to an animal or human decreases, the uniformity of its biodistribution, metabolism and pharmacokinetics will diminish. Drug concentrations are typically limited to avoid the complexity and costs associated with FDA approval. Additionally, many labeled compounds are intended as analogues of existing biological molecules which already exist at trace concentrations (e.g. hormones). Under conditions where the isotope linkage may be unstable, the ability to track both isotope and drug (e.g. by high performance liquid chromatography (HPLC) is essential to determine true drug biodistribution and pharmacokinetics. This cannot be done at tracer concentrations. The concentration range is enormous: technetium compounds are administered in the picomolar range; fluorinated drugs, nanomolar range; iodinated compounds, sub-micromolar range; immunohistochemicl studies, 100 μM range. Thus we are faced with interpretation of data based on administered drugs whose concentration represents an untested variable spanning roughly 6 orders of magnitude.

2.1.vii A few representative isotopes used for present imaging applications are shown below (Table 1). Obvious tradeoffs are apparent: of particular note is that fluorine has the shortest isotopic half-life, but the best imaging resolution.

Table 1: Properties of Representative Imaging Isotopes

	PET		SPECT	
	^{18}F	^{124}I	^{123}I	^{99m}Tc
half-life	1.83 hr	100.1 hr	13.1 hr	6.0 hr
Ligand Stability	high	medium-low	medium-low	low
Isotope Target	Bone	Thyroid	Thyroid	Various
Linear Resolution	2.0 mm	4.0 mm*	12-15 mm	8-12 mm
Voxel	8 mg	>64 mg	2500 mg	1000 mg
Attenuation	Yes	Yes	Possible	Possible
Absolute DPM	Yes	Yes	No	No
Patient Dose	Low	High*	Medium	Low
Clearance	Renal Possible	Gut	Gut	Gut
Labeling Ease	Often Difficult	Variable	Variable	Easy
Availability	Fair	Poor	Good	Excellent

* Only 25% of the decays of ^{124}I result in positrons. The extra gamma emissions and high positron energy lower the possible image resolution and substantially increase the patient radiation dose.

In addition to the inherent resolution of the specific isotope/camera used, we know very little about the absolute spatial extent of hypoxic regions of human tumors. Some data is now being generated by the studies using invasive assays of 2-nitroimidazoles[18-21]. Though tissue pieces larger than a few mm of extent are seldom assessed in detail, it seems unlikely that large contiguous areas of high binding will be found. Thus, the resolution of the technique as a whole may be limited by the biology as well as the physics.

CONCLUSIONS

Use of 18F-EF5 as A Non-Invasive Hypoxia Imaging Agent

Although EF5 has been quite successful in invasive applications of hypoxia imaging (drug aducts are detected by immunohistochemistry after tissue biopsy) it runs against the grain of recent drug development for non-invasive image applications. For example, it is the most lipophilic 2-nitroimidazole yet tested, and thus its plasma half-life is much longer (12 hr[42]) than the isotope half life of fluorine 18 (110 min). On the one hand, this amplifies a general problem for fluorine; since the drug half-life is almost always longer than the isotope half-life, one is faced with the problem of resolving non-metabolized parent drug from hypoxia-dependent binding. On the other hand, the lipophilic character of EF5 provides for a very uniform biodistribution of the drug, and we reasoned that there could be real advantages in emphasizing even biodistribution while minimizing drug clearance. This has the effect of increasing the background from non-metabolized drug in all tissues, without obliterating the imaging possibilities in gut, kidney, etc. EF5 has an extremely stable structure *in vivo*. No metabolites or breakdown products have been observed in any of the mammalian species investigated to date. Furthermore, the chemical stability of the isotope linkage to the sidechain is outstanding, because of the general C-F bond stability and as enhanced by having a multiply-fluorinated bond. In fact, the drug is stable in neat boiling trifluoroacetic acid, present during synthesis[43]. Despite the short half life of 18F, the linear resolution of the isotope is several-fold greater than that of other isotopes, particularly those used for single photon emission computed tomography (SPECT) which has much reduced spatial resolution. In view of our observation that hypoxia in human tumors may be less extensive and less severe than in rodent tumors, it appears essential to emphasize resolution.

REFERENCES

1. M. Hockel, C. Knoop, K. Schlenger, B. Vordran, E. Baussmann, M. Mitze, P. G. Knapstein, And P. Vaupel, Intratumor pO_2 predicts survival in advanced cancer of the uterine cervix. Radiotherapy and Oncology 26, 45-50 (1993).
2. D. M. Brizel, S. P. Scully, J. M. Harrelson, L. J. Layfield, J. M. Bean, L. R. Prosnitz, And M. W. Dewhirst, Tissue oxygenation predicts for the likelihood of distant metastases in human soft tissue sarcoma. Cancer Res. 56, 941-943 (1996).
3. M. Nordsmark, M. Overgaard, And J. Overgaard, Pretreatment oxygenation predicts radiation response in advanced squamous cell carcinoma of the head and neck. Radioth. and Oncol. 41, 31-39 (1996).
4. A. W. Fyles, M. Milosevic, R. Wong, M. C. Kavanagh, M. Pintilie, W. Chapman, W. Levin, L. Manchul, T. J. Keane, And R. P. Hill, Oxygenation predicts radiation response and survival in patients with cervix cancer. Radioth. and Oncol. 48, 149-156 (1998).
5. S. D. Young, R. S. Marshall, And R. P. Hill, Hypoxia induces DNA overreplication and enhances metastatic potential of murine tumor cells. Proc. Nat. Acad. Sci. USA 85, 9533-9537 (1988).
6. T. G. Graeber, C. Osmanian, T. Jacks, D. E. Housman, C. J. Koch, S. W. Lowe, And A. J. Giaccia, Hypoxia-mediated selection of cells with diminished apoptotic potential in solid tumours. Nature 379, 88-91 (1996).
7. T. Y. Reynolds, S. Rockwell, And P. M. Glazer, Genetic instability induced by the tumor microenvironment. Cancer Res. 56, 5754-5757 (1996).
8. G. E. Adams, In: Selective Activation of Drugs by Redox Processes (Adams, G.E., Breccia, A., Fielden, E.M. and Wardman, P., eds). NATO Series A 198, Plenum Press, NY (1990).
9. C. J. Koch, S. M. Evans, And E. M. Lord, Oxygen dependence of cellular uptake of EF5 [2-(2-nitro-1H-imidazol-1-yl)-N-(2,2,3,3,3-pentafluoroproply)acetamide] : analysis of drug adducts by fluorescent antibodies vs bound radioactivity. Br. J. Cancer 72, 869-874 (1995).
10. C. J. Koch, C. C. Stobbe, And K. A. Baer, Metabolism induced binding of ^{14}C-misonidazole to hypoxic cells: kinetic dependence on oxygen concentration and misonidazole concentration. Int. J. Radiat. Oncol. Biol. Phys. 10, 1327-1332 (1984).

11. C. J. Koch, Measurement of Absolute Oxygen Levels in Cells and Tissues using Oxygen Sensors and the 2-nitroimidazole EF5. Meth. Enzymol. - Antioxidants and Redox Cycling In Press (2002).
12. A. J. Varghese, S. Gulyas, And J. K. Mohindra, Hypoxia-dependent reduction of 1-(2-nitro-1-imidazolyl)-3-methoxy-2-propanol by Chinese hamster ovary cells and KHT tumor cells in vitro and in vivo. Cancer Res. 36, 3761-3765 (1976).
13. J. D. Chapman, K. A. Baer, And J. Lee, Characteristics of the metabolism-induced binding of misonidazole to hypoxic mammalian cells. Cancer Res. 43, 1523-1528 (1983).
14. B. M. Garrecht, And J. D. Chapman, The labelling of EMT-6 tumors in Balb/c mice with ^{14}C-misonidazole. Brit. J. Radiol. 56, 745-753 (1983).
15. J. A. Raleigh, G. G. Miller, A. J. Franko, C. J. Koch, A. F. Fuciarelli, And D. A. Kelley, Fluorescence immunohistochemical detection of hypoxic cells in spheroids and tumours. Br. J. Cancer 56, 395-400 (1987).
16. R. J. Hodgkiss, G. Jones, A. Long, J. Parrick, K. A. Smith, And M. R. L. Stratford, Flow cytometric evaluation of hypoxic cells in solid experimental tumours using fluorescence immunodetection. Br. J. Cancer 63, 119-125 (1991).
17. E. M. Lord, L. W. Harwell, And C. J. Koch, Detection of hypoxic cells by monoclonal antibody recognizing 2-nitroimidazole adducts. Cancer Res. 53, 5271-5276 (1993).
18. A. S. Kennedy, J. A. Raleigh, G. M. Perez, D. P. Calkins, D. E. Thrall, D. B. Novotny, And M. Varia, Proliferation and hypoxia in human squamous cell carcinoma of the cervix: first report of combined immunohistochemical assays. Int. J. Radiat. Oncol. Biol. Phys. 37, 897-905 (1997).
19. K. I. Wijffels, J. H. Kaanders, P. F. Rijken, J. Bussink, F. J. Van Den Hoogen, H. A. Marres, P. C. De Wilde, J. A. Raleigh, And A. J. Van Der Kogel, Vascular architecture and hypoxic profiles in human head and neck squamous cell carcinomas. Brit. J. Cancer 83, 674-683 (2000).
20. S. M. Evans, S. Hahn, D. R. Pook, W. T. Jenkins, A. A. Chalian, P. Zhang, C. Stevens, R. Weber, I. Benjamin, N. Mirza, M. Morgan, S. Rubin, W. G. Mckenna, E. M. Lord, And C. J. Koch, Detection of hypoxia in human squamous cell carcinoma by EF5 binding. Cancer Res. 60, 2018-2024 (2000).
21. S. M. Evans, S. Hahn, D. R. Pook, P. J. Zhang, W. T. Jenkins, D. Fraker, R. A. Hsi, W. G. Mckenna, And C. J. Koch, Hypoxia in human intraperitoneal and extremity sarcomas. Int. J. Radiat. Oncol. Biol. Phys. 49, 587-596 (2000).
22. W. T. Jenkins, S. M. Evans, And C. J. Koch, Hypoxia and necrosis in rat (L glioma and Morris 7777 hepatoma tumors: comparative measurements using EF5 binding and Eppendorf needle electrode. Int. J. Rad. Onc. Biol. Phys. 46, 1005-1017 (2000).
23. M. C. Kavanagh, A. Sun, Q. Hu, And R. P. Hill, Comparing techniques of measuring tumor hypoxia in different murine tumors: Eppendorf pO_2 histograph, [3H]misonidazole binding and paired survival assay. Radiat. Res. 145, 491-500 (1996).
24. M. I. Saunders, S. Dische, P. J. Anderson, And J. R. Flockhard, The neurotoxicity of misonidazole and its relationship to dose, half-life and concentration in the serum. Br. J. Cancer 37: , 268-270 (1978).
25. J. S. Rasey, K. A. Krohn, And S. Freauff, Bromomisonidazole: Synthesis and characterization of a new radiosensitizer. Radiat. Res. 91, 542-554 (1982).
26. D. C. Jette, L. I. Wiebe, And J. D. Chapman, Synthesis and in vivo studies of the radiosensitizer 4-(^{82}Br)bromomisonidazole. Int. J. Nuc. Med. and Biol. 10, 205-210 (1983).
27. M. B. Parliament, J. D. Chapman, R. C. Urtasun, A. J. Mcewan, L. Golberg, J. R. Mercer, R. H. Mannan, And L. I. Wiebe, Non-invasive assessment of human tumour hypoxia with I-iodoazomycin arabinoside: preliminary report of a clinical study. Br. J. Cancer 65, 90-95 (1992).

28. J. S. Rasey, W.-J. Koh, M. L. Evans, L. M. Peterson, T. K. Lewellen, M. M. Graham, And K. A. Krohn, Quantifying regional hypoxia in human tumors with positron emission tomography of [^{18}F]fluoromisonidazole: a pretherapy study of 37 patients. Int. J. Radiat. Oncol. Biol. Phys. 36, 417-428 (1996).
29. J. E. Moulder, And S. C. Rockwell, Hypoxic fractions of solid tumors: experimental techniques, methods of analysis and a survey of existing data. Int. J. Radiat. Oncol. Biol. Phys. 10, 695-712 (1984).
30. J. D. Chapman, A. J. Franko, And C. J. Koch, The fraction of hypoxic clonogenic cells in tumor populations. In: Biological Bases and Clinical Implications of Tumor Radioresistance 61-73, Masson Publishing, NY (1983).
31. S. M. Evans, W. T. Jenkins, B. Joiner, E. M. Lord, And C. J. Koch, 2-nitroimidazole (EF5) binding predicts radiation sensitivity in individual 9L subcutaneous tumors. Cancer Res. 56, 405-411 (1996).
32. P. L. Olive, R. E. Durand, J. A. Raleigh, C. Luo, And C. Aquino-Parsons, Comparison between the comet assay and pimonidazole binding for measuring tumour hypoxia. Brit. J. Cancer 83, 1525-1531 (2000).
33. M. R. Woods, E. M. Lord, And C. J. Koch, Detection of individual hypoxic cells in multicellular spheroids by flow cytometery using the 2-nitroimidazole EF5, and monoclonal antibodies. Int. J. Radiat. Oncol. Biol. Phys. 34, 93-101 (1996).
34. P. Vaupel, K. Schlenger, C. Knoop, And M. Hockel, Oxygenatiuon of human tumors: evaluation of tissue oxygen distribution in breast cancers by computerized O_2 tension measurements. Cancer Res. 51, 3316-3322 (1991).
35. J. D. Chapman, E. L. Engelhardt, C. C. Stobbe, R. F. Schneider, And G. E. Hanks, Measuring hypoxia and predicting tumor radioresistance with nuclear medicine assays. Radioth. & Oncol. 46, 229-237 (1998).
36. L. L. Wiebe, & D. Stypinski, Pharmacokinetics of SPECT radiopharmaceuticals for imaging hypoxic tissues. Quart. J. Nuc. Med. 40, 270-284 (1996).
37. A. Nunn, K. Linder, And H. W. Strauss, Nitroimidazoles and imaging hypoxia. Eur. J. Nuc. Med. 22, 265-280 (1995).
38. T. J. Tewson, Synthesis of [^{18}F]fluoroetanidazole: a potential new tracer for imaging hypoxia. Nuc. Med. & Biol. 24, 755-760 (1997).
39. J. S. Rasey, P. D. Hofstrand, L. K. Chin, And T. K. Tewson, Characterization of [^{18}F]Fluoroetanidazole, a new radiopharmaceutical for detecting tumor hypoxia. J. Nuc. Med. 40, 1072-1079 (1999).
40. M. M. Graham, L. M. Peterson, J. M. Link, M. L. Evans, J. S. Rasey, W.-J. Koh, J. H. Caldwell, And K. A. Krohn, Fluorine-18-fluoromisonidazole radiation dosimetry in imaging studies. J. Nuc. Med. 38, 1631-1636 (1997).
41. B. G. Siim, W. T. Laux, M. D. Rutland, B. N. Palmer, And W. R. Wilson, Scintigraphic imaging of the hypoxia marker (99m)technetium-labeled 2,2'-(1,4-diaminobutane)bis(2-methyl-3-butanone) dioxime (99mTc-labeled HL-91; prognox): noninvasive detection of tumor response to the antivascular agent 5,6-dimethylxanthenone-4-acetic acid. Cancer Res. 60, 4582-4588 (2000).
42. C. J. Koch, S. M. Hahn, K. J. Rockwell, J. M. Covey, W. K. Mckenna, And S. M. Evans, Pharmacokinetics of the 2-Nitroimidazole EF5 [2-(2-nitro-1-H-imidazol-1-yl)-N-(2,2,3,3,3-pentafluoropropyl)acetamide] in Human Patients: Implications for Hypoxia Measurements In Vivo. Cancer Chemoth. Pharmacol. 48, 177-187 (2001).
43. W. R. Dolbier, A.-R. Li, C. J. Koch, C.-Y. Shiue, And A. V. Kachur, ^{18}F-EF5, a marker for PET detection of hypoxia: Synthesis of precursor and a new fluorination procedure. Appl. Rad. and Isot. 54, 73-80 (2001).

DIFFUSE OPTICAL MEASUREMENT OF HEMOGLOBIN AND CEREBRAL BLOOD FLOW IN RAT BRAIN DURING HYPERCAPNIA, HYPOXIA AND CARDIAC ARREST

Joseph P. Culver*, Turgut Durduran*, Cecil Cheung*, Daisuke Furuya#, Joel H. Greenberg#, and A. G. Yodh*

1. Introduction

Tissue viability and function are often manifest in hemodynamic signatures, because metabolism and vascular regulatory mechanisms modulate the concentration, oxygenation and flow characteristics of blood cells. Assessment of tissue viability and function from hemodynamic measures can be complicated however, by the interplay between vascular supply, tissue oxygen consumption and regulatory effects. It is therefore desirable to independently monitor hemoglobin concentration, oxygenation and flow. In this contribution we present diffuse optical measurements of hemoglobin concentration, oxygenation and flow during global modulations in rat brain, through the intact skull.

Several different multi-parameter measurement schemes have already been employed in rat brain models of functional activation. For example functional magnetic resonance imaging (fMRI) measures of blood oxygen (BOLD), cerebral blood volume (CBV), and cerebral blood flow (CBF) have been combined to calculate oxygen metabolism $CMRO_2$ [6] during rat forepaw stimulation. In a different vein, near surface optical techniques have been used in exposed cortex preparations, for example combining optical spectroscopy and laser Doppler flowmetry [7], or combining phosphorescence oxygen sensors with laser Doppler flowmetry [1]. In our work we employ deep tissue near infrared (NIR) technologies [8] and a novel correlation spectroscopy measure of blood flow. The near infrared methodologies are minimally invasive and performed with the skull intact. We present data derived from models of

* Department of Physics and Astronomy, # Cerebrovascular Research Center, Department of Neurology, University of Pennsylvania, Philadelphia, PA 19104

hypercapnia, hypoxia and cardiac arrest; changes in rCBF, hemoglobin concentration and oxygenation are discussed.

2. Methods

We briefly review our methodology. A more detailed description is found in [3].

2.1. Instrument

Near infrared spectroscopy with diffuse photon density waves (DPDW) was used to measure tissue oxy- and deoxy-hemoglobin concentrations, and diffuse light correlation spectroscopy was used to measure tissue blood perfusion. The diffuse photon density wave measurements employed two intensity modulated (70Mhz) laser diodes (786nm and 830nm). Light reflected from the tissue was detected by four avalanche photodiodes and demodulated to deduce amplitude and phase variations. The diffuse light correlation flowmetry measurements utilized a single mode, large coherence length (>1m) laser source operating at 790nm. Single mode fibers were used to relay image single speckles to avalanche photodiodes, and a correlator chip analyzed the time variant speckle intensity to obtain the temporal intensity autocorrelation functions. All three lasers were multiplexed onto a two-dimensional 3x3 grid (8mmx10mm) of nine optical fibers located in the image plane of a camera. A relay lens projected the probe grid onto the sample (e. g. rat cranium) permitting non-contact measurements. DPDW measurements were interlaced between successive diffuse correlation measurements.

2.2.Theory

We take the measurement plane to coincide with the xy-plane, which is normal to the z-axis. The transport of diffuse photon density waves ($\Phi(\rho)$) is modeled using an extrapolated zero boundary condition, semi-infinite media approximation with solutions [5] of the form

$$\Phi(\rho) = \frac{3\mu_s' S_o}{4\pi}\left(\frac{e^{ikr_1}}{r_1} - \frac{e^{ikr_2}}{r_2}\right)$$

Here $k^2 = 3\mu_s'(i\omega/\upsilon - \mu_a)$, k is complex wavenumber of the DPDW, μ_a is the tissue absorption coefficient, μ_s' is the tissue scattering coefficient, ω is the modulation frequency, and v is the speed of light in the medium. $r_1 = \sqrt{\rho^2 + z_o^2}$, $r_2 = \sqrt{\rho^2 + (z_o + 2z_B)^2}$, ρ is the source detector separation along the surface, z_o is $\sim 1/\mu_s'$, and z_B, the extrapolated zero boundary distance, corresponds to the distance above the tissue surface at which the light fluence extrapolates to zero. The data was fit to the expression above to extract wavelength-dependent tissue absorption, and then spectral decomposition was used to extract the hemoglobin concentrations ([Hbr], [HbO_2]) from the tissue absorption assuming a 50/50 water/lipid background absorption ratio.

The diffuse light electric field autocorrelation function ($G(\rho,\tau)$)was analyzed in a similar way. The primary difference is that the wavenumber, k_D, associated with $G(\rho,\tau)$, depends on the autocorrelation time, i.e., $k_D{}^2 = 3\mu_s{}'\mu_a + 6\mu_s^{'2}\mu_a^2 k_o^2 \alpha D_B \tau$; here τ is the autocorrelation time delay, k_0 is the optical photon's wavenumber in the medium. The probability that a photon is scattered by a moving "cell", α, is presumed proportional to cerebral blood volume [3]. The blood flow speed is parameterized by a Brownian diffusion constant D_B and the product αD_B parameterizes blood flow. The correlation measurement gives αD_B.

2.3. Experiments

Adult male Sprague-Dawley rats weighing 300-325 g were fasted overnight prior to measurements. The animals were anesthetized (Halothane 1-1.5%, N_2O 70%, O_2 30%) and catheters were placed into a femoral artery to monitor the arterial blood pressure and into a femoral vein for drug delivery. The body temperature was maintained at 37±2C. The animal was tracheotomized, mechanically ventilated and fixed on a stereotaxic frame. The scalp was reflected since the fur introduced a slight distortion onto the near-infrared signals. The probe was placed symetrically about midline and covered a region from 2mm anterior to 6mm posterior of the rhinal fissure. Following the surgical procedures, halothane was withdrawn and anesthesia was maintained with α-chloralose (60 mg/kg), followed by supplemental doses of 30 mg/kg every hour. The hypercapnic challenge consisted of elevating blood p_aCO_2 by adding carbon dioxide (8%) into the breathing mixture. The hypoxic modulation involved decreasing the arterial p_aO_2 by reducing the inspired O_2 to 10%. Cardiac arrest was achieved with a 1 mL, 3 M solution of KCL.

3. Results

Figure 1 shows the percent increase from the baseline level of αD_B in one experiment where the animal breathed 8% CO_2 for 10 minutes. The numbers shown in the figure were averaged over the area of the imaged brain, which comprised an 8x10 mm ^{2} area of cortex. During the CO_2 inhalation, the p_aCO_2 increased from 38 mmHg to 75 mmHg, and the average relative increase of αD_B was about 0.77 (77%) from the baseline. Figure 1 also shows the concurrent changes in the hemoglobin concentration averaged over the same cortical area. A 0.17 (17%) increase in the hemoglobin concentration was observed. The hemoglobin concentration apparently increases at a slower rate than αD_B and also decays more slowly after the CO_2 was switched off. The average relative increase of hemoglobin oxygenation in the cortex during the same CO_2 stimulation is shown in the bottom panel of Figure 1. The increase was about 16%; the oxygenation dynamics followed the flow changes more closely than the total hemoglobin concentration. We also obtained time courses for hypoxia and cardiac arrest (not shown). In Figure 1, the relative changes averaged over the last 10 minutes of the hypercapnia are plotted along with the relative changes during hypoxia and cardiac arrest. The values are obtained by averaging the signal from 5 to 15 minutes following the perturbation.

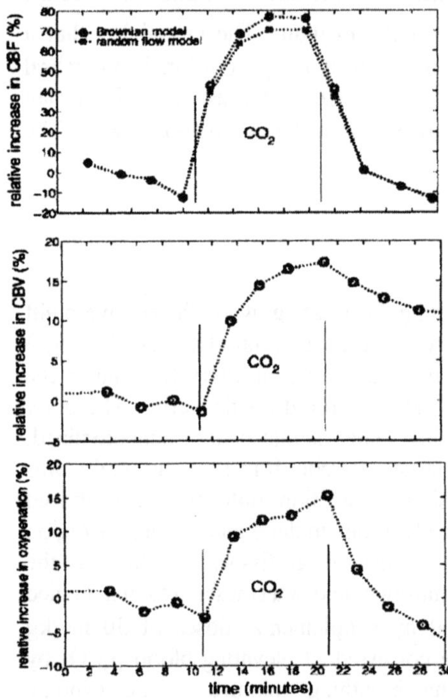

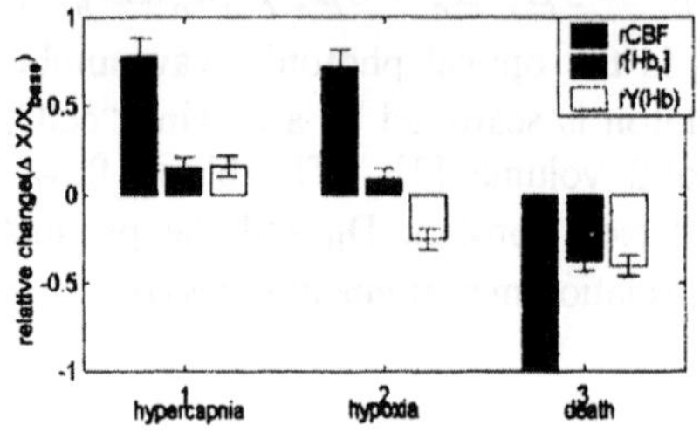

Fig. 1
(left) Time series of rCBF, rCBV and rYt during hypercapnia,
(right) relative changes of three quantities during three different physiological states.

4. Discussion

Vasodilatation increased blood flow (CBF) and blood volume (CBV) in both hypercapnia and hypoxia. Since oxygen metabolism remains roughly constant during hypercapnia and hypoxia, the flow increases are countered by a reduced oxygen extraction fraction. The measured tissue saturation (Y_t) is a combination of arterial, capillary, and venous tissue compartments. During hypercapnia the arterial saturation is constant so that the reduced oxygen extraction fraction results in increased tissue saturation, while during hypoxia, the drop in arterial saturation results combined with a decreased oxygen extraction fraction results in decreased tissue saturation. Cardiac arrest completely stops blood flow although the oxygen saturation does not go to zero. From these three modulations we have measured three different trends in the hemoglobin saturation and blood flow.

5. Summary

We have demonstrated the ability to concurrently measure relative changes in cerebral blood flow, hemoglobin concentration, and hemoglobin oxygenation with a single non-contact, non-invasive instrument. Our measurements from rat hypercapnia, hypoxia and cardiac arrest models are in reasonable agreement with the literature, and offer the possibility for further growth and quantification. The optical techniques used in this study are attractive also because they enable experimenters to measure vascular response of deep tissues. The new instrument and concept may also be applicable to human studies especially in infants and neonates permitting noninvasive monitoring of cerebral hemodynamics and oxygen (see [4] and [2] for examples of NIR spectroscopy).

We gratefully acknowledge support from grants NIH RO1-HL57835-01.

References

[1] Ances, B.M., Wilson, D.F., Greenberg, J.H. and Detre, J.A., Dynamic changes in cerebral blood flow, O-2 tension, and calculated cerebral metabolic rate of O-2 during functional activation using oxygen phosphorescence quenching, *Journal of Cerebral Blood Flow and Metabolism*, 21 (2001) 511-516.

[2] Benaron, D.A., Hintz, S.R., Villringer, A., Boas, D., Kleinschmidt, A., Frahm, J., Hirth, C., Obrig, H., van Houten, J.C., Kermit, E.L., Cheong, W.F. and Stevenson, D.K., Noninvasive functional imaging of human brain using light, *Journal of Cerebral Blood Flow and Metabolism*, 20 (2000) 469-477.

[3] Cheung, C., Culver, J.P., Yodh, A.G., Takahashi, K. and Greenberg, J.H., In vivo cerebrovascular measurement combining diffuse near-infrared absorption and correlation spectroscopies, *Physics in Medicine and Biology*, 46 (2001) 2053-2065.

[4] Danen, R.M., Wang, Y., Li, X.D., Thayer, W.S. and Yodh, A.G., Regional imager for low-resolution functional imaging of the brain with diffusing near-infrared light, *Photochemistry and Photobiology*, 67 (1998) 33-40.

[5] Haskell, R.C., Svaasand, L.O., Tsay, T.T., Feng, T.C. and McAdams, M.S., Boundary-Conditions For the Diffusion Equation in Radiative- Transfer, *Journal of the Optical Society of America a-Optics Image Science and Vision*, 11 (1994) 2727-2741.

[6] Mandeville, J.B., Marota, J.J.A., Ayata, C., Moskowitz, M.A., Weisskoff, R.M. and Rosen, B.R., MRI measurement of the temporal evolution of relative CMRO2 during rat forepaw stimulation, *Magnetic Resonance in Medicine*, 42 (1999) 944-951.

[7] Mayhew, J., Johnston, D., Berwick, J., Jones, M., Coffey, P. and Zheng, Y., Spectroscopic analysis of neural activity in brain: Increased oxygen consumption following activation of barrel cortex, *Neuroimage*, 12 (2000) 664-675.

[8] Villringer, A. and Chance, B., Non-invasive optical spectroscopy and imaging of human brain function, *Trends in Neurosciences*, 20 (1997) 435-442.

MULTIORGAN MONITORING OF HEMODYNAMIC AND MITOCHONDRIAL RESPONSES TO ANOXIA AND CARDIAC ARREST IN THE RAT

Ari Kraut*, Efrat Barbiro-Michaely, Yehuda Zurovsky, and Avraham Mayevsky†‡

1. INTRODUCTION

All tissues in the body are dependent upon the continuous supply of energy to perform their various functions. Glucose and oxygen are brought to the tissue through the bloodstream to fulfill this demand. The basic need of glucose as an energy substrate and oxygen for respiration is common to all tissues. However, the different organs in the body have various levels of demand for these substances. For this reason, the body distributes its resources differently to different organs[1].

The transfer of electrons through the respiratory chain is involved with ionization of mitochondrial NADH, driven from glycolysis procedures. The energy elicited from this electron transfer is used to maintain an intact cellular homeostasis. A decrease in supply of blood, as well as oxygen, may affect the balance between O_2 demand and supply. As a result, the accumulation of reduced NADH in the tissue will occur[2]. Under normal conditions, the flow of blood through the body is regulated in order to maintain a steady supply of blood to various parts of the body under varying cardiovascular conditions. However, each organ has its own internal autoregulation mechanism to maintain required perfusion rates regardless of the circulatory state of the entire organism[3]. As mentioned above, each of the organs in the body has unique hemodynamic and metabolic characteristics. Consequently, the response of each organ to the lack of O_2 induced by respiration of pure N_2 or cardiac arrest, is expected to be different. As of today, only in one study the brain, liver, kidney and testis were continuously monitored simultaneously[4]. In this

* Deceased September 2000.
† All authors: Faculty of Life Sciences, Bar-Ilan University, Ramat Gan 52900, Israel.
‡ Correspondence: Prof. A. Mayevsky, Tel: 972-3-5318218; Fax: 972-3-5351561;
Email: mayevsa@mail.biu.ac.il

study, surface fluorescence was used as an indicator of mitochondrial NADH under hypoxia and anoxia in those organs. In both models, an increase in NADH level was observed in each of the organs monitored. In 1998, a single combined probe for laser Doppler flowmetry (LDF) and NADH fluorometry was developed and used for the measurement of tissue viability[5]. In other studies these two techniques were used separately to monitor various organs. During cardiac arrest, the heart muscle ceases to contract, mean arterial blood pressure (MABP) drops to zero, and blood ceases to flow through the body. As a result, ischemia is induced in all organs. Ischemia in the brain causes a decrease in CBF and an increase in NADH level[5]. Ischemia of the liver was shown to cause an increase in NADH [6] and a drop in LDF[7]. Occlusion of the renal artery led to a rise in renal NADH[8, 9]. In the present study, we monitored continuously monitor the rat brain, liver, kidney and testis, simultaneously. The significance of such a model is to investigate the effect of various pathophysiological conditions or different drugs on the hemodynamic and metabolic responses in different organs. In addition, such a model can reveal the order in which different organs are responding to the same pathophysiological event.

2. METHODS

2.1. Animal Preparation

Wistar male rats (250-300 g) were anesthetized by 0.3 ml/100 gr Equithesin solution (each ml contains: pentobarbital 9.72 mg; chloral hydrate 42.51 mg; magnesium sulfate 21.25 mg; propylene glycol 44.34% w/v; alcohol 11.5% and water). During the entire experiment, the rats were maintained anesthetized by addition of 0.1 ml Equithesin every half-hour. In addition, a heating pad was placed under the rat to maintain body temperature at 37°C.

A midline incision was made in the skin, exposing the skull. Two holes were drilled in the skull. A 3.5 mm hole was drilled in the left parietal bone for the fixation of a cannula in which the probe was inserted, and a second small hole, in which a screw was inserted for better fixation to the skull. The cannula was then fixated to the skull using acrylic cement. Then the rat was turned over on its back for further operation. A hole in the experimental table, beneath the head, allowed the insertion of the probe into the brain cannula. The femoral artery and vein were exposed and cannulated for blood pressure monitoring, via a UTR-disposable pressure transducer (Biometrix), and KCL injection to the heart respectively. For the exposure of the kidney and liver, an abdomen midline section below the rib cage was created. The central lobe of the liver was exposed. Additionally, the left kidney was isolated from the juxtaposed spleen and intestine by placing a clamp around it. Then the right testis was exposed and the probes were placed on each organ. All probes, except for the brain, were held in place with micromanipulators during the entire experiment. Parafilm was placed around the tip of these probes and glued to the tissue using cyanoacrylate adhesive. Parafilm was also used for the prevention of hydration. A black cloth was placed over the parafilm to avoid room light from entering the monitoring site and causing monitoring artifacts.

2.2. Multi Site-Multi Parametric Monitoring System (MSMP)

Monitoring of the brain, liver, kidney and testis in the rat was performed using the MSMP- Multi Site-Multi Parametric monitoring system[10]. This monitoring system includes 4 bundles of optical fibers for the NADH redox state measurement, by surface fluorometry, as well as for tissue blood flow (TBF), using Laser Doppler Flowmetry (LDF). Each probe was placed on the surface of the organ monitored. The principle of NADH monitoring from the surface of the brain is that excitation light (366 nm) passes from the fluorometer to the organ via a bundle of optic fibers made of quartz. The emitted light (450 nm), together with the reflected light, at the excitation wavelength (366 nm), was transferred to the fluorometer via another bundle of fibers. Monitoring of the TBF was accomplished using the Perimed® Periflux PF 2B. This device works on the physical principle of the Doppler shift[11, 12]. All the signals monitored during the experiment were transferred to a multi-channel computerized data acquisition and recording system (Labview A/D software, National Instruments Co., USA.) in addition to a Grass polygraph paper recorder for further analysis.

2.3. Experimental Protocol

One half-hour after surgery, short anoxia was induced by inhalation of pure N_2 until respiratory arrest. The rats were immediately resuscitated. Three hours after the induction of anoxia, 1 ml of KCL 1 M was injected i.v. to induce cardiac arrest.

2.4. Statistical Analysis

The ANOVA and Duncan's multiple range test were used to detect significant differences between the four organs. A value of $P<0.05$ was considered to be significant. Mean ± S.E. values were calculated for each parameter in a group of 18 rats.

3. RESULTS

The rate of change in both NADH and TBF during cardiac arrest, induced by injection of 1 ml of 1 M KCL, is presented in Figure 1. The initial basal level of 0% NADH and 100% TBF were set before the injection of KCL. Approximately 3 seconds after KCL injection, an increase in NADH was observed in all four organs. This increase in NADH was associated with a fall in TBF to its lowest level. The rate of fall in TBF in the testis was significantly slower as compared to all other organs ($P<0.001$). Figure 2 presents the maximum levels of NADH during anoxia from four organs. As seen, three significantly different groups were observed. The brain and liver showed NADH levels of 49±11% and 52±10.5%, respectively. A second group was the level of NADH in the kidney, 66±33%. The third group was the NADH level measured in the testis, which was the lowest, reaching values of 22±12%. An increase in NADH levels indicates imbalance between oxygen demand and supply. As NADH reached its maximum values, the lowest level of the TBF (tissue blood flow) is usually observed.

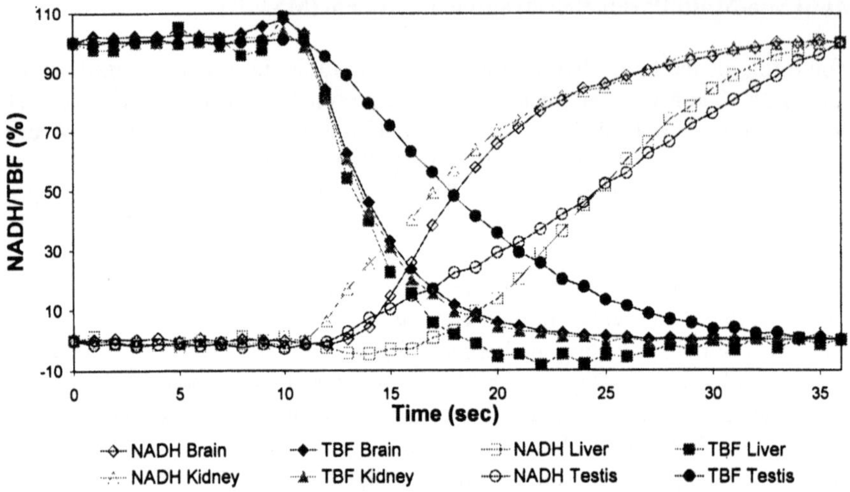

Figure 1. Changes in NADH and TBF during cardiac arrest. One experiment out of nine is displayed. Cardiac arrest was induced by injection of 1 ml of 1M KCl at time 10 sec. 0% NADH and 100% TBF were set as resting levels before induction of cardiac arrest. The rate of fall in TBF in the testis was found to be significantly different from the other organs using ANOVA (p<0.001).

Figure 3 shows the levels of minimum TBF in the four organs during anoxia (100% N_2). The use of ANOVA and Duncan's multiple range test exposed two significantly different groups regarding the lowest level of TBF. In the brain, TBF decreased to 66±28% of the basal level (100%). TBF in the liver, kidney and testis was 39±27, 22±27 and 29±21%, respectively. The time needed for the NADH to reach its maximum level showed three significant groups (Figure 4). In the brain, 38±18.5 seconds passed until maximum level of NADH was recorded. However, in the testis, a longer period of 60±16 seconds was needed to retain maximum NADH values. Maximum NADH levels were observed in the liver and kidney, 52±12.5 and 42±12.5 seconds after the induction of anoxia (Figure 5). The reduction in TBF during the anoxic

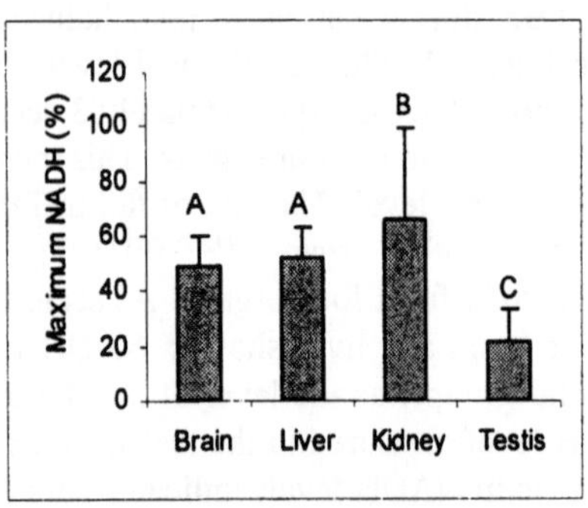

Figure 2. Maximum NADH values during anoxia (n=18). Values are expressed in percentage, where 0% is set as NADH level before anoxia induction. Three significantly different groups (A,B,C) were found using ANOVA and Duncan's multiple range test (*P*<<0.01)

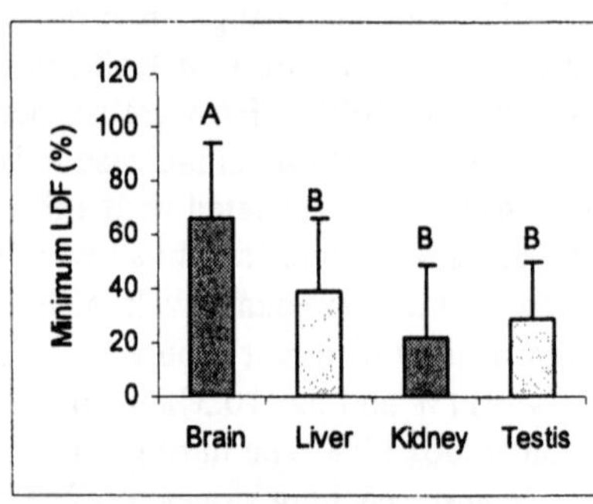

Figure 3. Minimum LDF values during anoxia (n=18). Values are expressed in percentage, where 100% is set as LDF level before anoxia induction and 0% is set as TBF level at death. Two significantly different groups (A,B) were found using ANOVA and Duncan's multiple range test (*P* <<0.01).

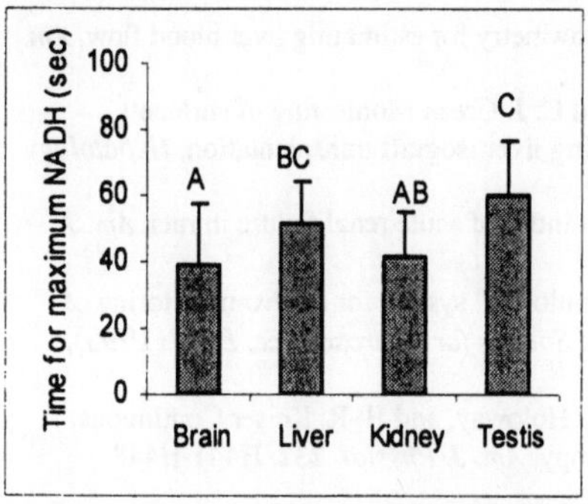

Figure 4. Time to Maximum NADH Values During Anoxia (n=18). The times from induction of anoxia until maximum NADH levels are shown. Three significantly different groups (A,C,BC/AB) were found using ANOVA and Duncan's multiple range test ($P<0.01$).

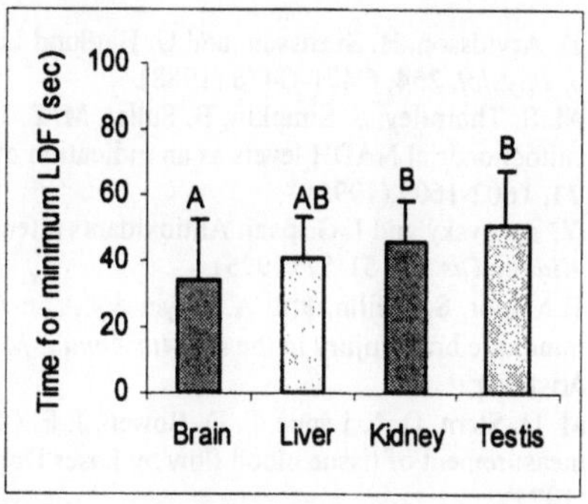

Figure 5. Time to minimum TBF values during anoxia (n=18). The times from induction of anoxia until minimum TBF levels are shown. Two significantly different groups (A,B) were found using ANOVA and Duncan's multiple range test ($P<0.01$).

episode showed two significantly different groups. In the brain, 33±20 seconds for minimum TBF were needed, while in the kidney and testis, 46±12 and 51±15 seconds, respectively, were needed for the TBF to get to its minimum values.

4. DISCUSSION

As seen, the brain, liver, kidney, and testis have distinctively different responses to anoxia. The brain was affected by anoxia more rapid than the testis, probably due to its massive consumption of O_2. Moreover, the brain apparently has the best mechanisms for autoregulation of blood flow, thus the decrease in CBF during anoxia was the smallest one. When cardiac arrest was induced, the testis had the most delayed reaction probably due to the fact that the testis require less blood than the brain, liver and kidney, which are highly perfuse organs.

Acknowledgements: This study was supported by the IHEL foundation and the HRDS.

REFERENCES

1. W. F. Ganong Circulation through special regions.*Review of Medical Physiology*. Appleton and Lange. 1991, 563.
2. K. E. van Holde and D. K. Mathews *Biochemistry* (The Benjamin/Cummings Publishing Company, Redwood City, CA, 1990).
3. C. Hinojosa-Laborde, B. H. Frohlich, and A. W. Cowley, Jr. Contribution of regional vascular responses to whole body autoregulation in conscious areflexic rats, *Hypertension* **17**, 1078-1084 (1991).
4. A. Mayevsky and B. Chance Intracellular oxidation reduction state measured *in situ* by a multichannel fiber-optic-surface fluorometer, *Science* **217**, 537-540 (1982).
5. A. Mayevsky, S. Meilin, T. Manor, E. Ornstein, N. Zarchin, and J. Sonn Multiparametric monitoring of brain oxygen balance under experimental and clinical conditions, *Neurol. Res.* **20**, S76-S80 (1998).
6. B. Chance and B. Schoener A correlation of absorption and fluorescence changes in ischemia of the rat liver, *in vivo*, *Biochemische Zeitschrift* **341**, 340-345 (1965).

7. D. Arvidsson, H. Svensson, and U. Haglund Laser Doppler flowmetry for estimating liver blood flow, *Am. J. Physiol.* **254**, G471-G476 (1988).
8. M. S. Thorniley, S. Simpkin, B. Fuller, M. Z. Jenabzadeh, and C. J. Green Monitoring of surface mitochondrial NADH levels as an indication of ischemia during liver isograft transplantation, *Hepatology* **21**, 1602-1609 (1995).
9. Y. Zurovsky and I. Gispaan Antioxidants attenuate endotoxin-induced acute renal failure in rats, *Am. J. Kidney Dis.* **25**, 51-57 (1995).
10. T. Manor, S. Meilin, and A. Mayevsky A new "multiprobe-multisite" system for *in vivo* monitoring of traumatic brain injury in the rat, *4th Annual Meeting of Israel Society for Neuroscience, Eilat* (1995), Abstract.
11. M. D. Stern, D. L. Lappe, P. D. Bowen, J. E. Chimosky, G. A. Holoway, and H. R. Keiser Continuous measurement of tissue blood flow by Laser Doppler spectroscopy, *Am. J. Physiol.* **232**, H441-H448 (1977).
12. U. Dirnagl, B. Kaplan, M. Jacewicz, and W. Pulsinelli Continuous measurement of cerebral cortical blood flow by Laser Doppler Flowmetry with a rat stroke model, *J. CBF Metab.* **9**, 589-596 (1989).

IMPORTANCE OF GLOMUS CELL PLASMA MEMBRANE AND MITOCHONDRIAL MEMBRANE POTENTIALS DURING ACUTE HYPOXIA SIGNALING IN THE RAT CAROTID BODY

Arijit Roy, Abu B. Al-mehdi, Anil Mokashi, and Sukhamay Lahiri*

1. INTRODUCTION

The mammalian carotid body (CB) is an oxygen addicted chemoreceptor organ, which consists of type I glomus, cells; the seat of oxygen sensing. There are two consensus hypotheses[1] regarding the CB hypoxia signaling pathway; (a) non-respiratory/ plasma membrane and (b) respiratory/ metabolic.

Plasma membrane depolarization, an essential element of non-respiratory pathway, is one of the major unresolved issues in CB physiology. Recording plasma membrane potential (E_m) with the patch clamp technique, hypoxia (Po_2 = 12-20 Torr) reversibly and significantly depolarized rat glomus cells on an average by 8.6 $\pm$ 0.9 mV (range +5 to +13 mV)[2]. However, it was also reported that hypoxia (Po_2 = 1-61 Torr) caused hyperpolarization in maximum of the glomus cells[3].

According to the respiratory model, depolarization of the glomus cell E_m is not considered to be the prime event in the hypoxia signaling cascade[4] and all agents known to impair mitochondrial oxidative phoshporylation lead to CB CSN excitation[5,6].

The aim of the present study was to elucidate the importance of glomus cell E_m and ψ_m (mitochondrial membrane potential) in the CB hypoxia (acute) signaling pathway. We measured indirectly E_m and ψ_m by a non-invasive fluorescence micro-imaging technique, using two different voltage sensitive probes namely: bis-oxonol and JC-1 respectively. Analysis of the fluorescence data revealed that hypoxia (Po_2= 8-10 Torr) caused an initial depolarization (not significant) followed by hyperpolarization ($P<0.05$) of the glomus cell E_m and preferentially depolarized glomus cell ψ_m ($P<0.05$).

* Sukhamay Lahiri, Department of Physiology, University of Pennsylvania, Philadelphia, PA 19104, USA.

Oxygen Transport to Tissue XXIII, edited by D. Wilson *et al.*
Kluwer Academic/Plenum Publishers, 2003

2. MATERIALS AND METHODS

The glomus cells were separated enzymatically from adult Sprague-Dawley rats (200-250 g) as described before[7]. Cultured glomus cells[7] on glass cover-slips were placed horizontally on a specially designed stainless-steel air-tight round coverglass window chamber (125 μl) connected with the superfusion tube at one end and an outlet at the other end. The chamber was mounted on a heated (37^0 C) plateform (Warner Instruments, Hamden, CT) on an inverted microscope and perfused by different solutions by gravity. The composition of the basic superfusate, the modified Tyrode with $CO_2 - HCO^-_3$ was (in mM): 112 NaCl, 4.7 KCl, 21.4 $NaHCO_3$, 2.2 $CaCl_2$, 1.1 $MgCl_2$, 22.0 sodium glutamate, 5.0 glucose plus 5 mM HEPES, and dextran 4g/L (mol. Wt. 74, 200). The osmolarity of the buffer was within normal range of 290-300 mosM and the pH was adjusted to 7.40. Hypoxic solution was prepared by proportionately mixing 100% O_2, 100% N_2, and 100% CO_2 solutions prepared in buffer, so as to get a required Po_2 of 8 to 10 Torr. Control normoxia – normocapnia solution was prepared from 21% O_2 and 5% CO_2. Pco_2 of all solutions were kept between 33-35 Torr. All solutions were measured for Po_2, Pco_2 and pH in a PHM73/Blood Gas monitor (Radiometer Copenhagen) prior to use. For elevated extracellular K^+ ($[K^+]_e$) solution, equimolar KCl was substituted for NaCl.

Glomus cell E_m change was measured in a non-invasive way using a non-toxic voltage sensitive fluorescent dye bis-oxonol ([bis-(1,3-dibutylbarbituric acid) trimethine oxonol; Molcular Probes, Eugene,OR). The lipophilic anionic bis-oxonol molecules undergo a potential–dependent distribution between the cytoplasm and the plasma membrane from the extracellular medium by the Nernst equilibrium[8]. The mechanism underlying the increase in bis-oxonol fluorescence with cellular membrane depolarization is usually ascribed to dye translocation or partitioning between extracellular free dye and plasma membrane, or cytosol and plasma membrane[9], or probably the orientation of dye molecules within the membrane[10]. Repolarization or hyperpolarization of the plasma membrane results in extrusion of the dye and thus a decrease in fluorescence[8]. The negative charges of bis-oxonol molecules ensure that the dye does not accumulate in mitochondria and the fluorescence is not affected by changes in ψ_m[11]. In the present study, aliquots of 1mM bis-oxonol solution (as stock) were prepared in ethanol and stored at -20° C. The glomus cells were incubated in an experimental concentration of 1 μM bis-oxonol for dye loading in Tyrode buffer under normoxic condition ($PO_2 \sim 130$ Torr) at 37° C for 30 min. For measuring E_m, fluorescence was excited at 490 nm (λ_{ex}) and measured at 520 nm (λ_{em}).

Glomus cell ψ_m was measured, using a voltage sensitive probe JC-1 (5,5`,6, 6`-tetrachloro-1,1`, 3,3`-tetraethylbenzimidazolyl-carbocyanine iodide; Molecular Probes, Eugene,OR). This lipophilic membrane permeant cationic probe has been shown to be distributed between the cytosol and mitochondria and exits in a monomeric form emitting green fluorescence at 527 nm after excitation at 490 nm (λ_{ex}). However, with the increase in average ψ_m (-130 to –180 mV), JC-1 is able to form aggregates emitting red fluorescence with a large shift in emission (λ_{em} = 590 nm). Since the dye is not significantly quenched in the cell, increase in JC-1 fluorescence is used to indicate relative hyperpolarization while decrease in fluorescence intensity is used to indicate mitochondrial depolarization[12].

In the present study, aliquots of JC-1 (1mM) stock solution were made in DMSO and stored in foil-wrapped eppendorf tubes in a refrigerator (-20° C). A working solution of 10 μM JC-1 was prepared in Tyrode buffer and the glomus cells were incubated in an experimental concentration of 1 μM JC-1 for 30 min at 37° C for effective loading.

Glomus cells were viewed with a Nikon Diaphot TMD epifluorescence microscope (60x, 100x oil-immersion objective), equipped with an optical filter changer (Lambda 10-2; Sutter Inst. Co., Novoto, CA) and illuminated with a 75–W mercury lamp. The epifluorescence illuminator was equipped with narrow bandpass filters and a triple-band dichroic mirror (D/F/R-BS&M; Chroma Technology Corp. Brattleboro, VT) for excitation. To prevent rapid bleaching of the fluorescent preparation a neutral density filter was used. The glomus cell images from the microscope were acquired during a 50-100 ms exposure time with a computer controlled 12 bit digital cooled CCD camera (Hamamatsu ORCA 100), using a graphics control software (MetaMorph Imaging system, Universal Imaging Corp.). Functional glomus cells were identified as more or less round cells with diffused green fluorescence for bis-oxonol and brightly stained red fluorescence for JC-1/ J-aggregates. The regions of interest were digitally marked and the pixel intensities within the region were then averaged together to obtain a measure of the fluorescence intensity of an individual cell. The background regions outside the cells were also digitally marked and the average pixel intensity was subtracted from each cell image. Changes in fluorescence intensity of the region of interest were acquired before, during and after stimulation. Image analysis was done with MetaMorph software (Universal Imaging Corp., USA).

Changes in the fluorescence intensity (arbitrary unit; a.u.) were expressed as percent of basal response (control), which was considered 100%. Results are expressed as mean ± SE of experiments for each condition. Replicate experiments were carried out using cells from separate isolation. Differences among groups were evaluated with one-way ANOVA using SigmaStat (Jandel Scientific, CA). Statistical significance was determined at $P < 0.05$.

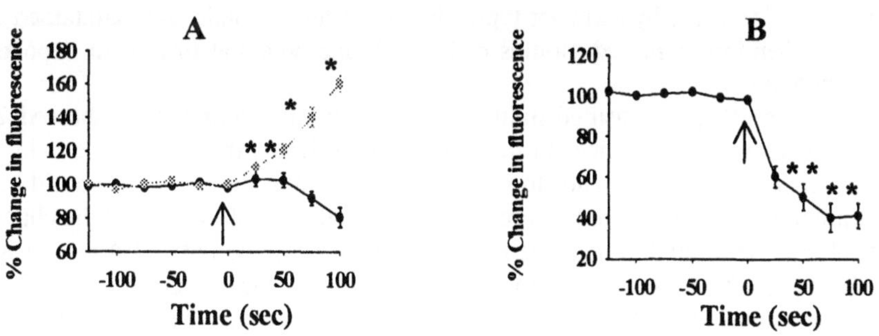

Figure 1. Summary of the time course of average fluorescence intensity responses (arbitrary unit) of rat glomus cells (n=8). **A:** Glomus cell E_m as measured by bis-oxonol, showing decrease in fluorescence (hyperpolarization) during acute hypoxia (Po_2= 8-10 Torr; black) and increase in fluorescence (depolarization) during 25 mM $[K^+]_e$ (Po_2 = 125-130 Torr; grey). **B:** Glomus cell ψ_m as measured by JC-1, showing decrease in fluorescence (depolarization) during acute hypoxia (Po_2 = 8-10 Torr). Arrows indicating point of application of hypoxia (A&B) and high K^+ (A). * P< 0.05.

3. RESULTS

Average bis-oxonol fluorescence intensities (mean ± SE; arbitrary unit) of eight glomus cells during normoxia (Po_2 =125 –130 Torr), hypoxia (Po_2 =8-10 Torr) and 25 mM $[K^+]_e$ (Po_2 =125 –130 Torr) are shown in Fig. 1A. Initially cells were treated with normoxia and images were taken every 25s for 125s (negative numbers). This was followed by abrupt change to hypoxia and images were taken every 25s for up to 100s (positive number). The same time sequence was followed for 25 mM $[K^+]_e$. Hypoxia produced an initial (25s) 2-3% increase (not significant) in average fluorescence, followed by a decrease in fluorescence to 80 ± 6% (P<0.05) at 100s. The same cells when exposed to 25 mM $[K^+]_e$ during normoxia, produced increase in fluorescence to 110 ± 3% (P<0.05) at 25s, followed by further increase to 160 ± 5% (P<0.05) at 100s. Thus, hypoxia caused mild depolarization initially, followed by significant hyperpolarization of the glomus cell E_m.

Average JC-1 fluorescence intensities (mean ± SE) of eight glomus cells during normoxia (Po_2 =125 –130 Torr) and hypoxia (Po_2 = 8-10 Torr) are shown in Fig. 1B. Steady-state images were taken during normoxia for 125s, followed by hypoxia for 100s. With hypoxia, fluorescence dropped to 60 ± 5% (P<0.05) at 25s and finally to 41± 6% (P<0.05) at 100s. Thus, the glomus cells mitochondria showed clear sign of depolarization with hypoxia.

4. DISCUSSION

In the present study, the glomus cell E_m and ψ_m responses to hypoxia were delayed. The possible reasons are; (a) the time taken for hypoxic solution to reach and exchange in the recording chamber obviated the ability to take images before 15-20s (the lag period) and (b) both bis-oxonol and JC-1 being slow response probes (optical response could take seconds to minutes) the actual changes in fluorescence signal was delayed. Therefore, it is highly probable that the fluorescence signals lag behind the true time course of glomus cell E_m and ψ_m. However, the time courses for the increase in glomus cell intracellular calcium ($[Ca^{2+}]_i$)[13], dopamine (DA)[14] release and increased CB CSN discharge[14] to acute hypoxia are typically within few seconds and maintained for > 1minute; we therefore expected glomus cell membrane potential to remain depolarized for that period too.

Glomus cell E_m measured indirectly by bis-oxonol fluorescence showed 2-3% increase in average fluorescence during initial (0 to 25s) exposure to acute hypoxia (Fig.1A). This increase corresponds to mild depolarization of the E_m, which could be due to inhibition of the leak K^+ current[15]. Previously, Donnelley[16] estimated that the change in glomus cell E_m due to inhibition of the leak K^+ current, could be only +1.4 mV. It would seem unlikely to effect the transduction pathway with that magnitude of depolarization. With further exposure to hypoxia, there was a 20% decrease (at 100s) in average fluorescence (Fig. 1A) which corresponds to significant hyperpolarization of the glomus cell E_m. Hypoxia induced membrane hyperpolarization, as reported by others in hippocampal neurons, is due to increased conductance of K_{ATP}[17] and K_{Ca} channels[18]. However, in rat glomus cells there are no functional expressions of K_{ATP} channels[19]; further K_{Ca} channel conductance decrease with hypoxia[20]. Thus, hypoxia induced hyperpolarization of the glomus cell E_m is rather unclear and doesn't comply with the plasma membrane hypothesis of hypoxia signaling. Is depolarization an essential element of acute hypoxia signaling? Due to subpopulations of glomus cells identified on the basis of size and on synapses to nerve endings, many glomus cells may not respond or respond differently to hypoxia[21]. However, there is no report on subpopulations based on electrophysiologic criteria[22]. Pang and Eyzaguiree[3] reported that isolated glomus cells most commonly hyperpolarized and clustered cells usually depolarize. According to them, isolated cells are not enveloped by sustenticular cells (type II cells) which release K^+ and thus the retro-effects of released K^+ would be either absent or much less pronounced. However, a number of isolated cells react to hypoxia with depolarization, indicating that the presence of sustenticular process is not the only factor affecting glomus cell behavior. Hypoxia has been reported to inhibit the O_2 - sensitive K^+ channels, leading to possible depolarization of the glomus cell membrane[2]. However, classical K^+ channel blockers (4-AP and TEA) fail to depolarize glomus cell membrane[15], increase in $[Ca^{2+}]_i$[23], dopamine release[23] and CSN activity[23].

Glomus cell ψ_m measured indirectly by JC-1 fluorescence showed a significant 40% decrease in average fluorescence during hypoxia initially (0 to25s) (Fig. 1B). With prolonged exposure to hypoxia there was a 60% fluorescence decrease at 100s. This clearly suggests mitochondrial depolarization. The possible mechanism could be: hypoxia preventing the cytochrome oxidase of the mitochondrial respiratory chain to reduce O_2, decreasing the intracellular phosphate potential and depolarizing the mitochondria. Mitochondrial uncouplers (DNP and FCCP) have been shown to depolarize type I cell mitochondria[24], leading to substantial increases in $[Ca^{2+}]_i$[4], neurosecretion[25] and CSN discharge[6]. Further, Lahiri et. al.[26] using high CO (PCO~550 Torr) as a metabolic inhibitor, have shown increase in CSN activity like hypoxia. Also, CO induced increase in glomus cell $[Ca^{2+}]_i$[27] and DA release[28] have been reported. Recently, our imaging study (unpublished) also showed glomus cell mitochondrial depolarization with high CO.

Finally, comparing the glomus cell E_m and ψ_m fluorescence data, it is apparent that acute hypoxia causing mitochondrial depolarization is a critical factor in the hypoxia signaling pathway even at hyperpolarized E_m. A similar observation was reported in rat hippocampal neurons, where hypoxia led to ψ_m depolarization, increased Ca^{2+} influx even at hyperpolarized E_m[29]. They suggested that with altered mitochondrial metabolism during hypoxia there might be a shift in the activation curve of calcium channels in the hyperpolarizing direction so that significant numbers of channels open at resting E_m and could provide increased influx of Ca^{2+}. More recently, Rizzuto et. al.[30] have reported close appositions between mitochondria with endoplasmic reticulum (ER) membranes and depolarization of the mitochondria likely to release calcium from ER. If, similar mechanisms exist in the glomus cell; then hypoxia causing mitochondrial depolarization and plasma membrane hyperpolarization would release Ca^{2+} from internal stores as well as influx through plasma membrane channels, elevating cytosolic Ca^{2+}, followed by neurotransmitter release and finally CSN excitation. The fact that acute hypoxia causes hyperpolarization of the glomus cell E_m and depolarization of the ψ_m, would add support to the hypothesis that acute hypoxia signaling is mediated through changes in mitochondrial respiration.

REFERENCES

1. S. Lahiri, C. Rozanov, A. Roy, B. Storey, and D.G. Buerk, Regulation of oxygen sensing in peripheral arterial chemoreceptor, *Int. J. Biochem.Cell Biol.* **33**, 755-774 (2001).
2. C.N. Wyatt, C. Wright, D. Bee, and C. Peers, O_2-sensitive K^+ currents in carotid body chemoreceptor cells from normoxic and chronically hypoxic rats and their roles in hypoxic chemotransduction., *Proc Natl Acad Sci* **92**, 295-299 (1995).

3. L. Pang, and C. Eyzaguirre, Different effects of hypoxia on the membrane potential and input resistance of isolated and clustered carotid body glomus cells., *Brain Res.* **575**, 167-173 (1992).
4. M.R. Duchen, and T.J. Biscoe, Relative mitochondrial membrane potential and $[Ca^{2+}]_I$ in type I cells isolated from the rabbit carotid body, *J Physiol.* **450**, 33-61 (1992).
5. S.S. Krylov, and S.V. Anichcov, The effects of metabolic inhibitors on carotid chemoreceptors, in *Arterial Chemoreceptors*,edited by R.W. Torrance (Blackwell, 1968), pp 103-109.
6. E. Mulligan, and S. Lahiri, Dependence of carotid chemoreceptor stimulation by metabolic agents on Pao_2 and $Paco_2$, *J. Appl. Physiol.* **50**, 884-891 (1981).
7. A. Mokashi, D. Ray, F. Botre, M. Katayama, S. Osanai, and S.Lahir, Effects of hypoxia on intracellular pH of glomus cells cultured from rat and cat carotid bodies, *J Appl Physiol.* **78**, 1875-1881 (1995).
8. H.J. Apell, and B. Bersch, Oxonol VI as an optical indicator for membrane potentials in lipid vesicles. *Biochem Biophys Acta,* **903**, 480-494 (1987).
9. T. Brauner, D.F. Hulser, and R.J. Strasser, Comparative measurements of membrane potentials with microelectrodes and voltage- sensitive dyes, *Biochem Biophys Acta.* **771**, 208-216 (1984).
10. A.B. Al-mehdi, H. Shuman, and A.B. Fisher, Oxidant generation with K^+- induced depolarization in the isolated perfused lung, *Free Radical Biol & Med.* **23**, 47-56 (1997).
11. C.F. Mohr, and C. Fewtrell, IgE receptor-mediated depolarization of rat basophilic leukemia cells measured with the fluorescent probe bis-oxonol, *J Immunol.* **138**, 1564 – 1570 (1987).
12. M. Reers, S.T. Smiley, C. Mottola-Hartshorn, A. Chen, M. Lin, and L.B. Chen, Mitochondrial membrane potential monitored by JC-1 dye, Methods in Enzymol., **260**, 406-417 (1995).
13. J. Urena, R. Fernandez-Chacon, A.R. Benot, G. Alvarez de Toledo, and J. Lopez-Barneo, Hypoxia induces voltage dependent Ca^{2+} entry and quantal dopamine secretion in carotid body glomus cells, *Proc. Natl. Acad. Sci.* **91**, 10208-10211 (1994).
14. D.G. Buerk, S. Osanai, A. Mokashi, and S. Lahiri, Dopamine, sensory discharge, and stimulus interaction with CO_2 and O_2 in cat carotid body, *J. Appl. Physiol.* **85**, 1719-1726 (1998).
15. K.J. Buckler, A novel oxygen-sensitive potassium current in rat carotid body type I cells, *J. Physiol.* **498**, 649-662 (1997).
16. D.F. Donnelly, K^+ currents of glomus cells and chemosensory functions of carotid body, *Respir. Physiol.* **115**, 151-160 (1999).
17. N. Fujimura, E. Tanaka, S. Yamamoto, M. Shigemori, and H. Higashi, Contribution of ATP-sensitive potassium channels to hypoxic hyperpolarization in rat hippocampal CA1 neurons in vitro, *J. Neurophysiol.* **77**, 378-385 (1997).
18. G. Erdemi, Y.Z. Xu, and K. Krnjevic, Potassium conductance causing hyperpolarization of CA1 hippocampal neurons during hypoxia, *J. Neurophysiol.* **80**, 2378- 2390 (1998).
19. C. Peers, and J. O'Donnell, Potassium currents recorded in type I carotid body cells from the neonatal rat and their modulation by chemoexcitatory agents, *Brain Res.* **572**, 259-266 (1990).
20. C. Peers, Hypoxic suppression of K^+ currents in type I carotid body cells: selective effect of Ca^{2+}-activated K^+ current, *Neurosci. Lett.* **119**, 253-256 (1990).
21. D.M. McDonald, and R.A. Mitchell, The innervation of glomus cells, ganglion cells and blood vessels in the rat carotid body: a quantitative ultrastructural analysis, *J. Neurocytol.* **4**, 177-230 (1975).
22. M.T. Perez-Garcia, A. Obeso, J.R. Lopez-Lopez, B. Herreros, and C. Gonzalez, Characterization of cultured chemoreceptor cells dissociated from adult rabbit carotid body, *Am. J. Physiol.* **263**, C1152-C1159 (1992).
23. A. Roy, C. Rozanov, D.G. Buerk, A. Mokashi, and S. Lahiri, Suppression of glomus cell K^+ conductance by 4-aminopyridine is not related to $[Ca^{2+}]_I$, dopamine release and chemosensory discharge from carotid body, *Brain Res.* **785**, 228-235 (1998).
24. K.J. Buckler, and R.D. Vaughan-Jones, Effects of mitochondrial uncouplers on intracellular calcium, pH and membrane potential in rat carotid body type I cells, *J. Physiol.* **513**, 819-833 (1998).
25. A. Rocher, A. Obeso, C. Gonzalez, and B. Herreros, ionic mechanisms for the transduction of acidic stimuli in rabbit carotid body glomus cells, *J. Physiol.* **433**, 533-548 (1991).
26. S. Lahiri, D.G. Buerk, D. Chugh, S. Osanai, A. Mokashi, Reciprocal photolabile O_2 consumption and chemoreceptor excitation by carbon monoxide in the cat carotid body: evidence for cytochrome a_3 as the primary O_2 sensor, *Brain Res.* **684**, 194-200 (1995).
27. A. Mokashi, A. Roy, C. Rozanov, S. Osanai, B.T. Storey, S. Lahiri, High Pco does not alter pH_I, but raises $[Ca^{2+}]_I$ in cultured rat carotid body glomus cells in the absence and presence of $CdCl_2$. *Brain Res.* **803**, 194-197 (1998).
28. D.G. Buerk, D.K. Chugh, S. Osanai, A. Mokashi, S. Lahiri, Dopamine increases in cat carotid body during excitation by carbon monoxide: Implications for a chromophore theory of chemoreception., *J Autonom Nerv Sys.* **67**, 130-136 1997).

29. A.V. Nowicky, and M.R. Duchen, Changes in $[Ca^{2+}]_I$ and membrane currents during impaired mitochondrial metabolism in dissociated rat hippocampal neurons, *J Physiol.* **507**, 131-145 (1998).
30. R. Rizzuto, P. Pinton, W. Carrington, F.S. Fay, K.E. Fogarty, L.M. Lifshitz, R.A. Tuft, and T. Pozzan, Close contacts with the endoplasmic reticulum as determinants of mitochondrial Ca^{2+} responses, Science, **280**, 1763-1766 (1998).

ESTIMATION OF HYDROXYL RADICALS BASED ON THE SALICYLATE TRAPPING METHOD IN HIPPOCAMPUS OF STROKE-PRONE SPONTANEOUSLY HYPERTENSIVE RATS (SHRSP) DURING TRANSIENT ISCHEMIA AND RECIRCULATION

[1]Hiroko Negishi, Katsumi Ikeda, Yasuo Nara, and Yukio Yamori

1. INTRODUCTION

Neuronal damage following transient cerebral ischemia is mediated by various mechanisms, among which oxygen radical-mediated processes play a central role[1, 2]. Ischemia and subsequent recirculation provide circumstances that favor their production. Free radicals are extremely reactive compounds that can interact with lipids, enzymes and DNA to produce various harmful effects. Several reports have suggested that hippocampal neurons in Stroke-prone Spontaneously Hypertensive rats (SHRSP) which are regarded as model animals for ischemic stroke in humans, were innately more vulnerable to ischemic damage than in the mother strain of SHRSP, Wistar Kyoto rats (WKY)[3, 4]. In a previous study, we demonstrated the important role of free radicals in the development of neuronal damage in the hippocampus of SHRSP subjected to cerebral ischemia and reperfusion[5]. However, direct evidence of hydroxyl radicals, which are very reactive and readily react with a number of compounds, in the course of cerebral ischemia and reperfusion in SHRSP has not been obtained.

In this study, we measured the levels of 2,3-dihydroxybenzoic acid (2,3-DHBA) following salicylate administration, as a biological marker of hydroxyl radicals in the hippocampus, to observe the role of hydroxyl radicals in SHRSP during ischemia and recirculation (see below).

[1] Hiroko Negishi and Katsumi Ikeda, Kyoto Univ., Kyoto 6068501, Japan. Yasuo Nara, Univ. of East Asia, Shimonoseki 7518503, Japan. Yukio Yamori, WHO Collaborating Center, Kyoto 6068413, Japan

Oxygen Transport to Tissue XXIII, edited by D. Wilson *et al.*
Kluwer Academic/Plenum Publishers, 2003

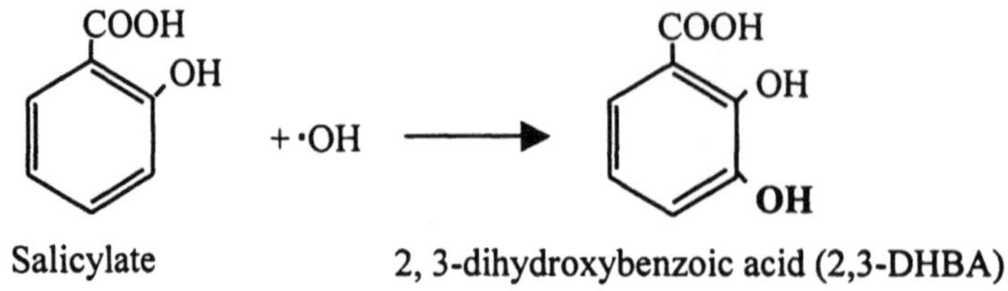

2. METHODS

2.1 Animals

Five male SHRSP/Izm, five male Wistar-Kyoto/Izm rats (WKY) which is the normotensive rat and the mother strain of SHRSP, and five male rats for sham-operated (SHRSP; n=3, WKY; n=2) aged 14 weeks fed a plain laboratory diet (SP; Funabashi Farm, Chiba, Japan) were used. We measured the production of hydroxyl radicals in the hippocampus of SHRSP and WKY. These rats were mounted on a stereotaxic frame under pentobarbital anesthesia, and a microdialysis probe (A-I-4-01, EICOM) was stereotaxically implanted in the CA1 layer of the hippocampus. The probes were subsequently perfused with Ringer's solution including salicylate (5 mM) at a flow rate of 2 μl/min using a microinfusion pump (AS-10, EICOM). The carotid arteries were exposed before ischemia, and samples of microdialysis perfusion were collected at 20-minute intervals, 20 min before ischemia; during the 40 min, the period of global cerebral ischemia by carotid artery occlusion; and during the 60 min of recirculation. In sham-operated rats, samples were collected at 20-min intervals after the carotid arteries were exposed without ischemia.

We used salicylate, a highly effective hydroxyl radical trap, which on scavenging hydroxyl radical forms the stable compound 2,3-DHBA[6, 7]. 2,3-DHBA was analyzed in microdialysis samples using an INERT-HPLC-ECD system (DG-100, EICOM, Kyoto, Japan). The tubes, injector and needle cylinder, in the HPLC-ECD system, were coated with inert material. The cylinder was steeped in nitric acid and hydrogen peroxide, and then coated with silicon. Non-specific production of hydroxyl radicals was inhibited by these treatments. Using this method, we confirmed that the formation

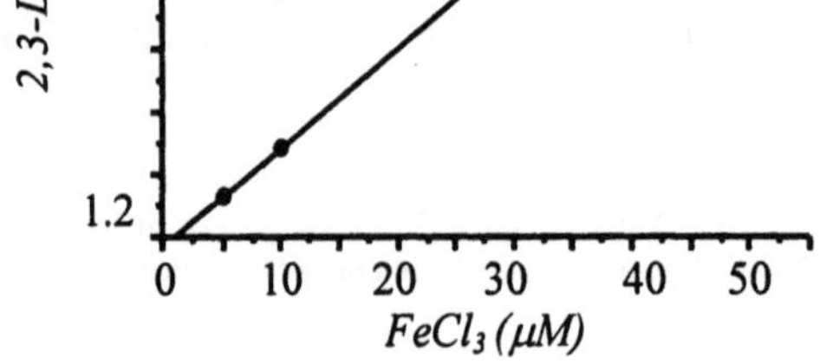

Fig. 1 Calibration curve of dihydrobenzoic acid (2, 3-DHBA), hydroxyl radicals (r=0.988). Known concentrations of $FeCl_3$, were added to a sodium phosphate buffer with 2'-deoxyribose, diethylenetriamine pentaacetic acid and hydrogen peroxide.

of 2,3-DHBA was dependent on the concentration of $FeCl_3$ which is the oxidizing agent, in 100 mM sodium phosphate buffer (pH 7.4), 10 μM diethylenetriamine pentaacetic acid, 1 mM 2'-deoxyribose, and 500 μM hydrogen peroxide, indicating the production of hydroxyl radicals in vitro (Fig. 1).

Statistical analyses were performed using the Stat View 5.0 program, SAS Institute Inc. The 2,3-DHBA levels in rats during the preischemia, ischemia and recirculation time periods were analyzed with Friedman's test. The changes in physiological variables among different groups were analyzed with Mann-Whitney's test. Values of $p<0.05$ were regarded as significant.

3. RESULTS AND DISCUSSION

The levels of 2,3-DHBA in the samples during the first 20 min of recirculation were the highest in SHRSP and WKY. However, increases in 2,3-DHBA levels were not observed in sham-operated rats. The 2,3-DHBA levels in SHRSP gradually increased during the 40 min of ischemia, but the levels during both ischemia and recirculation in SHRSP were significantly higher than in WKY (Figure 2). The peak production of hydroxyl radical rates in the hippocampus occurred during the first 20 min of recirculation, which was calculated from the change in hydroxyl radicals in the hippocampus during ischemia and recirculation.

In our current study, it was observed that hippocampal CA1 neurons in SHRSP were more vulnerable than those neurons in WKY in occlusion of carotid arteries and reperfusion. This is consistent with our observation that production of hydroxyl radicals during ischemia and recirculation was higher in the hippocampus of SHRSP compared to that of WKY.

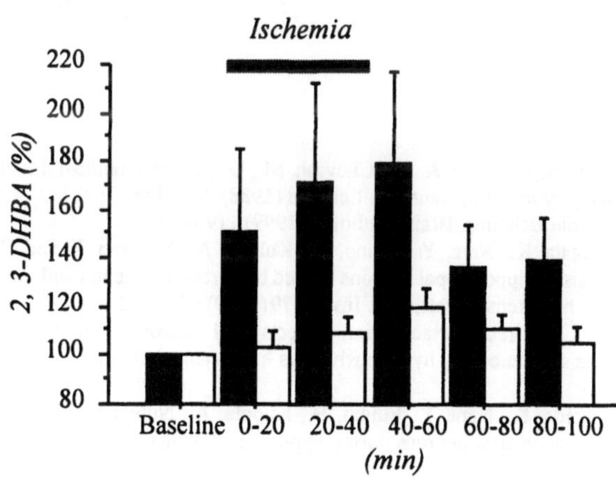

Fig. 2 The changes in the levels of 2, 3-DHBA in SHRSP (■)(n=5) and WKY (□)(n=5) hippocampal perfusates during 40 min of ischemia and 60 min recirculation. 2, 3-DHBA in SHRSP *vs* WKY, $p<0.05$. Baseline *vs* Recirculation (40-60 min), $p<0.05$ both SHRSP and WKY.

Several studies have demonstrated increased hydroxyl radicals in the rat brain during ischemia[8-11]. It was argued that a possible mechanism of hydroxyl radical production during ischemia is the "Fenton reaction", based on the iron- catalysed conversion of superoxide anion radicals and H_2O_2 into hydroxyl radicals, as ischemia induces a release of iron from ferritin and transferrin[8]. A recent study documented that lactoacidosis induced an increase in hydroxyl radical concentrations[12]. Our previous study showed increased lactate levels in acute ischemia by bilateral carotid artery occlusion in SHRSP, but not in WKY[13]. The increased lactate, therefore, may contribute to the production of hydroxyl radicals during ischemia. In the recirculation period, xanthine oxidase[14], mitochondorial electron transport and nitric oxide (NO) are known as sources of hydroxyl radical production. Our recent study observed direct evidence that enhanced nitric oxide production in the hippocampus of SHRSP during recirculation following ischemia (submitted). The increase in NO may be one of the sources of hydroxyl radical formation. It is known that reactive oxygen species, such as superoxides, and hydroxyl radicals, damage cells and DNA. 8-Hydroxyl-2'-deoxyguanosine (8-OHdG) is a product of oxidative DNA damage following specific enzymatic cleavage after 8-hydroxylation of a guanine base by hydroxyl radicals[15]. In relation to DNA damage by hydroxyl radicals, Cui et al. reported increased 8-OHdG formation in rat brains subjected to a focal ischemia and recirculation[16].

CONCLUSION

The production of hydroxyl radicals in the hippocampus highest during the first 20 min of recirculation. The changes in 2, 3-DHBA levels during ischemia and recirculation in SHRSP were significantly higher than in Wistar-Kyoto rats. These results suggest that hippocampal neuronal damage following ischemia and recirculation is, in part, caused by the increase in hydroxyl radicals during ischemia and recirculation.

REFERENCES

1. Cao, W., Carney, J.M., Duchon, A., Floyd, R.A. and Chevion, M., Oxygen free radical involvement in ischemia and reperfusion injury to brain, Neurosci. Lett., 88 (1988) 233--238.
2. Love, S., Oxidative stress in brain ischemia, Brain. Pathol. 9 (1999) 119--131.
3. Tagami, M., Ikeda, K., Yamagata, K., Nara, Y., Fujino, H., Kubota, A., Numano, F. and Yamori, Y., Vitamin E prevents apoptosis in hippocampal neurons caused by cerebral ischemia and reperfusion in stroke-prone spontaneously hypertensive rats, Lab. Invest. 79(1999) 609--615.
4. Yasui, M. and Kawasaki, K., CCK_B receptor activation protects CA1 neurons from ischemia-induced dysfunction in stroke-prone spontaneously hypertensive rats hippocampal slices, Neurosci. Lett. 191 (1995) 99--102.
5. Tagami, M., Yamagata, K., Ikeda, K., Nara, Y., Fujino, H., Kubota, A., Numano, F. and Yamori, Y., Vitamin E prevents apotosis in cortical neurons during hypoxia and oxygen reperfusion, Lab. Invest. 78 (1998) 1415--1429.
6. Gerlach, M., Xiao, A.Y., Heim, C., Lan, J., God, R., Feineis, D., Bringmann, G., Riederer, P. and Sontag, K.H, 1-Trichloromethyl-1,2,3,4-tetrahydro-beta-carbonline increases extracellular serotonin and stimulates hydroxyl radical production in rats, Neurosci. Lett., 257 (1998) 17--20.
7. Lancelot, E., Lacanu, L., Revaud, M.L., Boulu, R.G., Plotkine, M. and Callebert, J., Glutamate induces hydroxyl radical formation in vivo via activation of nitric oxide synthase in Sprague-Dawley rats, Neurosci. Lett., 242 (1998) 131--134.
8. Lancelot, E., Callebert, J., Revaud, M.L., Boulu, R.G. and Plotkine, M., Detection of hydroxyl radicals in rat striatum during transient focal cerebral ischemia: possible implication in tissue damage, Neurosci. Lett., 197 (1995) 85--88.

9. Ste-Marie, L., Vachon, P., Vachon, L., Bemeur, C., guertin, M.C. and Montgomery, J., Hydroxyl radical production in the cortex and striatum in a rat model of focal cerebral ischemia, Can. J. Neurol. Sci., 27 (2000) 152--159.
10. Wei, J., Huang, N.C. and Quast, M.J., Hydroxyl radical formation in hyperglycemic rats during middle cerebral artery occlusion / reperfusion, Free. Radic. Biol. Med., 23 (1997) 986--995.
11. Yamamoto, T., Yuki, S., Watanabe, T., Mitsuka, M., Saito, K.I. and Kogure, K., Delayed neuronal death prevented by inhibition of increased hydroxyl radical formation in a transient cerebral ischemia, Brain. Res., 762 (1997) 240--242.
12. Piantadosi, C.A. and Zhang, J., Mitochondrial generation of reactive oxygen species after ischemia in the rat, Stroke, 27 (1996) 327--332.
13. Nara, Y., Horie, R. and Yamori, Y., Effect of acute and chronic reduction of cerebral blood flow on glucose metabolitelevels in SHRSP, Stroke, 12 (1981) 693--695.
14. Remblier, C., Pontcharraud, R., Tallineau, C., Priou, A. and Huguet, F., Lactic acid-induced increase of extracellular dopamine measured by microdialysis in rat striatum: evidence for glutamatergic and oxidative mechanisms, Brain, Res., 837 (1999) 22--28.
15. Shigenaga, M. and Ames, B., Assays for 8-hydroxy-2'-deoxyguanosine: A biomarker of in vivo oxidative DNA damage, Free. Radic. Biol. Med., 10 (1991) 211—216.
16. Cui, J., Holmes, E.H., Greene, T.G. and Liu, P.K., Oxidative DNA damage precedes DNA fragmentation after experimental stroke in rat brain, FASEB. J., 14 (2000) 955--967.

EFFECT OF HYPOXIA AND ISCHEMIA ON EXPRESSION OF SELECTED GENES IN BRAIN OF NEWBORN PIGLETS

[1]Tatiana Zaitseva, Jennifer Creed, Gregory Schears, Jin Shen, Krystyna Knight, David F. Wilson, and Anna Pastuszko

1. INTRODUCTION

In the brain, function is acutely dependent on delivery of oxygen at a pressure sufficient to maintain the cellular metabolic status. Interruption of oxygen delivery leads to loss of functional integrity (consciousness) within seconds, with progressive deterioration occurring during periods of oxygen deprivation. This oxygen deprivation can result from either decreased oxygen content in the blood (hypoxic-hypoxia) or low blood flow (ischemia), although these names are frequently used interchangeably in the literature. It has been shown, however, that the physiological response and neuropathology of brain are different for hypoxia than for ischemia[1]. In the present study, we investigated the effects of hypoxic-hypoxia and of ischemic-hypoxia on cortical oxygenation and the expression of specific genes in neonatal piglet brain. Experimental conditions were selected such that the models had similar values for the mean cortical oxygenation. The results show that the changes in expression of the selected genes were significantly greater following ischemic-hypoxia than in hypoxic-hypoxia.

2. MATERIALS AND METHODS

2.1 Animal preparation:

Newborn piglets, age 3-5 days, were used for this study. Anesthesia was induced with halothane (Halocarbon Laboratories, Augusta, SC; 4% mixed with 96% oxygen),

[1] Dept. Biochemistry & Biophysics, University of Pennsylvania, School of Medicine and Dept. Anesthesiology & Critical Care, The Children Hospital, Philadelphia, PA 19104, U.S.A.

and 1.5% lidocaine-HCl (Abbott Laboratories, North Chicago, IL) was used as a local anesthetic. The halothane was reduced to 0.6-0.8% after tracheotomy and the femoral artery and femoral vein were cannulated. For ischemic-hypoxia models both carotid arteries were isolated and silk suture was placed loosely around each carotid artery. After the surgery was completed, the halothane was withdrawn entirely and Fentanyl-citrate (Elkins-Linn, Inc., Cherry Hill, NJ; 30 μg/kg) was injected intravenously at approximately one hour intervals throughout the experiments. The animals were paralyzed with pancuronium (0.3 mg/kg IV) and mechanically ventilated with a mixture of oxygen and nitrous oxide (in control conditions with 21-22% oxygen and 78-79% N_2O). The head was placed in a Kopf stereotaxic holder and an incision was made along the midline of the scalp. The scalp was removed to expose the skull and a hole approximately 5 mm in diameter was made in the skull over one parietal hemisphere, where the cortical oxygen pressure was measured through the intact dura. In all experiments, the blood pressure, body temperature and heart rate were continuously monitored. Arterial blood samples were taken every 15 min throughout the course of the experiment and the blood pH, $PaCO_2$ and PaO_2 were measured using a Model 178 pH/Blood Gas Analyzer (Corning). At the end of each experiment, the anesthetized animals were euthanized by intravenous administration of saturated KCl solution, the brains were perfused and used for gene expression.

All animal use procedures were in strict accordance with the NIH Guide for the Care and Use of Laboratory Animals and were approved by the local Animal Care Committee.

2.2 Experimental models

Hypoxic-hypoxia: In this model the FiO_2 was changed from 21% to 7% for one hour followed by reoxygenation for 2 hours.

Ischemic-hypoxia – Ligation of carotid arteries and bleeding: _Ischemic-hypoxia was induced by ligation of both carotid arteries by pulling the snares snugly around them. Blood, approximately 70 ml, was then withdrawn from the arterial catheter into syringes to reduce the systemic arterial pressure to about 40 Torr. After one hour of the ischemic-hypoxia, the carotid ligatures were released and the blood that had been previously withdrawn was reinfused. The animals were then maintained through a two hour period of recovery.

2.3 Brain preparation for the aRNA amplification analysis

At the end of the experimental protocol, the chest of the anesthetized animals was opened and a tube was inserted into the ascending aorta to perform perfusion of the brain. The brains were perfused *in situ* with 4% paraformaldehyde. After fixation, brains were embedded in paraffin and cut into 6-μm sections and slices were used for aRNA amplification as described by Eberwine et al[2] and Phillips and Eberwine[3].

2.4 Measurements of oxygen by the oxygen dependent quenching of phosphorescence

Cortical oxygen pressure was measured by the oxygen dependent quenching of phosphorescence[4-7]. This is a non-invasive optical method in which an oxygen sensitive phosphor (Oxyphor R2) was injected IV at approximately 22 mg/kg. The measurements were made with a frequency domain phosphorometer (PMOD 1000, Oxygen Enterprises, Ltd.). The calculated lifetime is approximately the mean for the phosphor in the tissue microvasculature and the measured oxygen pressures are in good agreement with values obtained using a time domain phosphorometer.

3. RESULTS

3.1 Effect of hypoxic-hypoxia and hypoxic-ischemia on physiological parameters and cortical oxygen pressure in newborn piglets

During hypoxic hypoxia and ischemia the pH of the arterial blood decreased at one hour from 7.42±0.01 and 7.39±0.02 to 7.23±0.04 ($p<0.001$) and 7.18±0.05 ($p<0.001$), respectively (Tables 1 and 2). At the end of the two hour recovery period from hypoxia or ischemia the blood pH increased to 7.37±0.02 and 7.36±0.02, values not significantly

Table 1. Physiological parameters in newborn piglet during hypoxia and recovery

Experimental conditions	Timc (hr)	pII	PaO_2	$PaCO_2$	Heart Rate	Blood Pressure (Torr)	PcO_2 (Torr)
Normoxia		7.42±0.01	128±6	35±3	166±9	90.8±4.8	56.8±4.3
Hypoxia	0.5	7.33±0.02[a]	22±1[b]	41±2	181±8	66.7±4.5[a]	7.0±0.6[b]
	1	7.23±0.04[b]	22±1[b]	41±3	163±9	55.8±5.0[b]	5.7±1.0[b]
Posthypoxic recovery	2	7.37±0.02	116±8	48±5	163±15	100.0±7.0	45.6±9.6

The newborn piglets (n=6) were subjected to one hour of hypoxia and 2 hours of posthypoxic recovery. The values are the means ± SEM for n experiments. [a]$p<0.05$; [b]$p<0.001$ for difference from normoxic conditions as determined by one-way analysis of variance withrepeated measures by Wilcoxon signed-rank test.

Table 2. Physiological parameters in newborn piglet during ischemia and recovery

Experimental conditions	Time (hr)	pH	PaO_2	$PaCO_2$	Heart Rate	MAB (Torr)	PcO_2 (Torr)
Normoxia		7.39±0.02*	100±6*	37.4±1.8*	173±13	86±2	47±2
Occlusion + bleeding	0.5	7.32±0.02*	76±11*	41.1±1.5*	163±13	49±10[b]	12±3[b]
	1	7.18±0.05*[b]	82±9*	39.2±2.2*	191±12	47±10[b]	9±2[b]
Postischemic recovery	2	7.36±0.02*	85±5*	38.9±1.5*	197±8	106±4	44±3

The newborn piglets (n=5) were subjected to one hour of occlusion and bleeding and 2 hours of postischemic recovery. The values are the means ± SEM for n experiments. [a]$p<0.05$; [b]$p<0.001$ for difference from normoxic condition as determined by one-way analysis of variance with repeated measures by Wilcoxon signed-rank test.
* indicates values taken from parallel experiments.

different from controls. The blood pressure decreased during hypoxia from 90.8±4.8 Torr to 55.8±14 Torr ($p < 0.001$). At the end of two hours recovery it had increased to 100±7 Torr, a value not different from control. Similar changes in blood pressure were observed during ischemia and postischemic recovery. Control oxygen pressure was between 56.8±4.3 Torr (control) and this decreased during hypoxia to 5.7±1.0 ($p<0.001$) and during ischemia to 9.6±1.5 Torr ($p<0.001$). During the recovery period, cortical oxygenation increased to values not significantly different from control values.

3.2 Effect of hypoxia and ischemia on expression of selected genes in striatum of newborn piglets

The changes in selected gene expression in striatum of newborn piglets following hypoxia or ischemia and two hours recovery are shown in Figure 1. As can be seen, hypoxia caused significant increase only in the level of mRNA for NMDAR1, by 40% ($p<0.05$). Ischemia increased levels of mRNA for NMDAR1 by 59% ($p<0.05$), DARPP-32 by 37% ($p<0.01$), CamKII by 41% ($p<0.01$) and GluR1 by 67% ($p<0.01$). The results showed that ischemic-hypoxia caused much larger alterations in genes expression in the signaling pathways tested than did hypoxic-hypoxia.

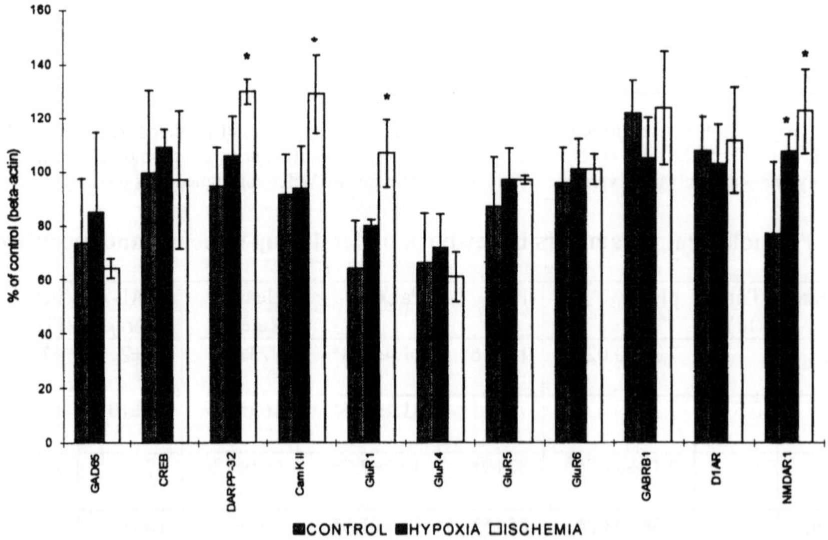

The piglets were subjected to hypoxic hypoxia or ischemic hypoxia and recovery as described in Material and Methods. The presented data are means for 5 experiments. * $p < 0.05$ for significant difference from baseline values as determined by one way analysis of variance, followed by the Wilcoxon signed-rank test.

Figure 1. Effect of hypoxic hypoxia and ischemic hypoxia on expression of selected genes in striatum of newborn piglets.

4. DISCUSSION

The oxygen dependence of the brain of newborns and adults has been extensively studied, but this work has suffered from a lack of effective methods for measuring oxygen in the tissue. Thus, despite great effort there remains much that we do not know about regulation of the vascular system and the quantitative relationships between oxygen pressure and neuronal function. The technology for optically measuring oxygen pressure, developed in our laboratory, now makes it possible to obtain more accurate oxygen measurements under physiological conditions, facilitating correlation of tissue oxygenation with other physiological/biochemical properties. The present study has shown that the changes in the gene expression induced by ischemic-hypoxia were significantly greater than by hypoxic-hypoxia, indicating that in the former there are local regions of tissue with substantially lower oxygen in the former than in the latter. This is consistent with the presence of greater oxygen extraction, and therefore larger intra-capillary gradients, under conditions with low flow (ischemic-hypoxia) as compared to conditions with high flow (hypoxic-hypoxia).

These changes in gene expression following ischemic-hypoxia can be of major importance in the development of long range dysfunction. Increases of expression of NMDAR1and GluR1 are consistant with other studies and may contribute to the excitotoxic effects of glutamate[8]. In addition to increasing glutamate receptors, ischemia also increases levels of DARPP-32 and CamKII. DARPP-32 is a neuronal phosphoprotein regulated by both glutamate and dopamine receptors. It contributes to control of the activity of protein phosphatase 1, major seroine/threonine phosphatase in brain. Increase in DARPP-32 and in CamKII expression may also reflect the changes in neuroplasticity in brain. It had been shown that CamKII can mediate plasticity at gluamatergic synapses by increasing single channel conductance of existing functional AMPA receptors or by recruiting new high conductance state AMPA receptors[9].

Thus, our data shows that ischemia produces a more complex pattern of gene expression profiles then does hypoxia with similar mean tissue oxygenation. It will be important to extend these studies with cellular analyses using histological methods to determine how the differences in gene expression correlate with neuronal and functional injury. Knowledge of causal interrelationship between extrinsic factors and intrinsic responses is a conceptual pre-condition for understanding pathogenic processes and ultimately understanding how to reverse them.

ACKNOWLEDGEMENTS

This research was supported in part by grants NS-31465 and HL-58669.

REFERENCES

1. Simon, R.P. (1999) Hypoxia versus ischemia. Neurology 5, 7-8.
2. Eberwine J., Yeh H., Miyashiro K., Cao Y., Nair S., Finnell R., Zettel M., and Coleman P. (1992) Analysis of gene expression in single live neurons. Proc. Natl. Acad. Sci. U.S.A. 89, 3010-3014.
3. Philips J. and Eberwine J.H. (1996) Antisense RNA amplification: a linear amplification method for analyzing the mRNA population from single living cells. Methods: A Companion to Methods in Enzymology 10, 283-288.

4. Wilson, D.F., Rumsey, W.L., Green, T.J., and Vanderkooi, J.M. (1988) The oxygen dependence of mitochondrial oxidative phosphorylation measured by a new optical method for measuring oxygen concentration J. Biol. Chem. 263, 2712-2718.
5. Pawlowski, M., and Wilson, D.F. (1992) Monitoring of the oxygen pressure in the blood of live animals using the oxygen dependent quenching of phosphorescence. Adv. Exptl. Med. Biol., 316, 179-185.
6. Rumsey W.L., Vanderkooi J.M., and Wilson D.F. (1988) Imaging of phosphorescence: a novel method for measuring oxygen distribution in perfused tissue. Science, 241, 1649-1651.
7. Vinogradov, S.A., Lo, L.-W., Jenkins, W.T., Evans, S.M., Koch, C., and Wilson, D.F. (1996) Non invasive imaging of the distribution of oxygen in tissue in vivo using near infra-red phosphors. Biophys. J. 70, 1609-1617.
8. Heurteaux C., Lauritzen I., Widmann C., and Lazdunski M. (1994) Glutamate-induced overexpression of NMDA receptor messenger RNAs and protein triggered by activation of AMPA/kainate receptors in rat hippocampus following forebrain ischemia. Brain Res. 659, 67-74.
9. Derkach V., Barria A., and Soderling T.R. (1999) Ca2+/calmodulin-kinase II enhances channel conductance of alpha-amino-3-hydroxy-5-methyl-4-isoxazolepropionate type glutamate receptors. Proc. Natl. Acad. Sci. U.S.A. 96(6), 3269-74.

BRAIN OXYGENATION DURING CARDIOPULMONARY BYPASS AND CIRCULATORY ARREST

[1]Gregory Schears*, Jin Shen, Jennifer Creed, Tatiana Zaitseva, David F. Wilson, William J. Greeley* and Anna Pastuszko

ABSTRACT

Quantitative measurements of oxygen distribution in the microcirculation of the brain cortex of newborn piglets were made during different modes of cardiopulmonary bypass. Three groups of animals, anesthetized and mechanically ventilated, were studied. The first group of animals were maintained on normothermic cardiopulmonary bypass (CPB) at a flow of 100ml/kg/min, while the second and third groups underwent low flow hypothermic cardiopulmonary bypass (40 ml/kg/min at 18°C) (LFCPB) and deep hypothermic (18°C) circulatory arrest (DHCA), respectively. After bypass, the piglets were monitored for a two hours post-bypass recovery period.

CPB caused a decrease in the cortical oxygen from 62 ± 3 mm Hg to 32 ± 7 mm Hg at the beginning of bypass and to 36 ± 5 mm Hg at the end of bypass. During the recovery period, cortical oxygenation steadily decreased, reaching 29 ± 8 mm Hg at the end of the experiment. With initiation of LFCPB, cortical oxygen decreased to 22 ± 7 mm Hg. Upon rewarming cortical oxygen increased to 37 ± 5 mm Hg and then decreased again to about 30 mm Hg at the end of two hours of post-bypass recovery. Similar changes in cortical oxygenation were observed during DHCA. In DHCA cortical oxygen decreased to 19 ± 4 mm Hg and during rewarming and recovery increased to 35 ± 6 mm Hg.

In conclusion, it has been shown that in newborn piglets recovering from CPB, LFCPB and DHCA, when the blood pressure remained above 55 mm Hg and therefore total blood flow should be well maintained, oxygen pressure in the microvasculature is significantly lower than for pre-bypass. It is suggested that the decreased oxygenation is due to increased heterogeneity in resistance in the microcirculatory units, resulting in broadened distribution of flow rates and oxygen levels.

---

[1] Dept. of Biochem. & Biophys., University of Pennsylvania, School of Medicine, and Dept. of Anesthesiology & Critical Care*, Children's Hospital of Philadelphia, Philadelphia, PA 19104.

Oxygen Transport to Tissue XXIII, edited by D. Wilson *et al.*
Kluwer Academic/Plenum Publishers, 2003

1. INTRODUCTION

The use of deep hypothermic cardiopulmonary bypass with and without circulatory arrest has substantially improved operating conditions for children undergoing congenital heart operations, resulting in increased survival and reduced cardiac morbidity. As overall surgical outcome has improved, neurologic dysfunction has become a more prominent concern. The pathogenesis of neuropsychologic dysfunction and brain injury has not been fully elucidated. One proposed mechanism is global impairment of cerebral blood flow and oxygenation during CPB. During cardiopulmonary bypass, factors such as hemodilution, hypocapnia, temperature, cerebral microembolization, vascular resistance, and acid-base status can impact cerebral metabolism[1-3]. However, the exact mechanisms affecting cerebral oxygenation during CPB and, consequently, their relationship to cerebral injury remain incompletely understood. An important limitation in the already published studies is the lack of direct measurements of cerebral oxygen levels during bypass and recovery.

The present study used oxygen dependent quenching of phosphorescence to examine the effects of cardiopulmonary bypass on cortical oxygen pressure in a newborn piglet model. This method was used previously to measure of oxygen distribution in brain and other tissue as the retina.[4-8]

2. METHODS

2.1 Animal model

Newborn piglets, 2-4 days of age were premedicated with ketamine (20 mg/kg IM) and acepromazine (1 mg/kg IM), intubated, and placed on mechanical ventilation (Sechrist Infant Ventilator, model IV - 100 B). Anesthesia was maintained with fentanyl (100 mcg/kg IV bolus, and 50 mcg/kg/hr infusion) and neuromuscular blockade with pancuronium (0.3 mg/kg IV). The ventilator was set with a positive inspiratory pressure of 25 mm Hg and positive and expiratory pressure of 3 mm Hg. Respiratory rate and inspiratory oxygen fraction were titrated and maintained on arterio $PaCO_2$ of 35 - 45 mm Hg and a PaO_2 of 200 - 300 mm Hg. Sodium bicarbonate (8.5 %) was used to maintain a base excess between -3 and 3 mmol/L. A jugular venous cannula and femoral arterial cannula were placed for collection of samples for measuring oxygen pressure, hemoglobin desaturation, pH and for monitoring blood pressure. The animal's temperature was monitored throughout the study by nasopharyngeal temperature probe (Yellow Spring Instrument, Inc., Yellow Spring, OH), cortical and deep brain temperature probes (Shiley, Inc., 15mm and 30mm myocardial probes). Temperature was maintained at 36°C except for the period of induced hypothermia. The head of animal was placed in a Kopf stereotaxic holder and an incision was made along the midline of the scalp. The scalp was removed to expose the skull and a hole approximately 5 mm in diameter made in the skull over one parietal hemisphere for measuring the oxygen pressure. After one hour of stabilization period CPB was performed. Sham operated animals served as the control. In all experiments, the blood pressure, body temperature and respiratory rate were continuously monitored. Blood samples were routinely taken for the measurement of blood pH, $PaCO_2$ and PaO_2 using a Beckman blood gas machine. The cortical oxygen pressure was measured at regular intervals during the experiment. All animal use procedures were in strict accordance with the NIH Guide for the Care and Use of Laboratory Animals and have been approved by the local Animal Care Committee.

2.2 CPB Technique and Experimental Protocol

The protocol and techniques used in this investigation were designed to duplicate those practiced in the clinical setting. The CPB circuit was consist of two Stocker Shiley Roller Pumps (Model 10 - 10-00, Shiley, Inc.), a membrane oxygenator and a biomedical 370 heat exchanger (Biomedicus). The pump was primed with whole donor blood and normal saline solution, to maintain a CPB hematocrit value between 20 - 25%. Oxygen and carbon dioxide flows were used to maintain blood gases within preset limits using alpha stat blood gas management. For cardiopulmonary bypass, the common carotid artery and interior jugular vein were exposed through a right neck incision. After intravenous heparin (300 units/kg) was administered, an 8 French arterial and venous cannula was inserted to the level of the aortic arch and the right atrium, respectively. CPB was then begun. Full CPB flow was set at 100 ml/kg/min. The animals were maintained on normothermic, full flow CPB for 120 min and then were weaned off CPB for 240 min. For LFCPB the bypass was begun as described above and then animals were immediately cooled to a nasopharyngeal/brain temperature of 18°C during a 20-25 minute period. When the temperature reached 18°C, the CPB circuit flow was reduced to 25-40 ml/kg/min. After 90 minutes of LFCPB, the flow was increased again to 100 ml/kg/min and the piglets rewarmed to a temperature of 36°C during a 40 min period. For DHCA the temperature was decreased to 18°C (as described above) and then the CPB circuit was turned off. After circulatory arrest (60 minutes) CPB was resumed at 100 ml/kg/min, and the piglet rewarmed as described above.

2.3 Measurements of oxygen by oxygen dependent quenching of phosphorescence

Cortical oxygen pressure was measured by the oxygen dependent quenching of phosphorescence[7, 9]. This is an optical method in which an oxygen sensitive phosphor (Oxyphor G2) was injected IV at approximately 2 mg/kg[10]. The measurements were made with a frequency domain phosphorometer (PMOD 2000, Oxygen Enterprises, Ltd.)[9]. The excitation light (635 nm) was carried to the tissue through one branch of a bifurcated light guide and illuminated an approximately 5 mm diameter area. The phosphorescence (790 nm maximum) was returned through the second branch of the bifurcated light guide, filtered to remove the excitation light and measured. The phosphorescence lifetime was determined from the phase relationship relative to the excitation light measured at a phase shift of 25 ± 2 degrees. The calculated lifetime is approximately the mean for the phosphor in the tissue microvasculature.

Statistical evaluation: All values are expressed as means for n experiments ± SEM. Statistical significance was determined using one-way analysis of variance with repeated measures by Wilcoxon signed-rank test. $p < 0.05$ was considered statistically significant.

3. RESULTS

3.1 Effect of normothermic CPB on cortical oxygen pressure

The changes in cortical oxygen pressure during normothermic bypass and post-bypass recovery are shown in Fig.1. The cortical oxygen decreased from 62 ± 3 mm Hg to 32 ± 7 mm Hg ($p<0.001$) at the beginning of bypass and to 36 ± 5 mm Hg ($p<0.001$) at the end of bypass. During the recovery period, cortical oxygenation steadily decreased, reaching 29 ± 8 mm Hg ($p<0.001$) at the end of the experiments.

The blood pressures was significantly decreased compared to pre-bypass values. It decreased from the pre-bypass value of 104 ± 3 mm Hg to between 70-80 mm Hg ($p<0.001$) during normothermic bypass and further to about 60 mm Hg ($p<0.001$) during post-bypass recovery.

3.2 Effect of LFCPB on cortical oxygen pressure

Cortical O_2 decreased from baseline with the initiation of LFCPB (Fig. 2). Pre-bypass cortical pressure was 61 ± 4 mm Hg and decreased on bypass, before cooling, to 29 ± 5 mm Hg ($p<0.001$) and further decreased during the cooling to 22 ± 7 mm Hg ($p<0.001$). During rewarming cortical oxygen increased to 37 ± 5 mm Hg, but was still statistically significant below pre-bypass value ($p<0.001$) and then decreased to about 30 mm Hg and stayed at this level during the entire recovery period. Changes in cortical oxygenation were generally parallel to changes in the arterial blood pressure.

During LFCPB, blood pressure decreased significantly from 109 ± 4 mm Hg to 62 ± 14 mm Hg ($p<0.001$) before cooling, to 46 ± 8 mm Hg ($p<0.001$) on cooling and to 68 ± 12 mm Hg ($p<0.001$) after cooling. During warming, blood pressure increased to 71 ± 5 mm Hg, a value statistically significant lower than pre-bypass ($p<0.001$) and then, during post bypass recovery, steadily decreased to 57 ± 3 mm Hg ($p<0.001$) by the end of the experiment.

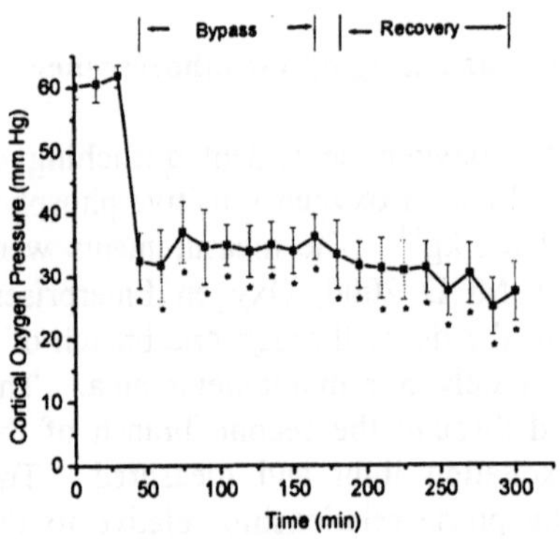

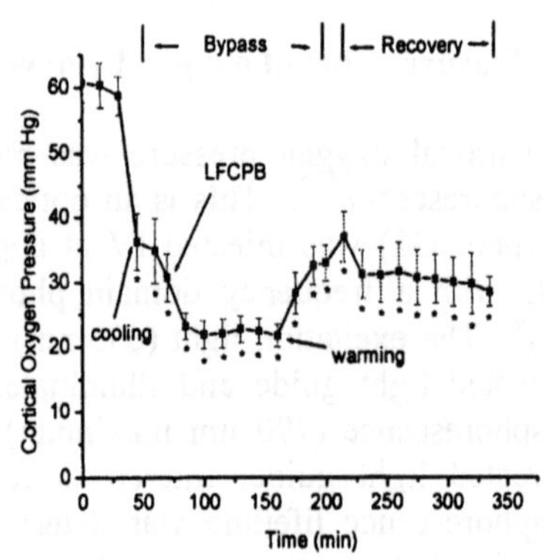

Figure 1. Effect of normothermic cardiopulmonary bypass and post-bypass recovery on cortical oxygen pressure in newborn piglets. The results are expressed as the mean ± SEM for 7 experiments.
* $p < 0.001$ for significant difference from pre-bypass values as determined by one-way analysis of variance, followed by the Wilcoxon signed-rank test.

Figure 2. Effect of LFCPB and post-bypass recovery on cortical oxygen pressure in newborn piglets. The results are expressed as the mean ± SEM for 7 experiments. * $p < 0.001$ for significant difference from pre-bypass values as determined by one way analysis of variance, followed by the Wilcoxon signed-rank test.

3.3 Effect of hypothermic circulatory arrest on cortical oxygen pressure

Cortical oxygen decreased during DHCA from 61.9 ± 4 mm Hg to 19.1 ± 4 mm Hg ($p < 0.001$) and during rewarming and recovery increased again to 34.8 ± 9 mm Hg ($p < 0.001$) (Fig. 3). The blood pressure during DHCA decreased from 102 ± 13.1 to 18.3 ± 2.5 mm Hg ($p < 0.001$). During rewarming and the two hours of recovery the blood pressure increased again to 53 ± 5 mm Hg ($p < 0.001$).

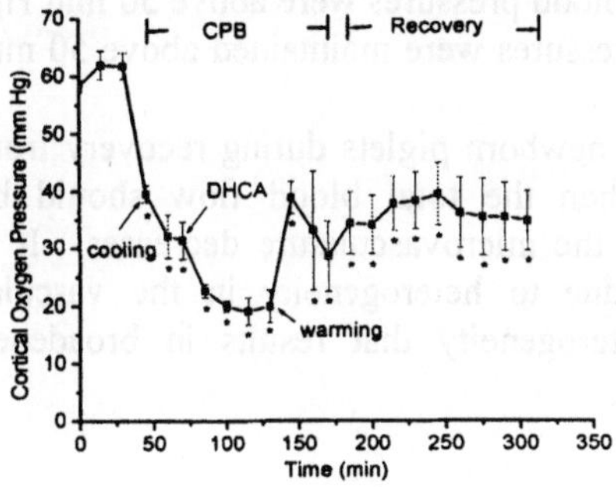

Figure 3. Effect of DHCA post-bypass recovery on cortical oxygen pressure in newborn piglets. The results are expressed as the mean ± SEM for 6 experiments. * $p < 0.001$ for significant difference from pre-bypass values as determined by one way analysis of variance, followed by the Wilcoxon signed-rank test.

4. DISCUSSION

The present study investigated the behavior of cortical oxygen pressure in newborn piglets during three different modes of cardiopulmonary bypass. It has been shown that normothermic bypass, LFCPB and DHCA all cause a decrease in cortical oxygenation. During LFCPB and DHCA, cortical oxygen decreased much more than during normothermic bypass. However, the recovery of cortical oxygenation from DHCA is significantly better than for the other two models. The changes in cortical oxygenation during LFCPB and DHCA are parallel to changes in blood pressure. The decrease in blood pressure immediately after going on bypass during LFCPB and DHCA is similar to that seen clinically. It is probably due to the effects of hemodilution and of nonpulsatile flow coupled with vasodilation due to substances released in response to exposure of the blood to non-endothelialized surfaces of the CPB circuit. Similar to LFCPB and DHCA, normothermic CPB caused a decrease in blood pressure and decrease of cortical oxygenation. However, the blood pressure during normothermic bypass was above 50 mm Hg, the lower limit for blood flow remaining constant (autoregulation) in newborn piglets[11-13]. The observation that, during normothermic bypass, oxygen pressure in the cortex decreases before there is any significant decrease in blood flow raises some important questions concerning the mechanism of regulation of the microcirculation in the brain. This finding could be explained if the total flow through the tissue were maintained, but flow in the microvasculature became more unevenly distributed, with a fraction of the vessels in the microcirculatory unit becoming much more dilated than others. This would allow total blood flow to be maintained while in local regions of the vasculature, the flow and thereby the oxygen pressure, would be significantly decreased. If both the patterns of flow and the total flow remained unchanged during autoregulation, the oxygen pressure at each point in the vasculature would remain unaltered, leaving no viable sensory system for regulating vasculature resistance. On the other hand, if decreasing blood pressure resulted in an unequal vasodilatation among the vessels, total flow could be maintained but the oxygen pressure in some vessels would decrease, the lower flow vessels providing the signal used to determine vasculature resistance. Our data suggests that measurements of microvessel blood flow or oxygen distributions (histograms) for the tissue vasculature would become broader as the blood pressure is lowered, with an increasing fraction at lower flow velocity and oxygen pressures.

The same phenomenon could explain the observation that cortical oxygen pressures during post-bypass recovery from LFCPB and DHCA were significantly lower, as compared to pre-bypass conditions, although the blood pressures were above 50 mm Hg. During post-bypass recovery the arterial blood pressures were maintained above 50 mm Hg with saline/packed red blood cells.

In conclusion, it has been shown that in the newborn piglets during recovery from normothermic bypass, LFCPB and DHCA, when the total blood flow should be reasonably well maintained, oxygen pressure in the microvasculature decreases. It is suggested that the decreased oxygenation is due to heterogeneity in the vascular resistance of the microcirculatory units, a heterogeneity that results in broadened distribution of flow rates.

ACKNOWLEDGEMENTS

Supported in part by grants HL-58669 and NS-31465.

REFERENCES

1. Nollert G, Mohnle P, Tassani-Prell P, Reichart B. Determinants of cerebral oxygenation during cardiac surgery. Circulation 1995; 92: 327-333.
2. Tanaka J, Shiki K, Asou T, Yasui H, Tokunaga K. Cerebral autoregulation during deep hypothermic nonpulsatile cardiopulmonary bypass with selective cerebral perfusion in dogs. J Thorac Cardiovasc Surg 1988; 95: 124-132.
3. Undar A, Andropoulos DB, Fraser CD Jr. Comparison of pH-stat and alpha-stat cardiopulmonary bypass on cerebral oxygenation and blood flow in relation to hypothermic circulatory arrest in piglets. Anesth 1999; 90: 926-927.
4. Huang ChCh, Lajevardi NS, Tammela O, Pastuszko A, Delivoria-Papadopoulos M, Wilson DF. Relationship of extracellular dopamine in striatum of newborn piglets to cortical oxygen pressure. Neurochem Res 1994; 19: 649-55.
5. Pastuszko A. Metabolic response of dopaminergic system during hypoxia in newborn brain. Biochem Med Metab Biol 1994; 51: 1-15.
6. Pastuszko A, Lajevardi SN, Chen J, Tammela O, Wilson DF, Delivoria-Papadopoulos M. Effects of graded levels of tissue oxygen pressure on dopamine metabolism in striatum of newborn piglets. J Neurochem 1993; 60: 161-6.
7. Wilson DF, Pastuszko A, DiGiacomo JE, Pawlowski M, Schneiderman R, Delivoria-Papadopoulos M. Effect of hyperventilation on oxygenation of the brain cortex of newborn piglets. J App Physiol 1991; 70: 2691-6.
8. Shonat RD, Wilson DF, Riva CE, Pawlowski M. Oxygen distribution in the retinal and choroidal vessels of the cat as measured by a new phosphorescence imaging method. Applied Optics 1992; 31: 3711-18.
9. Vinogradov SA, Fernandez-Seara MA, Wilson DF. Frequency domain instrument for measuring phosphorescence lifetime distributions in heterogeneous samples. Rev Sci Inst 2001; 72: 3396-3406.
10. Vinogradov SA, Lo LW, Wilson DF. Dendritic polyglutamic porphyrins: Probing porphyrin protection by oxygen dependent quenching of phosphorescence. Chem Eur J 1999; 5: 1338-47.
11. Yonetani M, Huang ChCh, Lajevardi SN, Pastuszko A, Delivoria-Papadopoulos M, Wilson DF. Effect of hemorrhagic hypotension on extracellular level of dopamine , cortical oxygen pressure and blood flow in brain of newborn piglets. Neuroscien Lett 1994; 180: 247-52.
12. Armstead WM, Leffler CW, Busija DW, Beasley DG, Mirro R. Adrenergic and prostanoid mechanisms in control of cerebral blood flow in hypotensive newborn pigs. Am J Physiol 1988; 254: H671-7.
13. Armstead WM, Mirro R, Busija DW, Leffler ChW. Opioids and Prostanoid System in the Control of Cerebral Blood Flow in Hypotensive Piglets. J Cereb Blood Flow Metab 1991; 11: 380-7.

EXPRESSION OF ANGIOPOIETIN-1 AND –2 IN THE RAT BRAIN DURING CHRONIC HYPOXIA AND DE-ADAPTATION

Paola Pichiule and Joseph C. LaManna

1. INTRODUCTION

Angiogenesis is defined as the formation of new vascular networks from pre-existing blood vessels. The mature vascular network of the adult is relatively stable and vascular endothelial cells exist in a differentiated, quiescent state and rarely proliferate. Angiogenic remodeling occurs only in certain situations, hypoxia and ischemia being primary inducers of angiogenesis. In the brain, angiogenesis is a critical compensatory strategy for brain adaptation to chronic hypoxia, which presumably results in shorter intercapillary diffusional distances and improvement of tissue oxygenation[1]. This adaptive response is reversible upon return to normoxia, although the mechanism controlling vascular regression is not known[2].

A variety of angiogenic factors have been identified, among them, vascular endothelial growth factor (VEGF) and angiopoietins act specifically on endothelial cells [3]. Angiopoietins include at least two secreted ligands for the endothelial cell tyrosine kinase receptor, Tie-2. Angiopoietin-1 (Ang1) is an agonistic ligand that induces Tie-2 activation whereas angiopoietin-2 (Ang2) binds to this receptor but does not stimulate its phosphorylation and may be an endogenous antagonist[4].

The actions of the angiopoietins appear to be different from those of VEGF. Analysis on cultured endothelial cells revealed that Ang1 did not elicit mitogenic responses, although it could stimulate endothelial cell survival, migration, and capillary-like tubule formation[5-8]. Ang1 appears to be widely expressed in adult tissues and Tie-2 has been found constitutively phosphorylated in quiescent endothelium[9]. On the other hand, Ang2 in adults is primarily expressed at sites of vascular remodeling[4].

Paola Pichiule and Joseph.C. LaManna, Case Western Reserve University, Department of Anatomy, Cleveland, OH 44106-4938; PHONE 216-368-1112; FAX 216-368-1144.

In contrast to the large information regarding the control of VEGF expression by metabolic stress such as hypoxia, much less is known about the molecular mechanism regulating the expression of angiopoietins. In this study, we investigated brain expression of Ang1 and Ang2 during chronic hypoxia. Considering that Ang2 has been implicated not only in angiogenesis but also in vascular regression, we also studied the expression of angiopoietins during normoxic recovery after chronic hypoxia (de-adaptation).

2. MATERIAL AND METHODS

2.1. Exposure to chronic hypoxia

Male Wistar rats (2-3 months) were exposed to hypoxia for periods up to 3 weeks in hypobaric chambers maintained at a pressure of 380 Torr (0.5 ATM, equivalent to 10% normobaric oxygen). Normoxic controls were kept outside the chambers but in the same location. When animals were kept for more than one day, chambers were opened for cage cleaning and food and water replenishment. A group of rats that were initially adapted for 3 weeks to hypobaric hypoxia were allowed to recover at normoxia for up to 7 days.

2.2. Immunoprecipitation and Western blot analysis of Ang1 and Ang2

Anesthetized animals were decapitated and brains were rapidly removed. Cortex was dissected and homogenized in ice-cold lysis buffer (1x PBS, 1% Nonidet P-40, 0.5% sodium deoxycholate, 0.1% SDS) supplemented with protease inhibitors (1 µg/ml leupeptin, 0.5µg/ml aprotinin, 1.5 µg/ml pepstatin, 0.1 mg/ml PMSF). Homogenates were centrifuged (10000g x10 min, 4°C) and supernatants stored at –80°C. Protein content in the supernatants was determined by Bradford protein assay with bovine serum albumin as standard (Bio-Rad).

Tissue lysates (200 µg) was incubated with either goat anti-Ang1 or anti-Ang2 IgG (Santa Cruz Biotech.) and rocked overnight at 4°C. Subsequently, 10 µl of Protein G-agarose was added, and the sample incubated for another 2h at 4°C. Pellet was collected by centrifugation (1,000xg for 5 min), washed with PBS, and boiled for 3 min in Laemmli sample buffer. The samples were separated on SDS-PAGE (7.5%; reducing conditions), proteins were transferred electrophoretically to nitrocellulose membrane and probed with indicated primary antibodies. This is followed by incubation with corresponding horseradish peroxidase-conjugated secondary antibody. Detection is performed through enhanced chemiluminescense reaction (ECL, Amersham).

2.3. Ang2 immunohistochemistry

Rats were deeply anesthetized and perfused transcardially with ice-cold phosphate buffered saline (pH 7.4) followed by 4% phosphate-buffered paraformaldehyde. Brains were removed, post-fixed in 2% paraformaldehyde for 24 hours and embedded in paraffin. Sections (6µm) were deparaffinized, hydrated and subjected to antigen retrieval at 90 °C for 15 minutes using Target Retrieval Solution (Dako). Subsequently, sections were incubated with 10% normal horse serum for 2 h and incubated overnight with goat polyclonal antibody against Ang2 (1:200, Santa Cruz). Immunodetection was accomplished by incubation with a biotinylated anti-goat antibody (1:250) and then with avidin biotin peroxidase complex (Vector Lab).

3. RESULTS

Expression of Ang1 and Ang2 in the brain cortex of rats exposed to prolonged hypoxia was analyzed by immunoprecipitation followed by Western blot analysis. In accordance with previous reports, our data showed basal expression of Ang1 protein in normoxic brain. Moreover, chronic hypoxia did not affect Ang1 expression. In contrast, no detectable levels of Ang2 protein were found in normoxic brain but it was strongly induced during chronic hypoxia. This increase was detected after 12 hours of hypoxic exposure, remained elevated for up to 14 days of hypoxia but returned to pre-hypoxic values at 3 weeks of hypoxia (Fig 1).

When rats were returned to normoxia for 3 weeks following 3 weeks of hypobaric hypoxia, capillary density returned to pre-hypoxic baseline density. In order to elucidate the mechanism involved in this process of capillary regression, we analyzed the expression levels of Ang1 and Ang2 during de-adaptation phase. Our results showed that Ang2 but not Ang1 expression was rapidly induced during de-adaptation and this induction lasted for at least 7 days (Fig 1).

In agreement with our immunoblot results, Ang2 expression was not detectable by immunohistochemistry neither in control nor in 21 days-hypoxia brain sections. At 7 days of hypoxia and 7 days of de-adaptation, Ang2 expression was found to be associated with vessels (Fig 2).

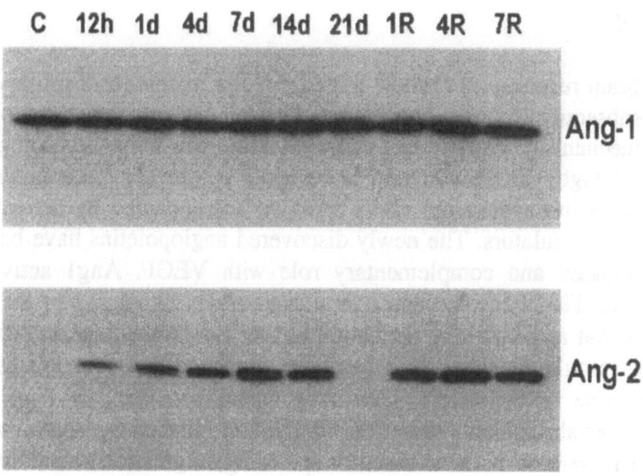

Figure 1. Western Blots showing Ang1 and Ang2 protein levels in brain cortex of rats subjected to normoxia (C), hypoxia (12 hours to 21 days) or hypoxia followed by 1, 4 and 7 days of normoxic recovery (R).

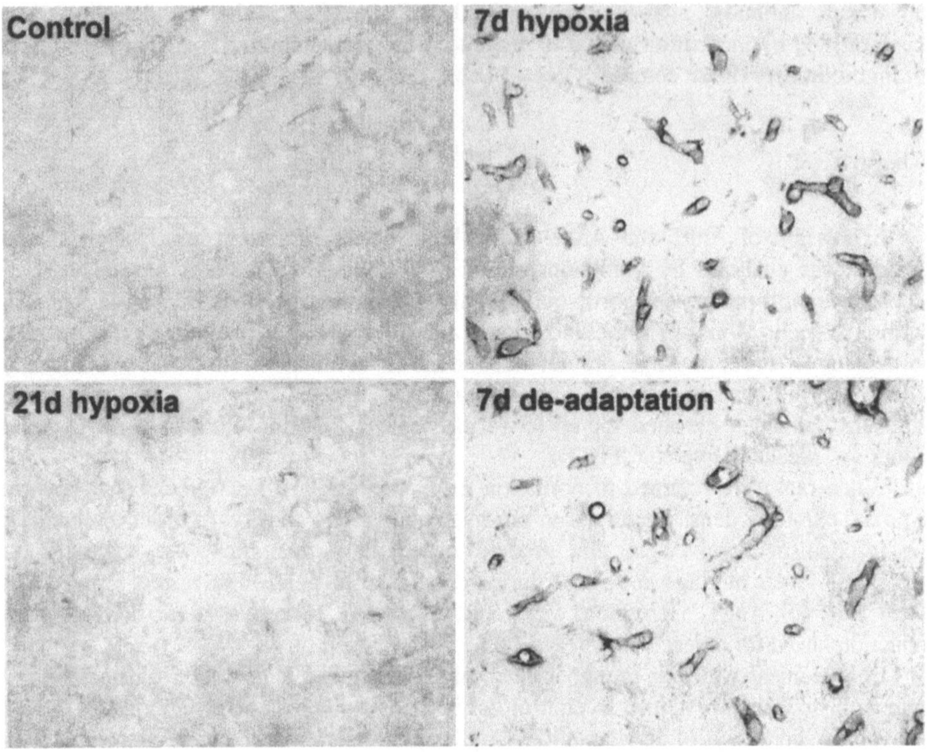

Figure 2. Immunohistochemical localization of Ang-2 in the brain cortex of rats exposed to normoxia, hypoxia (7 and 21 days) or hypoxia (21 days) followed by 7 days of normoxic recovery (de-adaptation).

4. DISCUSSION

A notable brain response to chronic hypoxia is the increase in capillary density that is reversed by subsequent return to normoxia. This suggests the existence of an intrinsic physiological mechanism controlling microvascular plasticity according to oxygen availability. Although VEGF induction is certainly a relevant mechanism in hypoxia-driven angiogenesis, its angiogenic effect must be accompanied by activation of other positive angiogenic regulators. The newly discovered angiopoietins have been suggested to play a coordinated and complementary role with VEGF. Ang1 activates tyrosine phosphorylation of Tie-2 receptor and it is constitutively expressed in the adult tissue, thereby it seems that a constitutive activation of the Tie-2 signaling pathway is required for maintaining vascular integrity[5,6]. It has been proposed that the expression of Ang2 blocks an otherwise constitutively stabilizing signal provided by Ang1, leading to localized vessel destabilization. Vessel destabilization induced by Ang2 may contribute to either vessel regression or new vascular sprouting; regression seems to occur when Ang2 is expressed in the absence of VEGF, although new sprouting occurs when Ang2 is expressed together with VEGF[10]. In agreement with this hypothesis, our results showed that Ang2 protein expression but not Ang1 was transiently induced during chronic hypoxia in the brain cortex. This is consistent with *in vitro* data showing that hypoxia and VEGF induced Ang2 but not Ang1 or Tie-2 expression in cultured endothelial cells[11]. It was previously reported that cerebral capillary density was increased by 33% after 1 week of hypobaric hypoxia and by 70% at 3 weeks[2]. In addition, VEGF protein expression was increased at 1-14 days of hypobaric hypoxia[12]. Thus, Ang2 upregulation occurred at times when angiogenesis was occurring and VEGF expression was increased. In addition, Ang2 protein expression was also induced during the de-adaptation phase when vessel regression was occurring. Further studies are required to establish when during de-adaptation Ang2 induction subsides and if this correlates with the return of vessel density to normoxic values.

In conclusion, the temporal induction of Ang2 expression is compatible with vascular remodeling involving either angiogenesis or capillary regression. These results suggest that the vascular effect of Ang2 activity depends on its interaction with other angiogenic regulators.

Acknowledgments

We thank Sue Foss for editing this manuscript. This work was supported by NINDS grants NS-38632 and NS-37111

References

1. J.C. LaManna and S.I Harik, Brain metabolic and vascular adaptations to hypoxia in the rat, in: Oxygen Transport to Tissue XIX, edited by Harrison and Delpy (Plenum Press, New York, 1997), pp163-167.
2. N. Harik, S.I Harik, N.T Kuo, K. Sakai, R.J. Przybylski, J.C. LaManna, Time-course and reversibility of the hypoxia-induced alterations in cerebral vascularity and cerebral capillary glucose transporter density, Brain Res **737**, 335-338 (1996).
3. N.W. Gale and G.D. Yancopoulos, Growth factors acting via endothelial cell-specific receptor tyrosine kinases: VEGFs, Angiopoietins, and ephrins in vascular development, Genes & Development **13**, 1055-1066 (1999).
4. P.C. Maisonpierre, C. Suri, P.F Jones, S. Bartunkova, S.J. Wiegand, C. Radziejewski , D. Compton, J. McClain, T.H. Aldrich, N. Papadopoulos, T.J Daly, S. Davis, T.N. Sato, G.D. Yancopoulos, Angiopoietin-2, a natural antagonist for Tie-2 that disrupts in vivo angiogenesis, Science **277**, 55-60 (1997).
5. S. Davis, T.H. Aldrich, P.F. Jones, A. Acheson, DL Compton, V. Jain, T.E. Ryan, Bruno J, C. Radziejewski, P.C. Maisonpierre, G.D. Yancopoulos, Isolation of angiopoietin-1,a ligand for the Tie2 receptor, by secretion-trap expression cloning, Cell **87**, 1161-1169 (1996).
6. A. Papapetropoulos, D. Fulton, Mahboubi K, R.G. Kalb, D.S. O'Connor, F. Li, D.C. Altieri, W.C. Sessa, Angiopoietin-1 inhibits endothelial cell apoptosis via the Akt/Survivin pathway, J. Biol. Chem **275,** 9102-9105 (2000).
7. B. Witzenbichler, P.C. Maisonpierre, P. Jones, G.D. Yancopoulos, J.M.Isner, Chemotactic properties of angiopoietin-1 and –2, ligands for the endothelial-specific receptor tyrosine kinase Tie2, J. Biol. Chem. **273,** 18514-18521(1998).
8. T.I. Koblizek, C.Weiss, G.D. Yancopoulos, U. Deutsch, W. Risau, Angiopoietin-1 induces sprouting angiogenesis in vitro, Curr Biol **8,** 529-532 (1998).
9. A.L. Wong, Z.A. Haroon, S. Werner, M.W. Dewhirst, C.S. Greenberg, K.G. Peters, Tie2 expression and phosphorylation in angiogenic and quiescent adult tissues, Circ Res **81,** 567-574 (1997).
10. J. Holash, P.C. Maisonpierre, D. Compton, P. Boland, C.R. Alexander, D. Zagzag, G.D. Yancopoulos, S.J. Wiegand, Vessel cooption, regression, and growth in tumors mediated by angiopoietins and VEGF, Science **284**,1994-1998 (1999).
11. O. Hideyasu, H. Takagi, K. Suzuma, A. Otani, M. Matsumura, Y. Honda, Hypoxia and Vascular endothelial growth factor selectively upregulate angiopoietin-2 in bovine microvascular endothelial cells, J. Biol. Chem. **274,** 15732-15739(1999).
12. J.C Chavez, F. Agani, P. Pichiule, J.C LaManna, Expression of hypoxia-inducible factor-1α in the brain of rats during chronic hypoxia, J. Appl. Physiol **89**,1937-1942 (2000).

HYPOXIA-INDUCIBLE FACTOR-1α ACCUMULATION IN THE RAT BRAIN IN RESPONSE TO HYPOXIA AND ISCHEMIA IS ATTENUATED DURING AGING

Juan C. Chavez and Joseph C. LaManna

1. INTRODUCTION

A common feature of senescent organisms is the decline in their ability to respond to different types of stress such as hypoxia or ischemia. Previously it was reported that aging was associated with a defect in compensatory angiogenesis in response to tissue ischemia[1]. This deficit was associated to a reduction of vascular endothelial growth factor (VEGF) expression, an endothelial-specific mitogen that is essential for angiogenesis during development and adulthood [2]. VEGF as well as other hypoxia-responsive genes share a common regulatory pathway. They are regulated by the transcription factor hypoxia inducible factor-1 (HIF-1), a heterodimeric protein consisting of two subunits, HIF-1α and HIF-1β. Both subunits are required for DNA binding and transactivation of target genes [3]. HIF-1α is unique to HIF-1 and its expression is primarily regulated by oxygen tension. During normoxia the HIF-1α gene is expressed continuously; however, the HIF-1α protein is rapidly degraded by the ubiquitin-proteosome system[4,5,6]. During hypoxia, degradation of HIF-1α is suppressed and thereby it accumulates in the nucleus almost instantly. In contrast, the HIF-1β subunit is constitutively expressed and its levels do not change significantly during hypoxia[3].

In this study, we asked if the brain response to hypoxia and transient global ischemia is impaired as a function of age. Our results indicate that in old animals HIF-1α accumulation in the brain in response to hypoxia and ischemia is impaired. Such reduced activation of HIF-1α might be responsible for the age-dependent increase in susceptibility of the brain to hypoxic-ischemic stress.

Juan C. Chavez and Joseph C. LaManna, Case Western Reserve University, Department of Anatomy, Cleveland, OH 44106-4938 Phone 216-368-1112 Fax 216-368-1144

Oxygen Transport to Tissue XXIII, edited by D. Wilson *et al.*
Kluwer Academic/Plenum Publishers, 2003

2. MATERIAL AND METHODS

2.1 Induction of Transient Global Ischemia by Cardiac Arrest and Resuscitation

Transient global cerebral ischemia was produced by a modification of the cardiac arrest model described by Crumrine and LaManna[7]. Male Fisher 344 rats (6 or 24 months old) were anesthetized with 2.5% halothane in a gas mixture of 70% nitrous oxide and 30% oxygen. The ventral tail artery was cannulated to monitor systemic arterial blood pressure and to obtain blood samples for blood gases and pH measurements. In addition, a catheter was inserted through the external jugular vein into the right atrium. Body temperature was maintained at 37°C by using a feedback controlled infrared heat lamp. Animals were allowed to recover completely from anesthesia, for at least 1 h, before the induction of cardiac arrest. Cardiac arrest was induced by sequential intra-atrial injection of D-tubocurare (0.3mg) and ice-cold potassium chloride solution (0.5 M, 0.12 ml/100 g of body weight). Resuscitation efforts began after 7 min of arrest. For this purpose rats were orotracheally intubated for mechanical ventilation accompanied by chest compression. Once a spontaneous heartbeat returned, a small dose of epinephrine (2 μg) was administered in order to achieve a mean arterial blood pressure of at least 80 mmHg.

The duration of ischemia was between 10 and 12 minutes and it was defined as the period between the decrease of blood pressure to zero and its return to 80% of pre-arrest value. Ventilation was adjusted to achieve normoxia and normocapnia until rats regained spontaneous respiration. In the sham-operated animals, the same surgical procedure was performed without cardiac arrest and resuscitation. Rats were sacrificed after 3 hours of recovery from cardiac arrest, the brains were rapidly removed and stored at –80°C.

Exposure to hypoxia

Male Fisher 344 rats (6 and 24 months old) were exposed to hypoxia (8% O_2, balance N2) for 4 hours in a plexiglass chamber. Controls were kept at normoxic conditions.

Western Blot Analysis

Brain cortical samples were homogenized in ice-cold lysis buffer (20mM HEPES, pH 7.5, 1.5 mM $MgCl_2$, 0.2mM EDTA, 0.1 M NaCl) supplemented with 0.2 mM dithiothreitol, and protease inhibitors (1 μg/ml leupeptin, 0.5 μg/ml aprotinin, 1.5 μg/ml pepstatin, 0.5mM PMSF). After addition of NaCl to a final concentration of 0.45 M, homogenates were centrifuged at 20000x g for 15 minutes at 4°C. Supernatants were collected and mixed with equal volume of homogenization buffer containing 40% (v/v) glycerol. Protein concentrations were determined by Bradford protein assay with bovine serum albumin as standard (Bio-Rad). Brain cortical lysates (200 μg of protein) were electrophoresed on 7.5% SDS-PAGE gels (HIF-1α and HIF-1β) under reducing conditions and transferred to nitrocellulose membrane by standard procedures. Membranes were blocked with 10% nonfat dry milk for 2 hours and incubated with HIF-1α mouse monoclonal antibody (Novus). After washing, membranes were incubated with the appropriate secondary horseradish peroxidase-conjugated antibodies. Antigen-antibody complexes were visualized by enhanced chemiluminescence detection (ECL,

Amersham). Crude nuclear extracts of Hep3B cells (ATTC, CRL-1830) exposed to 1% or 20% oxygen were used as positive controls (30 μg of protein) in HIF-1α Western Blot analysis.

3. RESULTS

3.1. Physiological Variables

The major pathophysiological effects of cerebral global ischemia induced by cardiac arrest and resuscitation have been described before[7]. These include an initial arterial acidosis, hypertension and hemoconcentration that returned to pre-ischemic levels between 30 and 180 min of reperfusion. In addition, a decreased cerebral blood flow during the first hours of recovery was reported[7]. Rats regained spontaneous respiration within 3 hours after resuscitation and regained consciousness before 12h. Usually, resuscitated young rats are able to move and feed themselves by 24 hours. In contrast to young animals, old rats do not survive more than 4 days after cardiac arrest and resuscitation.

3.2. HIF-1α Western Blot Analysis

Western blot analysis showed that HIF-1α migrates with a characteristic diffuse pattern that may correspond to posttranslational modifications of this protein. HIF-1α was hardly detected in brain cortical samples of normoxic or sham-operated animals both in young and old animals. Hypoxia (10% oxygen for 4 hours) and transient global cerebral ischemia induced HIF-1α accumulation in the brain cortex of young rats.

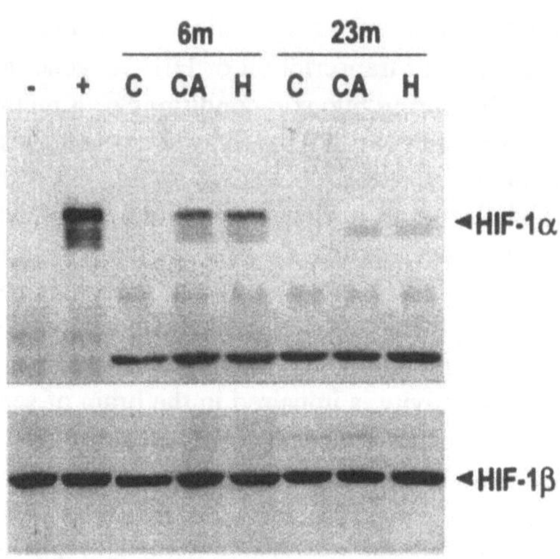

Figure 1. Western Blot showing HIF-1α and HIF-1β in the cerebral cortex of young (6m) and old (24m) F344 rats exposed to normoxia (C) 10% oxygen for 4 hours (H) or subjected to cardiac arrest and resuscitation (CA).

Conversely, HIF-1α accumulation after either hypoxia or ischemia was significantly reduced in 24 month rats compared to 6 month rats (Figure 1). The constitutive expression of HIF-1β in the cerebral cortex was not affected by either hypoxia or ischemia in young and old animals.

4. DISCUSSION

In this study we found a significant induction of HIF-1α in the cerebral cortex of young F344 rats exposed to a mild hypoxia or subjected to cardiac arrest and resuscitation. In contrast, HIF-1α accumulation was reduced in the cerebral cortex of old rats exposed to the same treatments. Our results suggest that HIF-1 activation in the brain in response to hypoxia or ischemia is attenuated as a function of age. Considering that HIF-1 regulates the activation of genes involved in oxygen homeostasis, its deficient activation might be responsible for an increased susceptibility of senescent organisms to a hypoxic/ischemic stress. For instance, a reduced expression of VEGF as a result of impaired HIF-1 activation during aging might preclude the formation of new vessels necessary for compensating insufficient tissue oxygenation. In addition, recently it has been shown that VEGF might have a neurotrophic role in the brain allowing survival of neurons during oxygen deprivation[8,9].

Although there are no reports describing the pathophysiological consequences of cardiac arrest in old rats, it is likely that changes similar to those described in young rats occur. In this model approximately a 50% survival rate of young rats is achieved after 24 hours of recovery. In contrast, old rats have a much higher mortality rate. Almost all animals die within 24 hours after cardiac arrest. Whether the impaired activation of HIF-1 in response to hypoxia/ischemia is the cause of the high mortality rate remains to be determined.

Potential mechanisms by which aging could lead to impaired HIF-1 response may include a decrease in the rate of transcription of HIF-1α gene, a reduced rate of HIF-1α protein synthesis especially during hypoxic conditions or a mutation of the HIF-1α gene that might prevent its expression. Alternatively, the oxygen sensing mechanism required for HIF-1 activation might be impaired during aging. Recently it was suggested that mitochondria might have a role in oxygen sensing and HIF-1 activation during hypoxia. Although the model of mitochondria as part of the oxygen sensing mechanism is controversial[10-13], it is tempting to speculate that mitochondrial impairment occurring during aging can potentially affect the oxygen sensing process and consequently affect the activation of HIF-1. Our findings are in agreement with a previous report showing that during hypoxia HIF-1 activity is impaired in the brain of senescent mice as indicated by a decrease in its DNA binding capacity[14]. Moreover, HIF-1 activity was also found attenuated in an in vitro model of aging[15]. Regardless of the mechanism responsible for a deficient HIF-1 response in the aged rat brain, our results as well as those reported before suggest that the lack of an adequate response to hypoxia might be a general feature of the aging process. Overcoming this deficiency might increase the resistance of old organisms to any condition that involves hypoxia.

Acknowledgments

We thank Sue Foss for her technical expertise and editing this manuscript. This work was supported by NINDS grants NS-38632 and NS-37111

References

1. Rivard, J-E. Fabre, M. Silver, D. Chen, T. Murohara, M. Kearney, M. Magner, T. Asahara, J. Isner, Age-dependent impairment of angiogenesis, Circulation **99**, 111-120 (1999).
2. G. Neufeld, T. Cohen, S. Gengrinovitch, Z. Poltoral, Vascular endothelial growth factor (VEGF) and its receptors, FASEB J. **12**, 9-22 (1999).
3. G.L. Wang, B-H Jiang, E.A. Rue, G.L. Semenza, Hypoxia inducible factor-1 is a basic-helix-loop-helix-PAS heterodimer regulated by cellular O2 tension, Proc. Natl. Acad. Sci. USA **92**, 5510-5514 (1995).
4. L.E.Huang, J. Gu, M. Schau, F.H. Bunn, Regulation of hypoxia-inducible factor 1α is mediated by an O_2-dependent degradation domain via the ubiquitin-proteosome pathway, Proc. Natl. Acad. Sci. USA **95**, 7987-7992 (1998).
5. S. Salceda, J. Caro, Hypoxia-inducible factor 1α (HIF-1α) protein is rapidly degraded by the ubiquitin-proteosome system under normoxic conditions, J. Biol Chem **272**, 22642-22647 (1997).
6. U.R. Jewell, I Kvietikova, A. Scheid, R. H. Wenger, M. Gassmann. Induction of HIF-1α in response to hypoxia is instantaneous. FASEB J. **15**, 1312-1314 (2001).
7. C. Crumrine J.C. LaManna Regional cerebral metabolites, blood flow, plasma volume, and mean transit time in total cerebral ischemia in the rat, J Cereb Blood Flow Metab. **11**, 272-282 (1991).
8. K.L. Jin, X.O. Mao, D.A. Greenberg. Vascular endothelial growth factor: direct neuroprotective effect in in-vitro ischemia, Proc.Natl. Acad. Sci. USA **97**, 10242-10247 (2000).
9. M. Sondell, G. Lundborg, M. Kanje. Vascular endothelial growth factor has neurotrophic activities and stimulates axonal autogrowth, enhancing cell survival and Schwann cell proliferation in the peripheral nervous system. J Neurosci **19**, 5731-5740 (1999)
10. N.S. Chandel, D.S. McClintock,C.E. Feliciano, T.M. Wood, J.A. Melendez, A. M. Rodrigues, PT. Schumacker. Reactive oxygen species generated at mitochondrial complex III stabilize hypoxia-inducible factor 1a during hypoxia. J Biol Chem **275** 25130-25138 (2000)
11. F. H. Agani, P. Pichiule, J.C. Chavez, JC. LaManna. The role of mitochondria in the regulation of hypoxia-inducible factor 1 expression during hypoxia. J Biol Chem **275**, 35863-35867 (2000).
12. E. C. Vaux, E. Metzen, K. M. Yeates, P. J. Ratcliffe. Regulation of hypoxia inducible factor is preserved in the absence of a functioning mitochondrial respiratory chain. Blood **98**, 296-302 (2001).
13. V. Srinivas, I. Leshchinsky, N. Sang, M.P. King, A. Minchenko, J. Caro. Oxygen sensing and HIF-1 activation does not require an active mitochondrial respiratory chain electron-transfer pathway. J Biol Chem. **276**, 21995-21998 (2001).
14. G. Frenkel-Denkberg, D. Gershon, A. P. Levy. The function of hypoxia-inducible factor 1 (HIF-1) is impaired in senescent mice. FEBS Letters 462: 341-344 (1999)
15. A. Rivardi, L Berthou-Soulie, N. Principe, M. Kearney, C. Curry, D. Branellec , G. L. Semenza, J. M. Isner. Age-dependent defect in vascular endothelial growth factor expression is associated with reduced hypoxia-inducible factor 1 activity, J. Biol Chem **275**, 29643-29647 (2000).

LUNG ISCHEMIA: ENDOTHELIAL CELL SIGNALING BY REACTIVE OXYGEN SPECIES

A progress report

Aron B. Fisher, Abu B. Al-Mehdi, Zhihua Wei, Chun Song, and Yefim Manevich*

1. INTRODUCTION

The major manifestations of tissue ischemia include cellular anoxia due to inadequate oxygen delivery and secondary acidosis. The reintroduction of oxygen upon reperfusion leads to increased production of reactive oxygen species (ROS) and subsequent oxidative tissue injury. This sequence of events has been called the ischemia/reperfusion syndrome. In contrast to other organs, ischemia of the lungs such as that which accompanies vascular occlusion is not expected to compromise lung ventilation or oxygen delivery and if so would not result in tissue anoxia or acidosis. However, the lungs do show increased generation of ROS during an ischemic episode as described below. Our studies have investigated the mechanisms and significance of ROS generation with non-hypoxic lung ischemia. We have determined that ROS generation represents one component of a complex signaling response of endothelium to ischemia and that this signaling cascade is triggered by the loss of shear stress. Further, we have determined that the stimulus for ROS production is endothelial membrane depolarization, presumably associated with inactivation with a shear stress K^+ channel. The following describes our experiments documenting this new paradigm of ischemic response.

2. THE ISOLATED PERFUSED LUNG MODEL OF LUNG ISCHEMIA

Studies were carried out with the isolated perfused lung preparation from rats or mice. Lungs were perfused in a temperature controlled (37°C) chamber with recirculating perfusate consisting of Krebs bicarbonate buffer plus added bovine serum albumin or dextran and were continuously ventilated with 5% CO_2 in air or oxygen. In our hands, this isolated preparation remains metabolically stable for several hr. Global ischemia was

*Aron B. Fisher, Abu B. Al-Mehdi, Zhihua Wei, Chun Song, and Yefim Manevich, University of Pennsylvania Medical Center, Institute for Environmental Medicine, Philadelphia, PA 19104-6068.

Oxygen Transport to Tissue XXIII, edited by D. Wilson *et al.*
Kluwer Academic/Plenum Publishers, 2003

produced by abrupt discontinuation of perfusion while ventilation continued. Tissue ATP content in this preparation showed essentially no change during one hr of ischemia indicating that tissue oxygenation is maintained adequately by lung ventilation[1, 2]. By contrast, ATP content decreases approximately 65% under these conditions if lungs are ventilated with N_2[2].

This report will emphasize studies of membrane potential and ROS generation, primarily utilizing fluorophores as reporter molecules with the following excitation (ex) and emission (em) wavelengths in nm. Cell membrane potential was measured with bis-oxonol (480 ex/520 em) or di-8-ANEPPS (440 ex/530 em), membrane localizing dyes that give increased fluorescence with depolarization. ROS production was monitored with dichlorofluorescein (490 ex/530 em) or hydroethidine (485 ex/585 em), radical traps that under our experimental conditions were localized intracellularly, or the extracellular reporters, Amplex red (545 ex, 610 em) and cytochrome c (absorbance at 550 nm).

2.1. Fluorescence At The Pleural Surface

Our initial studies to investigate the cellular response to acute ischemia utilized the technique of surface fluorometry[1, 3, 4]. A light guide was placed against the pleural surface and the fluorescence emission was recorded using specific fluorophores with the appropriate excitation. Changes in cell membrane potential during ischemia were monitored with the relatively slowly (seconds) responding dye bis-oxonol[1]. Intrinsic lung fluorescence before dye infusion was detected and was unchanged with ischemia. Fluorescence increased to reach a new plateau with infusion of the dye and with ischemia demonstrated a rapid increase that was reversible with reperfusion. These changes were reproducible with subsequent cycles of ischemia and reperfusion and are compatible with ischemia-mediated cell membrane depolarization.

ROS generation was monitored at the pleural surface by fluorescence of the dye hydroethidine which is converted to ethidium by superoxide anion[3]. We demonstrated a progressive increase in ethidium (the product) and decrease in hydroethidine (the substrate) fluorescence that was initiated shortly after the onset of lung ischemia. The increased ethidium fluorescence with ischemia was confirmed by fluorescence microscopy of fixed lung sections. To investigate potential linkage between membrane depolarization and ROS generation, lungs were perfused with a solution containing high K^+ or the K^+_{ATP}-channel blocker, glyburide[4]. These agents resulted in both membrane depolarization and ROS generation compatible with a cause and effect relationship.

2.2. Cellular Localization By Fluorescence Imaging

The signals recorded by surface fluorometry reflect changes in the lung tissue but obviously do not reflect changes in specific cell types in this cellularly heterogeneous organ. In order to determine the cellular localization of the fluorescence signals, we carried out fluorescence imaging of subpleural capillaries using an inverted fluorescence microscope, a CCD camera, and Metamorph® imaging software. Endothelial localization of bis-oxonol or DCF administered in the perfusate was confirmed by demonstrating both co-localization with acetylated LDL and clearly distinct localization from a type II epithelial cell marker[1,5]. Imaging during lung ischemia showed endothelial cell depolarization and ROS generation[1, 5]. The timing of these changes following onset of ischemia was determined by video recording at 18 frames per sec resolution[6]. Membrane potential was evaluated using the fast responding dye, di-8-ANEPPS. Increased fluorescence was noted immediately upon cessation of flow and reached an apparent plateau at approximately 1-2 sec. Calibration of the fluorescence signal through the use of solutions containing elevated K^+ indicated that membrane depolarization with ischemia was approximately 15-20 mv. Membrane depolarization with ischemia was prevented by pre-perfusion of lungs with lemakalim, a K^+ channel agonist. Amplex red, a highly sensitive dye that is localized to the intravascular space and interacts with H_2O_2 was used for imaging ROS production. A progressive increase of Amplex red fluorescence was observed at 2-3 sec of ischemia and was abolished by addition of catalase to the perfusate.

2.3. Subcellular Source of ROS

We utilized several inhibitors in order to determine the enzymatic pathway responsible for ROS generation by lung endothelium with lung ischemia. Pretreatment of lungs with allopurinol had no effect on ROS generation indicating that xanthine oxidase was not the responsible enzyme[2]. On the other hand, ROS generation was markedly inhibited by pretreatment with diphenyleneiodonium, an inhibitor of flavoproteins that has been widely used as an inhibitor of NADPH oxidase[2]. Pretreatment with PR39, a polypeptide inhibitor of sarc homology-3 (SH-3) domains, also inhibited ROS production with ischemia[5]. This latter effect is compatible with a role for NADPH oxidase since SH-3 domains are known to be involved in assembly of the oxidase complex. Confirmation of endothelial NADPH oxidase as a source of ROS was obtained by study of mice with "knock out" of $gp91^{phox}$, the flavoprotein component of NADPH oxidase[5]. Isolated perfused lungs from wild type mice demonstrated ROS production with ischemia similar to that observed with rats while there was no increase in ROS production with lungs from the gene-targeted mice. Lung imaging demonstrated that ROS production in mouse lungs with ischemia was endothelial and not from polymorphonucleor leukocytes (PMN) or macrophages.

3. ENDOTHELIAL CELL ISCHEMIA *IN VITRO*

The results of the isolated perfused lung studies indicated that acute cessation of flow was accompanied by rapid endothelial cell membrane depolarization followed by activation of endothelial NADPH oxidase and production of ROS. We next demonstrated that we were able to reconstitute these effects *in vitro* using isolated cell systems.

Bovine pulmonary artery endothelial cells (BPAEC) were selected as a readily available cell line that could provide a model of endothelial cellular response to altered shear stress (i.e., simulated ischemia). Cells were cultured on a plastic support that was placed in a parallel plate chamber in order to subject cells to laminar flow[7]. It is well known that cells exposed to shear stress demonstrate a variety of cellular changes including re-orientation of both the actin cytoskeleton and the long axis of the cells in the direction of flow, activation of enzymes and cell signaling mechanisms, and gene induction. We predicted correctly that flow adaptation by the cell would be required for them to subsequently demonstrate a signaling response to the loss of shear stress associated with cessation of flow. Flow adapted cells when subjected to simulated ischemia demonstrated endothelial cell membrane depolarization and ROS generation[7].

Membrane depolarization was detected by an increase in bis-oxonol fluorescence and ROS generation was detected by cytochrome c added to the extracellular perfusate. An increase in cytochrome c absorbance inhibitable by added SOD indicates the extracellular generation of superoxide anion. Neither membrane depolarization nor ROS generation was observed in cells that had not been flow adapted. Production of superoxide was markedly inhibited by the presence of DPI while extracellular catalase had no effect. On the other hand, the intracellular oxidation of DCF with simulated ischemia was inhibited by catalase and DPI while SOD had no effect. Our interpretation of these results is that with ischemia, endothelial membrane NADPH oxidase generates superoxide extracellularly where it is dismuted to H_2O_2 which in the absence of flow can really diffuse into the cells. These events were observed in the first several minutes of simulated ischemia and clearly occurred prior to oxygen depletion of the medium due to cellular metabolism. The adequacy of cellular oxygenation was confirmed by measuring medium PO_2 and by comparing air vs. oxygen saturated buffer which showed identical results during the first several minutes of ischemia.[7]

Subsequent experiments utilized an *in vitro* model of cells grown in an artificial capillary system[8, 9]. This system which permitted continuous oxygenation of cells during ischemia by perfusion of medium through abluminal ports simulates ischemia in the intact lung. Cells were removed from attachment to the artificial capillaries by mild trypsinization. BPAEC that had been flow adapted for 48 hr exhibited increased ROS production with acute cessation of flow[8]. ROS production was detected by increased DCF fluorescence using fluorometer analysis of cell sonicates, flow cytometric analysis of intact cells, and observation by fluorescence microscopy. Cells that had not been flow adapted did not show an ROS response to simulated ischemia. Subsequent experiments with this model demonstrated that simulated ischemia in flow adapted cells results in activation of ERK1/2 and the transcription factors NFκβ and AP-1[8, 9]. These responses were dependent on ROS generation and were prevented by pre-treatment of cells with DPI or N-acetylcysteine. Additional downstream events included Ca^{2+}/calmodulin dependent NO generation and evidence at 24 hr of increased endothelial cell division[7, 8, 10]. A schematic illustration of these events is shown in Figure 1.

4. SUMMARY

These studies using both intact lung and reconstituted cell systems have shown that pulmonary endothelial cells respond rapidly (within several seconds) to the acute cessation of perfusate flow, i.e., ischemia. These effects represent a response to the loss of shear stress and are unrelated to changes in cellular oxygenation. The immediate response is partial depolarization of the endothelial cell membrane followed by activation of endothelial NADPH oxidase and the extracellular generation of superoxide anion. Dismutation of superoxide to H_2O_2 generates a cell signaling molecule that results in the activation of protein kinases and transcription factors which in turn lead to NO generation and activation of endothelial cell division. The presumed physiological role of this signal cascade is the generation of a vasodilator (NO) and the formation of new capillaries in the effort to restore blood flow.

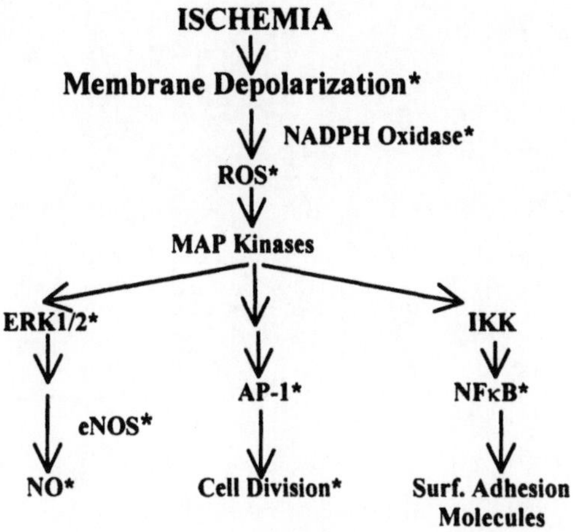

Figure 1. Schema of events in the lung circulation during ischemia. Cessation of flow leads to rapid endothelial cell membrane depolarization due to closure of K^+ channels, activation of membrane-associated NADPH oxidase, generation of ROS, and subsequent cell signaling leading to increased expression of surface adhesion molecules, NO generation and cell division. The asterisks indicate those reactions for which we have obtained experimental evidence as described in the text.

Acknowledgements

We thank Dr. Vladimir Muzykantov for helpful suggestions and Elaine Primerano for typing the manuscript. The research reported herein was supported by National Institutes Health grant HL 60290 (SCOR in ARDS).

References

1. A.B. Al-Mehdi, G. Zhao, and A.B. Fisher, ATP-independent membrane depolarization with ischemia in the oxygen-ventilated isolated rat lung. *Am. J. Respir. Cell. Mol. Biol.* **18,** 653-661 (1997).
2. G. Zhao, A.B. Al-Mehdi, and A.B. Fisher, Anoxia-reoxygenation versus ischemia in isolated rat lungs. *Am. J. Physiol. (Lung Cell. Mol. Physiol 17)* **273,** L1112-L1117 (1997).
3. A. Al-Mehdi, H. Shuman, and A.B. Fisher, Intracellular generation of reactive oxygen species during nonhypoxic lung ischemia. *Am. J. Physiol. (Lung Cell. Mol. Physiol 16)* **272,** L294-L300 (1997).
4. A. Al-Mehdi, H. Shuman, and A.B. Fisher, Oxidant generation with K^+-induced depolarization in the isolated perfused lung. *Free Rad. Biol. Med.,* **23,** 47-56 (1997).
5. A. Al-Mehdi, G. Zhao, C. Dodia, K. Tozawa, K. Costa, V. Muzykantov, C. Ross, F. Blecha, M. Dinauer, and A.B. Fisher, Endothelial NADPH oxidase as the source of oxidants in lungs exposed to ischemia or high K^+. *Circ. Res.,* **83,** 730-737 (1998).
6. C. Song, A.B. Al-Mehdi, and A.B. Fisher, An immediate endothelial cell signaling response to lung ischemia. *Am. J. Physiol. (Lung Cell. Mol. Physiol.)* in press (2001).
7. Y. Manevich, A. Al-Mehdi, V. Muzykantov, and A.B. Fisher, Oxidative burst and NO generation as the initial response to "ischemia" in flow-adapted endothelial cells. *Am. J. Physiol. (Heart Circ. Physiol.)* **280,** H2126-H2135 (2001).
8. Z. Wei, K. Costa, A.B. Al-Mehdi, C. Dodia, V. Muzykantov, and A.B. Fisher, Simulated ischemia in flow adapted endothelial cells leads to generation of reactive oxygen species and cell signaling. *Circ. Res.* **85,** 682-689 (1999).
9. Z. Wei, A.B. Al-Mehdi, and A.B. Fisher, Signaling pathway for nitric oxide generation with simulated ischemia in flow-adapted endothelial cells. *Am. J. Physiol. (Heart Circ. Physiol.)* in press (2001).
10. A.B. Al-Mehdi, C. Song, K. Tozawa, and A.B. Fisher, Ca^{2+}-and phosphatidylinositol 3-kinase-dependent nitric oxide generation in lung endothelial cells in situ with ischemia. *J. Biol. Chem.* **275,** 39807-39810 (2000).

NEAR INFRARED SPECTROSCOPY (NIRS) MEASUREMENTS DURING GLOBAL CEREBRAL ISCHEMIA IN SHEEP

Alfredo A. Martinez-Coll, Michael K. Morgan, and Hung T. Nguyen*

1.INTRODUCTION

With the onset of cerebral ischemia a complex chain of hemodynamic and metabolic events is set into motion. Initially, the diminished blood flow results in vasodilatation and an increase in tissue oxygen extraction to maintain oxygen metabolism. Depending on the duration and extent of the ischemia, the reduction in cerebral blood flow may lead to irreversible brain damage[1,2]. Near infrared spectroscopy (NIRS), allows continuous estimates of the concentrations of oxygenated and deoxygenated hemoglobin (HbO_2 and Hb, respectively) and the redox state of cytochrome aa3 in tissue. Addition of the signals for HbO_2 and Hb can be used as an indicator of changes in blood volume (BV), while their difference provides a measure of hemoglobin oxygen desaturation (DESAT)[3] or tissue oxygen extraction. The aim of this study was to compare changes in perfusion and oxygen extraction in brain (monitored by laser Doppler flowmetry and arteriovenous oxygen differences) with NIRS BV and DESAT signal attenuation during global cerebral ischemia and reperfusion in sheep.

2. MATERIALS AND METHODS

Following approval from the Royal North Shore Hospital Animal Care and Ethics Committee (ACEC), sixteen adult Merino Cross ewes with an average body weight of 48.4 (+/- 5.3) kg were used in the study.

* Alfredo A. Martinez-Coll and Hung T. Nguyen, Key University Research Strength in Health Technologies, University of Technology, Sydney, Broadway NSW 2007. Michael K. Morgan, Department of Neurosurgery, Royal North Shore Hospital, St.Leonards, NSW 2065, Australia.

2.1 Anesthesia

The animals were anesthetized with pentothal (1g/20 ml) as induction, intubated with a cuffed tracheal tube (10 mm ID), immobilized with Tracrium (0.6 mg/kg) and ventilated with oxygen (>40%) at 1 l/min, nitrous oxide (52%) at 2 l/min, and 1% halothane through a respirator (Bird model 8; Bird Australia, Pty. Ltd., Chatswood, NSW, Australia). Maintenance fluid was provided through peripheral venous access with 0.9% NaCl. Anesthesia was maintained with a continuous fentanyl infusion (10μg/ml) at 30 ml/kg for the first 30 minutes and then 20 ml/kg for the remainder of the procedure. Electrocardiography was monitored with electrodes clipped to the extremities. A rectal temperature probe monitored body temperature.

2.2 Surgery

After wool was removed from the neck and thigh area, incisions were made along both sides of the trachea and both carotid arteries were isolated and loosely fitted with rubber ties. A dual cannula for central venous pressure monitoring and IV drug administration was placed in the left jugular vein. Both femoral arteries were isolated and cannulas introduced for blood sampling and blood pressure monitoring (left side), as well as blood pressure reduction through the removal of blood (right side). The animals were turned over and placed in a prone position with the legs stretched and secured to the operating table; the head was also secured in a specially designed head holder. A midline incision in the head was made, the skin reflected and the skull exposed. Burr holes were drilled on opposite sides of the skull and adjusted to allow fixation of both the laser Doppler flowmeter and NIRS probes. A third burr hole was used for cannulation of the sagittal sinus for direct venous sampling.

2.3 Instrumentation

Changes in cerebral tissue perfusion were monitored with an ALF21 laser Doppler flowmeter (Advance Co., Ltd, Japan). It offers non-destructive and continuous monitoring of perfusion (range 0-100 ml/min/100g) at various sites in the body, relying on a low-power output semiconductor laser (780 nm with an optical power output of <2mW). Measurements provide fine tissue blood flow such as capillary flow, using a 1mm diameter half-spherical probe.

The research NIRS instrument consists of five 1 watt solid-state lasers (780, 800, 830, 850, and 980 nm) fired sequentially at 5μs pulses for a 1 ms cycle, and a 5 mm^2 photodiode receiver. In vitro assessment of the instrument has been previously reported[4]. A fiber optic adaptation was designed and implemented; two fiber optic bundles were used: one containing the signals from the five lasers, and the other for the photodetector. The optode separation was 0.5 cm.

The signals from the LDF and NIRS instrument, arterial and venous pressures, and ECG were fed to a MacLab 16S data acquisition module running Chart 3.5.6 software (AD Instruments, Castle Hill, Australia) at a sampling rate of 4 Hz. A fluid filled pressure transducer was used to monitor central venous pressure, while a 5F Millar catheter-tip manometer (Millar Instruments, Inc., Houston, TX) provided arterial pressure recordings. The pressure signals were amplified (System 6; Triton Technology, Inc., San Diego, CA) and recorded with the MacLab analog-to-digital converter. The MacLab hardware and software provided the option for continuous display and recording of all channels. The NIRS and LDF probes were placed on opposite hemispheres adjacent to each other directly on top of the brain over the intact dura matter.

2.4 Ischemia

Global cerebral ischemia was implemented through hypotension and bilateral carotid occlusion. Following a period of stabilization, blood gas samples (both arterial and venous) were taken and MAP lowered to < 50 mmHg through hemorrhage. The removal of blood was then stopped to maintain that level of arterial pressure and the carotid ties were pulled and tightened in place with alligator forceps to completely occlude flow. After a 4-minute period of ischemia blood gas samples were taken, the carotid ties released and the response to reperfusion observed for the next 5 minutes. The animals were sacrificed with an intravenous bolus of KCl (3g).

2.5 Statistical Analysis

Means +/-SD were calculated for measurements of all monitored parameters. Significant differences between control and test conditions between perfusion values (by LDF) and NIRS BV ($\lambda_{780}+\lambda_{850}$) signal attenuation, and venous oxygen saturation (SvO_2) and NIRS DESAT ($\lambda_{780}-\lambda_{850}$) signal attenuation were determined by paired t-tests with $\alpha = 0.05$. The relationship between oxygen extraction fraction (OEF- calculated from arteriovenous oxygen differences) and MAP to NIRS DESAT signal attenuation was also explored.

3. RESULTS

The response to global cerebral ischemia (Fig.1) was subdivided into two categories. In one group blood gas samples were obtained before hypotension and carotid occlusion as well as at the end of ischemia; in the other, the samples were drawn after blood removal and at the end of the ischemic period. In the first group of animals (n=10), there were significant decreases in SvO_2 (from 84.6 +/- 5.8 to 64.8 +/- 17 .1%), MAP (from 71.15 +/- 20.5 to 52.9 +/- 21.6 mmHg), and perfusion (from 41.6 +/- 10.7 to 29 +/- 6.37 ml/min/100g); $p < 0.01$. A significant increase in NIRS signal attenuation was found at various wavelengths (800, 830, 850, and 980 nm) coupled with a decrease in the NIRS DESAT signal ($p < 0.05$), but no significant difference in NIRS BV. NIRS signal attenuation at 780 and 850 nm increased by 10.6% and 23.6%, respectively. A strong correlation was observed between MAP and NIRS DESAT signal attenuation during blood removal leading to hypotension ($r^2=0.83$). In the second group of animals (n=6) there were minor changes in SvO_2 (from 59.4 +/- 13.1 to 54.4 +/- 16 %), MAP (47.82 +/- 5.1 to 63.6 +/- 15.7 mmHg), and perfusion (from 38.1 +/- 9.4 to 37.1 +/- 9.2 ml/min/100g). No significant differences in NIRS signal attenuation were found neither at individual wavelengths nor for BV or DESAT. After 5 minutes of reperfusion, SvO_2 returned to near pre-blood removal levels (81.71 +/- 9.65 % compared to 84.6 +/-5.8 % at baseline), whilst MAP and cerebral perfusion showed little change. NIRS signal attenuation also returned to near pre-blood removal levels.

The arterial and venous oxygen contents (CaO_2 and CvO_2, respectively) were calculated as O_2 content = $[Hb]*1.34*SO_2 + 0.03*pO_2$. OEF was calculated at various times using the following relationship: OEF= 1- (CvO_2/CaO_2). The effect of ischemia and reperfusion on the calculated OEF was: 1) an increase of 69% in the group of animals in which blood gas samples were taken before blood removal, 2) an increase of 14% in the group of animals in which blood gas samples were taken after removal of blood, and 3) a decrease of 51% after 5 minutes of reperfusion. A significant linear relationship (r^2~0.8) was found between DESAT and the calculated OEF for the majority of the animals (11 of a total of 16). In 4 of the 5 animals there was no correlation and in one animal the NIRS DESAT signal did not change significantly.

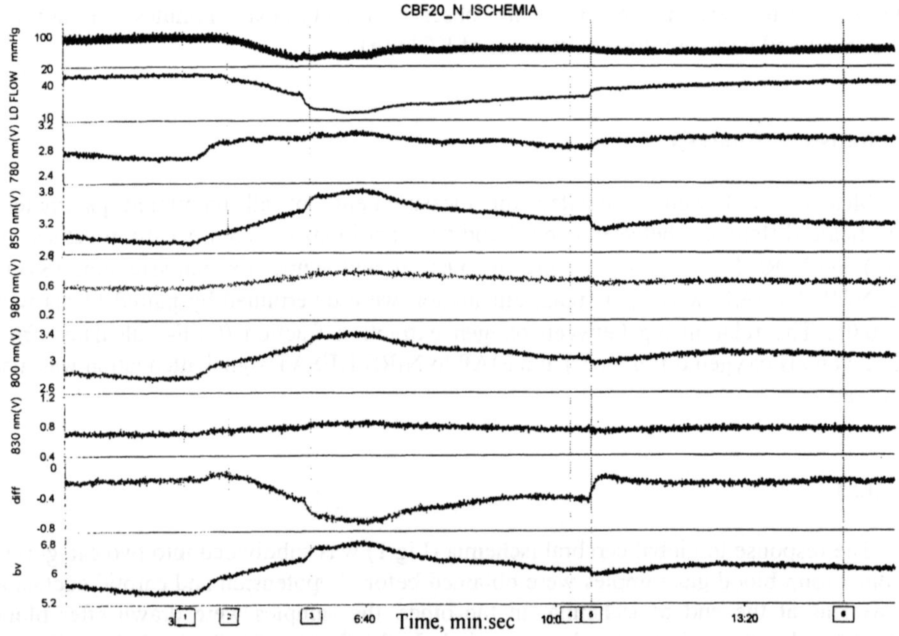

Figure 1. Characteristic trace during ischemia. Markers are: 1) baseline blood gas, 2) blood removal, 3) carotid occlusion, 4) blood gas at 4-min ischemia, 5) release of carotid occlusion 6) 5-min reperfusion.

4. DISCUSSION

The extent of ischemia (coupled with hypotension) would result in a significant reduction in cerebral blood flow, and hence, oxygen delivery. The tissue severely deprived of oxygen and beyond the lower limit of autoregulation increases OEF to compensate for the diminished flow. When blood gas samples where taken prior to blood removal, there was an increase in OEF during ischemia of 69%. These results are supported by observations of increased OEF to > 70% following reductions in flow[5].

In the group of animals in which blood gases were obtained after removal of blood, OEF during ischemia only showed a 14% increase. This result suggests that significant changes in metabolism take place during hypotension, and that the ischemic insult only marginally furthers changes in cerebral metabolism. In a similar study, Koga et al found significant decreases in brain pO_2 and a reduction in Cyt.aa3 during hypotension, but only slight additional changes following bilateral carotid occlusion[6]. In the same study, the average reduction in cortical CBF during ischemia was 27.5%, not dissimilar to the 29.2% change in perfusion reported here. Because changes in blood flow were measured by laser Doppler flowmetry, we report on cortical "perfusion" as percentage changes from a baseline, as it is widely accepted that LDF values are only relative units.

The change in NIRS signal attenuation at 850 nm (mainly due to HbO_2) from pre-blood removal and at the end of ischemia was 23.6 %. The extent of the ischemic insult correlates with results of NIRS changes during carotid endarterectomy where a ΔHb_{diff} of 6.8µm/L (equivalent to a 20% decrease in the level of HbO_2) provided 100% specificity for severe cerebral ischemia[7]. Interestingly, NIRS BV signal did not show a significant difference between baseline and ischemia; however, the changes in NIRS DESAT where significant ($p<0.05$). Tsuji et al found similar results with cerebral ischemia during hypotension in piglets, as well as a significant correlation between MAP and Hb_{diff}[8]; an observation also found in our study ($r^2=0.83$). Measurements after 4 minutes of reperfusion following ischemia showed a decrease in OEF of 51%. The results correlate well with the observations of an overshoot in flow during the early phase of reperfusion, known as post-ischemic hyperemia[9], which may also be coupled with metabolic depression[10]. We suggest that the marginal change in NIRS BV signal during ischemia is due in part to the presence of the carotid *rete mirabile* in the brain of ungulates (such as sheep). Although the function of the *rete* as heat exchanger to cool the arterial blood destined for the brain has been established, its role in cerebral blood flow regulation is not as clear[11]. The rete may serve as a reservoir for oxygented blood between the two hemispheres, and therefore, aid in the redistribution of blood during reductions or interruptions in flow.

In summary, NIRS signal attenuation shows good correlation with changes in perfusion and oxygen extraction. Simple NIRS instruments can be used successfully as a research tool in the study of global changes in cerebral blood flow and metabolism by providing accurate real-time information on the oxygen supply-demand relationship as well as hemodynamics.

REFERENCES

1. R. Griggs and A. I. Arieff, in: *Metabolic and brain dysfunction in systemic disorders,* edited by R. Griggs and A. I. Arieff (Little, Brown and Company, London, 1992), pp. 39-54.
2. R. Spetzler and D. Nehls, in: *Cerebral Blood Flow: Physiologic and Clinical Aspects,*edited by J. H. Wood (Mc-Graw Hill Book Company, Sydney, 1987), pp. 651-676.
3. P. Cooper, G. Wilson, D.T.A. Hardman, O. Kawaguchi, YF Huang, A. Martinez-Coll et.al, Blood oxygen desaturation heterogeneity during muscle contraction recorded by near infrared spectroscopy. *Adv. Exp. Med. Biol.* **428**, 285-292. (1997).
4. A. Martinez-Coll, P. Cooper, G. Murphy, and H. Nguyen, Assessment of a laser-poweredmultiwavelength near-infrared spectrometer.,*Proceedings to the 19th International Conference IEEE/EMBS*, Chicago, Il (1997), pp 696-699.
5. F. Finnerty, L. Witkin, and J. Fazekas, Cerebral hemodynamics during cerebral ischemia induced by acute hypotension, *J. Clin. Invest.* **33**, 1227-1232. (1954).

6. H. Koga and G. Austin, Cortical oxidative metabolism under conditions of ischemia, hypoxia, and asphyxia in the rabbit, *J. Neurosurg.* **59**, 57-62. (1983).
7. P. Kirkpatrick, J. Lam, P. Al-Rawi, P. Smielewsky, and M. Czosnyka, Defining thresholds for critical ischemia by using near-infrared spectroscopy in the adult brain, *J. Neurosurg.* **89**, 389-394. (1998).
8. M. Tsuji, A. Duplessis, G. Taylor, R. Crocker, and J. Volpe, Near infrared spectroscopy detects cerebral ischemia during hypotension in piglets, *Pediatr. Res.* **44**, 591-595. (1998).
9. E. Kagstrom, M. Smith, and B. Siesjo, Recirculation in the rat brain following incomplete ischemia, *J. Cereb. Blood Flow and Metab.* **3**, 183-192. (1983).
10. W. Pulsinelli, D. Levy, and T. Duffy, Regional cerebral blood flow and glucose metabolism following transient forebrain ischemia, *Ann. Neurol.* **11**, 499-509. (1982).
11. M. Baker, A brain-cooling system in mammals, *Scien. Amer.* **240**, 114-122. (1979).

SIMULATIONS OF CAPILLARY NETWORK OXYGEN TRANSPORT DURING TRANSIENT ISCHEMIA IN THE PRESENCE AND ABSENCE OF TISSUE MYOGLOBIN

Daniel Goldman*

1. INTRODUCTION

Capillaries are one of the principal sites of oxygen transport to tissue and are particularly important when the supply of oxygen is relatively low. The steady-state transport of oxygen in the microcirculation has been studied computationally by a number of researchers[1-6]. Some studies have been performed of the time-dependent case[7-9], however realistic biophysics and spatial heterogeneity in microvascular geometry and blood flow have not been included.

In many physiological situations microvascular oxygen delivery takes place under inherently time-varying conditions. Recently we used a biologically realistic computational model to study oxygen transport by capillary networks under oscillatory flow conditions[10]. Significant enhancement of oxygen delivery was shown to occur for a certain range of physiological parameters, however for realistic frequencies this enhancement was only found in the absence of myoglobin. These results indicated that vasomotion does not affect oxygen transport in striated and cardiac muscle and suggested that the characteristics of time-dependent oxygen transport are fundamentally different in tissues containing myoglobin compared to those where myoglobin is absent. The goal of the work described here is to begin investigating other physiological situations where time-dependent oxygen transport might be important and to determine how these are affected by the presence or absence of myoglobin.

Our previous study found that in the absence of myoglobin vasomotion can increase oxygen delivery to hypoxic tissue, on average, relative to steady flow with the same mean flow rate. In the present work we use a more accurate version of a computational model described previously[10] to study time-dependent oxygen transport from capillary networks. These networks are constructed using data on the

---

*Department of Mathematical Sciences, New Jersey Institute of Technology, Newark, NJ, 07102.

Oxygen Transport to Tissue XXIII, edited by D. Wilson *et al.*
Kluwer Academic/Plenum Publishers, 2003

hamster cheek pouch retractor muscle. This is a striated muscle containing myoglobin, however we believe our results are at least suggestive of what occurs in other tissues, including those without myoglobin. We first calculate steady-state tissue and capillary spatial oxygen distributions for fixed tissue oxygen supply and consumption rate. We then simulate transient ischemia using a step decrease in the convective oxygen supply followed by a restorative step increase. We follow the time evolution of the tissue oxygen distribution as it reaches an intermediate state and then returns to the original steady state. For the Mb-containing and Mb-free cases, we discuss and compare the time-course of tissue oxygen distributions and the characteristics of the ischemic steady state.

2. COMPUTATIONAL MODEL

A modified form of our existing computational model[5,6,10] was used to calculate steady-state blood flow and time-dependent oxygen transport in a realistically complex, three-dimensional capillary network. The network was based on a muscle-fiber construction using measured parameters for striated muscle and included several structural features such as anastomoses and tortuosity that have been shown to be important in producing oxygen transport heterogeneity[6]. For the work described here the network contained 25 parallel capillaries, which corresponds to one large functionally independent network or two interconnected networks of average size[11]. The tissue domain was 400μm in length and had a cross-section of 129μm by 123μm, giving a capillary density of 1576 per square millimeter[1].

A two-phase continuum flow model, based on the *in vivo* rheological model of Pries *et al.*[12], was used to calculate the steady-state distribution of total blood flow, Q, and discharge hematocrit, H_D, throughout the capillary network[10]. Flow resistance in each capillary segment is given by Poiseuille's Law with an apparent viscosity that depends on hematocrit and vessel radius (Fahraeus-Lindqvist effect). Also included in the model is the plasma skimming effect, which determines the relative distribution of hematocrit at vessel bifurcations. A uniform pressure difference between the arterial and venous ends of the network was used to establish a concurrent flow pattern in the parallel capillaries. For the reference case the average velocity in the network was 200 μm/sec, which is approximately twice that observed under resting conditions[1]. For all simulations inlet hematocrit to the network was fixed at 30%. The effect of the onset of transient ischemia on blood flow in the capillary network was modeled by a step-wise decrease of the velocity in all capillary segments to 50% of its reference value. The hematocrit distribution in the network was assumed to be unaffected by the velocity changes due to transient ischemia.

The oxygen transport model[10] included free and Hb-bound oxygen in the blood and facilitated diffusion of oxygen in the tissue due to the presence of Mb. The tissue was taken to be homogeneous and isotropic and blood was taken to be a continuous two-phase mixture. Since discrete erythrocytes and radial velocity and concentration gradients inside the capillaries were not considered, intravascular oxygen transport resistance was modeled using hematocrit-dependent mass transfer coefficients[6,13]. Oxygen transport inside the capillaries was described by a time-dependent equation for SO_2, while transport in the tissue was described by a time-dependent equation for PO_2 [10]. These equations were connected by a flux boundary condition determined by the local mass transfer coefficients and no-flux boundary conditions were used for PO_2 at all outer tissue boundaries. Capillary SO_2 was given by the Hill equation and first-order equilibrium expressions were used for tissue O_2 consumption and Mb-O_2 saturation.

The time-dependent oxygen transport equations were solved numerically using a finite-difference scheme based on our previously validated method[6,10]. The new solution technique has improved accuracy in both space and time and therefore is more suitable for studying realistic, time-dependent transport problems. For normal flow conditions, solutions in the tissue and capillary spaces were advanced from an arbitrary initial condition until they became constant in time. A step decrease in blood flow was then implemented and solutions were followed until they reached an ischemic steady state. Finally, flow was returned to its normal value and solutions were followed until they returned to the original steady state. For all simulations inlet SO_2 to the capillary network was 63%[1], and tissue oxygen consumption was fixed at approximately 8×10^{-4} ml O_2/ml/sec corresponding to moderately elevated metabolism. The biophysical parameters used in transport calculations are those of Goldman and Popel[10].

3. STEADY-STATE OXYGEN DISTRIBUTIONS

Figure 1 shows the steady-state distributions of tissue PO_2 for normal and ischemic blood flow, with and without tissue myoglobin. It can be seen that ischemia changes the shape of the distributions and shifts them toward lower PO_2 values. In

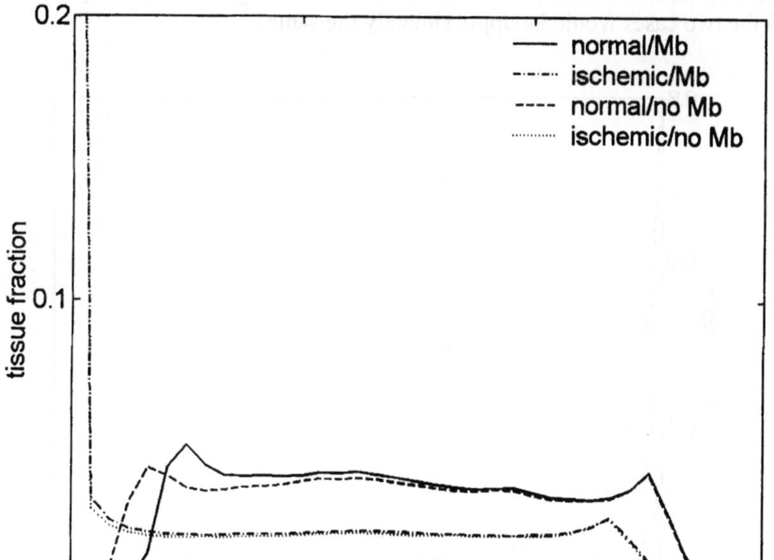

Figure 1. Steady-state tissue oxygen distributions for normal flow and ischemia, with and without tissue myoglobin.

addition, myoglobin is seen to have only a slight effect in decreasing oxygen transport heterogeneity under ischemic conditions. Without Mb, tissue PO_2 was 17.2±8.5 (mean±S.D) for normal flow and 8.6±10.2 for ischemia. With Mb, tissue PO_2 was 17.8±8.0 and 8.8±10.2. For normal flow the coefficient of variation (C.V.=S.D./mean) of tissue PO_2 was slightly lower with Mb than without (0.45 vs. 0.49), but for ischemia both CV's were 1.2 indicating a large amount of heterogeneity.

4. PROPERTIES OF TIME-DEPENDENT OXYGEN DELIVERY

Figure 2 shows the time course of tissue PO_2 distributions during transient ischemia when Mb is present in the tissue. It was found that once ischemia begins it takes tissue longer to become hypoxic when Mb is present compared to when it is absent, but it also takes longer for complete recovery from hypoxia once normal flow resumes. In both cases recovery time was significantly longer than the time required for hypoxia to develop. With Mb present, it took 8.5s from the beginning of ischemia for the onset of hypoxia, defined as the existence of tissue PO_2's less than 1 torr. In this case it took 15.0s for hypoxia to disappear once normal flow resumed. By contrast, in the absence of Mb hypoxia began after only 1s of ischemia and disappeared after 3.25s of normal flow. These results imply that for ischemic periods greater than about 8s, the average amount of tissue exposed to hypoxia will be greater when Mb is present than when it is absent. For example, a 5s period of ischemia would produce no hypoxia with Mb present but would produce about 6.75s of hypoxia without Mb. However, a 10s period of ischemia would produce 16.5s of hypoxia with Mb and only 11.75s without it. For significantly longer ischemic periods the two cases would be approximately the same.

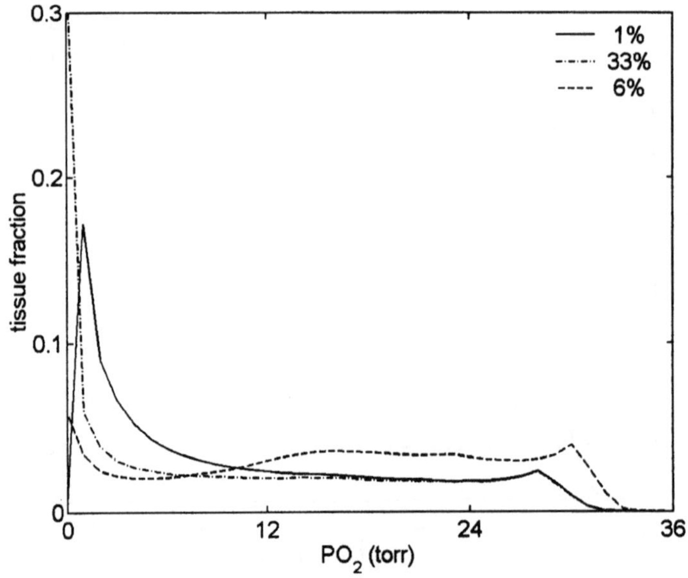

Figure 2. Tissue oxygen distributions showing different degrees of hypoxia as ischemia begins (1%) and continues for several seconds (33%), and as normal flow starts to return (6%).

These results are interesting because they imply that the oxygen buffering effect of Mb also has a potential drawback depending on the time course of ischemia. This finding is similar to that for vasomotion[10] where it was found that oxygen transport enhancement is hindered by the presence of Mb. In both cases, the capacity for short-term buffering of oxygen is accompanied by a delay in re-oxygenation under certain conditions. It is now of interest to compare these results to experimental findings, in particular to compare PO_2 distributions in critically perfused tissues containing Mb (e.g., heart) to those without Mb (e.g., brain).

ACKNOWLEDGMENTS

This work was supported by NJIT SBR grant 421860 and a grant from NPACI/SDSC.

REFERENCES

1. M. L. Ellsworth, A. S. Popel, and R. N. Pittman, Assessment and impact of heterogeneities of convective oxygen transport parameters in capillaries of striated muscle: experimental and theoretical, *Microvasc. Res.* **35**, 341-362 (1988).
2. K. Groebe, A versatile model of steady state O_2 supply to tissue. Application to skeletal muscle, *Biophys. J.* **57**, 485-498 (1990).
3. T. W. Secomb and R. Hsu, Simulation of O2 transport in skeletal muscle: diffusive exchange between arterioles and capillaries, *Am. J. Physiol.* **267**, H1214-H1221 (1994).
4. L. Hoofd and Z. Turek, Realistic modeling of capillary spacing in dog gracilis muscle greatly influences the heterogeneity of calculated tissue oxygen pressures, *Adv. Exp. Med. Biol.* **388**, 333-340 (1996).
5. D. Goldman and A. S. Popel, Computational modeling of oxygen transport from complex capillary networks: relation to the Microcirculation Physiome, *Adv. Exp. Med. Biol.* **471**, 555-564 (1999).
6. D. Goldman and A. S. Popel, A computational study of the effect of capillary network anastomoses and tortuosity on oxygen transport, *J. Theor. Biol.* **206**, 181-194 (2000).
7. D. D. Reneau, Jr., D. F. Bruley, and M. H. Knisely, A digital simulation of transient oxygen transport in capillary-tissue systems (cerebral grey matter), *A.I.Ch.E. J.* **15**, 916-925 (1969).
8. E. P. Salathe and C. Chen, The role of myoglobin in retarding oxygen depletion in skeletal muscle, *Math. Biosci.* **116**, 1-20 (1993).
9. T. D. Lagerlund and P. A. Low, Mathematical modeling of time-dependent oxygen transport in rat peripheral nerve, *Comput Biol. Med.* **23**, 29-47 (1993).
10. D. Goldman and A. S. Popel, A computational study of the effect of vasomotion on oxygen transport from capillary networks, *J. Theor. Biol.* **209**, 189-199 (2001).
11. B. R. Berg and I. H. Sarelius, Functional capillary organization in striated muscle, *Am. J. Physiol.* **268**, H1215-1222 (1995).
12. A. R. Pries, T. W. Secomb, T. Gessner, M. B. Sperandio, J. F. Gross, and P. Gaehtgens, Resistance to blood flow in microvessels in vivo, *Circ. Res.* **75**, 904-915 (1994).
13. C. D. Eggleton, A. Vadapalli, T .K. Roy, and A. S. Popel, Calculations of intracapillary oxygen tension distributions in muscle, *Math. Biosci.* **167**, 123-143 (2000).

CEREBRAL OXYGENATION DURING CHRONIC OBSTRUCTIVE PULMONARY DISEASE[+]

Raj Jasani, Mark Sanders, Edwin Nemoto,[*] Istvan A. Hajduk, Charles W. Atwood Jr., Patrick W. Strollo, Jr.[1]

1. INTRODUCTION

Cognitive dysfunction has been recognized in chronic obstructive pulmonary disease (COPD) patients who are hypoxic during wakefulness (1,2). COPD patients with acceptable arterial oxygen (PaO_2) during wakefulness may experience arterial oxyhemoglobin desaturation to ≤ 85% during sleep (3). It is unknown whether cerebral oxygenation is compromised in sleeping patients with COPD with or without arterial hypoxemia. Therefore, we studied the effect of spontaneous arterial desaturation in COPD patients during sleep on regional cerebral oxygenation (rSO2) to test the hypothesis that patients with COPD suffer decrements in rSO2 during sleep.

2. METHODS

All studies were done in the Sleep Laboratory at the Veterans Administration Pittsburgh Healthcare System following a protocol approved by the Research and Development Committee and Sub-committee on Human Studies of the Pittsburgh Veterans Affairs Healthcare System.

2.0 Study Design

The study is a prospective study employing simultaneous standard polysomography (PSG) which involves the simultaneous recording of electroencephalogram, electro-oculogram, submental electromyogram, airflow by thermistor, ventilatory effort by inductance plethysmography, electrocardiogram, pulse oximetry and in this study,

[1] Division of Pulmonary, Allergy and Critical Care Medicine, Department of Medicine and Department of Neurological Surgery[*], School of Medicine, and the Veterans Affairs Healthcare System, University of Pittsburgh, Pittsburgh, PA 15213

cerebral oximetry (INVOS4100, Somanetics, Inc., Troy, MI). A technologist scored the PSG recordings.

2.1 Screening Procedures

Study participants were referred to or managed by the Pulmonary Sleep Clinic or Pulmonary Medicine Clinic. Clinical probability of sleep/apnea/hypopnea and the presence or absence of COPD was determined by the patient's pulmonary or sleep physician or by clinically indicated pulmonary function test results.

2.2 Experimental Paradigm

As part of routine medical care, the patients had a medical examination, and pulmonary function tests. The patients' data were compared to the inclusion/exculsion criteria, and if eligible, invited to participate in the study. Signed informed consent was obtained. The patient spent one night in the sleep laboratory with continuous PSG, cerebral oximetry and pulse oximetry recording. Inclusion criteria: Severe COPD (FEV1/FVC ration <70% and FEV1:,50% of predicted values) and with sleep/apnea/hypopnea (Apnea + Hypopnea Index, <15). Exclusion criteria: Home oxygen dependent. Prior PSG to evaluate for sleep disordered breathing.

The definition of sleep/apnea/hypopnea was as follows: apnea: cessation of ventilation> 10 seconds; hypopnea: reduction of inspiratory airflow >30% for > 10 seconds; Apnea-Hypopnea Index: average number of apneas + hypopneas per hour of sleep.

2.3 Data Analysis

Patients with COPD without SAH (Apnea Hypopnea Index $\leq$ 15 by full-night PSG) were included in the analysis. Statistical analysis was performed by one-way ANOVA with a P value of ≤ 0.05 as significant. The data are expressed as mean $\pm$ sd.

3. RESULTS

Six males with an average age of 63.7 ± 10.6 years with a sleep efficiency of 51 ± 26 and an apnea index of 1.03 ± 0.02 were studied. Arterial oxygen parameters showed

Table 1. Arterial oxygenation parameters during sleep and awake

Parameters	Mean ± SD
Wake SpO2 (%)	95 ± 2.3
Nadir SpO2 (%)	89.2 ± 3.8
Avg. Sleep SpO2 (%)	94 ± 2.5
DEF (avg. # desaturations >4% per hr of sleep)	13.3 ± 31.3
% Sleep Time SpO2<90	6.8 ± 13.5
% Sleep Time SpO2 90-99%	93 ± 13.9

an average sleep arterial saturation of 94% and wake SaO2 of 95% (Table 1). The average nadir in SaO2 was 89% with approximately 7% of the time with SaO2 <90%.

Analysis of bilateral cerebral oximetry during periods of arterial desaturation during sleep, identified by pulse oximetry, revealed that in the 6 patients, 20 cerebral oxyhemoglobin desaturations of greater than 10% of baseline occurred coincidental with arterial desaturation. These periods of arterial desaturation resulted in an average decrease in SpO_2 from a mean of 93% to 89% and were associated with significant bilateral decreases in cerebral oxygenation from 62% to 55% (Table 2).

Table 2. Left ($LrSO_2$) and right ($RrSO_2$) hemispheric cerebral oxygenation and arterial oxygen saturation (SaO_2) in 6 patients with COPD during periods of arterial saturation and desaturation during sleep

Period (N)	$LrSO_2$	$RrSO_2$	SaO_2
Saturation (16)	$62 \pm 8^{*}$	$62 \pm 9^{\Psi}$	$93 \pm 3^{**}$
Desaturation (20)	55 ± 9	55 ± 10	89 ± 3

$^{**}p=0.0001$, $^{*}p=0.01$, $^{\Psi}p=0.03$ on comparison of saturation and desaturation episodes.

Especially interesting were observations in one of the 6 patients, of asymmetry between the hemispheres in level of cerebral oxygenation where one hemisphere may increase oxygenation while the other decreased (Fig. 1). This asymmetry is of special interest because it probably represents the development of a "steal" phenomenon which may contribute to the development of regional cerebral ischemia and stroke (5).

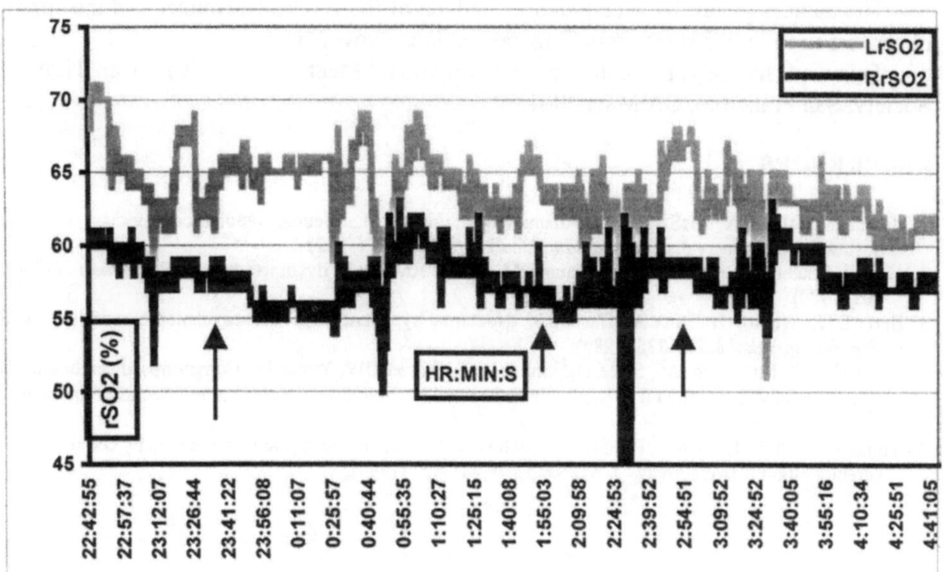

Figure 1. Cerebral oximetry on the left (LrSO2) and right (RrSO2) hemispheres in a patient during sleep. Hemoglobin oxygenation is consistently higher on the left compared to the right. Note that in specific instances (arrows) when oxygenation increases in the left hemisphere, it may either remain unchanged or decreases in the right hemisphere which may represent a "steal" phenomenon.

4. DISCUSSION

The studies suggest that significant cerebral desaturations occur during sleep in conjunction with arterial desaturation in patients with COPD who do not have PSG evidence of significant SAH. These desaturations during sleep are important in that they may contribute to the cognitive decline seen in these patients with COPD (1,2).

The observed asymmetries in cerebral oxygenation especially during episodes of desaturation is a clinically important finding in that it may represent a spontaneous test of cerebrovascular reserve which may reveal cerebrovascular hemodynamic failure (4). In addition to the desaturations, it may also account for the cognitive decline frequently observed in these patients. The presence of hemodynamic failure reflected by a decrease in cerebral blood flow due to the "steal" phenomenon may indicate a high risk for stroke. Thus, it may be prudent to monitor cerebral oxyenation in these patients as part of the monitoring armamentarium in the sleep diagnostic lab.

CONCLUSIONS

Cerebral oxygenation decreased during arterial desaturation episodes in sleeping patients with COPD. Arterial desaturations or apnea with accentuation of hemispheric asymmetries suggest compromised cerebrovascular reserve and stroke risk.

ACKNOWLEDGEMENTS

The authors gratefully acknowledge the loan of the cerebral oximeter and the supply of sensors used in this study from Somanetics, Inc., Troy, MI.

This work has been presented to the International Meeting of the American Thoracic Society, San Francisco, CA May, 2001.

REFERENCES

1. Grant I, Heaton RK, McSweny AJ, Adams KM, Timms RM. Neuropsychological findings in chronic obstructive pulmonary disease. *Arch Int. Med* 142,1470-1476 (1982)
2. Grant I, Heaton RK, McSweny AJ, Adams KM, Timms RM. Brain dysfunction in COPD. *Chest* 77, 308-309 (1980).
3. Berry DTR, Webb WB, Block AJ, Bauer RM. Nocturnal hypoxemia and neuropsychologic variables. *J Clin Exp Neuropsychol* 8, 229-238 (1986).
4. Webster MW, Makaroun MS, Steed DL, Smith HA, Johnson DW, Yonas H: Compromised cerebral blood flow reactivity is a predictor of stroke in patients with symptomatic carotid artery occlusive disease. *J Vasc Surg* 21, 338-45 (1995).
5. Yonas H, Smith H, Durham S, Pentheny SL, Johnson DW. Increased stroke riskpredicted by compromised cerebral blood flow reactivity. *J. Neurosurg*;79,483-489 (1993)

PHYSIOLOGY OF FUNCTIONAL ACTIVATION

John A. Detre*

Nearly all studies of task-specific activation using functional neuroimaging rely on the existence of a close coupling between regional changes in brain metabolism and regional cerebral blood flow (CBF), herein referred to as activation-flow coupling (AFC). Regional CBF changes are used as a surrogate marker for changes in regional brain function, and uncoupling of blood flow and metabolism or neuronal function may result in false negative or false positive activation. Changes in blood flow and metabolism occur with excitatory or inhibitory neurotransmission, both of which are energy consuming processes[1].

Studies in both animal models and human subjects using a variety of modalities have shown that in normal brain, blood flow changes occur following a latency of 0.5 to 1.5 seconds and build to a peak in approximately 4-8 seconds, even for stimuli of much shorter duration[2]. The AFC response is so stereotyped in normal subjects that it may be taken as a constant in statistical models used to analyze fMRI data. However, there are likely to be both regional variations and pathological alterations in AFC as well as individual variations in AFC, even in normals[3]. For example, in patients with severe stenosis of arteries supplying a given brain region, delayed vascular transit times may be observed along with an attenuation of functionally induced hemodynamic changes[4,5].

Currently, the most popular means of visualizing regional brain activation is using functional magnetic resonance imaging (fMRI) with blood oxygenation level dependent (BOLD) contrast. BOLD contrast reflects a complex interaction between blood flow, blood volume, and hemoglobin oxygenation[6-8]. Functional contrast is obtained because the iron present in hemoglobin becomes paramagnetic only when it is deoxygenated[9], producing a local susceptibility increase manifested as a change in T2*, among other effects . This change in hemoglobin oxygenation can be observed using a variety of pulse sequences, including routine gradient echo sequences and gradient echo echoplanar sequences[10] which particularly emphasize T2* effects. With regional brain activation, a reduction in T2* is observed, reflecting a decrease in regional deoxyhemoglobin which has been attributed to increases in CBF which exceed increases in oxygen metabolism,

* John A. Detre, M.D., Department of Neurology, University of Pennsylvania School of Medicine, 3400 Spruce St., Philadelphia, PA 19104

though the precise basis for this mismatch is uncertain. Some modeling studies based on the physiology of blood flow changes and oxygen diffusion suggest that large changes in flow may be required to increase oxygen delivery even to a small extent[11]. There are several shortcomings to the use of BOLD contrast including the lack of absolute quantification and difficulties in measuring changes in BOLD contrast in brain regions with high static susceptibility.

BOLD contrast is not the only approach available for fMRI. Over the past decade, noninvasive methods for direct measurement of cerebral blood flow (CBF) using MRI have also been developed[12-14]. This class of techniques is termed arterial spin labeled (ASL) perfusion MRI, and utilizes electromagnetically labeled arterial blood water as endogenous tracer for measuring CBF. This approach is capable of providing sequential multislice CBF images of human brain which are suitable for studying regional brain metabolism in response to cognitive and sensorimotor tasks, pharmacological intervention, and other long-term effects such as recovery or response to therapy. Preliminary data demonstrate that task-activation measured by ASL shows less inter-subject variability than BOLD contrast, that ASL contrast shows stable noise characteristics over the entire frequency spectrum, making it suitable for studying sequential changes in resting or activated brain function over periods greater than a few minutes. Functional signal changes detected by ASL also have superior spatial and temporal resolution as compared to BOLD contrast[15]. Furthermore, the effects of ASL can be measured using imaging sequences that are insensitive to static susceptibility effects. Over the next several years, an increased use of ASL approaches for fMRI is anticipated.

Surprisingly little is known about the mechanism of AFC. The phenomenon was originally described in 1890 by Roy and Sherrington[16]. Studies since that time have failed to identify a specific hormonal or metabolic factor that mediates AFC[2]. The recent identification of nitric oxide gas as a highly vasoactive and rapidly diffusible substance has suggested the possibility of this neuro-chemical as a mediator, but a variety of other metabolites and even direct synaptic connections have also been proposed. Even the purpose of AFC remains undetermined. While it is often assumed that a regional CBF increase is required to supply oxygen and nutrients, studies of brain energy metabolism in response to functional activation have not conclusively supported this idea[17], and more recent data confirm that cerebral blood flow changes are not regulated by glycolytic demands[18]. Additionally, blood flow effects can be pharmacologically decoupled from brain activation without loss of at least electrophysiological responses[19]. An alternative possibility is that regional blood flow increases to remove toxic waste products of metabolism. Several lines of evidence suggest that the oxygen to glucose ratio decreases during functional activation, resulting in lactate accumulation that must be cleared. It has been speculated that this is at least partially attributable to the use of glycogen as an energy source during functional activation[20].

The extent to which non-neuronal constituents of brain parenchyma contribute to the overall metabolic rate is also uncertain, and may be variable. An extensive body of research now suggests that much of the glucose utilized during glutamatergic neurotransmission is actually consumed by astrocytes[21]. These cells have glucose transporters on their end-feet directly abutting the capillaries, and terminate neurotransmission through glutamate uptake from the synaptic cleft. Glutamate is cycled back to neurons as glutamine, and lactate is at least partially cycled back to neurons to be oxidized. Blood flow in functionally activated brain can be increased by lactate infusion and decreased by pyruvate infusion, suggesting that redox state may be an important mediator of AFC[22].

Changes in tissue oxygen and oxygen consumption during functional activation are also poorly understood. Early PET studies during sustained visual stimulation in humans suggested an uncoupling of glycolytic and oxidative metabolism, with both CBF and CMRGlu increasing about 50% with stimulation and CMRO2 remaining minimally changed[17]. However, near infrared optical spectroscopy studies suggested an early increase in deoxyhemoglobin which is then washed out by a much larger and less well localized increase in oxyhemoglobin due to increased CBF[23]. Similar early increases in deoxyhemoglobin have been observed with BOLD fMRI, primarily using visual stimulation in animals and humans at high magnetic field strengths[24,25]. Evidence for an early decrease in deoxyhemoglobin and therefore an increase in oxygen consumption has also been obtained using phosphorescence quenching methods[26,27], and more recently by direct oxygen electrode measurements in tissue[28]. This latter study also demonstrated even larger decreases in tissue oxygen following prolonged stimulation, which were not accompanied by increases in flow, suggesting that decreased tissue oxygen is not the signal for blood flow increases. This has also been confirmed in recent PET studies in human subjects in which CBF with visual stimulation did not increase with hypoxia as compared to normoxia[29].

The spatial resolution limits for functional imaging have begun to be characterized in brain regions with known functional organization. Ocular dominance columns and iso-orientation "pinwheels" have been imaged in primary visual cortex[30,31], and discrete activation due to stimulation of adjacent vibrissae has been observed in whisker barrel cortex[32]. Comparison of fMRI of motor cortex with intraoperative mapping by electrocorticography has revealed accuracy of fMRI to within about 1 cm, which has also been confirmed in a recent primate study[33]. Analysis of microelectrode recordings obtained during BOLD fMRI suggests that BOLD signal changes correlate best with local field potentials, rather than axonal firing patterns[34].

REFERENCES

1. Nudo, R.J. & Masterson, R.B., Stimulation-induced [14C]2-deoxyglucose labeling of the synaptic activity in the central auditory system. *J. Comp. Neurol.* **245**, 553-565 (1986).
2. Villringer, A. & Dirnagl, U., Coupling of brain activity and cerebral blood flow: basis of functional neuroimaging. *Cereb. Brain Metab. Rev.* **7**, 240-276 (1995).
3. Aguirre, G.K., Zarahn, E. & D'Esposito, M., The variability of human BOLD hemodynamic responses. *Neuroimage* **8**, 360-9 (1998).
4. Powers, W.J., Fox, P.T. & Raichle, M.E., The effect of carotid artery disease on the cerebrovascular response to physiologic stimulation. *Neurology* **38**, 1475-8 (1988).
5. Stoll, M., Seidel, A., Schimrigk, K. & Hamann, G.F., Hand gripping and acetazolamide effect in normal persons and patients with carotid artery disease. *J. Neuroimaging* **8**, 27-31 (1998).
6. Kwong, K.K., Belliveau, J.W., Chesler, D.A., Goldberg, I.E., Weisskoff, R.M., Poncelet, B.P., Kennedy, D.N., Hoppel, B.E., Cohen, M.S. & Turner, R., Dynamic magnetic resonance imaging of human brain activity during primary sensory stimulation. *Proc. Natl. Acad. Sci. USA* **89**, 5675-5679 (1992).
7. Ogawa, S., Menon, R.S., Tank, D.W., Kim, S.G., Merkle, H., Ellerman, J.M. & Ugurbil, K., Functional brain mapping by blood oxygenation level-dependent contrast magnetic resonance imaging. A comparison of signal characteristics with a biophysical model. *Biophys. J.* **64**, 803-12 (1993).
8. Mandeville, J.B., Marota, J.J.A., Ayata, C., Moskowitz, M.A., Weisskoff, R.M. & Rosen, B.R., MRI measurement of the temporal evolution of relative CMRO2 during rat forepaw stimulation. *Magn. Reson. Med.* **42**, 944-51 (1999).

9. Thulborn, K.R., Waterton, J.C., Matthews, P.M. & Radda, G.K., Oxygenation dependence of the transverse relaxation time of water protons in whole blood at high field. *Biochim. Biophys. Acta* **714**, 265-70 (1982).
10. Stehling, M.K., Turner, R. & Mansfield, P., Echo planar imaging: magnetic resonance imaging in a fraction of a second. *Science* **254**, 43-50 (1991).
11. Buxton, R.B. & Frank, L.R., A model for the coupling between cerebral blood flow and oxygen metabolism during neural stimulation. *J. Cereb. Blood Flow Metab.* **17**, 64-72 (1997).
12. Detre, J.A., Leigh, J.S., Williams, D.S. & Koretsky, A.P., Perfusion imaging. *Magn. Reson. Med.* **23**, 37-45 (1992).
13. Kwong, K.K., Chesler, D.A., Weisskoff, R.M., Donahue, K.M., DAvis, T.L., Ostergaard, L., Campbell, T.A. & Rosen, B.R., MR perfusion studies with T1-weighted echo planar imaging. *Magn. Reson. Med.* **34**, 878-887 (1995).
14. Kim, S.G. & Tsekos, N.V., Perfusion imaging by a flow-sensitive alternating inversion recovery (FAIR) technique: application to functional brain imaging. *Magn. Reson. Med.* **37**, 425-35 (1997).
15. Silva, A.C., Lee, S.-P., Yang, G., Iadecola, C. & Kim, S.-G., Simultaneous blood oxygenation level-dependent and cerebral blood flow functional magnetic resonance imaging during forepaw stimulation in the rat. *J. Cereb. Blood Flow Metab.* **19**, 871-9 (1999).
16. Roy, C.S. & Sherrington, C.S., On the regulation of the blood-supply of the brain. *J. Physiol.* **11**, 85-108 (1890).
17. Fox, P.T. & Raichle, M.E., Focal physiological uncoupling of cerebral blood flow and oxidative metabolism during somatosensory stimulation in human subjects. *Proc. Natl. Acad. Sci. USA* **83**, 1140-1144 (1986).
18. Powers, W.J., Hirsch, I.B. & Cryer, P.E., Effect of stepped hypoglycemia on regional cerebral blood flow response to physiological brain activation. *Am. J. Physiol.* **270**, H554-9 (1996).
19. Lindauer, U., Megow, D., Schultze, J., Weber, J.R. & Dirnagl, U., Nitric oxide synthase inhibition does not affect somatosensory evoked potentials in the rat. *Neurosci. Lett.* **216**, 207-10 (1996).
20. Shulman, R.G., Hyder, F. & Rotheman, D.L., Cerebral energetic and the glycogen shunt: Neurochemical basis of functional imaging. *Proc. Natl. Acad. Sci. USA* **98**, 6417-22 (2001).
21. Magistretti, P.J. & Pellerin, L., Cellular mechanisms of brain energy metabolism and their relevance to functional brain imaging. *Phil. Trans. Soc. Lond. B* **354**, 1155-63 (1999).
22. Ido, Y., Chang, K., Woolsey, T.A. & Williamson, J.R., NADH: sensor of blood flow need in brain, muscle, and other tissues. *FASEB* **15**, 1419-21 (2001).
23. Malonek, D. & Grinvald, A., Interactions between electrical activity and cortical microcirculation revealed by imaging spectroscopy: implications for functional brain mapping. *Science* **272**, 551-4 (1996).
24. Menon, R.S., Hu, X., Andersen, P., Ugurbil, K. & Ogawa, S., Cerebral oxy/deoxy hemoglobin changes during neural activation: MRI timecourse correlates to optical reflectance measurements. *SMR 2nd Meeting, Book of Abstracts* 68 (1994).
25. Yacoub, E., Le, T.H., Ugurbil, K. & Hu, X., Further evaluation of the initial negative response in functional magnetic resonance imaging. *Magn. Reson. in Med.* **41**, 436-41 (1999).
26. Vanzetta, I. & Grinvald, A., Increased cortical oxidative metabolism due to sensory stimulation: implications for functional brain imaging. *Science* **286**, 1555-8 (1999).
27. Ances, B.M., Wilson, D.F., Greenberg, J.H. & Detre, J.A., Dynamic changes in CBF, O2 tension and calculated CMRO2 during functional activation using oxygen phosphorescence quenching. *J. Cereb. Blood Flow Metab.* **21**, 511-6 (2001b).
28. Ances, B.M., Buerk, D.G., Greenberg, J.H. & Detre, J.A., Temporal dynamics of the partial pressure of brain tissue oxygen during functional forepaw stimulation in rats. *Neurosci. Lett.* **306**, 106-10 (2001a).
29. Mintun, M.A., Lundstrom, B.N., Snyder, A.Z., Vlassenko, A.G., Shulman, G.L. & Raichle, M.E., Blood flow and oxygen delivery to human brain during functional activity: THeoretical modeling and experimental data. *Proc. Natl. Acad. Sci. USA* **98**, 6859-64 (2001).
30. Menon, R.S., Ogawa, S., Strupp, J. & Ugurbil, K., Ocular dominance in human V1 demonstrated by functional magnetic resonance imaging. *J. Neurophys.* **77**, 2780-7 (1997).
31. Kim, D.S., Duong, T.Q. & Kim, S.G., High-resolution mapping of iso-orientation columns by fMRI. *Nat. Neurosci.* **3**, 164-9 (2000).
32. Yang, X., Hyder, F. & Shulman, R.G., Activation of single whisker barrel in rat localized by functional magnetic resonance imaging. *Proc. Natl. Acad. Sci. USA* **93**, 475-8 (1996).
33. Disbrow, E.A., Slutsky, D.A., Roberts, T.P. & Krubitzer, L.A., Functional MRI at 1.5 tesla: a comparison of the blood oxygenation level-dependent signal and electrophysiology. *Proc. Natl. Adcad.Sci. USA* **97**, 9718-23 (2000).
34. Logothesis, N.K., Pauls, J., Augath, M., Trinath, T. & Oeltermann, A., Neurophysiological investigation of the basis of the fMRI signal. *Nature* **412**, 150-7 (2001).

NONLINEAR ANALYSIS OF HEART RATE VARIABILITY IN INFANTS WITH APPARENT LIFE-THREATENING EVENTS

Philippe Lemmerling, Sabine Van Huffel, Gunnar Naulaers, Hans Daniels, and Hugo Devlieger*

1. ABSTRACT

Some studies have shown that neonates having experienced an Apparent Life-Threatening Event (ALTE) have an increased risk of dying of the Sudden Infant Death Syndrome (SIDS)[1]. In this study the heart rate variability in neonates has been analyzed through a study of the long range correlations and scale invariance (self-similarity) in their cardiac rhythm. The goal is to see whether it is possible to distinguish normal infants from infants with ALTEs using the previously mentioned nonlinear measures of the heart rate variability.

Twelve ALTE infants are included in this study. To determine the degree to which these neonates suffer from ALTEs two examinations are performed: a polysomnography[1] and home monitoring. Using the nonlinear measures obtained from the nonlinear analysis of the heart rate variability, the group suffering most from ALTEs can clearly be separated from the other neonates.

* Philippe Lemmerling and Sabine Van Huffel, SISTA/COSIC Division, Dept. of Electrical Engineering (ESAT), K.U.Leuven, Kasteelpark Arenberg 10, 3001 Leuven, Belgium. Gunnar Naulaers, Hans Daniels and Hugo Devlieger, Dept. of Paediatrics, University Hospital Leuven, Herestraat 49, 3000 Leuven, Belgium

[1] i.e. the simultaneous recording of physiological signals, such as ECG, heart rate, peripheral oxygen saturation, respiration, EEG, EOG and EMG, during 8 to 12 hours.

Oxygen Transport to Tissue XXIII, edited by D. Wilson *et al.*
Kluwer Academic/Plenum Publishers, 2003

2. INTRODUCTION

Some studies have indicated that the occurrence of ALTEs is a risk factor for SIDS. Therefore, a measure predicting the degree to which a neonate suffers from ALTEs could be used as a measure to predict the risk for a neonate to die of SIDS.
In this study twelve ALTE infants are included. Two examinations are performed in order to determine in an objective way the degree to which a neonate suffers from ALTEs: a polysomnography in the hospital and home monitoring as follow-up. In this way the group of ALTE neonates can be split up into four classes: AA (abnormal polysomnography, abnormal follow-up (3 babies)), NA (normal polysomnography, abnormal follow-up (1 baby)), AN (abnormal polysomnography, normal follow-up (2 babies)) NN: normal polysomnography, normal follow-up (6 babies)). A polysomnography is defined as being abnormal when one of the following events occur during the examination:

- a bradycardia of less than 50 bpm for at least 4 seconds
- an obstructive apnea of at least 15 seconds with a concurrent bradycardia of less than 60 bpm and/or concurrent oxygen desaturation of less than 80%.

An abnormal follow-up (i.e. home monitoring) is defined as the occurrence of a bradycardia of less than 50 bpm during the follow-up.

In addition a control group of 7 healthy infants is included. It is clear that the neonates most at risk for ALTEs are those in class AA. The heart rate variability is analyzed by means of the Detrended Fluctuation Analysis (DFA)[2,3] of the RR time series, which consists of the duration of the time intervals between successive R peaks in the electrocardiogram (ECG). As explained in the next section, four numbers are derived from the obtained DFA function. They correspond to the very short term (α_0), the short term (α_1), the long term (α_2) and the very long term (α_3) scale-invariant (fractal) correlation property of the RR time series. The rationale behind this approach is the following. Recent research[4] indicated that a deficit in the arcuate nucleus may be partially responsible for some SIDS deaths. Further research[5] suggested the existence of cellular connections between the arcuate nucleus and the caudal raphe, a region implicated in cardiorespiratory control. Deficiencies in cardiorespiratory control will be reflected in the RR time series and thus it makes sense to analyze this time series as possible source of information about SIDS.

3. METHODS

3.1. Derivation of RR time series

Due to the long duration of the ECG recordings (on average 8 hours) the RR time series has to be extracted in an automated way from the ECG. The R peaks are detected using the second order derivatives of the ECG followed by an automated validation/correction of the obtained RR time series.

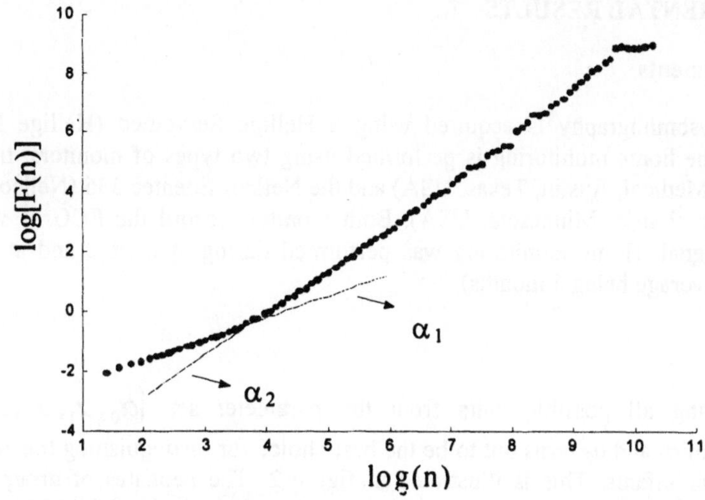

Figure 1. This figure shows the DFA function F(n) as a function of the frame length. Straight lines indicate self-similarity.

3.2. DFA

The DFA analysis derives a function F(n) from the RR time series r(i), i=1,...,N in the following way. First r(i) is integrated over time and shifted by its average r_{ave} over the N data points: $y(k)=\sum_{i=1}^{k} r(k) - r_{ave}$. Next the integrated time series y is divided into frames of equal length n. In each frame of length n, a straight line is fitted in least squares sense to y, yielding a new function $y_n(k)$, representing the local trend in each frame. Next the DFA function F(n) is obtained as the mean square fluctuation of the detrended integrated time series: $F(n)=\sqrt{\sum_{k=1}^{N}(y(k)-y_n(k))^2/N}$. Note that the different steps described before are necessary to obtain information on the self-similarity of the RR time series while taking into account its nonstationary character. Furthermore, a linear relation on a double log graph (see figure 1) indicates the presence of self-similarity (i.e. fluctuations over small frames are related to fluctuations over large frames in a power-law fashion). The determination of α_1 and α_2 is done by selecting an index n_{break} such that the global mean square error of two straight lines fitted to F(n) over the intervals log(n)=[2 ... n_{break}] and log(n)=[n_{break} ...6] is as small as possible, where log stands for the natural logarithm. The slope of the straight line fitted on the first interval is α_1 whereas the slope of the straight line fitted on the second interval is α_2 (see figure 1). Two additional lines are fitted to the DFA function over the interval to the left of log(n)=2 and to the interval log(n)=[7 ... 9.5], since their slopes α_0 and α_3 tend to deviate from respectively α_1 and α_2. Note that these different α's are called the scale-invariant (fractal) correlation properties of the RR time series.

4. EXPERIMENTAL RESULTS

4.1. Measurements

The polysomnography is acquired using a Hellige Servomed (Hellige Freiburg, Germany). The home monitoring is performed using two types of monitors: the Arvee 4800 (Arvee Medical, Austin, Texas, USA) and the Nellcor Edentec 336 (Nellcor Puritan Bennett, Eden Prairie, Minnesota, USA). Both monitors record the ECG as well as a respiratory signal. Home monitoring was performed during at least 2 and at most 18 months (the average being 5 months).

4.2. Results

Considering all possible pairs from the parameter set $\{\alpha_0, \alpha_1, \alpha_2, \alpha_3\}$, the combination of α_1 and α_2 turns out to be the best choice for distinguishing the AA group from the other infants. This is illustrated in figure 2. The neonates of group AA are represented by a "*", those of the NA group by a white disk, those of the NN group by a "+", those of the AN group by a "x" and those of the control group by a grey disk. The AA group is clearly characterized by a combination of higher α_1 values and lower α_2 values when compared to infants belonging to the other groups. Note however that there is one neonate of the AA group that has a significantly lower α_1 value than the other members of the AA group. It turned out that this neonate suffers from the long QT syndrome and should thus be considered as a special case. It should be mentioned that although some authors have indicated a long QT interval as one of the causes of SIDS[6] this hypothesis is rather controversial[7,8,9].

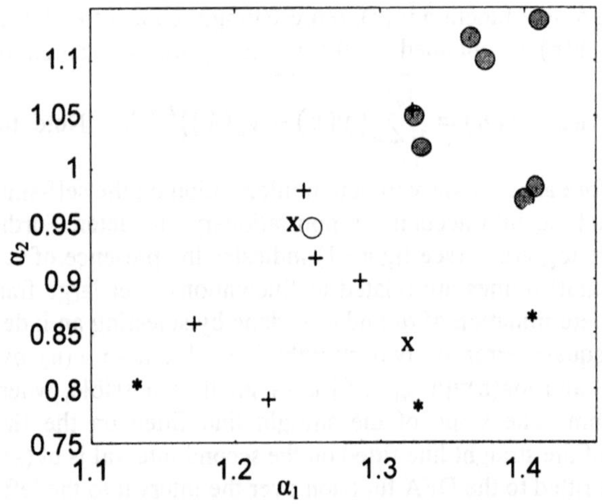

Figure 2. This figure illustrates the position of the members of the AA group ("*"), the NA group (white disk), the NN group ("+"), the AN group ("x") in the α_1- α_2 plane, where α_1 and α_2 are nonlinear measures derived from the RR time series.

CONCLUSIONS

Although the group of patients is too small for a statistical analysis, it is clear that a combination of α_1 and α_2 enables to distinguish healthy infants from infants suffering to a high degree from ALTEs (the AA group). While long-term correlations are most persistent for normal infants (α_2 close to 1), ALTE babies have a smaller α_2 index, pointing out a decreased long-term correlation, in particular for the AA group, which might indicate a higher risk for SIDS. Within the ALTE group, the NN group can be clearly separated from the AA group. The distinction between the NA, AN and AA groups is less clear. A larger follow-up study is needed to confirm the performance of the nonlinear measures, and to evaluate the predictive value of the different measures with respect to SIDS.

ACKNOWLEDGMENTS

Philippe Lemmerling is a post doctoral researcher of the FWO (Fonds voor Wetenschappelijk Onderzoek). This paper presents research results of the Belgian Programme on Interuniversity Poles of Attraction (IUAP P4-02 and P4-24), initiated by the Belgian State, Prime Minister's Office - Federal Office for Scientific, Technical and Cultural Affairs, of the Brite Euram Programme, Thematic Network BRRT-CT97-5040 'Niconet', of the Concerted Research Action (GOA) projects of the Flemish Government MEFISTO-666 (Mathematical Engineering for Information and Communication Systems Technology), of the IDO/99/03 project (K.U.Leuven) "Predictive computer models for medical classification problems using patient data and expert knowledge", and of the FWO (Fund for Scientific Research Flanders) projects G0200.00 and G078.01.

REFERENCES

1. Platt M.W., Blair P.S., Fleming P.J., Smith I.J., Cole T.J., Leach C.E., Berry P.J., Golding J., 2000, A clinical comparison of SIDS and explained sudden infant deaths: how healthy and how normal? CESDI SUDI Research Group. Confidential Inquiry into Stillbirths and Deaths in Infancy study. Arch. Dis. Child. (England), 82(2), pp. 98-106.
2. Peng C. K., Buldyrev S.V., Havlin S., Simons M., Stanley H.E., Goldberger A.L., 1994, On the mosaic organization of DNA sequences, Phys. Rev. E, 49, pp.1685-1689.
3. Peng C.K., Havlin S., Stanley H.E., Goldberger A.L., 1995, Quantification of scaling exponents and crossover phenomena in nonstationary heartbeat time series, Chaos, 5, pp. 82-87.
4. Kinney H.C., Filiano J.J., Sleeper L.A., Mandell F., Valdes-Dapena M., White W.F., 1995, Decreased muscarinic receptor binding in the arcuate nucleus in sudden infant death syndrome, Science (United States), 269(5229), pp. 1446-1450.
5. Zec N., Filiano J.J., Kinney H.C., 1997, Anatomic relationships of the human arcuate nucleus of the medulla: a DiI-labeling study, J. Neuropathol. Exp. Neurol., 56(5), pp. 509-22
6. Schwarz P.J., Stramba-Badiale M., Segantini A., Ausoni P., Bosi G., Giorgetti R., Grancini F., Marni E.D., Perticone F., Rosti D., Salice P., 1998, Prolongation of the QT interval and the sudden infant death syndrome, N Engl J Med., 338(24), pp. 1709-1714.
7. Guntheroth W.G., Spiers P.S., 1999, Prolongation of the QT interval and the sudden infant death syndrome, Pediatrics, 103, pp. 813-814.
8. Hoffman J.I., Lister G., 1999 The implications of a relationship between prolonged QT interval and the sudden infant death syndrome, Pediatrics, 103, pp. 815-817.
9. Shannon D.C., 1999, Method of analyzing QT interval can't support conclusions, Pediatrics, 103, p. 819.

THE EFFECT OF NICOTINAMIDE & HYPEROXIC GASES ON BLOOD GLUCOSE

Rachel A. Richardson, and Mark W. Dewhirst*

1. INTRODUCTION

Hypoxia plays a major role in tumor clinical behavior and response to therapy.[1,2] Many methods have been used to try to increase tumor pO_2 including Nicotinamide (Nic) ± hyperoxic gases.[3] In a recent study reported by Robinson et al., carbogen/Nic induced hyperglycemia[4]. This phenomenon had not been previously reported.

Hyperglycemia can be used to decrease O_2 consumption thereby increasing tumor pO_2. This process is known as the Crabtree effect. The Crabtree effect is the process where tumor cells utilize excess glucose via glycolysis and reduce their oxygen consumption rate even in the presence of O_2.[5] It was not clear in the study by Robinson et al., whether the increase in blood glucose was due to Nic, carbogen, or a combination of the two. Therefore, the intent of this study was to further investigate the effects that Nic and hyperoxic gases have on blood glucose.

2. MATERIALS & METHODS

Six different treatment regimens (n=10-14) were investigated in female Fischer 344 rats including: 1) Nic, 2) saline, 3) Nic & carbogen, 4) Nic & O_2, 5) saline & carbogen, and 6) saline & O_2. Rats were anesthetized with Nembutal (50 mg/kg i.p.) and the femoral artery was cannulated for monitoring blood pressure with Digital Manometer (Fiber optic Sensor Technologies Inc., Ann Arbor MI) and obtaining samples for blood glucose. Blood glucose levels were determined with an Accu-chek III blood glucose monitoring system (Mannheim Boehringer, Indianapolis, IN). Rats were placed on a heating pad and body temperature was maintained at 37°C. Baseline measurements were taken after the animal was anesthetized and placed on air for ten minutes. Then, animals were given either Nic (200 mg/kg) or an equivalent volume of saline and two blood glucose samples were taken at 5 and 15 minutes post injection, while still breathing air.

Mark W. Dewhirst, Department of Radiation Oncology, Duke University Medical Center, Duke University, Durham, NC, 27710, U.S.A. FAX 919-684-8718, Email: dewhirst@radonc.duke.edu

Oxygen Transport to Tissue XXIII, edited by D. Wilson *et al.*
Kluwer Academic/Plenum Publishers, 2003

Then they were switched to either carbogen or O_2 and additional blood samples were taken after 10, 20, and 30 minutes.

3. RESULTS

No treatment affected blood pressure. Blood glucose concentrations increased by approximately 9% in response to Nic ($p < 0.015$). Conversely, saline injection had no effect ($p > 0.051$). Although Nic was significantly different from its own baseline, when it was compared to saline there was no significant difference ($p > 0.05$).

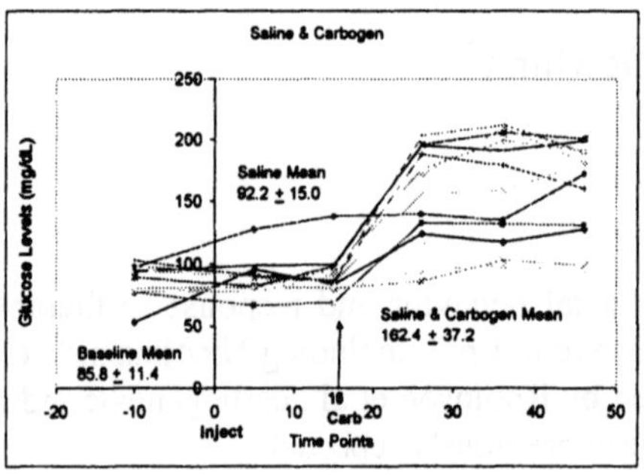

1a. Saline & Carbogen

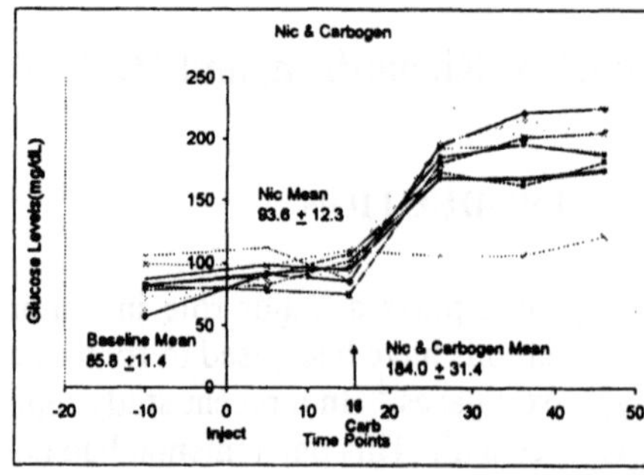

1b. Nic & Carbogen

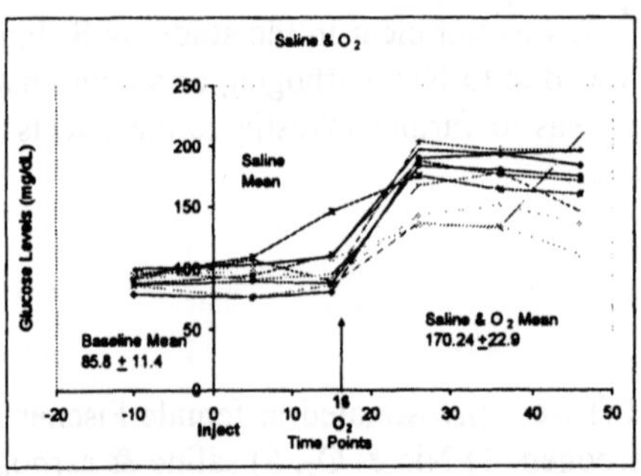

1c. Saline & O_2

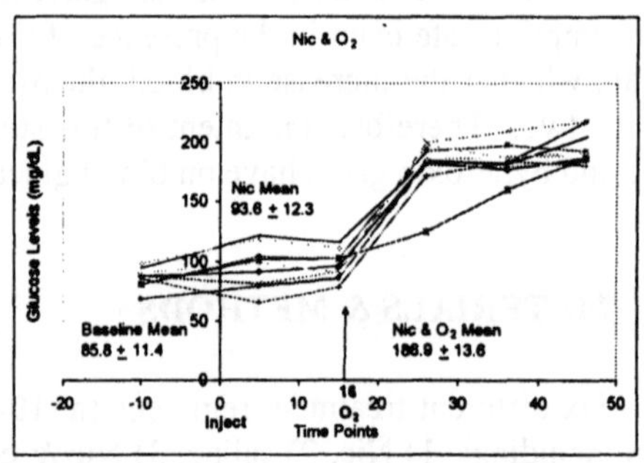

1d. Nic & O_2

Figure 1. Effects of Nic & Carbogen/O_2 on Blood glucose.

There was an approximately two-fold increase in blood glucose with both carbogen and 100% O_2. Nic did not affect the increase in blood glucose.

Both hyperoxic gases induced a significant increase in blood glucose compared to Nic or saline alone ($p < 0.01$), and there was no significant difference between hyperglycemia induced by carbogen vs. 100% O_2 independent of the presence of Nic ($p > 0.10$).

3. DISCUSSION

The results of our study indicate that the rise in blood glucose post-Nic is likely not biologically meaningful because it was not significantly different from saline alone, suggesting that it may be the stress of injection that caused the increase in blood glucose not the Nic itself.

The relative changes in glucose due to carbogen and 100% O_2 were not different, suggesting it is O_2 and not CO_2 that induces hyperglycemia. The mechanism by which hyperglycemia occurs after hyperoxic gas treatment is most likely glycogenolysis due to a stress response. It is most likely glycogenolysis because it is this process in the liver that regulates the minute-to-minute levels of blood glucose, and the increase in blood glucose levels after treatment with hyperoxic gas was very rapid (within 10 minutes).[6]

Even though there is a significant increase in blood glucose with this method it may not be an optimal way to induce the Crabtree effect and increase tumor pO_2. In studies by Brizel et al., carbogen breathing did not improve tumor pO_2 or response to radiation, suggesting that a carbogen induced increase in blood glucose concentration probably had no effect on O_2 consumption and hence tumor pO_2.[7,8] In contrast, in a recent study we demonstrated that glucose administration (1g/kg i.v.) 20 minutes prior to O_2 breathing yielded supra-additive effects on tumor pO_2, changing it an average of 27mmHg.[9] In contrast, glucose or O_2 alone only increased tumor pO_2 by 3-5mm Hg. This raises an interesting conundrum since the hyperglycemia associated with carbogen/O_2 breathing appears to be less effective in increasing tumor pO_2 than pretreatment with glucose followed by O_2 breathing. There are two possible reasons why this difference exists. First, it could be the magnitude of hyperglycemia. In prior work we achieved a four-fold increase, whereas hyperoxic gas alone only increased glucose by 1.8-fold.[9] Second, it could be that timing of hyperglycemia induction is important in inducing the Crabtree effect.

ACKNOWLEDGMENTS

This work was supported by a grant from the NIH/NCI CA40355

REFERENCES

1. D. Brizel, S.P. Scully, J.M. Harrelson, L.J. Layfield, J.M. Bean, L.R. Prosnitz, and M.W. Dewhirst, Tumor oxygenation predicts for the likelihood of distant metastases in human soft tissue sarcoma, *Canc Res* **56**, 941-943 (1991).
2. M. Höckel,C. Knoop, K. Schlenger, B. Vorndran, E. Baussman, M. Mitze, M. Knapstein and P. Vaupel, Intratumoral pO_2 predicts survival in advanced cancer of the uterine cervix, *Radiother Onc* **26**, 45-50 (1993).
3. D.K. Kelleher, and P.W. Vaupel, Nicotinamide exerts different acute effects on microcirculation function and tissue oxygenation in rat tumors, *Int J Radiat Oncol Biol Phys* **26**, 95-102 (1993).
4. S.P.Robinson, F.A. Howe, M. Stubbs and J.R. Griffiths, Effects of nicotinamide and carbogen on tumour oxygenation, blood flow,energetics, and blood glucose levels, *Br J Cancer* **82,** 2007-20014 (2000).
5. H.G. Crabtree, Observations of the carbohydrate metabolism of tumors, *Biochem J* **23**, 536-545 (1929).
6. A.L. Lehninger, D.L. Nelson, and M.M. Cox, in: *Principles ofBiochemistry,* (Worth, New York: 1993).
7. D.M. Brizel, W.D. Hage, R.K. Dodge, M.T. Munley, C.A. Piantadost and M.W. Dewhirst, Hyperbaric oxygen improves tumor radiation response significantly more than carbogen/nicotinamide, *Rad Res* **147**, 715-720 (1997).

8.D.M. Brizel, S. Lin, J.L. Johnson, M.W.Dewhirst and C.A. Piantadosi, The mechanisms by which hyperbaric oxygen and carbogen improve tumor oxygenation, *Br J Cancer* 72 1120-1124 (1995).
9.S.A. Snyder, J.L. Lanzen, R.D. Braun, G. Rosner, T. Secomb, J. Biaglow and M.W. Dewhirst, Simultaneous administration of glucose and hyperoxic gas achieves greater improvement in tumor oxygenation than by hyperoxic gas alone, *Int J Radiat Oncol Biol Phys,* In Press.

MEASUREMENT OF TISSUE OXYGENATION INDEX DURING THE FIRST THREE DAYS IN PREMATURE BORN INFANTS

Gunnar Naulaers, Geert Morren, Sabine Van Huffel, Paul Casaer, and Hugo Devlieger[1]

1. ABSTRACT

No normal values of tissue oxygenation index (TOI) of the brain are known regarding premature born infants. We measured TOI, a measure for the cerebral hemoglobin oxygen saturation, on the head of 15 preterm infants with a median postmenstrual age of 28 weeks (interquartile range (IQR) between 26-29 weeks) with spatially resolved spectroscopy (NIRO 300, Hamamatsu®) during the first three days of life. Infants with intra-ventricular hemorrhage or periventricular leucomalacia before the first measurement, as shown by ultrasound, were excluded. The first measurement was done within the first 6 hours of life, the second and third measurement at, respectively, 24 and 48 hours after this first measurement. The mean TOI was calculated if saturation did not change by more than 5% for at least 30 minutes. Other parameters measured were P_aO_2, PCO_2, pH, mean arterial blood pressure, heart rate, hemoglobin, glycemia and peripheral oxygen saturation.

There was a significant increase of TOI after 24 ($p< 0.05$) and 48 ($p < 0.001$) hours. The median TOI on the first day was 57% (95% CI : 54-65.7), 66.1% on the second day (95%CI : 61.9-82.3 %) and 76.1% on the third day (95%CI 67.8-80.1 %). No correlation was found between TOI and peripheral oxygen saturation, blood pressure, PaO_2, $PaCO_2$ and hemoglobin concentration after multiple regression analysis.

TOI increases in the first three days in premature born babies. The increase of TOI is not due to an increase of oxygenation or mean arterial blood pressure. In our opinion, it reflects the increase in cerebral blood flow during the first three days.

---

*G.Naulaers, P. Casaer and H. Devlieger, Department of Paediatrics, University Hospital Leuven, Belgium.
G. Morren and S. Van Huffel, ESAT/SISTA-COSIC, Dept of Electr Engineering, KU Leuven, Belgium.

2. INTRODUCTION

Spatially resolved spectroscopy (SRS) is a new NIRS method that measures cerebral hemoglobin oxygen saturation. The measurement of the Tissue Oxygenation Index (TOI) is realized by means of a light detector with three sensors at different distances from the light source. If light with an intensity I_0 is sent into tissue, the light is attenuated in the tissue due to scatter loss and absorption loss. If the distance between the light source and the sensor is large enough, the isotropy of scatter distribution becomes so homogeneous that the scatter loss becomes the same at the three sensors. Therefore any differences in intensity measured at the three different sensors can be interpreted as differences in absorption loss. Thus the local absorption change can be seen as a function of the distance between the light source and the sensors. The further mathematical approach has been previously reported (1,2).

TOI values were measured in healthy adult subjects by different groups and with different NIRS instruments (3-7) and no differences between the right and left forehead TOI values were found. Quaresima et al (3) compared the TOI values with the measurement of cerebral venous oxygen saturation (SvO_2) measured by near-infrared spectroscopy in healthy adult volunteers and concluded that TOI mainly reflected the intracranial venous compartment of circulation. As far as we know, no normal TOI values of premature born babies have been reported till now. We measured the TOI at days 1,2 and 3 to get a first impression of the "normal" values in this population and to see if it changes with age.

3. PATIENTS AND METHODS

3.1 Patients

Fifteen patients with a postmenstrual age of less than 31 weeks participated in the study. The median postmenstrual age was 28 weeks (range 25-30 weeks). A brain ultrasound was performed before starting up the measurement in all patients. Exclusion criteria were : an abnormal brain ultrasound before the measurement was started (i.e. intraventricular bleeding or periventricular leucomalacia), severe pulmonary hypertension as evidenced by echocardiography and congenital malformations.

3.2 Methods

NIRO 300 (Hamamatsu®) was used for spatially resolved spectroscopy. Only one optode was used and placed at the right frontoparietal side with 4 cm interoptode distance. All patients were measured within six hours after birth for at least 30 minutes. The second and third measurements were done 24 and 48 hours later.

The data were recorded in an analogous way with a sampling frequency of 100 Hz by the data acquisition system Codas (Dataq Instruments®, USA) and stored on a PC. The NIRO-300 signals are digital with a sampling rate of 6 Hz. They are converted to analogue signals with a sample-and-hold function before their introduction in the Codas system. At the same time ECG, pulse rate and peripheral oxygen saturation (beat-to-beat by Nelcor-2000® monitor), and mean arterial blood pressure (Siemens Sirecust®) were recorded by the Codas system. $PaCO_2$, PaO_2, pH, glycemia and hemoglobin were measured by an arterial blood sample on a blood gas analyzer (Radiometer, Copenhagen, Denmark) before and after the measurement. All measurements were done by the the first author.

The mean and 95% CI of the mean were calculated for the TOI values if the peripheral oxygen saturation did not change by more than 5 %. The mean and 95% CI of the mean of heart rate, peripheral oxygen saturation and mean arterial blood pressure over this period were also calculated. A Kolmogorov-Smirnov test was used to test for a normal distribution of the TOI for the whole group and the three subgroups (different days).

The ANOVA-test was applied to see if there was a significant difference. The Student-Newman-Keuls test was used for all pairwise comparisons. A p-value of less than 0.05 was considered to be significant. Multiple regression analysis was used to detect the important influences of the other parameters on TOI.

4. RESULTS

The median TOI was 57% (95% Confidence Interval (CI) for the median was 54 to 65.7 %) on day 1, 66.1% (95 CI was 61.9 to 82.2 %) on day 2 and 76.1% (95% CI was 67.8 to 80.1%) on day 3. There was a normal distribution for the whole group as well as for the three different subgroups (Kolmogorov-Smirnov test for Normal distribution was $p > 0.2$). An ANOVA-test showed a significant difference ($p<0.001$) between the groups. A Student-Newman-Keuls test for all pairwise comparisons showed that there was a significant increase in TOI from day 1 to day 2 and from day 1 to day 3, but not from day 2 to day 3.

To test the inclusion criteria an ANOVA-test was performed to find an eventual difference in peripheral oxygen saturation for the different days. No significant difference was found ($p = 0.066$). A significant increase ($p = 0.003$) was found in blood pressure from day 1 (mean blood pressure of 33.8 mmHg) to day 3 (mean blood pressure of 41.2 mmHg). The increase from day 1 to day 2 was not significant. There is also a significant difference in $PaCO_2$ on the different days. Mean $PaCO_2$ on day 1 was 31.7 mmHg, on day 2 45.2 mmHg and on day 3 37.6 mmHg. This is a significant difference ($p < 0.001$) between the three days, but there is an increase from day 1 to 2 and a decrease from day 2 to 3. There was no significant difference between PaO_2, HCO_3 and pH on the different days.

The increase of TOI from day 1 to day 3 remained significant after multiple regression analysis. Multiple regression analysis did not show any correlation between TOI and blood pressure, $PaCO_2$ or peripheral oxygen saturation.

5. DISCUSSION

The measurement of TOI in premature born infants with spatially resolved spectroscopy is a new tool to evaluate brain oxygenation. Near-infrared spectroscopy is a well validated technique to measure brain oxygenation, cerebral blood flow and cerebral blood volume. Until now, it is used only in research because it is very sensitive for movement and it can only measure values relative to the starting point in a continuous way. TOI, in contrast, is an absolute value and can be measured on different occasions in the same patient. Several studies show TOI values in healthy adult volunteers (3-7). Quaresima et al. (3) compared TOI with the cerebral venous oxygen saturation (S_vO2), measured by near-infrared spectroscopy (NIRO 300, Hamamatsu) , and concluded that TOI reflected mainly the saturation of the intracranial venous compartment of circulation. Other studies with different NIRS instruments also found this correlation, but between TOI and jugular bulb oximetry (8,9). In contrast, other studies comparing jugular bulb oximetry with TOI did not find a correlation (10-14) Normal values between 65% and 85% were reported. Teller et al. (15) reported the measurement of TOI of the liver and the brain in 25 infants, aged 1 to 54 days with a postmenstrual age between 29 and 40 weeks during feeding. Mean TOI of the brain was 62.1 ± 9.7%, which is lower than the values reported in adults. This is, to our knowledge, the only report regarding TOI values in premature born babies. They are comparable with our results on day 1 (59.2% ± 6.4%), although the TOI values on day 2 and 3 were higher and more compatible with the results in adult volunteers. There is an important increase from day 1 to day 2 and this was significant from day 1 to day 3. It can be concluded that TOI is not a stable factor over the first three days in premature born infants. We believe normal values will have to be studied for different postnatal and postmenstrual ages, as we already do for other physiological parameters such as mean arterial blood pressure.

There are different possible explanations for this increase in TOI. First, an increase in oxygenation was ruled out by assessing the results only when peripheral oxygen saturation remained stable. Also arterial oxygen content did not change significantly on the different days. Whether a decrease in fetal hemoglobin plays an important role has not been investigated.

TOI can increase due to an increase in cerebral blood flow most probably by a decreased oxygen extraction by better perfused cerebral tissue (6,16). Carlin et al.(6) found signs and symptoms of cerebral ischemia in two patients during awake carotid endarterectomy with TOI values of 40 and 48%. Decreases to 63% were not accompanied by any signs of cerebral ischemia. Meek et al.(17) described a significant increase in cerebral blood flow during the first 72 hours in premature born infants, independent of mean arterial blood pressure, $PaCO_2$ and hematocrit. However, this increase is explained by an increase in cerebral metabolism, so a "luxury perfusion" should be hypothesised to explain the increase in TOI. An important bias in our study is the low $PaCO_2$ values on the first day. This could partly explain the lower cerebral blood flow and hence the lower TOI values, although no correlation could be found between $PaCO_2$ and TOI in our study. In adults no correlation was found between TOI and $PaCO_2$ (12,14,). However, in contrast to adults, children and term babies, there is a statistical association between hypocapnia and brain damage and developmental deficit in preterm babies (18). The combination of low $PaCO_2$ with low blood pressure can explain a lower cerebral blood flow (CBF) and hence a larger oxygen uptake with a lower venous saturation. Further studies on the relation between CBF and TOI have to be done.

In conclusion, we found an increase in TOI over the first three days in premature born babies. This increase might reflect an increase in cerebral blood flow during the first three days in these patients. TOI is not a stable parameter in the first days of life and we believe that further studies should be done before this parameter can be mused for clinical purposes. Other studies for determining "normal values" and "critical values" in this particular population are certainly warranted.

REFERENCES

1. J. Matcher, P. Kirckpatrick, K. Nahid, M. Cope and D.T. Delpy. Absolute quantification methods in tissue near infrared spectroscopy. *Proc SPIE* **2389**, 486-495 (1995).
2. S. Suzuki, S. Takasaki, T. Ozaki and Y. Kobayashi. A tissue oxygenation monitor using NIR spatially resolved spectroscopy. *Proc SPIE* **3597**, 582-592 (1999).
3. Quaresima V, Sacco S, Totaro R and Ferrari M. Noninvasive measurement of cerebral hemoglobin oxygen saturation using two near infrared spectroscopy approaches. *J Biomed Optics*. **5**(2) : 201-205 (2000).
4. Samra S, Dorj P, Zelenoek GB and Stanley JC. Cerebral oximetry in ptients undergoing carotid endarterectomy under regional anesthesia. *Stroke* **27**(1), 49-55 (1996).
5. Misra M, Stark J, Dujovny M, Widman R and Ausman J.I.. Transcranial cerebral oximetry in random normal subjects. *Neurol. Res* **20**(2), 137-141 (1998).
6. Carlin RE, McGraw DJ, Calimlim JR and Maxcia MF. The use of near-infrared cerebral oximetry in awake carotid endarterectomy. *J Clin Anesth* **10**(2), 109-113 (1998).
7. Nielsen HB, Boushel R, Madsen P and Secher NH. Cerebral desaturation during exercise reversed by O_2 supplementation. *Am J Physiol* **277**(3), H1045-H1052 (1999).
8. Daubeney P.E.F, Pilkington SN, Janke E, Charlton GA, Smith DC and Webber SA. Cerebral Oxygenation Measured by Near-Infrared Spectroscopy: Comparison With Jugular Bulb Oxymetry. *Ann Thorac Surg* **61**(4), 930-4 (1996).
9. Sapire KJ, Gopinath SP, Farhat G, Thakar DR, Gabrielli A, Jones JW, Robertson CS and Chance B. Cerebral oxygenation during warming after cardiopulmonary bypass. *Crit Care Med* **25**(10), 1655-1662 (1997).
10. Buunk G, van der Hoeven G and Meinders AE. A comparison of near-infrared spectroscopy and jugular bulb oximetry in comatose patients resuscitated from a cardiac arrest. *Anaesthesia* **53**(1), 13-19 (1998).
11. Minassian AT, Poirier N, Pierrot M, Menei P, Granry JC, Ursino M and Beydon L. Correlation between Cerebral Oxygen Saturation Measured by Near-Infrared Spectroscopy and Jugular Oxygen Saturation in Patients with Severe Closed Head Injury. *Anesthesiology* **91**(4), 985-90 (1999).
12. Germon TJ, Young AER, Manara AT and Nelson RJ. Extracerebral absorption of near infrared light influences the detection of increased cerebral oxygenation monitored by near infrared spectroscopy. *J Neurol Neurosurg Psychiatry* **58**(4), 477-479 (1995)
13. Lewis SB, Myburgh JA, Thornton EL, Reilly PL. Cerebral oxygenation monitoring by near-infrared spectroscopy is not clinically useful in patients with severe closed-head injury: A comparison with jugular venous bulb oximetry. *Crit Care Med* **24**(8), 1334-1338 (1996)
14. Ali MS, Harmer M, Vaughan RS, Dunne JA and Latto IP. Spatially resolved spectroscopy (NIRO-300) does not agree with jugular bulb oxygen saturation in patients undergoing warm bypass surgery. *Can J Anesth* **48**(5), 497-501 (2001).
15. Teller J, Schwendener K, Wolf M, Keel M, Bucher HU, Fanconi S and Baenziger O. Continuous monitoring of liver oxygenation with near infrared spectroscopy during naso-gastric tube feeding in neonates. *Schweiz Med Wochenschr* **130**(18), 652-6 (2000).
16. Hueber DM, Franceschini MA, Ma HY, Zhang Q, Ballesteros JR, Fantini S, Wallace D, Ntziachristos V and Chance B. Non-invasive and quantitative near-infrared haemoglobin spectrometry in the piglet brain during hypoxic stress, using a frequency-domain multidistance instrument. *Phys Med Biol* **46**(1), 41-62 (2001).
17. Meek JH, Tyszczuk L, Elwell CE and Wyatt JS. Cerebral blood flow increases over the first three days of life in extremely preterm neonates. *Arch Dis Child Fetal Neonatal Ed* **78**(1), F33-F37 (1998)
18. Greisen G and Vannuci RC. Is Periventricular Leucomalacia a Result of Hypoxic-Ischaemic Injury? Hypocapnia and the Preterm Brain. *Biol Neonate* **79**(3-4) : 194-200 (2001)

SIMULTANEOUS MEASUREMENT OF CEREBRAL TISSUE OXYGENATION OVER THE ADULT FRONTAL AND MOTOR CORTEX DURING REST AND FUNCTIONAL ACTIVATION

Terence S. Leung, Clare E. Elwell, Julian R. Henty, and David T. Delpy*[1]

1. INTRODUCTION

Near-infrared spectroscopy (NIRS) has been shown to measure subtle haemodynamic changes in the human brain. Differential spectroscopy has been used to record changes in oxy-haemoglobin concentration $\Delta[HbO_2]$ and deoxy-haemoglobin concentration $\Delta[HHb]$, following various kinds of stimulation including motor [1], visual [2], and cognitive [3]. NIRS based topography systems have also been used to map cerebral haemodynamic responses [4,5] around the evoked region. NIRS provides an alternative to more expensive techniques such as positron emission tomography and functional magnetic resonance imaging, in the understanding of brain function.

Using multiple closely spaced detectors, the technique of spatially resolved spectroscopy [6], is based on the measurement of attenuation slope as a function of source detector separation. From these measurements it is possible to de-couple the absorption (μ_a) and scattering coefficients (μ_s), resulting in the calculation of scaled absolute haemoglobin concentrations, i.e. $k\cdot[HbO_2]$ and $k\cdot[HHb]$. The Tissue Haemoglobin Index (THI) is defined as $k\cdot[HbO_2]+k\cdot[HHb]$ while the Tissue Oxygenation Index (TOI) as $(k\cdot[HbO_2]/(k\cdot[HbO_2]+k\cdot[HHb]))*100\%$ [7]. The constant k here is a scaling factor for μ_s. TOI is a measure of oxygenation saturation in cerebral tissues whereas THI reflects the total amount of haemoglobin.

Using a continuous-wave type spectrometer with SRS capability and fast sampling rate of 6 Hz (NIRO 300, Hamamatsu Photonics KK), we aim to (1) establish the normal ranges of TOI and THI over both frontal and motor cortex, (2) compare the slow spontaneous oscillations (~0.1Hz) found in $\Delta[HbO_2]$ signals simultaneously measured over both sites, and (3) investigate the responses of $\Delta[HbO_2]$, $\Delta[HHb]$, TOI and THI in the resting and activated human adult brain.

*[1] *Department of Medical Physics & Bioengineering, University College London, London, WC1E 6JA, U.K.*

Oxygen Transport to Tissue XXIII, edited by D. Wilson *et al.*
Kluwer Academic/Plenum Publishers, 2003

2. METHODS

2.1 Experiment

Simultaneous measurements of cerebral oxygenation were made over the right frontal cortex and left motor cortex of ten normal adult volunteers (median age=36±12 years) using a NIRO 300 spectrometer. An optode spacing of 5 cm was used for the optical probe over the frontal cortex while a smaller spacing of 4 cm was used for that over the motor cortex to optimise signal to noise ratio. Both optical probes were covered with a black cloth. Whilst seated in a quiet, dark room, with their eyes closed, subjects were asked to relax and 5 minutes of resting data were recorded. Subjects were then asked to perform a finger opposition task with their right hand. The task was performed for 43 seconds followed by a 30 second rest period and measurements were made for a total of ten task/rest repetitions. Respiration was measured using a capnogram and the pulse plethysmograph (PPG) was recorded from a pulse oximeter probe on the left earlobe (Novametrix, USA). Raw data, including the optical densities at four wavelengths and attenuation slopes at three wavelengths, were collected from the spectrometer via the serial port onto a PC and other physiological signals recorded via an analogue-to-digital conversion card. The sampling rate was set to 6 Hz. All data were saved onto a hard disk for further analysis. The raw NIRS data were converted to $\Delta[HbO_2]$, $\Delta[HHb]$, TOI and THI with a Matlab program (Mathworks Ltd.) written based on a SRS literature from the manufacturer[7]. A differential pathlength factor of 6.4 was used [8].

3. RESULTS

3.1 TOI and THI in the resting brain

The mean and standard deviation of TOI and THI for 10 resting subjects over 5 minutes are shown in Table 1. TOI over the frontal cortex was found to be significantly higher than that over the motor cortex (paired t-test, $p<0.002$). THI over the frontal cortex was found to be significantly lower than that over the motor cortex (paired t-test, $p<0.003$).

3.2 Cross-correlation between $\Delta[HbO_2]$ measured over the frontal and motor cortex

The $\Delta[HbO_2]$ signals measured over the frontal and motor cortex were found to contain signal components, including those due to heart beat (~1 Hz), respiration (~0.25Hz) and slow spontaneous oscillations (~0.1Hz), as previously reported [9]. The signals sometimes also contained a slow drift (<0.005Hz). One possible origin of the slow spontaneous oscillations is the vasomotion in the capillary level [9]. In order to investigate this signal component more closely, the slow drift was first removed (detrended) by subtracting the lowpass filtered $\Delta[HbO_2]$ signal at a cut-off frequency of 0.005Hz from the raw $\Delta[HbO_2]$ signal. The resulting signal was again lowpass filtered at a cut-off frequency of 0.5Hz to remove the strong heart beat signals. The filtering was carried out by a 5th order lowpass Butterworth digital filter in forward-backward directions to avoid introducing a phase delay. The filtered signals were dominated by the slow spontaneous oscillations (~0.1Hz). An example of the filtered $\Delta[HbO_2]$ signals over the motor and frontal cortex is shown in Figure 1. For the 10 subjects at rest, a mean (± S.D.) correlation coefficient (r^2) of 0.57±0.18 was found between the filtered $\Delta[HbO_2]$ signals measured over the frontal and motor cortex.

3.3 Motor stimulation

The on-off period for the stimulation was 43-30 seconds, corresponding to approximately 0.014Hz. The NIRS signals, including $\Delta[HbO_2]$, $\Delta[HHb]$, TOI and THI, were detrended and lowpass filtered at 0.08 Hz as described above so that effects of other signal components were minimised. The filtered NIRS signals were then ensemble averaged over 8 cycles. The response to the stimulation was calculated as the difference between the average of 10 seconds worth of data just before the onset of the stimulation, and the average of 30 seconds worth of data 10 seconds after the onset of the stimulation. For the 10 subjects, the mean and standard deviation of responses of the NIRS signals over the two sites are summarised in Table 2. The t-test was conducted to assess the significance of the responses in the NIRS signals. Using a significant level of 0.03 as a threshold, Table 3 shows that there are no significant responses in any of the NIRS signals measured over the frontal cortex, whereas all of the NIRS signals measured over the motor cortex show significant responses to motor stimulation. Using a p value of 0.05, the $\Delta[HbO_2]$ response at the frontal cortex just achieved significance. To see if this could be due to a change in central circulatory parameters, the heart rate signals were examined. Figure 2 depicts the grand averaged heart rate signal. In the 10 subjects, the heart rates show significant increases during activation (mean = 4.1±3.1 beats/min., t-test:$p<0.003$). Figure 3 shows the grand averaged responses of the NIRS signals measured over the two sites for all subjects. It should be noted that although the grand averaged TOI and THI measured over the motor cortex showed responses starting slightly earlier than the stimulation, this behaviour can only be found in 5 of the 10 subjects and is likely due to variations unrelated to the stimulations.

4. CONCLUSIONS

It has been found that TOI and THI measured by the SRS based technique may be affected by the location of probes which prompts the need to specify the measurement site when these quantities are to be compared. The relatively high correlation coefficient found between $\Delta[HbO_2]$ signals over the frontal and motor cortexes may suggest a centrally mediated system playing a role in the production of the slow spontaneous oscillations around 0.1Hz. It can be seen that the heart rate increases during stimulation which may cause a systemic change in the brain[5]. The slight (but barely statistically significant) response in the NIRS signals measured over the frontal cortex may be due to this systemic change. SRS based measures such as TOI and THI were also shown to be able to detect evoked responses over the motor cortex during motor stimulation experiments. The availability of two simultaneous measurements, the relatively high sampling rate of 6 Hz and the absolute TOI measurements offered by the spectrometer will allow more detailed analyses of spontaneous and activation related changes in cerebral haemodynamics.

Table 1 Mean and standard deviation of raw TOI and THI for 10 resting subjects (recording duration = 5 minutes)

	TOI (%)		THI (a.u.)	
	(frontal)	(motor)	(frontal)	(motor)
Mean	68.49	59.25	37.65	91.27
S.D.	3.49	8.02	9.52	46.97

Table 2 Responses of NIRS signals over the frontal and motor cortex during motor stimulation experiments (n=10)

	Over Frontal Cortex		Over Motor Cortex	
	Mean	S.D.	Mean	S.D.
$\Delta[HbO_2]$ (μ molar)	0.112	0.177	0.571	0.289
$\Delta[HHb]$ (μ molar)	-0.052	0.099	-0.228	0.182
TOI (%)	0.051	0.246	1.28	0.527
THI (a.u.)	0.052	0.218	0.628	0.438

Table 3 Evoked responses found in NIRS signals measured over the motor cortex as determined by t-tests {H_0: response =0, H_1: response >0 (for Δ[HbO2],TOI,THI) or H_1: response < 0 (for Δ[HHb])}

	Over Frontal Cortex		Over Motor Cortex	
	p	Evoked response ($p<0.03$)	p	Evoked response ($p<0.03$)
Δ[HbO2]	0.04	×	<0.0001	√
Δ[HHb]	0.07	×	0.0017	√
TOI	0.26	×	<0.0001	√
THI	0.24	×	<0.001	√

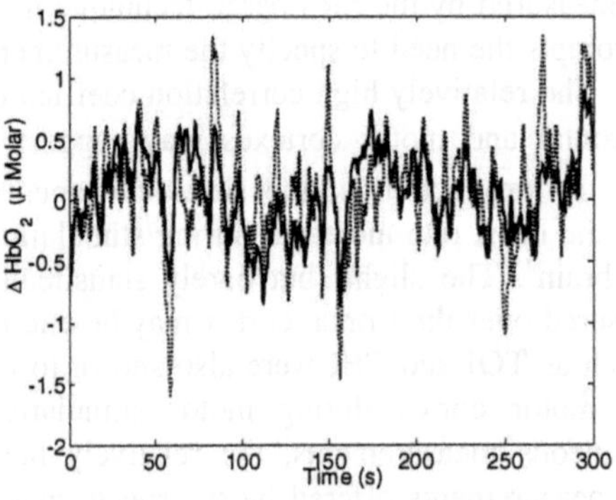

Figure 1. Filtered $\Delta[HbO_2]$ signals measured over the frontal (solid line) and motor (dotted line) cortex.

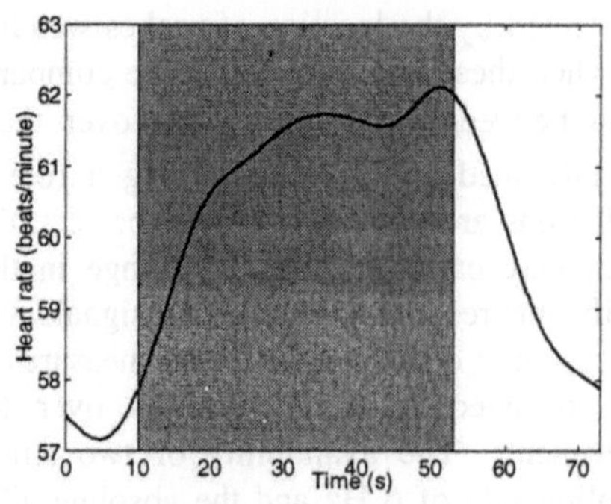

Figure 2. Grand averaged heart rate signal (n=10).

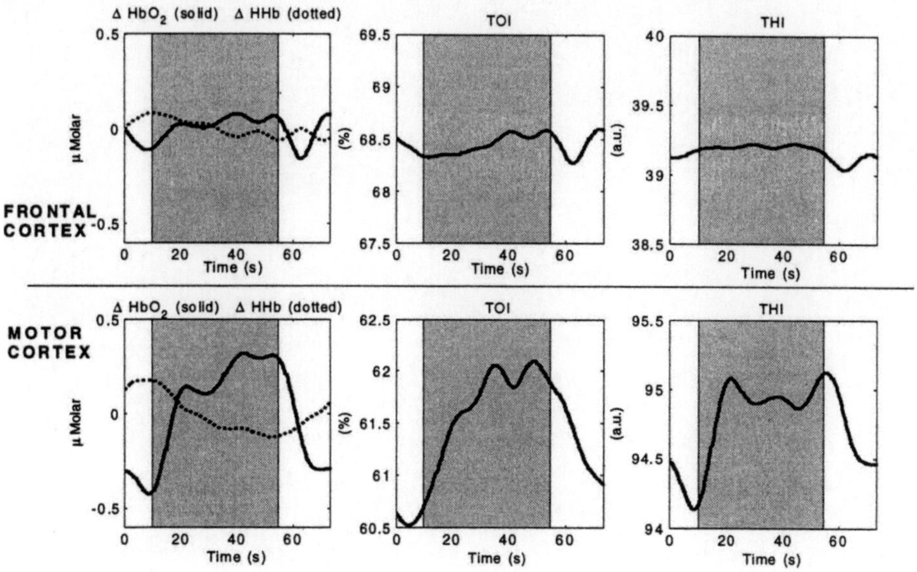

Figure 3. Grand averaged responses of NIRS signals in motor stimulation experiments (n=10).

ACKNOWLEDGEMENTS

The authors would like to thank Dr. M. Smith, Dr. M. Thavasothy and all the volunteers who participated in this study. This work was funded by the Wellcome Trust and supported by Hamamatsu Photonics KK.

REFERENCES

1. H.Obrig, C.Hirth, J.G.Junge-Hulsing, C.Doge, T.Wolf, U.Dirnagl, A.Villringer, "Cerebral oxygenation changes in response to motor stimulation," *J.Appl. Physiol.*, **81**, pp.1174-1183, 1996.
2. J.H.Meek, C.E.Elwell, M.J.Khan, J.Romaya, J.S.Wyatt, D.T.Delpy, S.Zeki, "Regional changes in cerebral haemodynamics as a result of a visual stimulus measured by near infrared spectroscopy," *Proc. R. Soc. Lond,* **B261**, pp.351-356, 1995.
3. C.Hock, F.Muller-Spahn, S.Schuh-Hofer, M.Hofmann, U.Dirnagl, A.Villringer, "Age dependency of changes in cerebral hemoglobin oxygenation during brain activation: a near-infrared spectroscopy study," *J. Cerebral Blood Flow and Metabolism*, **15**, pp.1103-1108, 1995.
4. A.Maki, Y.Yamashita, E.Watanabe, Y.Mayanagi, H.Koizumi,"Spatial and temporal analysis of human motor activity using non-invasive NIR topography," *Med. Phys.*, **22**, pp.1997-2005, 1995.
5. M.A.Franceschini, V.Toronov, M.E.Filiaci, E.Gratton, S.Fantini, "On-line optical imaging of the human brain with 160-ms temporal resolution," *Optical Express*, **6**(3), pp.49-57,2000
6. S.J.Matcher, O.Kirkpatrick, K.Nahid, M.Cope, D.T.Delpy, "Absolute quantification methods in tissue near infrared spectroscopy," *Proc. SPIE*, **2389**, pp.486-495, 1993.
7. S.Suzuki, S.Takasaki, T.Ozaki, Y.Kobayashi, "A Tissue Oxygenation Monitor using NIR Spatially Resolved Spectroscopy", *Proc. of SPIE*, **3597**, pp.582-592, 1999.
8. A.Duncan, J.H.Meek, M.Clemence, C.E.Elwell, L.Tyszczuk, M.Cope, D.T.Delpy, "Optical pathlength measurements on adult head, calf and forearm and the head of the new born infant using phase resolved optical spectroscopy," *Phys. Med. Biol.*, **40**, pp.295-304, 1995.
9. C.E.Elwell, R.Springett, E.Hillman, D.T.Delpy, "Oscillation in Cerebral Haemodynamics: Implications for Functional Activation Studies," *Adv. Exp. Med. Biol.*, **471**, pp.57-65, 1999.

DYSFUNCTIONAL MICROVASCULAR CONDUCTED RESPONSE IN IRRADIATED NORMAL TISSUE

M. Waleed Gaber, Michael D. Naimark, and Mohammad F. Kiani*

1. INTRODUCTION

Ionizing radiation is used widely to treat many conditions including cancer, arteriovenous malformations (AVM), macular degeneration, and intimal hyperplasia. Damage to the microvasculature constitutes one of the most important components of the late effects of radiation damage to many organs in clinical applications [1-3]. While the effects of ionizing radiation on microvascular structure and function of normal tissue have been studied, the mechanisms by which ionizing radiation interferes with the normal microvascular control processes are not well understood. An important question is why normal microvasculature is not able to repair radiation damage as efficiently as it is able to repair other forms of damage (e.g. wounds). Understanding the mechanisms by which ionizing radiation damages the microvasculature has important clinical implications.

Microvascular conducted response is the mechanism by which local vasomotor signals are conducted electrically through cell-cell coupling by as yet unknown mechanisms through gap junctions of endothelial and/or smooth muscle cell layers of the arterioles and capillaries [4]. For example, changes in capillary pressure may lead to a distal conducted vasoactive response in arterioles [5] and the magnitude of the vasoactive conducted response in daughter branches is integrated at bifurcations [6]. Since conducted response has been shown to play an important role in the coordination of microvascular blood flow control and microvascular network adaptation, its disruption may be an important mechanism of normal tissue microvascular damage after radiation exposure. For example, microvessels in irradiated tissue requiring increased blood flow may not be able to communicate their demand to the feed arterioles which may then lead to localized ischemia. The latter may be the reason why new capillaries that are formed after radiation exposure may not function as effectively as the initial microvasculature [3]. Further evidence is provided by our recent studies in which we have shown that functional abnormalities such as reduced blood flow in irradiated microvascular networks are apparent long before the manifestation of structural changes such as reduced capillary diameter[7].

The inducible form of nitric oxide synthase (iNOS) generates significant amounts of nitric oxide and is induced in response to inflammatory stimuli including radiation exposure [8]. Recent studies have suggested that a rise in the level of iNOS may be responsible for the reported dysfunction in the microvascular conducted response mechanism in sepsis [9-11]. This is based on the

* M. Waleed Gaber, Department of Radiation Oncology, St. Jude Children's Research Hospital, Memphis, TN 38105. Michael D. Naimark and Mohammad F. Kiani, School of Biomedical Engineering and Department of Radiation Oncology, University of Tennessee Health Science Center, Memphis, TN 38163.

finding that proinflammatory mediator-induced increase in iNOS suppresses the production of the constitutive form of nitric oxide synthase (cNOS) which then leads to an overall reduction in the formation of biologically active nitric oxide [12;13]. Due to the important role that microvascular conducted response plays in blood flow control and regulation, we have studied the early effects of ionizing radiation on conducted response in the microcirculation of normal tissue. Furthermore, we have tested the hypothesis that disruption of the nitric oxide pathway is in part responsible for the dysfunctional conducted response in the irradiated normal tissue.

2. MATERIALS AND METHODS

A cremaster muscle preparation [14] of male Golden Syrian Hamsters (Harlan, Indianapolis, IN) was used to study the effects of ionizing radiation on conducted response in normal microvasculature at 2-5 days post-irradiation. Animals were irradiated at 7-8 weeks of age, and age-matched nonirradiated animals were used as controls. All protocols were approved by the Animal Care and Use Committee of the University of Tennessee Health Science Center and followed guidelines of NIH.

Prior to irradiation (IR) animals were lightly anesthetized with an IP injection of 25mg/kg of Sodium Pentobarbital. A local single dose of 10 or 20 Gy was delivered using a Siemens MD-2 linear accelerator (2 Gy/min, 6 mV X-rays). A collimator, 3.5cm in diameter, was used to deliver a localized dose of radiation bilaterally to the testicle regions. Tissue equivalent bolus material was placed above the scrotum to establish electronic equilibrium and to insure the prescribed dose was delivered uniformly to the cremaster muscle. Experiments were performed 2-5 days post-IR. Prior to surgery animals were anesthetized with an IP injection consisting of 600mg/kg of urethane and 90mg/kg of α-chloralose. Body temperature was maintained at approximately 37°C by convective heating. Animals were intubated, catheterized (left femoral vein), and placed on a surgical board where the right cremaster muscle was pinned as a flat sheet with minimal disruption to the tissue. Preparations were maintained at a temperature of 35±0.5°C and superfused at a rate of 5ml/min with a bicarbonate buffered salt solution equilibrated with 5%CO_2-95%N_2.

Adenosine (0.1mM) dissolved in saline was topically applied via superfusate to assess the maximum diameter in each studied vessel. Other vasoactive agents were dissolved in saline and were applied locally via a pipette (~3 μm tip) by pressure ejection. Vessel diameter was measured approximately 500μm upstream of the pipette tip. Suffusate flow direction was in the direction of blood flow to eliminate diffusion of drugs upstream; FITC-dextran was added to the vasoactive substance in the pipette to verify the flow direction of vasoactive agent under epi-illumination.

Effects of locally applied endothelium dependent vasodilator acetylcholine (ACH, 10mM concentration in the pipette) and smooth muscle cell dependent vasoconstrictor phenylephrine (NE, 10mM concentration in the pipette) on control and irradiated animals were compared. These vasoactive substances diffuse over a distance of approximately 20μm from the pipette tip to the vessel wall and using a forced-convection mass transfer calculation [15] we estimate that their concentration near the vessel is 100 times lower than that in the pipette (i.e. 0.1mM). Effects of the presence of a selective blocker of iNOS (AG, 10mM aminoguanidine) in the superfusate on the magnitude of conducted response was also tested in both control and irradiate animals.

Arteriolar diameters were measured under bright field illumination using a chilled CCD camera (Hamamatsu C5985, Japan). All diameters were measured using a 20X objective lens. Flow direction of suffusate and vasoactive agent were verified under epi-illumination using a 4x objective lens.

All data are presented as percentage of the maximum diameter (Mean±SEM). Number of vessels tested (N) are given for each group. Effects of IR dose was tested using ANOVA. A multiple comparison procedure (Fisher's least significant difference) was used to discriminate among the means. In each experiment involving aminoguanidine, data were collected in a single vessel in the absence and presence of aminoguanidine in the superfusate and the paired t-test was used to compare the means. Differences were considered to be significant if $P<0.05$.

3. RESULTS

No microvascular lesions were observed in the irradiated arterioles. Average resting diameter of irradiated arterioles (25.4±0.8μm, Mean±SEM) was not significantly different from that of control (24.4±0.7μm). In the absence of the vasoactive agents, no significant changes in vessel diameters were observed when the pipette contained only saline and FITC-dextran (data not shown). Conducted response induced by local application of the endothelium-dependant vasodilator acetylcholine (ACH) was significantly (P<0.01) attenuated in arterioles in a radiation dose-dependent manner and was significantly different (P<0.01) from control in both 10 and 20 Gy irradiated arterioles; see Figure 1. Conducted vasoconstrictory response to locally applied phenylephrine was not significantly different between irradiated and control groups. The magnitude of conducted response in either control or irradiated arterioles was not significantly correlated with the resting vessel diameter (r^2<0.2).

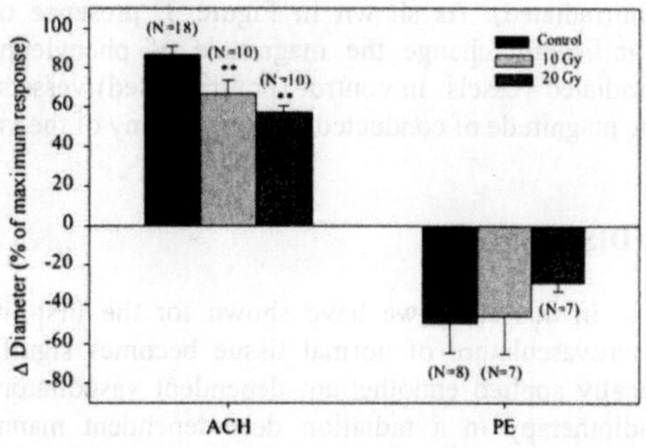

Figure 1. Conducted vasodilatation in response to the local application of acetylcholine (ACH) was significantly (P<0.01,**) lower in irradiated arterioles as compared to control. Conducted vasoconstrictory response to locally applied phenylephrine (PE) was not significantly different compared to control. Data are presented as Mean ± SEM.

As discussed in the Introduction, upregulated production of iNOS in inflammatory situations such as irradiation may lead to an overall decrease in nitric oxide production in the microvasculature which would then result in arterial vasoconstriction. Therefore, effects of the presence of a selective blocker of iNOS (aminoguanidine) in the superfusate on conducted response was investigated with the expectation that inhibition of iNOS would result in arterial vasodilation. Aminoguanidine (AG) alone did not significantly change the diameter of arterioles (data not shown).

Conducted vasodilatory response to acetylcholine in 10 Gy irradiated vessels was significantly (P<0.01) enhanced in the presence of aminoguanidine, see Figure 2; this response was restored to the control level in 10Gy irradiated arterioles and was not significantly different from control

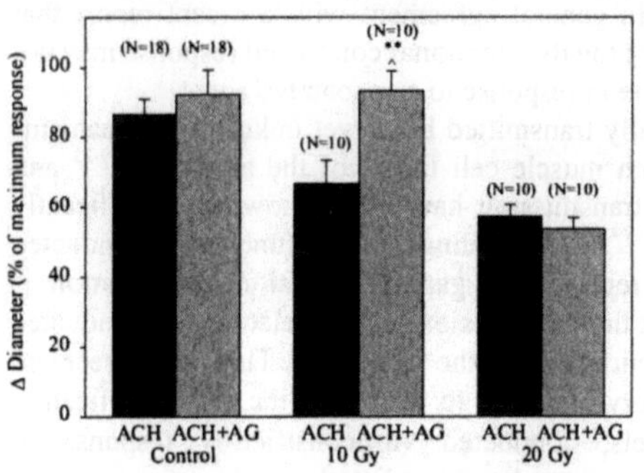

Figure 2. Presence of a selective blocker of iNOS (aminoguanidine, AG) in the superfusate significantly (P<0.01,**) increased the conducted vasodilatory response to acetylcholine (ACH) in 10Gy irradiated arterioles and restored it to the control level (^, not significantly different from control). Data are presented as Mean ± SEM.

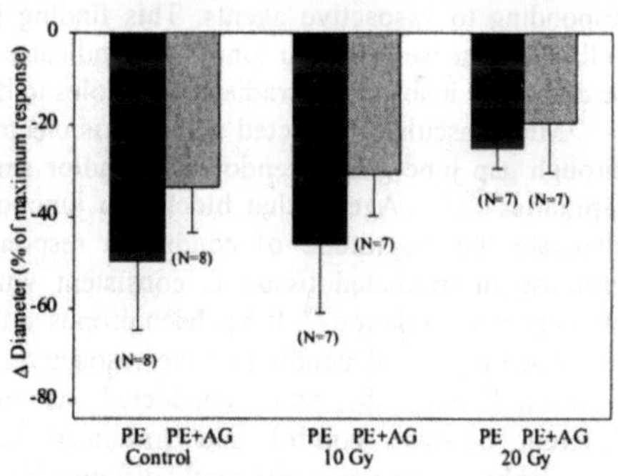

Figure 3. Presence of a selective blocker of iNOS (aminoguanidine, AG) in the superfusate did not significantly change the magnitude of conducted vasoconstrictory response to locally applied phenylephrine (PE). Data are presented as Mean ± SEM.

(unirradiated). As shown in Figure 3, presence of aminoguanidine in the superfusate did not significantly change the magnitude of phenylephrine conducted vasoconstrictory response in irradiated vessels. In control (nonirradiated) vessels, aminoguanidine did not significantly change the magnitude of conducted response to any of the vasoactive agents used in this study.

4. DISCUSSION

In this study we have shown for the first time that conducted response in the irradiated microvasculature of normal tissue becomes significantly dysfunctional. Conducted response to locally applied endothelium dependent vasodilator acetylcholine was significantly reduced after radiotherapy in a radiation dose-dependent manner. However, conducted response to locally applied smooth muscle cell dependent vasoconstrictor phenylephrine was not significantly different between irradiated and control groups. Since conducted response has been shown to play an important role in the coordination of microvascular blood flow control and microvascular network adaptation, its disruption may be an important aspect of normal tissue microvascular damage after radiation exposure. For example, microvessels in irradiated tissue requiring increased blood flow may not be able to communicate their demand to the feed arterioles which may then lead to ischemic tissue. In a mathematical modeling study we have shown that the lack of conduction and coordination of vasomotor signals during microvascular network formation and adaptation may lead to pathological microvascular network patterns and characteristics [16;17].

There are at least three possible mechanisms by which ionizing radiation can disrupt microvascular conducted response. First, ionizing radiation may damage the mechanisms by which microvessels respond to vasoactive agents. Second, the pathway for the transmission of the electrical signal may be disrupted by irradiation. Third, transduction mechanisms at the distal (upstream) location which cause the vasoactive response may be damaged by irradiation. In the following paragraphs we use our findings to speculate on these three mechanisms of radiation damage. Further studies are needed to conclusively establish whether any of these mechanisms or other mechanisms are in fact responsible for the disruption of conducted response in irradiated normal tissue.

Our data indicate that the mechanisms responsible for initiation of conducted response may not be damaged by radiation, since irradiated arterioles exhibited an ability to respond normally to topically applied adenosine compared to control. Therefore, irradiated vessels are capable of responding to vasoactive agents. This finding is in general agreement with a recent report that radiation increases vascular tone [18] and indicates that the dysfunctional conducted response may not be due to the inability of irradiated arterioles to dilate in response to a vasoactive agent.

Microvascular conducted response is electrically transmitted by as yet unknown mechanisms through gap junctions of endothelial and/or smooth muscle cell layers of the arterioles [19-21] and capillaries [22;23]. Agents that block gap junction transmission have been shown to significantly attenuate the magnitude of conducted response [19]. Our findings of dysfunctional conducted response in irradiated tissue is consistent with reports that gap junctional communication is modulated by radiation [24]. It has been proposed that the transmission path for electrically conducted response may vary depending on the vasoactive agent initiating the response [4]. This is in agreement with our findings that while conducted vasodilatory response to acetylcholine was significantly different between control and irradiated vessels, conducted vasoconstrictroy response to phenylephrine was not significantly different between control and irradiated vessels.

Type II NOS (iNos) is induced in response to sepsis and other inflammatory stimuli including radiation exposure. In the present study, a selective blocker of iNOS (aminoguanidine) was found to restore the acetylcholine conducted vasodilatory response in 10Gy irradiated vessels to the control level. These findings indicate that nitric oxide dependent mechanisms may be involved in the dysfunctional conducted response in irradiated normal tissue microvasculature. In general agreement with our findings, aminoguanidine is reported to restore acetylcholine conducted vasodilatory response to the control level in sepsis affected microvessels [9;10].

In summary, our results indicate that conducted arterial response is dysfunctional in irradiated normal tissue and Ionizing radiation preferentially damages endothelial cells in the microvasculature. Damage to the nitric oxide pathway may play an important role in normal tissue microvascular damage after radiation therapy but the precise mechanisms need to be further investigated. We are currently investigating the long term effects of ionizing radiation on microvascular conducted response.

We thank Dr. Karel Tyml (University of Western Ontario) for helpful discussions. Funding for this study was provided for by the National Institutes of Health (CA68154).

REFERENCES

1. Rubin P: Clinical Oncology. Philadelphia, PA, W.B. Saunders Company, 1993.
2. Dimitrievich GS, Fischer-Dzoga K, Griem ML: Radiosensitivity of vascular tissue. Radiat.Res. 99: 511-535, 1984.
3. Fajardo LF, Berthrong M: Vascular lesions following radiation. Pathol.Annu. 23: 297-230, 1988.
4. Segal SS: Cell-to-cell communication coordinates blood flow control. Hypertension 23(2): 1113-1120, 1994.
5. Rivers RJ: Remote effects of pressure changes in arterioles. Am.J.Physiol. 268: H1379-H1382, 1995.
6. Segal SS, Damon DN, Duling BR: Propagation of vasomotor responses coordinates arteriolar resistances. Am.J.Physiol. 256: H832-H837, 1989.
7. Roth NM, Sontag MR, Kiani MF: Early effects of ionizing radiation on the microvascular networks in normal tissue. Radiat.Res. 151: 270-277, 1999.
8. MacNaughton WK, Aurora AR, Bhamra J, Sharkey KA, Miller MJ: Expression, activity and cellular localization of inducible nitric oxide synthase in rat ileum and colon post-irradiation. Int.J.Radiat.Biol. 74: 255-264, 1998.
9. Tyml K, Yu J, McCormack DG: Capillary and arteriolar responses to local vasodilators are impaired in a rat model of sepsis. J.Appl.Physiol. 84: 837-844, 1998.
10. Hollenberg SM, Cunnion RE, Zimmerberg J: Nitric oxide synthase inhibition reverses arteriolar hyporesponsiveness to catecholamines in septic rats. Am.J.Physiol. 264: H660-H663, 1993.
11. Scott JA, Machoun M, McCormack DG: Inducible nitric oxide synthase and vascular reactivity in rat thoracic aorta: effect of aminoguanidine. J.Appl.Physiol. 80: 271-277, 1996.
12. Kessler P, Bauersachs J, Busse R, Schini-Kerth VB: Inhibition of inducible nitric oxide synthase restores endothelium- dependent relaxations in proinflammatory mediator-induced blood vessels. Arterioscler.Thromb.Vasc.Biol. 17: 1746-1755, 1997.
13. Bassenge E: Endothelial function in different organs. Prog.Cardiovasc.Dis. 39: 209-228, 1996.
14. Roth NM, Kiani MF: A "Geographic Information Systems" based technique for the study of microvascular networks. Ann.Biomed.Eng. 27: 42-47, 1999.
15. Bird RB, Stewart WE, Lightfoot EN: Concentration Distributions in Solids and in Laminar Flow (Transport Phenomena , Bird, Stewart, Lightfoot, eds). New York, John Wiley & Sons, pp519-553, 1960.
16. Hudetz AG, Kiani MF: The role of wall shear stress in microvascular network adaptation. Adv.Exp.Med.Biol. 316: 31-39, 1992.
17. Kiani, M. F. Mathematical modeling of microvascular growth and adaptation. Ph.D. Dissertation, Louisiana Tech University, Ruston, LA, 1990.
18. Bourlier V, Diserbo M, Joyeux M, Ribuot C, Multon E, Gourmelon P, Verdetti J: Early effects of acute gamma-radiation on vascular arterial tone. Br.J.Pharmacol. 123: 1168-1172, 1998.
19. Segal SS, Duling BR: Conduction of vasomotor responses in arterioles: a role for cell-to- cell coupling? Am.J.Physiol. 256: H838-H845, 1989.
20. Welsh DG, Segal SS: Endothelial and smooth muscle cell conduction in arterioles controlling blood flow. Am.J.Physiol. 274: H178-H186, 1998.
21. Xia J, Little TL, Duling BR: Cellular pathways of the conducted electrical response in arterioles of hamster cheek pouch in vitro. Am.J.Physiol. 269: H2031-H2038, 1995.
22. Dietrich HH: Effect of locally applied epinephrine and norepinephrine on blood flow and diameter in capillaries of rat mesentery. Microvasc.Res. 38: 125-135, 1989.
23. Dietrich HH, Tyml K: Capillary as a communicating medium in the microvasculature. Microvasc.Res. 43: 87-99, 1992.
24. Trosko JE, Chang CC, Madhukar BV: Modulation of intercellular communication during radiation and chemical carcinogenesis. Radiat.Res. 123: 241-251, 1990.

OXYGENATION AND BLOOD CONCENTRATION CHANGES IN HUMAN SUBJECT PREFRONTAL ACTIVATION BY ANAGRAM SOLUTIONS

Britton Chance, Shoko Nioka, Sajid Sadi, Connie Li *

1. INTRODUCTION

The goal of this five year project has been to better understand learning and problem solving, and eventually to make a classroom "Cognoscope" using Near Infrared (NIR) imaging. This will let us measure cognitive function online with students. Our imager measures cognitive activated oxygenation and blood concentration of changes in the prefrontal cortex. The scan time is one second, with 50 millisecond per light flash. We use a 3 wavelength LED, the advantage of which is FDA approval is not needed for power higher than 100 microwatts, while a laser diode at over 100 microwatts requires Class III regulations and switches, yellow stickers, and dead man contacts. Small silicon diode detectors with IC are used for detectors which are followed by sample and hold, ADC and signal processing giving a simultaneous time course of 18 outputs.

2. BIOLOGICAL SIGNALS

Incremental changes in oxy and deoxyhemoglobin are measured by a dual wavelength optical imaging method (1). The imager is calibrated in a liquid model with appropriate absorption (μ_a) and scattering (μ_s') where hemoglobin varies through the physiological range and incremental changes of oxygenation are possible. One can measure blood volume with a single wavelength or from a computed sum of these two adjacent wavelengths by simple algorithms. Only small optical changes are caused by the

* Britton Chance, Shoko Nioka, University of Pennsylvania, Department of Biochemistry and Biophysics, Philadelphia, PA 19104-6059

Oxygen Transport to Tissue XXIII, edited by D. Wilson *et al.*
Kluwer Academic/Plenum Publishers, 2003

relative difficulty of solving anagrams, about 1μM and small increments of absorption are involved. (An anagram is a multi-letter scrambled word which can be recognized by the high school juniors, about 1 every 10 seconds.) We present easy, medium and hard anagrams, for example, the number of correct solutions by students for 7 and 9 letter anagrams, where 7 is solvable, 9 is too hard. So just by adding two letters to the anagram, we can significantly change the success of the prefrontal activation. In image processing, we can subtract the image obtained with 7 letters from that with 9 letters. We also can use as a baseline, one or two letters less than the actual test. Thus our test is incremental with respect to an easier or more difficult test. Therefore, we image changes of oxygenation and blood volume in response to a change in problem solving difficulty.

In anagram solutions, Kounios and others (2,3), suggest that when you read the anagram, short-term memory is employed and thereafter a series of processes occur involving searches. Then a solution causes what is called an "Insight Event", there is insight into the problem that achieves a solution.

3. EXPERIMENTAL STUDY OF COGNITIVE FUNCTION

3.1. Experimental Protocol

A series of activations by anagram presentations over a 5-minute interval is recorded. When the difficulty matches ability there are frequent successes. The responses are analyzed so that we may make histograms of 100 to 400 events in the histograms which then have a Gaussian fit corresponding to a small increment of oxygenation in only a few of the 18 voxels that may be activated for a human subject solving a difficult 8 letter anagram. We take the mean value of the histogram, multiply it by the number of successes in that voxel relative to the total successes divided by the FWHM of the histogram. When difficulty and ability are matched, one or two voxels will predominate. There are 18 voxels, from ear to ear and hairline to eyebrow and each voxel can be displayed relative to the brain anatomy.

We also show images during training and post-training for 5 letter anagrams (Fig. 1), for example, the subject can make 614 correct solutions post-training in the given time as compared to 326 pre-training. The voxels form a chaotic pattern for the pre-training "search" pattern, but post-training solutions are obtained in a few discrete localizations. When we use 3 letter anagrams and then 8 letter anagrams (Fig. 2) for a trained subject, the subject is not interested in 3 letter easy anagrams but is challenged by the 8 letter anagrams. So a challenge is necessary to evoke a cognitive response and to focus attention on the problem solution (for details, see Summary).

DISCUSSION

We have found a robust prefrontal signal which seems to be predominantly an oxygenation signal. The blood volume signal being small and often negative for reasons that may be related to the questions that have been brought up about the Bold signal (4).

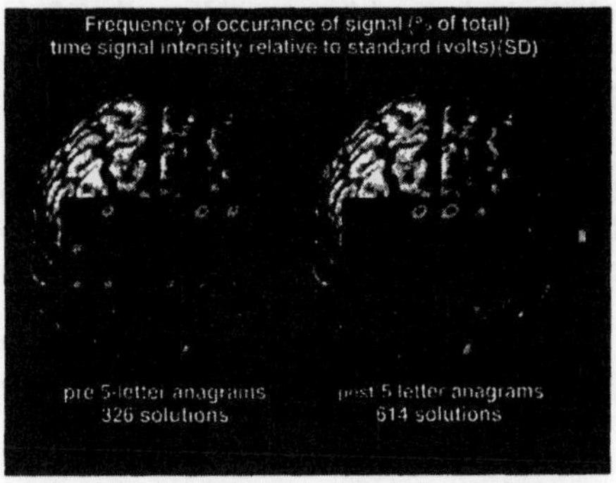

Fig. 1. Superimosed images on an average anatomical brain model exhibiting the prefrontal cortex and illustrating the effect of training of the subject. After three weeks she was able to double the number of solutions and to focus on two voxels in the middle upper portion of the brain cortex, whilst during training the subject was searching for the appropriate way to solve anagrams using most parts of the prefrontal region and only minor signals at the correct positions of the prefrontal cortex for effective and rapid anagram solution.

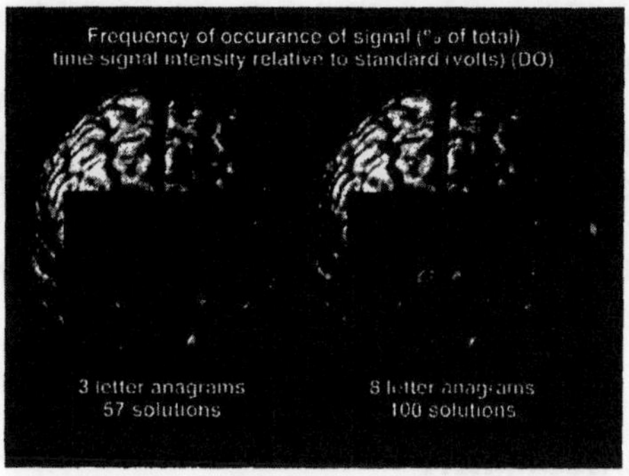

Fig. 2. The comparison of Figs. 1 and 2 gives the changes from individual to individual, and the importance of a match of difficulty and ability. This subject who is highly skilled in anagram solution and capable of 8 letter solutions did 100 solutions of a difficult problem and did not appreciate, did not enjoy the "too easy" 3 letter anagrams and presented weak signals in a chaotic pattern. Please note that the different individuals (Fig. 1 vs Fig. 2) gave different patterns.

Taking into account the paper of Magistretti and colleagues (5) the hypothesis is that astrocyte activation involves activation of glycolysis rather than respiration, we must consider that activation of glycolysis requires either a localized hypoxia or an inhibitor of respiration causing an activation of glycolysis by an increase of ADP and PI over and above the level required to activate mitochondria, i.e. 20 micromolar vs. roughly 200 micromolar (6).

Nevertheless, the evidence for activation of glycolysis in brain function seems solid. Thus, we are saddled with the hypothesis that glycolysis is activated by failure of respiration to supply the ATP necessary for the glutamate → glutamine conversion and the contact at activation of glycolysis. The usual hypothesis of compartmentation of glycolysis from respiration is always applicable but has never actually been proven in the models usually used for studying respiration/glycolysis interaction, particularly yeast and the ascites tumor cells (6). There the remarkable Crabtree effect is explained by an inhibition of the adenylate transferase blocking mitochondrial activity. And this indeed may be occurring in the astrocytes. Alternatively, lactic acidosis may block respiration, but this seems unlikely; the mitochondria are relatively insensitive to pH until 6.5 is reach where Davies and I found 50% inhibition of mitochondrial activity.

Further pursuing Raichle's idea that the blood volume changes may not be important to cell respiration (4), a sufficiency of oxygen being already present, we can take note that our cognition-related oxygenation is not always accompanied by blood volume changes. This is not inconsistent with the "BOLD" effect which measures a decrease of Hb whilst the NIR optical method measures actual concentration changes.

4. SUMMARY

Continued studies in cognitive function in problem solving of graded difficulties has now proceeded through several thousand tests of high school students in a six-week period last summer (2000). The optical imager used 9 sources and 23 detectors operated in time multiplex to give an image of the prefrontal region in several seconds. Thus, the image can be acquired in the state prior to activation, during and following activation. The images are acquired at two wavelengths, 750 and 850 nm, from which the changes of blood concentration (blood pooling) and relative oxygenation/deoxygenation state (metabolic activity) can be obtained and with suitable calibrations converted into micromolar changes of total Hb concentration and fractional oxygenation, respectively. The imager has a wearable pad applied to the forehead and the electronics now presents the running average of the two quantities mentioned above in the 16 voxels which cover the prefrontal region in 1 sec. Thus the protocol involves a scanning through anagrams of graded difficulty which are known to include the range of maximal activation (>3 solutions in the 30 sec test interval or are too easy (>10), or too difficult. Anagrams are presented in a sequence of three, four, five, etc. letter anagrams to the maximal level of difficulty and then back down again to the starting point. Each anagram is displayed for approximately one minute. Computer scanning of the results gives histograms of the several hundred tests per individual in the "training" interval (for three weeks) and in the post training interval about 400 tests, each for 4 levels of difficulty (3 weeks). Usually two or three complete trials per day were achieved with the result that the group of 7 students produced over 2600 test results in 2000. The histogram displays are under study to determine : a) the most fruitful voxels, b) the fraction of the total tests that appeared in those voxels, and c) the maximum signal level observed in the histogram display in units of micromolar hemoglobin. The product of these two was taken as the Figure of Merit and their displays of the 16 voxels gave patterns for trained and untrained responses. The preliminary conclusions of this study were: a) that the training effect was very large, pre-training exhibited a chaotic voxel distribution for all difficulty levels while trained students gave a higher output and activated only one or two of the 16 voxels. If the tests were too easy, too hard, i.e., an ability/difficulty mismatch, frustration or disattention gave similar chaotic patterns (7); b) a match between difficulty and ability activated only a few voxels in similar locations. Since attention and success measures appear to be of interest in school room studies, a wearable imager with local

LCD display can be supplied to the Senior class of the local high school for those individuals who have already had last Summer's training in anagrams. A wearable NIR measurement of metabolic activation and blood flow may be a useful educational aid. In summary, the optical method focuses attention on cognition activated oxygenation of prefrontal cortex.

ACKNOWLEDGEMENTS

Data acquisition depended upon the Summer Student Minority Program and instrument development by the Medical Diagnostic Research Foundation (MDRF), Philadelphia, Instrument team. This work was supported in part by MDRF, The Johnson Research Foundation, Department of Biochemistry and Biophysics, University of Pennsylvania, Philadelphia and by NIH Grant NS36633 and ONR (K. Pourrezaei).

REFERENCES

1. B. Chance, E. Anday, S., Nioka, S., Zhou, *et al*, A Novel Method for Fast Imaging of Brain Function, Non-Invasively, with Light. *Optics Express* **2**:411-423 (1998).
2. J. Kounios, R.W., Smith, W., Yang, P., Bachman, *et al.*, Cognitive Association Formation in Human Memory Revealed by Spatiotemporal Imaging. *Neuron.* **29**:297-306 (2001).
3. J. Kounios, P.J., Holcomb, S. and Kotz, (2000) On the Locus of the Semantic Satiation Effect: Evidence from Event-Related Brain Potentials. *Memory & Cognition* **28**:1366-1377 (2000).
4. M.A. MinTun, B.N. Lundstrom, A.Z. Snyder, A.G. Vlassenko, G.L. Shulman, and M.E. Raichle, Blood Flow and Oxygen Delivery to Human Brain during Functional Activity: Theoretical Modeling and Experimental Data. *Proc. Natl. Acad. Sci. USA* **98**:6895-6964 (2001).
5. P.J. Magistretti, L. Pellerin, D.L., Rothman. and R.G. Shulman, Neuroscience – Energy on Demand. *Science* **283**:496-497 (1999).
6. B. Chance, B. and Hess Metabolic Control Mechanism. IV. The Effect of Glucose upon the Steady State of Respiratory Enzymes in the Ascites Cells. *J. Biol. Chem.* **243**:2421-2427 (1959).
7. Gur, Ruben (*this volume*).

QUANTITATION OF THE CONCORDANCE BETWEEN CEREBRAL INTRAVASCULAR OXYGENATION AND MEAN ARTERIAL BLOOD PRESSURE FOR THE DETECTION OF IMPAIRED AUTOREGULATION

Geert Morren, Gunnar Naulaers, Philippe Lemmerling, Sabine Van Huffel, Paul Casaer, and Hugo Devlieger*

1. ABSTRACT

Since some important forms of brain injury in premature infants are caused in considerable part by disturbances in cerebral blood flow (CBF), it is important to be able to detect whether the cerebrovascular autoregulation, the mechanism by which CBF is maintained constant despite alterations in mean arterial blood pressure (MAP), is working properly. A recent study suggested that concordant changes in MAP and cerebral intravascular oxygenation (HbD), measured non-invasively by near-infrared spectroscopy (NIRS) as the difference between the concentration changes of oxygenated hemoglobin (HbO_2) and deoxygenated hemoglobin (Hb), reflect impaired cerebrovascular autoregulation[1]. Consequently, premature infants with impaired cerebrovascular autoregulation could be identified by simultaneous, continuous measurements of HbD and MAP.

From several premature babies, MAP, HbD and arterial oxygen saturation (SaO_2) were measured simultaneously at the University Hospital Leuven, Belgium. The concordance between MAP and HbD was quantitated using three different measures, among which a

* Geert Morren, Philippe Lemmerling and Sabine Van Huffel, SCD/SISTA Division, Dept. of Electrical Engineering (ESAT), K.U.Leuven, Kasteelpark Arenberg 10, 3001 Leuven, Belgium. Gunnar Naulaers, Paul Casaer and Hugo Devlieger, Dept. of Paediatrics, University Hospital Leuven, Herestraat 49, 3000 Leuven, Belgium

Oxygen Transport to Tissue XXIII, edited by D. Wilson *et al.*
Kluwer Academic/Plenum Publishers, 2003

newly developed measure that looks for similarity of the dynamics between signals. Some preliminary results obtained from the measured data are given.

2. INTRODUCTION

The most important forms of brain injury in premature infants (germinal matrix-intraventricular hemorrhage (GMH-IVH) and periventricular leukomalacia (PVL)) are caused in considerable part by disturbances in cerebral blood flow (CBF). Premature infants have a particular propensity for development of disturbances in CBF for 2 major reasons. First, because alterations in mean arterial blood pressure (MAP) are very common in such infants. And second, because cerebrovascular autoregulation, the mechanism by which CBF is maintained constant despite alterations in MAP, is either defective or absent in at least some infants. So, there is a relationship between impaired autoregulation and the occurrence of brain injury.
Two studies showed a relationship between CBF and MAP and concluded that autoregulation may be defective in some premature infants [2, 3]. Different authors did not find any correlation between CBF and MAP. However, they only measured at discrete time instants and not continuously. A recent study suggests that concordant changes in cerebral intravascular oxygenation (HbD) and MAP reflect impaired cerebrovascular autoregulation in continuous measurements. A previous study showed a strong correlation between HbD, measured non-invasively by near-infrared spectroscopy (NIRS) as the difference between the concentration changes of oxygenated hemoglobin (HbO_2) and deoxygenated hemoglobin (Hb), and volemic CBF, determined by radioactive microspheres provided the arterial oxygen saturation (SaO_2) does not change appreciably during the measurement [4]. Thus, under such conditions, HbD is a measure of CBF.
Consequently, premature infants with impaired cerebrovascular autoregulation could be identified by simultaneous, continuous measurements of HbD and MAP, and it might be possible to correct the cerebral circulatory disturbance and prevent the lesions.

3. METHODS

3.1. Correlation

The most straightforward way to quantitate the concordance between two signals *x(t)* and *y(t)* is the correlation coefficient *COR*, estimated as:

$$COR = \frac{\sum_{t=1}^{N}(x(t)-\bar{x})(y(t)-\bar{y})}{\sqrt{\sum_{t=1}^{N}(x(t)-\bar{x})^2 \sum_{t=1}^{N}(y(t)-\bar{y})^2}} \qquad (1)$$

where N is the number of samples considered and $\bar{x}$ and $\bar{y}$ stand for the mean of $x(t)$, respectively $y(t)$.

3.2. Coherence

Another way to quantitate the correlation in a frequency specific manner is based on the coherence function, defined as:

$$C_{xy}(f) = \frac{|P_{xy}(f)|^2}{P_{xx}(f)P_{yy}(f)} \tag{2}$$

where $P_{xy}(f)$ is the cross power spectral density of $x(t)$ and $y(t)$, and $P_{xx}(f)$, $P_{yy}(f)$ the power spectral densities of $x(t)$, respectively $y(t)$. The spectral density functions are estimated using Welch's method [5]. The coherence function is a measure of the linear dependence between the signals $x(t)$ and $y(t)$ at a given frequency. A coherence of 1 indicates perfect frequency specific correlation, a coherence 0 indicates a complete lack of frequency specific correlation. To obtain a measure of the coherence over a specific frequency band, the coherence function is averaged over that frequency band.

3.3. HTLS-SEP

More details about this recently proposed method can be found in [6]. The idea is to fit one linear model to two signals. Only if the two signals have the same dynamical properties (or, in other words, have common signal poles), the determined system will be able to explain both signals sufficiently well. The signals are modeled as a sum of K exponentially damped sinusoids:

$$x(t) = \sum_{k=1}^{K} a_k^x \, e^{j\varphi_k^x} \, e^{(j2\pi f_k - d_k)\Delta t} + n_x(t), \quad t = 0,1,\ldots N-1 \tag{3}$$

where f_k, $k=1,\ldots,K$ represent the frequencies, d_k, $k=1,\ldots,K$ the dampings, a_k^x, $k=1,\ldots,K$ the amplitudes, φ_k^x, $k=1,\ldots,K$ the phases, $n_x(t)$ the noise and Δt the constant sampling interval. The other signal $y(t)$ is modeled with the same frequencies f_k and dampings d_k, (thus the same signal poles $z_k \equiv e^{(j2\pi f_k - d_k)}$) but other complex amplitudes a_k^y and phases φ_k^y. The common poles are estimated using a subspace-based algorithm called HTLS [7]. The phases and amplitudes are estimated for $x(t)$ and $y(t)$ separately as the least squares solutions to the model equations, with the frequencies and dampings replaced by their estimates. Note that it is implicitly assumed that both signals are related by means of a linear model. However if the relation is almost linear, the nonlinear model can be approximated by a linear one and the technique can still be applied.

Using all estimated parameters, the part of $x(t)$ and $y(t)$ common to both signals can be reconstructed, giving $x_c(t)$ and $y_c(t)$. $x_c(t)$ will only be a good fit of $x(t)$, when $x(t)$ and $y(t)$

have common signal poles. The ratio of the energy of the common part to the sum of the energy of the common part and the energy of the residual $x_r(t)$ $(=x(t) - x_c(t))$ is used as measure of the importance of the common part:

$$CPC_x = \frac{\sum_{t=0}^{N-1}(x_c(t))^2}{\sum_{t=0}^{N-1}(x_c(t))^2 + \sum_{t=0}^{N-1}(x_r(t))^2} \tag{4}$$

If $x(t)$=HbD (CBF) and $y(t)$=MAP, then CPC_x is a measure of how much of the dynamics of HbD (CBF) can be explained by the MAP.

4. EXPERIMENTAL RESULTS

4.1. Measurements

From several premature infants, MAP, SaO_2 and HbD were measured simultaneously at the University Hospital Leuven. HbD was calculated as the difference between the concentration changes of HbO_2 and Hb, acquired using NIRS (NIRO-300®, Hamamatsu). SaO_2 was measured by means of a pulse oximeter. From all signals, the average values for non-overlapping 5-second intervals were calculated (0.2 Hz). This sampling frequency is high enough since major physiological importance can be attributed to the frequency band of 0-0.01Hz (i.e. changes occurring over several minutes).

4.2. Results

Since the concordance between the signals varies as a function of time, a sliding window approach was used: the measures were calculated on 30-minute recordings (i.e. 360 data points), and plotted against the starting time of the 30-minute window. Fig. 1 shows an example of the measurements on a premature infant, together with the corresponding measures. Between the 40th and the 60th minute, there was a decrease in SaO_2. Since the condition of constant SaO_2 does not hold, HbD can not be assumed to be proportional to CBF during this period, and therefore this period should be excluded from the analysis. During the last two hours of the recording, large (spontaneous) fluctuations in MAP are associated with parallel changes in HbD, while SaO_2 remains constant. This is a case of impaired cerebrovascular autoregulation. During this period, CPC (close to 1) and COR (about 0.8) are higher than COH, suggesting that CPC and COR are more sensitive in order to detect impaired autoregulation.

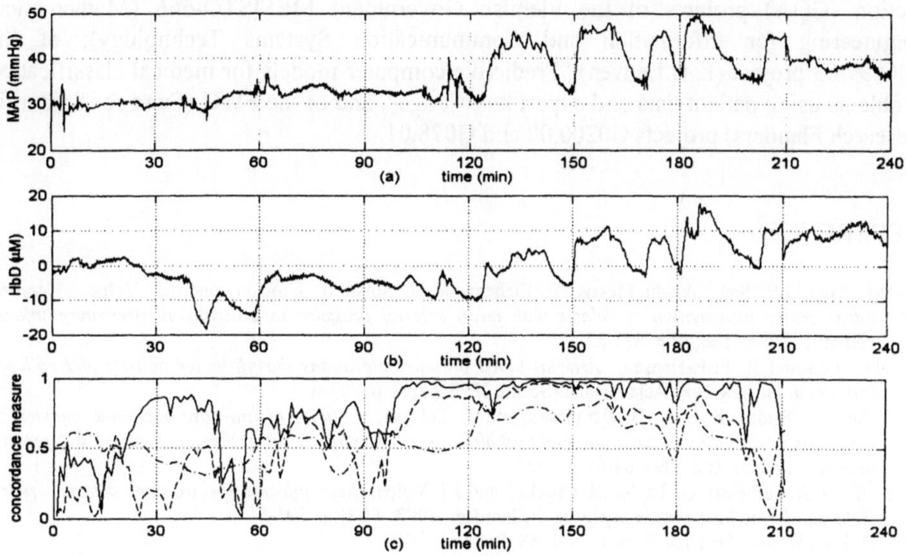

Figure 1. Simultaneous changes in MAP (a) and HbD (b) in a premature infant, and the corresponding measures (c): correlation (dashed line), coherence (dash-dotted line) and CPC (solid line).

CONCLUSIONS

The concordance between HbD and MAP, which reflects impaired cerebrovascular autoregulation, was quantified using three different measures. One of these was specifically developed for this application and looks for similarity of the dynamics between signals. The recordings analyzed so far indicate that the new method and correlation are better measures to detect impaired autoregulation than coherence analysis. Coherence analysis is sensitive to the parameter values used for the spectral density estimation. A problem with the new method is the determination of the model order K, which is twice the number of frequency components. A larger follow-up study is needed to confirm the performance of the new measure, and to evaluate the predictive value of the different measures with respect to possible brain damage statistically. Another advantage of the new method is that it allows the analysis of the dynamical properties of more than two signals simultaneously. This will allow us in the future to include other relevant physiological parameters such as arterial carbon dioxide tension (pCO_2) in the analysis.

ACKNOWLEDGMENTS

This paper presents research results of the Belgian Programme on Interuniversity Poles of Attraction (IUAP V-10-29), initiated by the Belgian State, Prime Minister's Office - Federal Office for Scientific, Technical and Cultural Affairs, of the Brite Euram Programme, Thematic Network BRRT-CT97-5040 'Niconet', of the Concerted Research

Action (GOA) projects of the Flemish Government MEFISTO-666 (Mathematical Engineering for Information and Communication Systems Technology), of the IDO/99/03 project (K.U.Leuven) "Predictive computer models for medical classification problems using patient data and expert knowledge", and of the FWO (Fund for Scientific Research Flanders) projects G0200.00 and G078.01.

REFERENCES

1. M. Tsuji, J.P. Saul, A. du Plessis, E. Eichenwald, J. Sobh, R. Crocker, and J.J. Volpe, *Cerebral intravascular oxygenation correlates with mean arterial pressure in critically ill premature infants.* Pediatrics, 2000. **106**(4): p. 625-32.
2. H.C. Lou and B. Friis-Hansen, *Arterial blood pressure elevations during motor activity and epileptic seizures in the newborn.* Acta Paediatr Scand, 1979. **68**(6): p. 803-6.
3. I. Seri, G. Rudas, Z. Bors, B. Kanyicska, and T. Tulassay, *Effects of low-dose dopamine infusion on cardiovascular and renal functions, cerebral blood flow, and plasma catecholamine levels in sick preterm neonates.* Pediatr Res, 1993. **34**(6): p. 742-9.
4. M. Tsuji, A. duPlessis, G. Taylor, R. Crocker, and J.J. Volpe, *Near infrared spectroscopy detects cerebral ischemia during hypotension in piglets.* Pediatr Res, 1998. **44**(4): p. 591-5.
5. S.M. Kay, *Modern Spectral Estimation.* 1988: Prentice Hall.
6. G. Morren, P. Lemmerling, S. Van Huffel, G. Naulaers, H. Devlieger, and P. Casaer. *Detection of autoregulation in the brain of premature infants using a novel subspace-based technique.* in *Proceedings of the 23rd Annual International Conference of IEEE Engineering in Medicine and Biology Society.* 25-28 Oct 2001. Istanbul, Turkey.
7. S. Van Huffel, H. Chen, C. Decanniere, and P. Van Hecke, *Algorithm for time-domain NMR data fitting based on total least squares.* J Magn Res A, 1994. **110**: p. 228-237.

EFFECTS OF FLUID PERCUSSION INJURY ON RAT BRAIN HEMODYNAMICS, IONIC, ELECTRICAL ACTIVITY, AND ENERGY METABOLISM *IN VIVO*

T. Manor, G. Rogatsky, N. Zarchin, and A. Mayevsky

1. INTRODUCTION

The most commonly used evaluation method for severity of brain injury in patients is the level of consciousness, which is measured by the Glasgow Coma Scale (GCS). The majority of cases admitted to hospitals alive are classified as "mild" (GCS 13-15), 10% as moderate (GCS 9-12) and 10% are classified as severe (GCS≤8) [1].

Several experimental models for traumatic brain injury have been developed [2]. The fluid percussion model, first described by Sullivan et al. [3] is performed by striking a fluid filled reservoir with a pendulum which creates a pressure pulse that is diverted to the pre-exposed Dura mater and brain tissue. The transient high intracranial pressure which is created, is similar to the pressure measured in the demicranium that simulates sudden trauma and causes the same neurological symptoms of behavioral depression in animals and humans [2]. The physiological effects of fluid percussion trauma on animals, such as altered blood flow, cerebrovascular unresponsiveness to changes in pCO_2, disruption of the blood brain barrier, increased ICP and vasogenic edema are similar to those observed in humans. Ionic balance is disturbed and pathophysiological phenomena, such as ischemia, spreading depression, epileptic activity and irreversible brain death are observed in cases of severe trauma [4]. To date most studies on the effects of fluid percussion injury as a model for traumatic brain injury have measured only one or two parameters simultaneously in real time, or have been preformed *in vitro* [5]. The aim of the present study was to perform, a qualitative and quantitative analysis of the fluid percussion model effects using simultaneous, on-line monitoring of metabolic, hemodynamic ionic and electrical parameters from the *in vivo* rat brain.

2. MATERIALS AND METHODS

2.1 Multiprobe Assembly (MPA)

This approach was developed several years ago by Mayevsky [6] and Mayevsky et al. [7] and has been used successfully in rats and gerbils. The MPA used in this study (Fig. 1) contained a fiber optic probe connected to a fluorometer/reflectometer for monitoring NADH and a laser-Doppler flowmeter (LDF) for monitoring CBF. The MPA used in this study also contained three ion selective electrodes capable of detecting millimolar changes in extracellular ion concentrations of K^+, Ca^{2+} and H^+ [8, 9]. Direct current (DC) potential was measured by a highly stable Ag-AgCl electrode located around the K^+ electrode. A reference electrode (also Ag-AgCl) was located below the skin in the animal's neck. Two polished stainless steel rods were inserted into the MPA for electrocorticograph (ECoG) measurement. Parenchymal pressure was measured with an ICP probe (Camino Lab, USA) which was integrated into the MPA and enabled measurement of intracranial pressure levels in mmHg. All the above components were mounted in a Plexiglass cannula. The signals obtained from each experiment were stored in a computer using the CODAS data acquisition program (DATA-Q Co. USA), which enabled receiving 16 channels simultaneously, analog to digital signal conversion, and analysis of the changes in the signals.

2.2 Animal Preparation

Male Wistar rats (180-280g) were anesthetized with 0.3 ml/100g body weight IP injection of Equithesin (each ml contains: pentobarbital 9.72 mg; chloral hydrate 42.52 mg; alcohol 11.5%; propylene glycol 44.34%; magnesium sulfate 21.25 mg and completed with distilled water). The skull was exposed by a midline incision in the skin. A 6mm hole was drilled in one parietal bone for the MPA and a 3.5mm hole was drilled in the contralateral parietal bone for the fluid percussion cannula. The Dura mater was gently removed and the MPA was located on the parietal cortex with the aid of a micromanipulator such that undue pressure on the brain was avoided. The fluid percussion cannula was screwed into the bone contralaterally. Both were cemented to the bone using four stainless steel screws and dental acrylic cement. Traumatic Brain Injury was induced using a lateral fluid percussion model manufactured at the Bar-Ilan University workshop. A low, moderate or severe hydraulic shock (amplitude of 2-4 atm, duration approx. 20 msec) was administered to the left parietal cortex at an angle of 30° to the plane of the lambda and bregma. The intensities were controlled by weights attached to a pendulum, which falls from various angles and hits a syringe. The physiological solution inside the syringe was pushed through a tube connected by a 2.8mm inner diameter metal cannula to the cranium. Monitoring of all parameters was performed contralaterally to the pressure induced.

2.3 Experimental protocol

The animals were allowed to awaken from anesthesia and two hours after surgery measurements recording begun. After stabilization, the brain was exposed to lateral fluid percussion and monitoring was performed for additional four hours. The animals were divided into three study groups (10 animals each). Each intensity of percussion resulted in max ICP pressure registered by the Camino probe at the time of insult: 1. mild trauma - 230±38mmHg; 2. moderate trauma - 279±33mmHg; 3. severe trauma - 315±27mmHg. These values are distorted by the slow response time of the Camino device. Control animals (n=6) were monitored for the same period of time but did not undergo fluid percussion.

2.4 Data evaluation & Statistical analysis

Samples of the analog signals were taken at 15, 30 and 60 min after TBI and then every hour. The results are presented as mean±SEM of the changes from baseline values (0 or 100%). A two-tailed t-test was used to compare each parameter at the various sampling times to its baseline pre-TBI value. ANOVA analysis was performed for determining the correlation between the various parameters at each time point under the different intensities of fluid percussion.

3. RESULTS

The animals were divided into 3 groups exposed to mild, moderate or severe fluid percussion injury as detailed in the methods. No changes or shifts of the monitored parameters were observed in the control animals.

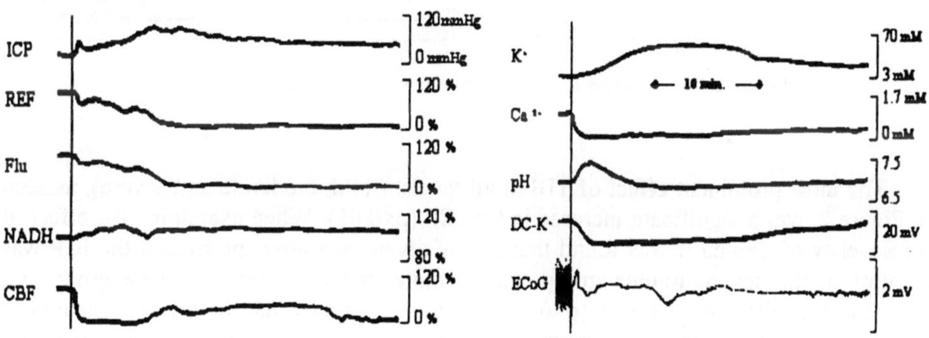

Figure 1: Analog responses of a rat exposed to severe injury. REF, FLU, DC-K^+ = reflectance, fluorescence and dc steady potential around potassium electrode. other abbreviations as in Fig 1.

Figure 1 presents typical recorded responses of a rat exposed to severe injury. Immediately following percussion (TBI), an ischemic depolarization (ID) occurred, as expressed by a decrease in CBF, elevation of extracellular K^+ and depression of EcoG. ICP increased and a negative shift of the DC steady potential was observed.

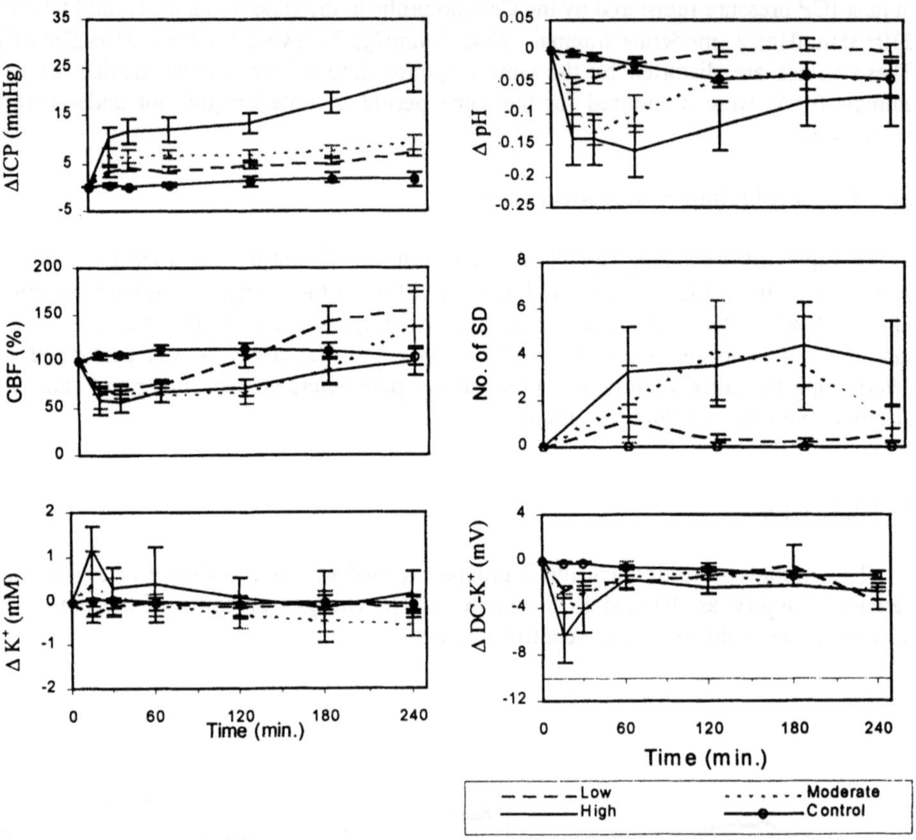

Figure 2. The responses of various parameters in the four models of TBI used.

The most prominent effect of TBI at all levels, (mild, moderate and severe), as seen in Figure 2, was a significant increase in the ICP ($p<0.01$). When examining the effect of the severity of trauma it was found that at any given time after the trauma the ICP was greatest in the severe trauma group, intermediate in the moderate trauma group and lowest in the mild trauma group ($p<0.05$). It should be noted that the greatest changes in ICP were recorded within 2-3 minutes of the trauma (as a result of the percussion) and immediately decreased. TBI caused a rapid decrease in CBF at all levels of trauma ($p<0.05$). This decrease in CBF was maintained for 60 minutes. After this time period an increase was observed but the reactions were heterogenous and no statistical significance was found. Extracelullar K^+ was elevated in the severe injury group, during the first 120 minutes (not significant). Extracelullar pH and the DC steady potential demonstrated a significant decrease during the first 60 minutes following all three levels of trauma ($p<0.05$), after which they increased and were not statistically different from pre-trauma levels. The averaged number of spreading depression (SD) waves per hour was maximal at 2-3h post trauma, and usually appeared after seizures. Among parameters that are not presented, ECoG was depressed proportionally to the injury level, CBV decreased most in the moderate trauma group, intermediately in the severe trauma group and least in the mild trauma group ($p<0.05$) during the first 60 minutes. NADH levels and extracelullar Ca^{2+} increased slightly during the first hour, and then decreased (not significant).

4. DISCUSSION

This study was performed in order to examine whether the multiprobe assembly (MPA) developed in our laboratory can supply additional information as to the state of the brain immediately following trauma and to try and shed some light on processes occurring immediately after trauma.

In this study, we found that ICP increases immediately following trauma (Fig. 2). This increase was found regardless of the severity of the trauma, but was correlated to the severity of the trauma in that it increased more as the trauma increased in severity. This indicates that ICP should indeed be used as a clinical tool for monitoring the severity of TBI. Cerebral blood flow decreased as the ICP increased, indicating impairment in the oxygen supply to the brain tissue. This apparently resulted in ischemia, as indicated by tissue acidosis. However, extracellular potassium ion concentrations did not increase significantly. This indicates that the ion pumps maintaining ionic homeostasis were only slightly affected by the decrease in blood flow or increase in acidosis. SD waves as well as seizures are apparent following trauma, and may aggravate the metabolic stress. In conclusion, it was found that the MPA can simultaneously monitor numerous parameters during and following traumatic brain injury. Furthermore, it was found that the most significant changes occur in ICP and that these induce decreases in CBF which result in tissue acidosis. Energy metabolism however, is not compromised, since NADH levels did not change and neither was there an increase in extracellular potassium ion concentrations.

REFERENCES

1. J. F. Kraus, D. L. McArthur, T. A. Silverman, and M. Jayaraman Epidemiology of brain injury. in: Narayan, R. J., Wilberger, J. E., and Povlishock, J. T. eds. *Neurotrauma*. New York, McGraw-Hill. 1996, 13-30.
2. J. T. Povlishock An overview of brain injury models. in: Narayan, R. K., Wilberger, J. E., and Povlishock, J. T. eds. *Neurotrauma*. McGraw-Hill. 1996, 1325-1336.
3. H. G. Sullivan, J. Martinez, D. P. Becker, J. D. Miller, R. Griffith, and A. O. Wist Fluid-percussion model of mechanical brain injury in the cat, *J. Neurosurg.* **45**, 521-534 (1976).
4. A. Mayevsky, A. Doron, T. Manor, S. Meilin, N. Zarchin, and G. E. Ouaknine Cortical spreading depression recorded from the human brain using a multiparametric monitoring system, *Brain Res.* **740**, 268-274 (1996).
5. A. Mayevsky, T. Manor, S. Meilin, N. Razon, G. E. Ouaknine, and E. Ornstein Multiparametric monitoring of tissue vitality in clinical situations, *Proc. SPIE* **4255**, 33-39 (2001).
6. A. Mayevsky Multiparameter monitoring of the awake brain under hyperbaric oxygenation, *J. Appl. Physiol.* **54**, 740-748 (1983).

7. A. Mayevsky, K. Frank, M. Muck, S. Nioka, M. Kessler, and B. Chance Multiparametric evaluation of brain functions in the Mongolian gerbil *in vivo*, *J. Basic Clin. Physiol. Pharmacol.* **3**, 323-342 (1992).
8. C. M. Friedli, D. S. Sclarsky, and A. Mayevsky Multiprobe monitoring of ionic, metabolic and electrical activities in the awake brain, *Am. J. Physiol.* **243**, R462-R469 (1982).
9. A. Mayevsky, R. B. Duckrow, E. Yoles, N. Zarchin, and D. Kanshansky Brain mitochondrial redox state, tissue hemodynamic and extracellular ion responses to four-vessel occlusion and spreading depression in the rat, *Neurol. Res.* **12**, 243-248 (1990).

AUTHOR INDEX

SUBJECT INDEX

MIX
Papier aus verantwortungsvollen Quellen
Paper from responsible sources
FSC® C105338

If you have any concerns about our products,
you can contact us on
ProductSafety@springernature.com

In case Publisher is established outside the EU,
the EU authorized representative is:
Springer Nature Customer Service Center GmbH
Europaplatz 3, 69115 Heidelberg, Germany

Printed by Libri Plureos GmbH
in Hamburg, Germany